W9-CSP-483

36 Springer Series in Solid-State Sciences
Edited by Hans-Joachim Queisser

Springer Series in Solid-State Sciences

Editors: M. Cardona P. Fulde H.-J. Queisser

Volume 40 **Semiconductor Physics** - An Introduction By K. Seeger

Volume 41 **The LMTO Method** Muffin-Tin Orbitals and Electronic Structure
By H.L. Skriver

Volume 42 **Crystal Optics with Spatial Dispersion, and Excitons**
By V.M. Agranovich and V.L. Ginzburg

Volume 43 **Resonant Nonlinear Interactions of Light with Matter**
By V.S. Butylkin, A.E. Kaplan, Yu.G. Khronopulo, and E.I. Yakubovich

Volume 44 **Elastic Media with Microstructure II** Three-Dimensional Models
By I.A. Kunin

Volume 45 **Electronic Properties of Doped Semiconductors**
By B.I. Shklovskii and A.L. Efros

Volume 46 **Topological Disorder in Condensed Matter**
Editors: F. Yonezawa and T. Ninomiya

Volume 47 **Statics and Dynamics of Nonlinear Systems**
Editors: G. Benedek, H. Bilz, and R. Zeyher

Volume 48 **Magnetic Phase Transitions**
Editors: M. Ausloos and R.J. Elliott

Volume 49 **Organic Molecular Aggregates,** Electronic Excitation and Interaction Processes
Editors: P. Reineker, H. Haken, and H.C. Wolf

Volume 50 **Multiple Diffraction of X-Rays in Crystals**
By Shih-Lin Chang

Volume 51 **Phonon Scattering in Condensed Matter**
Editor: W. Eisenmenger, K. Laßmann, and S. Döttinger

Volume 52 **Magnetic Superconductors and Their Related Problems**
Editors: T. Matsubara and A. Kotani

Volumes 1 - 39 are listed on the back inside cover

A. A. Chernov

Modern Crystallography III

Crystal Growth

With Contributions by
E. I. Givargizov, K. S. Bagdasarov, V. A. Kuznetsov,
L. N. Demianets, A. N. Lobachev

With 244 Figures

Springer-Verlag
Berlin Heidelberg New York Tokyo 1984

Professor Dr. *Alexander A. Chernov*
Professor Dr. *E. I. Givargizov*
Professor Dr. *K. S. Bagdasarov*
Dr. *V. A. Kuznetsov*
Dr. *L. N. Demianets*
Dr. *A. N. Lobachev* (†)

Institute of Crystallography, Academy of Sciences of the USSR, 59 Leninsky prospect, SU-117333 Moscow, USSR

Series Editors:

Professor Dr. Manuel Cardona
Professor Dr. Peter Fulde
Professor Dr. Hans-Joachim Queisser

Max-Planck-Institut für Festkörperforschung, Heisenbergstrasse 1
D-7000 Stuttgart 80, Fed. Rep. of Germany

Title of the original Russian edition:
Sovremennaia kristallografiia; Obrazovanie kristallov

ISBN 3-540-11516-1 Springer-Verlag Berlin Heidelberg New York Tokyo
ISBN 0-387-11516-1 Springer-Verlag New York Heidelberg Berlin Tokyo

Offset printing: Beltz Offsetdruck, 6944 Hemsbach/Bergstr. Bookbinding: J. Schäffer OHG, 6718 Grünstadt
2153/3130-543210

Modern Crystallography

in Four Volumes*

I Symmetry of Crystals. Methods of Structural Crystallography

II Structure of Crystals

III Crystal Growth

IV Physical Properties of Crystals

Foreword

Crystallography—the science of crystals—has undergone many changes in the course of its development. Although crystals have intrigued mankind since ancient times, crystallography as an independent branch of science began to take shape only in the 17th–18th centuries, when the principal laws governing crystal habits were found, and the birefringence of light in crystals was discovered. From its very origin crystallography was intimately connected with mineralogy, whose most perfect objects of investigation were crystals. Later, crystallography became associated more closely with chemistry, because it was apparent that the habit depends directly on the composition of crystals and can only be explained on the basis of atomic-molecular concepts. In the 20th century crystallography also became more oriented towards physics, which found an ever-increasing number of new optical, electrical, and mechanical phenomena inherent in crystals. Mathematical methods began to be used in crystallography, particularly the theory of symmetry (which achieved its classical completion in space-group theory at the end of the 19th century) and the calculus of tensors (for crystal physics).

---

* Published in *Springer Series in Solid-State Sciences,* I: Vol. 15; II: Vol. 21; III: Vol. 36; IV: Vol. 37

Early in this century, the newly discovered x-ray diffraction by crystals made a complete change in crystallography and in the whole science of the atomic structure of matter, thus giving a new impetus to the development of solid-state physics. Crystallographic methods, primarily x-ray diffraction analysis, penetrated into materials sciences, molecular physics, and chemistry, and also into many other branches of science. Later, electron and neutron diffraction structure analyses became important since they not only complement x-ray data, but also supply new information on the atomic and the real structure of crystals. Electron microscopy and other modern methods of investigating matter–optical, electronic paramagnetic, nuclear magnetic, and other resonance techniques–yield a large amount of information on the atomic, electronic, and real crystal structures.

Crystal physics has also undergone vigorous development. Many remarkable phenomena have been discovered in crystals and then found various practical applications.

Other important factors promoting the development of crystallography were the elaboration of the theory of crystal growth (which brought crystallography closer to thermodynamics and physical chemistry) and the development of the various methods of growing synthetic crystals dictated by practical needs. Man-made crystals became increasingly important for physical investigations, and they rapidly invaded technology. The production of synthetic crystals made a tremendous impact on the traditional branches: the mechanical treatment of materials, precision instrument making, and the jewelry industry. Later it considerably influenced the development of such vital branches of science and industry as radiotechnics and electronics, semiconductor and quantum electronics, optics, including nonlinear optics, acoustics, etc. The search for crystals with valuable physical properties, study of their structure, and development of new techniques for their synthesis constitute one of the basic lines of contemporary science and are important factors of progress in technology.

The investigation of the structure, growth, and properties of crystals should be regarded as a single problem. These three intimately connected aspects of modern crystallography complement each other. The study, not only of the ideal atomic structure, but also of the real defect structure of crystals makes it possible to conduct a purposeful search for new crystals with valuable properties and to improve the technology of their synthesis by using various techniques for controlling their composition and real structure. The theory of real crystals and the physics of crystals are based on their atomic structure as well as on the theoretical and experimental investigations of elementary and macroscopic processes of crystal growth. This approach to the problem of the structure, growth, and properties of crystals has an enormous number of aspects, and determines the features of modern crystallography.

The branches of crystallography and their relation to adjacent fields can be represented as a diagram showing a system of interpenetrating branches which have no strict boundaries. The arrows show the relationship between the branches, indicating which branch influences the activity of the other, although, in fact, they are usually interdependent.

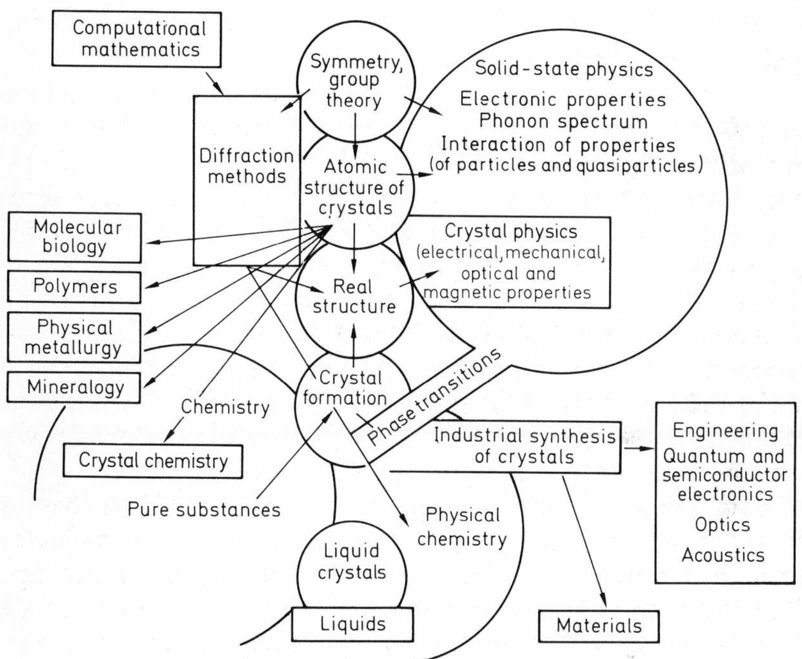

Branches of crystallography and its relation to other sciences

Crystallography proper occupies the central part of the diagram. It includes the theory of symmetry, the investigation of the structure of crystals (together with diffraction methods and crystal chemistry), and the study of the real structure of crystals, their growth and synthesis, and crystal physics.

The theoretical basis of crystallography is the theory of symmetry, which has been intensively developed in recent years.

The study of the atomic structure has been extended to extremely complicated crystals containing hundreds and thousands of atoms in the unit cell. The investigation of the real structure of crystals with various disturbances of the ideal crystal lattices has been gaining in importance. At the same time, the general approach to the atomic structure of matter and the similarity of the various diffraction techniques make crystallography a science not only of the structure of crystals themselves, but also of the condensed state in general.

The specific applications of crystallographic theories and methods allow the utilization of structural crystallography in physical metallurgy, materials science, mineralogy, organic chemistry, polymer chemistry, molecular biology, and the investigation of amorphous solids, liquids, and gases. Experimental and theoretical investigations of crystal growth and nucleation processes and their development draw on advances in chemistry and physical chemistry and, in turn, contribute to these areas of science.

Crystal physics deals mainly with the electrical, optical, and mechanical properties of crystals closely related to their structure and symmetry, and adjoins solid-state physics, which concentrates its attention on the analysis of laws defining the general physical properties of crystals and the energy spectra of crystal lattice.

The first two volumes are devoted to the structure of crystals, and the last two, to the growth of crystals and their physical properties. The authors present the material in such a way that the reader can find the basic information on all important problems of crystallography. Due to the limitation of space the exposition of some sections is concise, otherwise many chapters would have become separate monographs. Fortunately, such books on a number of crystallographic subjects are already available.

The purpose of such an approach is to describe all the branches of crystallography in their interrelation, thus presenting crystallography as a unified science to elucidate the physical meaning of the unity and variety of crystal structures. The physico-chemical processes and the phenomena taking place in the course of crystal growth and in the crystals themselves are described, from a crystallographic point of view, and the relationship of properties of crystals with their structure and conditions of growth is elucidated.

This four-volume edition is intended for researchers working in the fields of crystallography, physics, chemistry, and mineralogy, for scientists studying the structure, properties, and formation of various materials, for engineers and those engaged in materials science technology, particularly in the synthesis of crystals and their use in various technical devices. We hope that this work will also be useful for undergraduate and graduate students at universities and higher technical colleges studying crystallography, solid-state physics, and related subjects.

Modern Crystallography is written by a large group of authors from the Institute of Crystallography of the USSR Academy of Sciences, who benefited from the assistance and advice of many other colleagues. The English edition of all four volumes of *Modern Crystallography* is being published almost simultaneously with the Russian edition. The authors have included in the English edition some of the most recent data. In several instances some additions and improvements have been made.

B.K. Vainshtein

Preface

This book is intended for those who are investigating or plan to investigate processes of crystallization. It will also be handy as a reference on the basic concepts and techniques of crystal growth. In writing it, we did not expect the reader to have specialized knowledge and sought to explain the fundamentals of the science and practice of crystal growth consistently and systematically. Our objective was to put together a text from which the reader could get an idea of all that we consider to be the principal approaches to the up-to-date analysis of crystallization phenomena and the techniques of growing single crystals.

The processes of crystal formation are analyzed in their entirety from the unified standpoints of macroscopic and statistical thermodynamics and physicochemical kinetics, in correspondence with the methods and ideology of the physics and chemistry of solids.

Mutual enrichment of the different lines of investigation is particularly important in the study of the theoretical fundamentals, which are covered in Part 1, and in practical applications, to which Part 2 is devoted. When dealing with theory we tried, wherever possible, to touch upon practical problems, and when discussing practice, to proceed from the essence of fundamental phenomena. All of the caprices of growing crystals cannot be predicted as yet, but some are becoming understandable.

We attempted to achieve a maximum simplicity of presentation and create a qualitatively clear physical and physicochemical picture; for this reason many interesting details had to be sacrificed. After mastering the fundamentals, the reader can find relevant information in the primary literature.

The material is divided in two parts. The first part is an analysis of crystal nucleation and growth processes; Part 2 deals with the techniques of crystal growing. In Chap. 1, the thermodynamic equilibrium between a crystal and its environment is discussed, with attention given to the conditions of phase equilibrium, the measure of deviation from it, the energy of the interface between the crystal and its environment, the structure of this interface under various conditions, and the equilibrium shape of crystals. Chapter 2 describes the principal concepts of crystal nucleation in the bulk of the metastable phase and on foreign surfaces. This is followed by an analysis in Chap. 3 of the molecular kinetics involved in the motion of the phase boundary, as well as a discussion of its morphology. Chapter 4 is devoted to the effect of impurities on growth kinetics, the thermodynamics and kinetics of impurity trapping, and the physical causes leading to the formation of certain impurity inhomogeneities. Heat and mass transfer in crystallization, and the shapes and stability of growing crystals are the subject of Chap. 5. In Chap. 6, defects – inclusions, phase inhomogeneities, dislocations, and internal stresses – are discussed. Chapter 7 gives an idea of certain fundamental problems in industrial crystallization and of the behavior of an ensemble of crystals. Part 2 is divided into three chapters, which discuss crystal growth from the gas (vapor) phase (Chap. 8), from solutions (Chap. 9), and from melts (Chap. 10). Each section treats the physicochemical, procedural, and crystallographic aspects of crystal growing.

Before each group of methods is described, information is given on the physicochemical fundamentals upon which they are based. The choice of these methods depends on the properties of the substances which are to be obtained. For instance, crystals melting congruently at moderate temperatures can usually be most easily and rapidly obtained from melts. Crystals with high melting points ($T \geqslant 200°C$) and good solubility in some liquids, or with sufficiently high vapor pressure at moderate temperatures, are often grown from solutions or the gas phase. With compounds which decompose when heated it is often advantageous to use methods based on crystal formation in chemical reactions. In designing equipment it is also very important to select inert materials for parts that come into contact with the crystallizing substance, the solvent or its vapors, residual gases, the crystallization atmosphere, etc.

The method and regime and the growing techniques in general should ensure that the crystals produced have the assigned dimensions and degree of perfection. This is impossible to achieve unless a procedure is developed that is based to a large extent upon data pertaining to the mechanism and kinetics of growth and to the formation of defects, i.e., the crystallographic foundations of crystal growth. Specific types of defects correspond to each group of techniques and regimes. The relationship between growing conditions and structures that appear during growth is discussed after the respective techniques.

Part 1 (Chaps. 1–7) was written by A.A. Chernov. In Part 2, Chap. 8, on growing from the gas phase, was prepared by E.I. Givargizov. Sections 9.1,2, on growing from aqueous solutions, were written by V.A. Kuznetzov; Sect. 9.3, on hydrothermal solutions, by L.N. Demianets, V.A. Kuznetzov, and A.N. Lobachev; and Sect. 9.4, on growing from high-temperature solutions, by K.S. Bagdasarov, who also prepared Chap. 10, on growing from melts.

The authors are grateful to many colleagues for valuable advice and materials. Our special thanks for aid in preparation of the manuscript are due to L.A. Solomentseva and L.N. Obolenskaya.

The comments and advice of D.E. Temkin, editor of the original edition in Russian, were of great value to us. The extremely thorough and creative work done by A.M. Mel'nikova greatly exceeded her function as scientific editor and improved the text considerably. We feel the deepest gratitude to them.

Moscow, December 1983 *The Authors*

Contents

PART 1: CRYSTALLIZATION PROCESSES
(A.A. Chernov)

PART 2: THE GROWING OF CRYSTALS

(E.I. Givargizov, K.S. Bagdasarov, V.A. Kuznetsov, L.N. Demianets, A.N. Lobachev)

1. Equilibrium

Crystal growth and decrystallization (evaporation, melting, dissolution, etching, etc.) take place under the influence of a thermodynamic driving force determined by deviations in temperature, pressure, concentration, and external field intensity from their equilibrium values. Growth and decrystallization occur on interfaces whose structure and properties are close to equilibrium. Thus the basic principles governing phase equilibrium, driving force, and equilibrium interface structure and properties should be considered before entering into the kinetics of phase transformation. These aspects of equilibrium form the subject of this chapter.

1.1 Phase Equilibrium

1.1.1 One-Component Systems

Let us consider a one-component system containing two phases, such as the crystal and vapor or crystal and melt phases. If there is an exchange of particles between the two phases, a phase equilibrium will ultimately be established between them. The temperature T and the pressures P_S, P_M in the crystal and the medium (i.e., vapor or melt) must then be related by the equality of the chemical potentials of the crystalline solid (S) and the medium (M):

$$\mu_S(P_S, T) = \mu_M(P_M, T) .$$

This must be supplemented by mechanical equilibrium at the phase boundary. If we neglect the effect of surface energy and assume that no other forces are appliced to the boundary, then $P_S = P_M = P$ (P is the total pressure in the system) [Ref. 1.1, Sect. 81].

The chemical potential of a given phase is equal to the work which has to be performed to change the number of particles (atoms or molecules) in this phase by unity. Therefore μ_j, where j = S (solid), M (medium), V (vapor), and L (liquid), is also called the chemical potential of a particle in the corresponding phase. The quantity μ_j can be written as the sum

$$\mu_j = \varepsilon_j - Ts_j + P\Omega_j , \tag{1.1}$$

where ε_j is the potential energy, S_j is the entropy, and Ω_j is the specific volume of the phase per particle.

To get a general idea of the structure of the chemical potential, we estimate the terms of (1.1) using the simplest concepts.

We compute the potential energy ε_j from the internal energy of the atom or molecule, which usually changes little on transition of a particle from one phase to another. The simplest measure of ε_j determined in this manner is the heat of evaporation, i.e., the change of energy on transition of a particle to a rarefied vapor, in which particle interactions are very weak and ε_V can be taken as zero. For most substances the heat of evaporation varies from less than $\simeq 10$ kcal/mol for highly volatile, say organic substances with relatively weak Van der Waals bonds to more than $\simeq$ 100 kcal/mol for substances with a metallic, covalent, or ionic bond. Per particle, $\varepsilon_j \simeq (10\text{–}100) \times 4.18 \times 10^{10}/6.02 \times 10^{23} \simeq (0.7\text{–}7) \times 10^{-12}$ erg.

In condensed phases with molecules that are not all too complex (such as those of biopolymers), the interatomic distances a are $\simeq (1\text{–}5) \times 10^{-8}$ cm, i.e., $\Omega \simeq a^3 \simeq 10^{-22}\text{–}10^{-24}$ cm^3. At atmospheric pressure $P \simeq 10^6$ dyn/cm^2 and $\Omega \simeq 10^{-23}$ cm^3, $P\Omega \simeq 10^{-17}$ erg, which is much less than ε_j. Therefore the free energy $\varepsilon_j - Ts_j$ is sometimes used instead of the chemical potential. In vapors, the value of $P\Omega_V$ may be higher: at $T = 1000$ K in the ideal gas, $P\Omega_V = kT \simeq 1.4 \times 10^{-13}$ erg (here $k = 1.38 \times 10^{-16}$ erg/deg is the Boltzmann constant).

The entropy s_S of a crystal includes several terms. The configuration entropy is related to the number of ways in which the atoms can be accommodated at the different sites in the body and to the probabilities of their realization. The vibration and rotation components of s_S depend on the probabilities of finding the system at certain levels of vibration and rotation energy. For a defect-free crystal consisting of one kind of atoms, the vibrational entropy, $s_S = 3k \ln(kTe/h\bar{\nu})$, where $\bar{\nu}$ is the "mean geometric" frequency of the thermal vibrations of the atoms in the crystal, $h = 6.62 \times 10^{-27}$ ergs is the Planck constant, and e is the base of the natural logarithms [Ref. 1.1, Sect. 62]. At $\bar{\nu} = 3 \times 10^{13}$ s^{-1}, $T = 1000$ K we have $Ts_S \simeq 3kT \simeq 4 \times 10^{-13}$ erg. If X is the molecular fraction of the defects (e.g., vacancies or impurity atoms) relative to the number of sites which they can occupy in the lattice (e.g, the number of points in it), the configurational entropy of the crystal atoms $\simeq -kX$ Ref. 1.1, Sect. 88] and is low at $X \ll 1$. Conversely, the configurational entropy contributes substantially to the chemical potential of the defects themselves and plays an important role in their behavior.

Thus, at near-atmospheric and lower pressures the decisive role in (1.1) is played by the first two terms, and the energy term appreciably exceeds the entropy. However, transformations from one phase to another depend on the difference between the values of ε_j and Ts_j in these phases, rather than on their absolute values. For instance, from the equality of the chemical potentials of the crystalline solid (S) and liquid or gaseous medium (M), $\varepsilon_S - Ts_S = \varepsilon_M - Ts_M$, it follows that $\varepsilon_M - \varepsilon_S = T(s_M - s_S) = \Delta H$, where ΔH is the heat of fusion or evaporation. The quantity ΔH thus determined is positive, because the entropy of the less-ordered liquid or vapor exceeds that of the more ordered crystal. The heats of fusion usually do not exceed 10% of the heat of evaporation.

The phase equilibrium, i.e., the equality of chemical potentials of the phases, defines the temperature equality of the equilibrium vapor pressure (in a crystal–vapor system, the V–S curve in Fig. 1.1) or the pressure dependence of the melting point (in the crystal–melt system, the S–L curve in Fig. 1.1). At pressures and temperatures corresponding to the points on the equilibrium curves V–S, V–L, and S–L, equilibrium among the phases (vapor, liquid, and solid) is attained, with each of the phases stable in the corresponding temperature and pressure ranges in Fig. 1.1. The point O, at which the three phases coexist, is called the triple point. The V–L line terminates at the critical point.

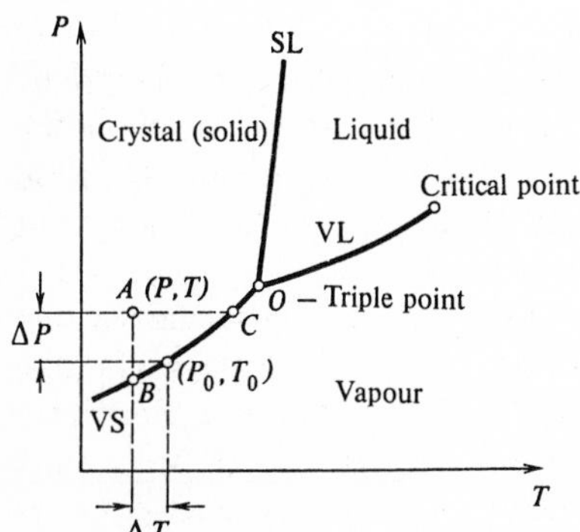

Fig.1.1. Phase diagram of vapor (V), crystal (S), and melt (L). The vapor in the state $A(P, T)$ is supersaturated relative to the crystal

If the pressure and temperature in a system correspond to some point with coordinates (P, T) (Fig. 1.1) outside the equilibrium curves, then $\mu_S \neq \mu_M$, and a driving force for phase transformation appears. The most general measure of this force is the difference $\Delta\mu$ between the chemical potentials of the corresponding phases or the ratio $\Delta\mu/kT$. For vapor–solid transformation $\Delta\mu = \mu_V(P, T) - \mu_S(P, T)$, for melt–solid transformation $\Delta\mu = \mu_L(P, T) - \mu_S(P, T)$, and for the condensation of vapor into liquid $\Delta\mu = \mu_V(P,T) - \mu_L(P, T)$. If the differences $\Delta\mu$ are negative, the processes must reverse.

In place of the difference in chemical potentials it is customary to use the directly measured values, i.e., the deviations ΔP and ΔT in pressure and temperature from their equilibrium values corresponding to some point (P_0, T_0) on the V–S, L–S, or V–L curve, where $\Delta P = P - P_0$, $\Delta T = T_0 - T$. If both these deviations are small, the deviation from equilibrium can be represented by

$$
\begin{aligned}
\Delta\mu &= \mu_V(P_0 + \Delta P, T_0 - \Delta T) - \mu_S(P_0 + \Delta P, T_0 - \Delta T) \\
&= \left(\frac{\partial \mu_V}{\partial P} - \frac{\partial \mu_S}{\partial P}\right)\Delta P - \left(\frac{\partial \mu_V}{\partial T} - \frac{\partial \mu_S}{\partial T}\right)\Delta T \\
&= (\Omega_V - \Omega_S)\Delta P + (s_V - s_S)\Delta T,
\end{aligned}
\tag{1.2}
$$

where Ω_V is the specific volume per particle in the vapor, Ω_S is the specific volume per identical particle in the crystal, and s_V and s_S are the respective entropies. The expression for deviation from equilibrium in the melt is quite similar to (1.2) and is obtained by replacing the subscript V with L.

If we choose a reference point such that $\Delta T = 0$, the difference ΔP is called the absolute supersaturation, and $\Delta P/P_0 \equiv \sigma$ the relative supersaturation. Assuming a vapor to be an ideal gas and neglecting the pressure-independent term, we have $\mu_V = kT \ln P$ and

$$\Delta\mu = kT \ln P/P_0 . \tag{1.3}$$

At small deviations from equilibrium, when $\sigma \ll 1$, we have $\Delta\mu \simeq kT \Delta P/P_0$.

If (as is often the case) supersaturation is achieved by cooling a vapor, which is saturated at $T = T_0$ and hence has a pressure $P = P_0$ (Fig. 1.1), then since $\Delta P = 0$, expression (1.2) gives the relationship between $\Delta\mu$ and the supercooling ΔT:

$$\Delta\mu = \Delta H \, \Delta T/T_0 , \tag{1.4}$$

where $\Delta H = T(s_V - s_S)$ is the heat of evaporation at a temperature T_0. In the two particular cases indicated, the deviations from equilibrium correspond to segments AB (at $\Delta T = 0$) and AC ($\Delta P = 0$). Expressing ΔP in terms of ΔT with the aid of the *Clapeyron-Clausius* equation, we can easily ascertain the equivalence of (1.3) and (1.4) at small deviations from equilibrium. The deviation from equilibrium in a supercooled liquid at a given pressure is defined by (1.4), where ΔH now implies the heat of fusion.

1.1.2 Multicomponent Systems

The equilibrium state in a two-component (binary) system made up of components A and B is described by the concentrations of the components, and by the pressure and temperature. The relative molar concentration $X_{\alpha j}$ of component $\alpha(\alpha = \text{A, B})$ in phase $j(j = \text{V, L, S})$ is defined as the fraction of molecules of type α with respect to the total number of molecules in phase j, therefore $X_{Aj} + X_{Bj} = 1$. Equilibrium in a two-component system is achieved when the chemical potentials of each of the components in the two contacting phases, e.g., a solid and a liquid, are identical:

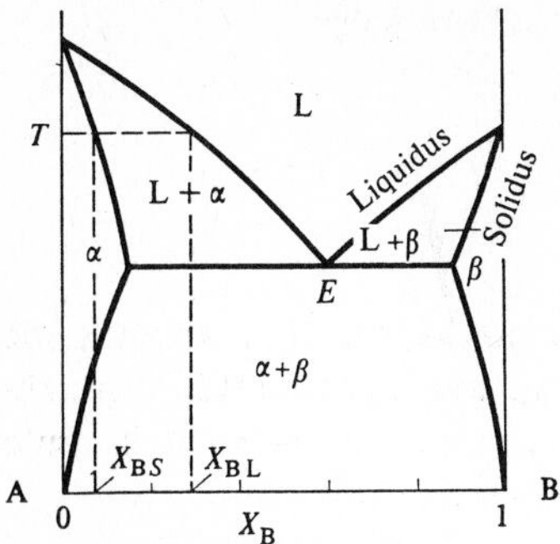

Fig.1.2. Phase diagram of a binary system with a eutectic point E. In the α and β areas the solid phases α and β are the only stable ones; $\alpha + \beta$ is a mixture of phases α and β; L is a liquid melt; L + α is a liquid with crystals of α; and L + β is a liquid with crystals of β

$$\begin{aligned} \mu_{AS}(P, T, X_{AS}) &= \mu_{AL}(P, T, X_{AL}), \\ \mu_{BS}(P, T, X_{BS}) &= \mu_{BL}(P, T, X_{BL}). \end{aligned} \tag{1.5}$$

Equalities (1.5) define the three-dimensional phase diagram, i.e., functions $X_{BL}(P, T)$ and $X_{BS}(P, T)$, which are represented in coordinates (P, T, X) by two surfaces, those of the liquidus and the solidus. A planar cross section (P = const) of one of the simplest diagrams is presented in Fig. 1.2. Curve $X_{BL}(P, T)$ at P = const in this two-dimensional (T, X)diagram is called the liquidus line, and curve $X_{BS}(P, T)$ at P = const the solidus line. The points on the liquidus and solidus lines, which correspond to a given temperature, define the concentrations of component B in the liquid (X_{BL}) and solid (X_{BS}) phases, which are at equilibrium with one another at this temperature. The ratio

$$K_0 = X_{BS}/X_{BL} \tag{1.6}$$

is called the equilibrium coefficient of the distribution of component B between the liquid and solid phases. Numerous complex phase diagrams are discussed in special literature. Extended reviews are given in *Rosenberger's* book [1.2a].

Deviation from equilibrium in a binary system is described by two quantities:

$$\Delta\mu_A = \mu_{AL} - \mu_{AS} \quad \text{and} \quad \Delta\mu_B = \mu_{BL} - \mu_{BS}\,.$$

A particular instance of a two-component system is a crystal–solution system in which the solvent (B) does not enter the crystal, i.e, $X_{BS} = 0$. Then, of the two relations in (1.5), only the first remains, which determines equilibrium solely with respect to the transition of the crystal substance (A) to solution and back. Thus this situation is quite similar to that in the solid-vapor system, and amounts to equilibrium in a one-component system. The concentration of the solute in solutions where the solvent does not enter the crystal is often denoted C. The deviation $\Delta C = C - C_0$ of the concentration from the equilibrium value C_0 is called the absolute supersaturation, and the ratio $\sigma = \Delta C/C_0$

the relative supersaturation. The difference between the chemical potentials of the crystallizing substance in solution and in the crystal is given by an equation similar to (1.2). For ideal (dilute) solutions

$$\mu_{AL} = kT \ln C + \psi(P, T), \qquad \Delta\mu = kT \ln C/C_0, \tag{1.7}$$

where ψ is a function independent of C. In the more general case of nonideal solutions, the concentrations C and C_0 in (1.7) must be replaced by the actual and equilibrium activities of the solute [1.2b, c]

When perfect crystals are grown from vapor, the typical numerical values of the relative supersaturations σ strongly depend on the type of substance and the growing conditions, but may reach unity (i.e., 100 %) or even more, whereas growth from solutions usually requires $\sigma < 0.1$. However, transparent crystals of potash alum can be obtained from aqueous solutions at room temperatures and at large supersaturations; $\sigma \lesssim 0.2$.

Crystals can also grow in chemical reactions that take place in the gas, liquid, and solid phases (in the solid phase the reaction rate is much lower). One of the numerous reactions used is so-called disproportionation:

$$2GeI_2 \rightleftarrows Ge + GeI_4 . \tag{1.8}$$

If the reaction is shifted to the right, crystalline germanium is obtained (GeI_2 and GeI_4 are gases). The replacement of the symbol $\rightleftarrows$ with $=$ means equilibrium. The equilibrium state of a gas mixture in a possible chemical reaction between its constituent substances A_α ($A_1 = GeI_2$, $A_2 = Ge$, $A_3 = GeI_4$) can be written in the generalized form

$$\sum_\alpha \nu_\alpha A_\alpha = 0 . \tag{1.9}$$

Here, ν_α are stoichiometric coefficients [for reaction (1.8) $\nu_1 = -2$, $\nu_2 = 1$, $\nu_3 = 1$, because the coefficients in the right side of the reaction are usually assumed to be positive], and the sum is taken over all the components involved in the reaction.[1] If δn acts of reaction (1.9) occur, this results in the formation of $\nu_\alpha \delta n$ particles of type A_α. If $\nu_\alpha < 0$, the number of A_α particles decreases by ν_α as a result of a single act. Hence δn reaction acts change the thermodynamic potential by

$$\delta\Phi = \sum_\alpha \nu_\alpha \mu_\alpha \delta n .$$

At equilibrium, the thermodynamic potential Φ must reach its minimum, i.e., $\delta\Phi = 0$, whence we obtain the equilibrium condition

[1] Reversing all the signs does not affect the result.

$$\sum_{\alpha} \nu_\alpha \mu_\alpha = 0 . \tag{1.10}$$

If all the reactants are ideal gases, the chemical potential of the αth reactant $\mu_\alpha = kT \ln P_\alpha + \psi(T)$, where P_α is the partial pressure of the reactant, and ψ is pressure independent, as before. Substituting these chemical potentials into (1.10), we have

$$\prod_{\alpha} P_{\alpha 0}^{\nu_\alpha} = \exp\left(-\frac{1}{kT} \sum_{\alpha} \nu_\alpha \psi_\alpha\right) \equiv \mathscr{K}_0(T) . \tag{1.11}$$

Function $\mathscr{K}_0(T)$ is called the equilibrium constant of reaction (1.9), and $P_{\alpha 0}$ stands for the equilibrium vapor pressure of the α component at a given temperature.

If one of the phases in the system is crystalline ($\alpha = \text{S}$) and the equilibrium vapor pressure over it is low, than without expressing the crystal potential in terms of the pressure of its vapors in the system, we get from (1.10):

$$\prod_{\alpha \neq \text{S}} P_{\alpha 0}^{\nu_\alpha} = \exp\left[-\frac{1}{kT}\left(\sum_{\alpha \neq \text{S}} \nu_\alpha \psi_\alpha + \nu_\text{S} \mu_\text{S}\right)\right] = \mathscr{K}_0(P, T) , \tag{1.12}$$

where $P = \Sigma_\alpha P_\alpha$ is the total pressure in the system. Taking into account the fact that the chemical potential of the crystal, μ_S, depends only slightly on the pressure, the right-hand side of (1.12) and of (1.11) can be assumed to be only a function of the temperature.

The deviation of the system from equilibrium in the case at hand is, by analogy with (1.7),

$$\sum_{\alpha} \nu_\alpha \mu_\alpha = kT\left[\ln \prod_{\alpha} P_\alpha^{\nu_\alpha} - \ln \mathscr{K}_0(T)\right]$$

or, in a system with solid phases,

$$\sum_{\alpha \neq \text{S}} \nu_\alpha \mu_\alpha = kT\left[\ln \prod_{\alpha \neq \text{S}} P_\alpha^{\nu_\alpha} - \ln \mathscr{K}_0(P, T)\right] .$$

Hence the analog of the above-introduced supersaturation $\ln(P/P_0)$ in the case under consideration is

$$\ln\left(\prod_{\alpha \neq \text{S}} P_\alpha^{\nu_\alpha} \Big/ \prod_{\alpha \neq \text{S}} P_{\alpha 0}^{\nu_\alpha}\right) = \ln\left[\prod_{\alpha \neq \text{S}} P_\alpha^{\nu_\alpha} \Big/ \mathscr{K}_0(P, T)\right] . \tag{1.13}$$

This is the definition of supersaturation during the growth of crystals from the gas phase in the course of a chemical reaction. In solutions, the definition is the same, but P_α is replaced by C_α or activities a_α in (1.11–13).

1.1.3 Crystallization Pressure

So far, the pressure in the crystal has been assumed to be equal to that in its environment. Let us now subject the crystal to the action of external forces, with the result that it becomes stressed. External forces can be created simply by placing a weight on the crystal. It turns out that a thin film of liquid (solution, melt)[2] exists between weight and crystal (and also between vessel bottom and crystal), and the crystal can exchange particles with the medium [1.3c, d].

Let us find the condition for phase equilibrium between a loaded crystal and its environment. To do this, we transfer δN particles from the medium into the crystal so that these particles become deposited at the crystal–weight interface. If S is the contact area, Ω_S the specific atomic volume of the crystal, and F the weight, then after the deposition of δN particles the weight will have risen by a height $\delta H = \Omega \delta N/S$, i.e., the thermodynamic potential of the crystal–medium–weight system in the field of gravity will have increased by $F\delta H$. Moreover the transfer of particles will increase the crystal potential by $\mu_S \delta N$ and reduce the environment potential by $\mu_M \delta N$. The phase equilibrium is characterized by the fact that the thermodynamic potential of the system is minimal with respect to possible transfers of particles from one phase to another. Therefore we must have $\mu_S \delta N - \mu_M \delta N + F\delta H = 0$, or

$$\mu_M(P, T) = \mu_S(P, T) + \Pi \Omega_S , \tag{1.14}$$

where $\Pi = F/S$ is the pressure of the load on the crystal. It should be emphasized that (1.14) holds only for those areas where the external pressure Π is applied. For the lateral surfaces of the crystal, condition (1.1) remains in force, the only difference being that the expression for the chemical potential of the crystal must take its stressed state into account. Since the elastic energy is proportional to the square of the stresses, the shift in equilibrium caused by this effect is much weaker (by a factor of $\sim \Pi/E$, where E is the Young modulus) than the shift

$$\Delta T_0/T_0 = \Pi \Omega_S/\Delta H , \tag{1.15}$$

which is found from (1.14). Here, ΔT_0 is the change in equilibrium temperature due to the external pressure Π. Relation (1.15) can also be read from right to

---

[2] The film does not pour out, because of the action of molecular (Van der Waals) forces [1.4–6], the repulsion forces of like-charged outer shells of the Debye layers near the surfaces, etc. The tendency of these forces to increase the thickness of the film, h, is equivalent to the existence in it of a so-called disjoining pressure and to a reduction in the chemical potential of the film as compared to that of the bulk liquid. In the case of Van der Waals forces, the disjoining pressure is equal to $B_n h^{-n}$, where $n = 3$–4, $B_3 = 10^{-15}$–10^{-14} erg, and $B_4 = 10^{-20}$–10^{19} erg cm. In addition to the Van der Waals and electrostatic (Debye) components, there is also a so-called structural (entropy) component of the disjoining pressure. This component results from partial ordering of the liquid near solid surfaces. For further references see [1.7b, c].

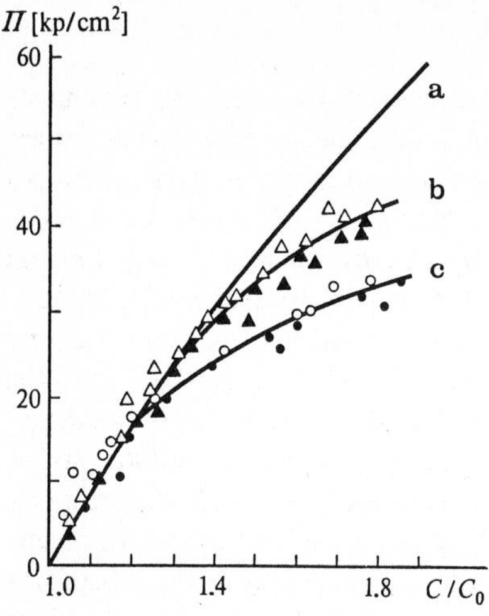

Fig.1.3a–c. Pressure which a crystal of potash alum can develop at different supersaturations on a load lying on it [1.7]. **(a)** Theoretical curve taking into account terms of the second order with respect to pressure; **(b)** experiment, (111) face; **(c)** experiment, (110) face. The *shaded circles* and *triangles* indicate that the crystal is growing, the *open symbols* that growth has terminated

left: the maximum load which can be lifted by a crystal in contact with a melt supercooled by ΔT_0 degrees is equal to ΠS. The pressure Π developed by the crystal is called the crystallization pressure. The dependence of the maximum load which can be lifted by a crystal of potash alum at different supersaturations in solution is given in Fig. 1.3 [1.8]. The shift in the equilibrium concentration of solution is found from condition $\mu_S(PT) + \Pi\Omega_S = \mu_M(P, T, C)$ at $P, T =$ const. For the ideal solution

$$\Delta C/C_0 = \Pi\Omega_S/kT ,$$

and at $\Delta C/C_0 \equiv\ \simeq 10^{-3}(0.1\%)$, $T \simeq 300$ K, $\Omega_S \simeq 3 \times 10^{-23}$ cm^3, the result is $\Pi \simeq 14$ kp/cm^2.

1.2 Surface Energy and Periodic Bond Chains

1.2.1 Surface Energy

The minimum work required to build a unit interface at a constant volume and temperature in the system is called the specific free-surface energy and denoted α. A new interface can be created (1) by stretching (elastically deforming) the initial surface, (2) by transferring a certain number of particles from one phase to another and forming a bump or depression on the surface, or (3) by "cutting" each of the initially homogeneous phase along the future interface and gluing the "halves" of the different phases together along the resulting surfaces. If the two phases are liquid, the minimum work is equal in all three cases. The work depends exclusively on the state of the surface layer, and this in turn depends solely on the temperature and chemical potential of the contacting phases, i.e., on parameters which remain unchanged after the manipulations described. Extension or compression of a liquid film in a classical experiment to determine the surface tension does not affect the state of the surface layer of this film, because the new molecules readily shift to the surface layer on extension or return into the bulk of the film on compression.

The situation is different when one of the phases is solid (crystalline). On extension of the surface (together with the body, of course) no such transfer of molecules is possible, and therefore not only the surface area, but also the density and mutual arrangement of the particles in the surface layer undergo changes. Therefore the work required to build a unit surface of the crystal by this method differs from the work expended in adding particles or "cutting and gluing," which do not distort the state of the surface layer in the sense mentioned above. Hence the surface energy and tension equal one another numerically for the liquid–liquid interface, but differ substantially for the interface between any phase and the solid. Hereafter we shall only use the surface energy α.

The surface energy of the solid–vacuum interface at absolute zero ($T = 0$) is approximately equal to the half-sum of the energies of all the interatomic bonds[3] which must be ruptured in order to obtain the surface under consideration (a crystal rupture results in the formation of two free surfaces). In actuality, the surface atoms regroup after the bonds break, and the surface energy reduces, but this change probably does not exceed 10%–20%. Such relaxation (reconstruction) effects will hereafter be ignored.

1.2.2 Periodic Bond Chains and Estimates of the Surface Energy

The binding energy of a pair of atoms (or molecules) reduces with increasing distance between them.[4] The bond between the atoms (or molecules) of each pair can be represented as a segment of the line joining the atoms, assuming that the segment length is proportional to the binding energy. Then all the bonds in a crystal may be classified in groups represented by segments of equal length. All segments of a specific length and orientation can be considered to be lying on chains, i.e., straight lines with crystallographic indices characterizing segment orientations. Chains consisting of segments of a given length and direction are arranged periodically in the crystal and are therefore called periodic bond chains (PBCs) [1.9–11]. In a simple cubic lattice with spacing a, the PBCs between the first nearest neighbors form three systems, in each of which the chains are parallel to one of the cube edges and are spaced at a. The PBCs between the second nearest neighbors are parallel to various direc-

[3] The energy of a bond is the work that must be expended to rupture the bond.

[4] That is, of course, if the distance itself exceedsthat corresponding to the minimum potential energy of the pair under consideration.

tions of the type [110] and are spaced at $a/\sqrt{2}$. The PBCs between the third nearest neighbors are parallel to [111] and $a/\sqrt{3}$ distant from each other, etc. The PBC diagram for a two-dimensional square lattice is shown in Fig. 1.4.

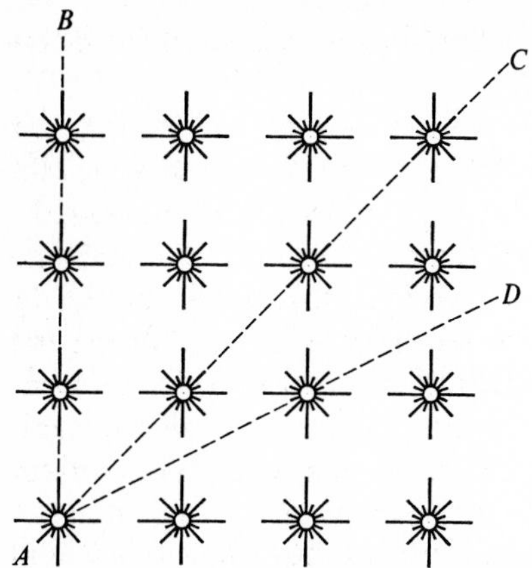

Fig.1.4. Periodic bond chains (PBCs) in a simple cubic lattice. Each *circle* denotes an atom or molecule. Each segment is directed towards the neighbor with which the given atom is bound; the segment length is proportional to the binding energy. A chain is made up of equivalent segments lying on the same straight line

The surface energy is determined by the energies of those bonds which belong to PBCs intersecting a given surface. The number of such PBCs is infinitely large, and the contribution from PBCs of more and more distant neighbors is ever-reducing due to the decrease in the energy of the corresponding bonds. Therefore in rough estimates for crystals with nonionic bonds, one usually restricts oneself to the first terms of this infinite sum, which correspond to the interaction of the nearest neighbors. For instance, in the first-nearest-neighbor approximation, the surface energy of the (100) face of a simple cubic lattice is $\alpha_{100} = \varepsilon_1/2a^2$, where ε_1 is the binding energy of the nearest neighbors. Taking into account the second nearest neighbors with a binding energy ε_2 gives $\alpha_{100} = \varepsilon_1/2a^2 + 4\varepsilon_2/2a^2$, because each surface atom has four uncompensated bonds ε_2.

The binding energy can be estimated from the heat of evaporation of the crystal, assuming that the number of bonds ruptured upon evaporation of each building block (atom, molecule, or aggregate) is half the number of bonds between this block and its neighbors in the crystal bulk (Sect. 1.3). For example, in the fcc lattice, which is typical of many metals, each atom in the bulk has $Z_1 = 12$ nearest neighbors, and the heats of evaporation of the metals are $\Delta H = 20$–60 kcal/mol, i.e., 1.3–4×10^{-12} erg per atom. At $\Delta H = 40$ kcal/mol, the energy of one uncompensated bond, i.e., half the energy of an unbroken bond, is $\varepsilon_1/2 = \Delta H/Z_1 = 3.3$ kcal/mol $= 2.3 \times 10^{-13}$ erg, see (1.22). Each atom in the surface layer (111) is associated with three uncompensated bonds, i.e., three PBCs of the type [110], and a surface $a^3\sqrt{3/8} \simeq 4 \times 10^{-16}$ cm^2 (if the lattice parameter is $\sim$4 Å, the interatomic distance is $\sim$ 3 Å). Therefore in the nearest-neighbor approximation for metals, $\alpha_{111} = 8\sqrt{3}\,\Delta H/a^2 Z_1 \simeq 2 \times 10^3$ erg/cm^2. In molecular crystals the heats of evaporation equal 10–

20 kcal/mol, and the intermolecular distances are 2–3 times larger; accordingly, the surface energies are 1–2 orders of magnitude lower, and amount to several tens of ergs per square centimeter.

At the crystal–liquid or crystal–solid interface, the bonds of the surface atoms are partially saturated, and therefore the surface energies are much lower than at the boundary with the vacuum. When estimating the surface energy of the crystal–melt interface, use is made of the above-mentioned relation, where ΔH is the heat of fusion, typically only 2–3 kcal/mol for metals. Accordingly, in this case $\alpha \simeq 100$ erg/cm^2.

Because of the long-range nature of Coulomb interaction, calculation of the surface energy of ionic crystals requires summation over a large number of neighbors. The relevant calculations [1.10] for the (100) face of a simple cubic lattice whose points are occupied by ions with a charge $\pm Ze$ lead to the expression

$$\alpha = 0.0326\,(Ze)^2/a^3\,,$$

where $e = 4.8 \times 10^{-10}$ cgs units is the electron charge, and a is the distance between the nearest positive and negative ions along the $\langle 100 \rangle$ directions (i.e., $a = 2.82$ Å for NaCl).

Experimental techniques and the results of measurements of interfacial energies, primarily for metallic and ionic systems, are reviewed in [1.1.2]. Tables 2.2, 3 summarize some results for metals.

The surface energy of the boundary between the crystal and any medium is anisotropic, i.e., it depends on the crystallographic orientation of the boundary. Since the main contribution to the surface energy is made by the PBC chains of the strongest bonds, the values of α are minimal for those surfaces in which the greatest number of strong-bond chains lie and which are thus intersected by the smallest number of such chains. For a simple cubic lattice these are primarily the faces of cube (100). They are parallel to two of the three systems of bond chains between the first nearest neighbors, and perpendicular to the chains of the third system, and are also intersected by all four systems of chains of second nearest neighbors. The faces of rhombododecahedron (110) contain only one of the systems of the first bonds, but also one system of the second. Accordingly, a contribution to their surface energy is made by two systems of chains of the first nearest neighbors and three systems of the second. The faces of octahedron (111) do not contain a single chain of the first nearest neighbors, but they contain three systems of the second.

Faces of any lattice which are parallel to at least two systems of chains of the strongest bonds are called F faces, faces parallel to one such system, S faces, and faces containing none, K faces. The surface density of atoms is usually highest for F faces (Sect. 5.2.3), and therefore they are said to be close packed. The symbols F, S, and K stand for flat, stepped, and kinked.

Estimating the surface energy is more complicated for faces which are not parallel to the symmetry planes in multicomponent crystals. Indeed, when the bond between dissimilar atoms is ruptured, it is not clear which part of the energy of this bond should be referred to which of the two surfaces formed. In such cases dividing into equal parts can only yield an approximate value.

1.2.3 Surface Energy Anisotropy

Let us now establish the nature of surface energy anisotropy. We first consider the energies of surfaces only slightly deflected from the F faces. Such surfaces are called vicinals. In Fig. 1.5a this is plane $C_1C_4C_4C_1$ with indices (1, 10, 0), which was obtained by a "rotation" about the x axis by the angle θ (in Fig. 1.5a, θ = arctan 0.1). Out of geometric necessity the vicinal consists of flat terraces, i.e., areas made up of a close-packed F face and the end faces of broken-off flat nets parallel to this face (in Fig. 1.5a the end faces, i.e., the step rises, are hatched). The edges of uncompleted nets are called steps. The work α_l that must be performed in order to build a step of unit length at constant volume and temperature is the specific free energy of the step (erg/cm). At $T = 0$ this energy depends on the number of PBCs intersecting the step rise, as is the case with surfaces.

In the simplest case, in which a vicinal face is formed by equidistant steps of elementary height a, its specific free surface energy $\alpha(\theta)$ is the sum of the energies of the steps and flat terraces:

$$\alpha(\theta) = (\alpha_l/a)\sin\theta + \alpha(0)\cos\theta\,, \tag{1.16}$$

where $\alpha(0)$ is the surface energy of a close-packed face.

The term reflecting the interaction between the steps is omitted in this expression for $\alpha(\theta)$, and the sparser the steps are, i.e., the smaller θ is, the more correct is the omission.

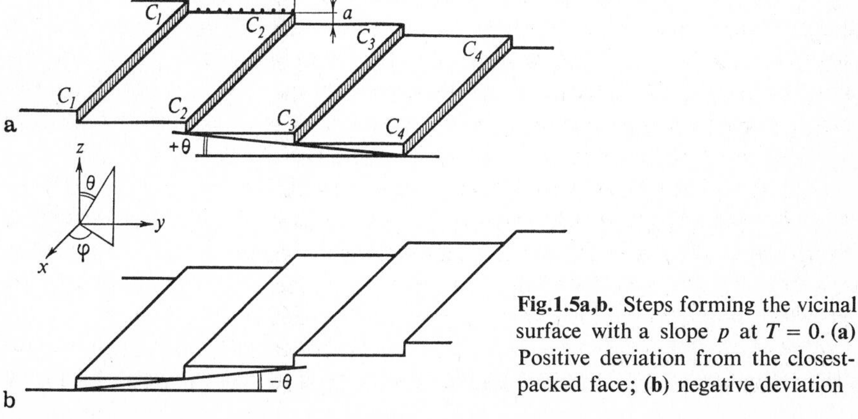

Fig.1.5a,b. Steps forming the vicinal surface with a slope p at $T = 0$. **(a)** Positive deviation from the closest-packed face; **(b)** negative deviation

The step interaction is of an electromagnetic and elastic nature, and is attributable to weak Van der Waals forces and forces due to electric dipoles of even charges generally associated with the steps. Each step gives rise to an elastic field in the crystal. The overlapping of these fields causes the elastic interaction [1.13, 14].

We now consider a vicinal obtained by deflection from the F face to the opposite side, i.e., by angle $-\theta$. Its energy is

$$\alpha(-\theta) = (\alpha_{\bar{l}}/a)\sin\theta + \alpha(0)\cos\theta\,, \tag{1.17}$$

where $\alpha_{\bar{l}}$ stands for the energy of the steps facing the opposite side (Fig. 1.5b). If the (x, z) plane is the symmetry plane of the crystal, then $\alpha_{\bar{l}} = \alpha_l$. Function $\alpha(\theta)$, defined by relations (1.16, 17), is continuous at all values of θ, but its derivatives with respect to θ at $\theta \geq 0$ and $\theta \leq 0$ are different: $\partial\alpha/\partial\theta = \alpha_l/a$ at $\theta \leq 0$, and $\partial\alpha/\partial\theta = -\alpha/\alpha_{\bar{l}}$ at $\theta \leq 0$. In other words, derivative $\partial\alpha/\partial\theta$ experiences a jump $(\alpha_l + \alpha_{\bar{l}})/a$ at $\theta = 0$. Graphically, this means that for values of θ corresponding to F faces (for a cubic and a tetragonal lattice the value is $\theta = 0, \pm\frac{\pi}{2}, \pi, \ldots$), the surface energy has sharp minima, as can be seen from Fig. 1.6a. In the vicinity of such a minimum, function $\alpha(\theta)$ cannot be expanded into a series in powers of θ because of the absence of a derivative at the point of minimum; in other words, the orientations corresponding to such faces are singular. The faces themselves are also called singular.

We shall now see how the energy of a surface depends on its orientation outside the vicinity of singular faces, i.e., what the anisotropy of the surface energy is in general. In other words, we shall establish the function $\alpha(\boldsymbol{n})$, where $\boldsymbol{n}$ is the unitary normal to a surface element whose surface energy is equal to α. The direction of the normal is assigned by two angles, φ and θ, which are indicated in Fig. 1.5:

$$\boldsymbol{n} = \{\sin\theta\cos\varphi, \sin\theta\sin\varphi, \cos\theta\}\,.$$

Let us consider the case of a simple rhombic lattice with the strongest bonds along three mutually perpendicular directions; we restrict ourselves to the corresponding PBCs and assume that the bond strengths in each direction are different. We first investigate the two-dimensional cross section $\varphi = \frac{\pi}{2}$, i.e., the energy change only upon rotation of the surface about the x axis by an arbitrary angle θ (Fig. 1.5). In this case, the steps are parallel to the x axis, i.e., to the $\langle 100\rangle$ direction, and $\alpha_l = \alpha_{010}$ and $\alpha(0) = \alpha_{001}$. Here, α_{001} and α_{010} are assumed to be equal to the specific surface energies of the (001) and (010) faces. Then (1.16) and (1.17) can be joined:

$$\alpha(\theta) = \alpha_{010}|\sin\theta| + \alpha_{001}|\cos\theta|\,. \tag{1.18}$$

The signs of the modulus physically reflect the fact that the contributions to

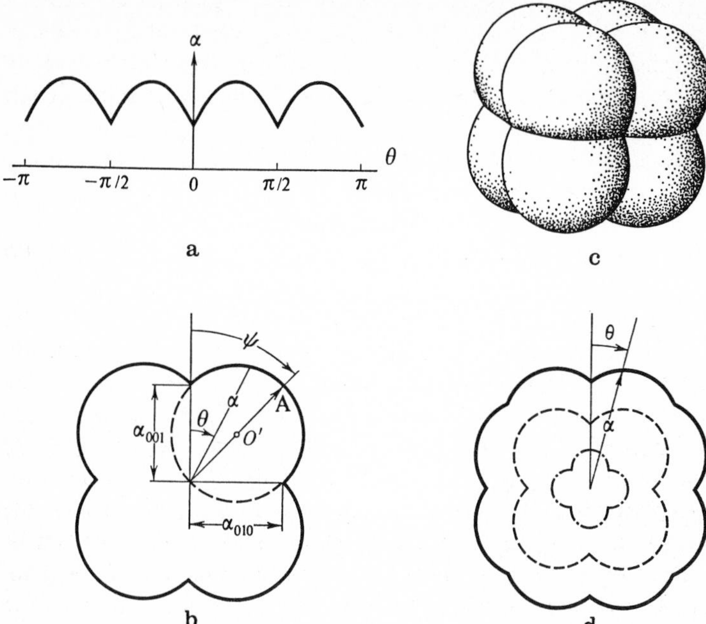

Fig.1.6a–d. Anisotropy of the surface energy $\alpha(\vartheta)$ for a cubic crystal at $T = 0$. (**a**) In rectangular coordinates; (**b**) in polar coordinates in a two-dimensional section; (**c**) spatial polar diagram in the first-nearest-neighbor approximation; (**d**) two-dimensional diagram in the second-nearest-neighbor approximation. The *dashed line* denotes the contribution of bonds with the first nearest neighbors, and the *dot-dashed line* that of bonds with the second nearest neighbors

the surface energy from the terraces and steps are always positive. In the region $0 \le \theta \le \frac{\pi}{2}$, where the sine and cosine are positive, the modulus signs can be omitted, giving (1.18) the form:

$$\alpha(\theta) = \sqrt{\alpha_{001}^2 + \alpha_{010}^2}\cos(\psi - \theta)\,, \quad \tan\psi = \alpha_{010}/\alpha_{001}\,. \tag{1.19}$$

Let us regard α and θ as the polar coordinates of a point on a plane, with α being the distance to the origin, i.e., the length of the radius vector, and θ the angle formed by the radius vector and the vertical axis and measured in a clockwise direction (Fig. 1.6b). In these coordinates (1.19) is the equation of a circle with diameter $\sqrt{\alpha_{001}^2 + \alpha_{010}^2}$ and center at point O′ ($\alpha = \alpha_{001}^2 + \alpha_{010}^2/2$, $\theta = \psi$). Such a circle is constructed in Fig. 1.6b for a cubic crystal, $\alpha_{001} = \alpha_{010}$. Recall that only the part of the circle enclosed in the first quadrant $0 < \theta < \frac{\pi}{2}$ has physical meaning. In accordance with (1.18), function $\alpha(\theta)$ is symmetric about reflections in the horizontal ($\theta = \frac{\pi}{2}$) and vertical ($\theta = 0$) planes, i.e., it is invariant to replacements $\theta \to -\theta$ and $\theta \to \pi + \theta$. Reproducing the circular arc (1.19) located in the first quadrant with the aid of these operations, we obtain the figure depicted by the solid line in Fig. 1.6b. This gives the anisotropy of the surface energy due to the selected PBCs, at $\varphi = \frac{\pi}{2}$, i.e., for the zone of faces parallel to the x axis.

Note that in order to obtain the expression $\alpha(\theta, \varphi)$ in the three-dimensional case, when φ is arbitrary, (1.18) can be written as a scalar product

$$\alpha = (\boldsymbol{A}\boldsymbol{n}) , \tag{1.20}$$

where $\boldsymbol{A} = \{\alpha_{010}, \alpha_{001}\}$, $\boldsymbol{n} = \{\sin\theta, \cos\theta\}$, and θ corresponds to the first quadrant. If angle φ is arbitrary, i.e., the orientation of the steps in the projection onto the (x, y) plane is chosen arbitrarily, then

$$a\alpha_l = a\alpha_{\bar{l}} = \alpha_{100}|\cos\varphi| + \alpha_{010}|\sin\varphi| . \tag{1.21}$$

Joining (1.21) to (1.16 and 1.17), we convince ourselves of the validity of (1.20) for the three-dimensional case with a vector $\boldsymbol{A} = \{\alpha_{100}, \alpha_{010}, \alpha_{001}\}$. In three-dimensional polar coordinates $(\alpha, \varphi, \theta)$ expression (1.20) describes a sphere with diameter $\sqrt{\alpha_{100}^2 + \alpha_{010}^2 + \alpha_{001}^2}$ and center at point $\alpha = \sqrt{\alpha_{100}^2 + \alpha_{010}^2 + \alpha_{001}^2}/2$, $\varphi = \arctan \alpha_{010}/\alpha_{100}$, $\theta = \arctan \sqrt{\alpha_{100}^2 + \alpha_{010}^2}/\alpha_{001}$. For a simple cubic lattice, reproducing the part of this sphere located in the first quadrant with the aid of reflections in the coordinate planes results in Fig. 1.6c. We see that the surface energy has six sharp singular minima corresponding to the F faces. At the values of φ and θ related to these faces, the derivatives of α with respect to both these angles are discontinuous. Faces parallel to only one system of PBCs are singular solely with respect to one of the angles (φ or θ); these are the sharp neckings in Fig. 1.6c.

As mentioned above, the energy of any surface is defined by the sum of the contributions from the different PBC systems, each contribution being independent of the others. Therefore, to find the part of the surface energy ascribable to the next nearest neighbors, one must determine the corresponding new PBC system and follow the reasoning just given for the PBCs of the first nearest neighbors. For instance, in a simple cubic lattice the atoms located at the ends of the diagonals of a face of a unit cell are the second nearest neighbors as regards distances and bond strength. Here, the (111) faces contain three systems of PBCs of the second nearest neighbors, and the (100) faces two systems, while the (110) faces contain one system. Corresponding to these faces are the sharp minima and neckings in the polar space diagram for the part of the surface energy due exclusively to the second nearest neighbors. The intersection of such a diagram by the yz plane (Fig. 1.5) is represented by the dot-dashed curve in Fig. 1.6d. It has minima in the [110] directions and a smaller radial size than the similar curve for the first nearest neighbors shown as the dashed line in Fig. 1.6d. Accordingly, the absolute depth of the minima and the value of the jump $\partial\alpha/\partial\vartheta$ for singular faces in the second-nearest-neighbor diagram are less than in that of the first nearest neighbors. The bonds of the first nearest neighbors are directed along ⟨100⟩, and those of the second nearest neighbors along ⟨110⟩. This is reflected in the mutual turn of the energy diagrams by $\frac{\pi}{4}$ (Fig. 1.6d). The summary diagram, which represents the anisotropy of the surface energy with due regard for both the first and second nearest neighbors (the continuous curve in Fig. 1.6d) consequently has sharp minima for both the (100) and (110) faces in the case at hand. Taking the next nearest neighbors into account adds new minima (and new neckings in the three-dimensional diagram) corresponding to all the possible PBC systems, i.e., to all the rational orientations. The sharpness and depth of these minima, however, decrease abruptly as the face indices become more complicated. As we shall see in the next section, these shallow minima disappear altogether because of the thermal fluctuations of the steps at $T > 0$.

1.3 Atomic Structure of the Surface

1.3.1 Surface Configurations and Their Energies

The surface of a perfect crystal at $T > 0$ is schematically represented in Fig. 1.7 for a simple cubic lattice. In contrast to Fig. 1.5, the step is not straight here: atoms or molecules may leave the step end face (as well as the surface layer) because of thermal motion, and shift either into the environment or onto the surface, into one of the surface states with a different number of bonds with the lattice. The various configurations on the surface are numbered in Fig. 1.7 and listed in the first two columns of Table 1.1.

The relative amounts of particles in all the indicated positions depend on their energies in these positions and on the temperature.

In homopolar, molecular, and, less precisely, metallic crystals with short-range bonds, the energies of the particles in the various positions are estimated by the number of first, second, third, etc., nearest neighbors. The energy of

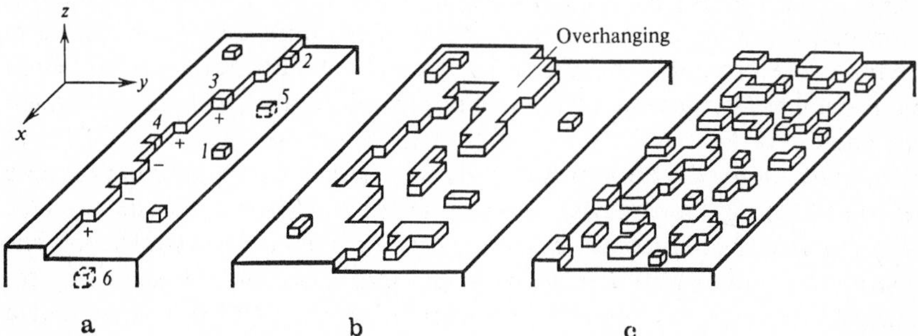

Fig. 1.7a–c. Surface of a crystal with a simple cubic lattice. **(a)** Different types of atomic positions at $T > 0$ (see Table 1.1). The *plus* and *minus signs* indicate positive and negative kinks on the step; **(b)** strongly developed step roughness, including "overhanging;" **(c)** an atomically rough surface with two possible levels of surface atoms

Table 1.1. Positions of the atoms in a crystal

Position no. in Fig. 1.7 (number of nearest neighbors)	Name of position	Numbers of neighbors (1st, 2nd, 3rd)
1	On the surface	1 4 4
2	At the step	2 6 4
3	In (at) the kink	3 6 4
4	In the step	4 6 4
5	In the surface layer	5 8 4
6	In the bulk	6 12 8

Table 1.2. Numbers of first three nearest neighbors for an atom in half-crystal ($\varepsilon_{1/2}$) and adsorption (ε_s) positions at different faces [1.15]

	Face	ε_s		Face	ε_s
Simple cubic	100	1 4 4	Hexagonal close-packed	0001	3 3 1
$\varepsilon_{1/2}=3\ 6\ 4$	110	2 5 2	$\varepsilon_{1/2}=6\ 3\ 1$	1010	4 4 0
	111	3 3 4		0112	4 2 1
	211	3 5 3		1011	5 2 0
				1120	5 2 1
Body-centered cubic	100	4 1 4			
$\varepsilon_{1/2}=4\ 3\ 6$	110	2 2 5	Diamond	100	1 5 5
	111	4 3 3	$\varepsilon_{1/2}=2\ 6\ 6$	110	2 4 6
	211	3 3 5		111	1 3 6
				311	2 5 4
Face-centered cubic	100	4 1 2		310	2 6 5
$\varepsilon_{1/2}=6\ 3\ 12$	110	5 2 10			
	111	3 3 9			
	210	6 2 10			
	311	5 3 10			
	531	6 3 0			

an atom which has Z_i neighbors of the ith order is conventionally written $Z_1Z_2Z_3 \ldots$ For instance, let us be satisfied with the accuracy corresponding to the interaction energy between neighbors of the third order inclusive. Then the energy of an atom adsorbed on the surface (Fig. 1.7, position *1*) should be written as 1 4 4. The energies of different positions are given in the third column of Table 1.1.

Of the positions, the one having the greatest importance in growth processes is position 3, in the kink, where the number of bonds is half that of an atom in the crystal bulk. This position is therefore called the half-crystal position, and the corresponding energy is denoted $\varepsilon_{1/2}$. The formulae of energy $\varepsilon_{1/2}$ and of the energy of an atom adsorbed on different faces, ε_s, are listed in Table 1.2.

A kink is called positive or negative depending on whether it starts or ends a new row at a step, i.e., widens or narrows the flat net ending in the step. At steps which on the average coincide in direction with one of PBCs, the number of positive and negative kinks is the same with respect to this PBC.

The importance of the kink in growth processes arises from the fact that the addition of a new particle to a single kink reproduces the initial configuration without changing the number of uncompensated bonds, i.e., the surface energy of the crystal. For this reason the change in internal energy caused by the detachment of an atom or molecule from the kink is equal to the heat of transformation (evaporation, fusion) per particle. The number of the ith neighbors in the kink on any step and any surface is always equal (as is easily verified) to half the number Z_i of bonds with identical neighbors in the crystal bulk. From this follows the estimate of the energy ε_1 of the bond between the nearest neighbors obtained with the aid of the experimentally measured heat of sublimation, ΔH:

$$\varepsilon_1 = 2\Delta H/Z_1 \,. \tag{1.22}$$

The fact that the surface energy is not affected by the detachment of an atom or molecule from the kink also means that the work to be performed for such a detachment is equal to the chemical potential of the crystal. Put differently, the chemical potential of a particle in the kink is equal to that of the crystal. Accordingly, the work of transferring the particle from the kink to the medium is equal to the difference between the chemical potentials of the medium and the crystal. The work of transferring particles from other positions on the surface to the medium is not equal to that difference, because the surface energy then changes.

In calculating the work required to detach ions (or complexes of ions) from different positions, and in determining the surface energy for a heteropolar crystal, we can no longer restrict ourselves to the approximation of even several nearest neighbors. The Coulomb interaction between ions, which slowly decreases with distance, necessitates summation over the entire lattice.

Let us consider, for instance, an ion in a kink at a step parallel to one of the two PBC systems forming the F face (100) of a simple cubic lattice. We assume that the kink is elementary, i.e., it is the end of a semiinfinite uncompleted atomic row forming the step rise. The energy of an ion in the kink must then be the sum of three energies of interaction: ε''' with the ions of the crystal half space bounded by the F face, ε'' with the ions of the half plane bounded by the step, and ε' with the ions of the half line bounded by the kink. Energy ε' is equal to the sum over the unlike-charged ions of the chain:

$$\begin{aligned}\varepsilon' &= \frac{(Ze)^2}{a}\left[1 - \frac{1}{2} + \frac{1}{3} - \cdots + (-1)^{k-1}\frac{1}{k} + \cdots\right] \\ &= \frac{(Ze)^2}{a}\ln 2 = 0.69\frac{(Ze)^2}{a}\,,\end{aligned}$$

where e is the electron charge, Z is the valence of the ion in a given lattice (so that Ze is the charge of the ion), and a is the distance between the ions in the chain.

Summation over the half plane yields the energy $\varepsilon'' = 0.114(Ze)^2/a$, and over the half space, the energy of adsorption on the surface $\varepsilon''' = 0.066(Ze)^2/a$ [1.10]. Thus the total energy of attraction of an ion in the kink is equal to $\varepsilon' + \varepsilon'' + \varepsilon''' = 0.87(Ze)^2/a$. The lower values of ε'' and ε''' as compared to ε' are due to the fact that for instance an ion on the (100) plane is attracted to the crystal primarily by the underlying ion of opposite sign, but is repulsed by its four neighbors in this plane.

With the aid of ε', ε'', and ε''' it is also easy to compute the energies of a number of other positions; for instance, for position 2 (at the step, Fig. 1.7), $\varepsilon'' + \varepsilon''' = 0.18(Ze)^2/a$. A similar calculation can be made not only for an individual ion, but also for a neutral molecule. For instance, a diatomic molecule consisting of a positive and a negative ion is attracted to the half line with an energy equal to $0.39(Ze)^2/a$, to the half plane with an energy equal to $0.29(Ze)^2/a$, and to the half space with an energy equal to $0.13(Ze)^2/a$. Consequently, for a diatomic molecule the energy of detachment from the half-crystal position is $0.75(Ze)^2/a$, i.e., less than for an individual ion. This computation leads to the conclusion that the adsorption energy on the F faces of ionic crystals constitutes a comparatively small fraction of the energy of evaporation, in this case about 1/6. The "standing" orientation of dipole molecules on ionic surfaces may be much more favorable than the "lying" one: in the case of KCl molecule on (100) KCl, $\varepsilon_s = 0.62$ eV (14 kcal/mol) for the former (cation down) and $\varepsilon_s = 0.3$ eV (7 kcal/mol) for the latter [1.19c]. On the F faces of crystals with short-range bonds $\varepsilon_s \simeq 0.3$–$0.5\Delta H$. Unfortunately, no experimental values of ε_s have been obtained so far for ionic crystals.

Another consequence of the repulsion of like-charged ions is that the binding energy of an ion located at a vertex of a crystalline polyhedron is higher than that of an ion on a polyhedron edge, which is higher than that of an ion on the surface. For molecular crystals, where no ionic repulsion occurs, the sequence is reversed.

The binding energies estimated above neglect mutual repulsion between ions. This (Born's) repulsion exists because the inert-gas-like electronic shells of the ions cannot penetrate one another. Calculations taking Born's repulsion into account give

$$[(Ze)^2/2a]\,M(1 - \rho/a) \qquad \text{and} \qquad [(Ze)^2/a]\,[M(1 - \rho/a) - 0.97]$$

for the energies needed to detach an ion or molecule, respectively, from the kink position on the surface. Here, $0.97(Ze)^2/a$ is the energy required to split the molecule into ions, M is the Madelung's constant, ρ is the distance characterizing the repulsion of ions, and $(1 - \rho/a)$ is the factor which takes repulsion in the lattice into account [1.16]. For alkali halide crystals, ρ varies from 0.29 to 0.35, so that $\rho/a \simeq 0.1$. For NaCl, $M = 1.7476$, $\rho = 0.321 \times 10^{-8}$, $a = 2.82 \times 10^{-8}$ cm, and the second formula yields 68 kcal/mol for the energy of molecular evaporation at 0 K.

The foregoing approach in analyzing the energy of atoms on the surface was developed in classical investigations by Kossel, Stransky, and Kaishev, and served as a basis for the molecular-kinetic theory of crystal growth.

1.3.2 Adsorption Layer

Let us now determine the probabilities of some of the above-discussed configurations on the surface of a crystal in equilibrium with another medium [1.17, 18]. We begin with particles which are adsorbed on the surface and form an adsorption layer. If the crystal is surrounded by a vapor with pressure P, then $P/\sqrt{2\pi mkT}$ particles of mass m in impinge on a unit surface per unit time. Some probably very small fraction of them can be reflected elastically and returned to the vapor, while the majority are adsorbed by the surface. On the other hand, each adsorbed particle executing thermal vibrations along the normal to the surface with a frequency $\nu_\perp \simeq 10^{12}$–10^{13} Hz can be desorbed on the average after $\exp(\varepsilon_s/kT)$ vibrations, having "lived" on the surface for a period

$$\tau_s = \nu_\perp^{-1} \exp(\varepsilon_s/kT) . \tag{1.23}$$

Suppose n_s to be the density of the particles adsorbed on the surface. Then, if the desorption flux n_s/τ_s equals the impinging flux $P/\sqrt{2\pi mkT}$, one has, making use of (1.23):

$$n_s = \frac{P}{\nu_\perp \sqrt{2\pi mkT}} \exp(\varepsilon_s/kT) . \tag{1.24}$$

Let us estimate n_s and τ_s for the (111) face of a silicon crystal surrounded by its own vapor saturated at a temperature $T = 1200$ K, i.e., having a pressure $P = 1.2 \times 10^{-7}$ mm Hg $= 1.6 \times 10^{-4}$ dyn/cm^2. The mass m of the silicon atom is $28 \times 1.7 \times 10^{-24}$ g. For the diamond lattice characteristic of silicon, in the nearest-neighbor approximation $\varepsilon_s^1 = 0.5\,\Delta H$ [see Table 1.2 and (1.22)], i.e., $\varepsilon_s = 55.5$ kcal/mol $= 55.5 \times 10^3 \times 4.18 \times 10^7/6.02 \times 10^{23}$ erg/atom $=$ 3.85×10^{-12} erg/atom. Assuming $\nu = 10^{13}$ Hz, we obtain $n_s \simeq 3 \times 10^{10}$ cm^{-2}. For one atomic site on the (111) face of silicon there is $\simeq 10^{-15}$ cm^2 of surface. Under the indicated conditions, therefore, the fraction of the sites occupied by atoms is only 3×10^{-5}. If the temperature is reduced to $T = 1000$ K, and the pressure of the surrounding vapor or the intensity of the molecular beam incident on the surface remains unaltered, the fraction of the occupied sites increases to $\simeq 3 \times 10^{-3}$.

The lifetime τ_s of the Si atom on the surface under the above-mentioned conditions is $\simeq 1.3 \times 10^{-3}$ s ($T = 1200$ K) and $\simeq 1.3 \times 10^{-1}$ ($T = 1000$ K). The adsorbed atoms execute thermal vibrations not only along the normal to the surface, but also parallel to it, which results in their jumping over to neighboring position, i.e., in diffusion along the surface. If the height of the potential barrier for jumping over into a neighboring well of the potential surface relief is U_D, the lifetime in each well between jumps is $\tau_D = \nu_{\parallel}^{-1} \exp(U_D/kT)$, where the vibration frequency $\nu_{\parallel}$ tangential to the surface may be several times lower than the frequency $\nu_{\perp}$ of normal vibrations, but is a value of the same order of magnitude, 10^{12}–10^{13} s^{-1}. In what follows we shall, as a rule, make no distinction between $\nu_{\parallel}$ and $\nu_{\perp}$, putting $\nu_{\parallel} = \nu_{\perp} = \nu$. Therefore the surface diffusion coefficient is

$$D_s \simeq a^2/4\tau_D \simeq (a^2\nu/4)\exp(-U_D/kT)\,. \tag{1.25}$$

During its lifetime τ_s on the surface a particle will have passed an average distance λ_s, which is called the diffusion mean free path[5]:

$$\lambda_s \simeq 2(D_s\tau_s)^{1/2} \simeq a\exp[(\varepsilon_s - U_D)/2kT]\,. \tag{1.26}$$

If $\varepsilon_s \simeq 0.5\,\Delta H$, $U_D \ll \varepsilon_s$, and $\Delta H/kT \simeq 25$, then $\lambda_s \simeq 510^2 a$.

If the activation energy of surface diffusion U_D is comparable to kT or even lower than the thermal energy, so-called nonlocalized adsorption takes place. Then adatoms or admolecules form a two-dimensional gas with nearly free motion of particles along the surface. Such a situation is most probable on the surfaces of organic crystals. In many cases of localized adsorption the relation $U_D \simeq (0.2$–$0.5)\ \varepsilon_s$ is justified.[6] For silicon atoms on (111) Si experiment yields $U_D \simeq 1.1$eV $= 25.3$ kcal/mol [1.19a]. Therefore according to (1.26), at $\varepsilon_s = 55.5$ kcal/mol and $T = 1000$ K the diffusion path along (111) Si is $\lambda_s \simeq 2 \times 10^3 a \simeq 1\,\mu$m (the distance a between the neighboring adsorption positions is $\simeq 4.5$ Å).

If there is a mobile adsorption layer on the surface, the particles arrive at each adsorption position either from neighboring adsorption positions or directly from the gas phase. On a (111) face of the diamond lattice, each adsorption position has six equivalent neighbors; on the (111) face of the fcc lattice there are three such neighbors; on the (100) face of the simple cubic lattice, four, etc. Accordingly, the frequency of arrival of mobile adatoms from neighboring sites at any adsorption position occupying on the average an area $\simeq a^2$ is $\simeq n_s a^2 \nu_{\parallel} \exp(U_D/kT)$ on the (111) face of the diamond lattice. The frequency of arrival directly from the gas phase is $\simeq Pa^2/\sqrt{2\pi mkT}$. With due consideration for (1.24) it is easy to see that the ratio of these frequencies is $(\nu_{\parallel}/\nu_{\perp}) \times \exp[(\varepsilon_s - U_D)/kT]$. As noted above, $(\nu_{\parallel}/\nu_{\perp})$ is of the order of unity, and ε_s usually appreciably exceeds U_D. Therefore the frequency of arrival of atoms at

[5] *Burton* et al. [1.17] adopted $\lambda_s = (D_s\tau_s)^{1/2}$ in their paper.

[6] Another useful empirical rule for estimating ε_s is known for metals: $\varepsilon_s/kT_0 \simeq 8$–$10$, where T_0 is the melting temperature [1.19b].

each adsorption position from neighboring positions in the adlayer is much higher than that of arrival directly from the gas phase: for (111) Si at $T = 1000$ K the ratio of the indicated frequencies $\simeq \exp(\varepsilon_s - U_D)/kT \simeq 4 \times 10^6 \gg 1$. This circumstance predetermines the leading role of surface diffusion in growth from vapors (Sect. 3.2.1) and molecular beams.

In crystal growth from the gas phase with the participation of chemical reactions (growing of films of Si, GaAs, refractory metals, etc), a quite different picture of the adsorption layer should be expected. The Si-H–Cl gas system contains molecules of H_2, $SiCl_2$, $SiCl_4$, SiCl, etc., and, in considerably smaller amounts, atoms of Si, Cl, and H. All these components can be adsorbed on the surface of a growing crystal, often with the formation of highly stable chemical bonds. For instance, the adsorption energy of hydrogen atoms on (111) Si is $\simeq$ 73 kcal/mol ([1.20]: see also [Ref. 1.21, p. F-215]), and the Si–Cl binding energy is $\simeq$ 105 kcal/mol [Ref. 1.21, p. F-215]. Consistent calculations of the adsorption coverages Θ of faces in contact with chemically complex gas systems are difficult even under equilibrium conditilons, because of the lack of accurate data on the binding energies on the surface, on the interactions of adparticles, on vibration states, and hence on partition functions of the adsorbed particles. Nevertheless, approximate estimates of the degree of coverage of the faces have been made [1.22]. The estimates obtained show that in Si–H–Cl and Ga–As–H–Cl systems, dense adsorption layers ($0.1 \lesssim \Theta \lesssim 1$) of faces (111) Si and (111) GaAs should be expected; the faces of both materials must be covered largely by atoms of hydrogen and chlorine and, in much smaller amounts, Si and Ga, and by molecules of $SiCl_2$, SiCl, GaCl, As_3, As_4. The large fraction of occupied adsorption sites in dense chemical adsorption layers is the reason why the values of the diffusion lengths λ_s are smaller there than in the dilute adlayers at the crystal–vapor interface of a one-component system. The value of λ_s is also reduced because the rotation of admolecules is stopped at the moments when they jump over to neighboring adsorption positions (i.e., because of the entropy activation barrier). So far there are few experimental data on the compositions and densities of the adsorption layers on growing crystals [1.23]. These data confirm the existence of the dense chemisorbed layer in the CVD process. Owing to the high strength of the chemical bonds between the surface and the particles which do not enter the crystal (Cl, H), these particles can be removed from the growing surfaces only by chemical reactions, but not by thermal desorption. Such reactions, together with those which produce the crystallizing substance, determine the growth process. The saturation of surface dangling bonds reduces the surface energy and the linear step energy, thus significantly enhancing the two-dimensional nucleation growth mechanism in chemical vapor deposition. The mechanisms of chemical vapor deposition (CVD) of Si and GaAs have been dicussed in many papers [1.22–24]. What follows refers predominantly to systems without chemical reactions.

1.3.3 Step Roughness

Because of the thermal vibrations of the atoms forming the step rise, these atoms may escape from the end face, leaving a "hole" in it, i.e., a pair of adjacent kinks of opposite sign. (Transitions of a particle from the step to the adsorption layer require less energy expenditure than escapes to the vapor phase, and are therefore much more frequent.) On the other hand, the adsorption layer particles migrating over the surface may join the step. As a result the step ceases to be straight, that is, smooth on the atomic scale, and acquires a certain number of kinks: it becomes rough. We shall restrict ourselves to the model of a step containing only kinks of atomic (unitary) size. Strictly speaking, this model is applicable only for low temperatures. We denote as n_+ and n_- the numbers of unitary positive and negative kinks per unit length of a step. Suppose that the density of the smooth sites on the step rise, where no kinks are present, is n_0. Then it is obvious that the sum

$$n_+ + n_- + n_0 = 1/a = n\,, \tag{1.27}$$

i.e., the linear density of the atoms on the step.

Assume that the appearance of one kink increases the step energy by w. In the nearest-neighbor approximation at the crystal–vapor interface $w = \varepsilon_1/2$, i.e., half of the binding energy between the nearest neighbors. Therefore the product of the ratios of the densities of sites with kinks of one sign or the other to the densitity of smooth sites is

$$n_+n_-/n_0^2 = \eta^2\,, \qquad \eta = \exp(-\,w/kT)\,. \tag{1.28}$$

If the average orientation of the step forms a small angle φ with the direction of the PBCs of the nearest neighbors, then

$$n_+ - n_- = \varphi/a\,. \tag{1.29}$$

Solving (1.27–29) for n_+ and n_-, we obtain the average distance between the kinks on the step with orientation [1.17]:

$$\lambda_k = \frac{1}{n_+ + n_-} = \lambda_{k0}\left[1 - \frac{1}{2}\left(\frac{\lambda_{k0}}{a}\right)^2\varphi^2\right], \tag{1.30a}$$

where

$$\lambda_{k0} = a\left[1 + \frac{1}{2}\exp(w/kT)\right] \tag{1.30b}$$

and actually determines the average distance between the kinks for a step coinciding with PBC in direction. If $\Delta H/kT \simeq 25$ and $w = \Delta H/6$, then $w/kT \simeq 4$

and $\lambda_{k0} \simeq 30a$. With an increase in temperature the kink density increases exponentially, and with a decrease it falls off. For silicon in contact with its vapor at $T = 1000$ K, $\lambda_{k0} \simeq 6 \times 10^5 a$.

The above reasoning is applicable, of course, not only to the crystal–vapor interface, but also to the crystal–melt and crystal–solution interfaces. In these latter cases, however, the estimates of w are less reliable, because the relation $w = \Delta H/Z_1$, where Z_1 is the number of bonds of an atom with the nearest neighbors in the bulk, remains only approximately true. For silicon, $\Delta H/kT_0 \simeq 3.3$($\Delta H \simeq 11.1$ kcal/mol, melting point $T_0 = 1685$ K), and $w/kT = 1.6$. Therefore at the crystal–melt boundary, $\lambda_{k0} = 3.6a$. Thus if an absolutely straight step which is smooth in the atomic scale is "made" on the surface, at $T > 0$ it will become curved locally, on the atomic scale, remaining straight only on the macroscale, i.e., on distances large as compared to the interatomic distances. This means that the appearance of kinks on a straight step is energetically favorable, because it reduces its free (and not total!) linear energy α_l. This reduction occurs in accordance with the expression

$$\alpha_l = U_l - TS_l \tag{1.31}$$

and is due to an increase in entropy with increasing intensity of fluctuations and with the corresponding increase in the number of kinks.

Here, $U_l = (n + n_+ + n_-)w$ is the total energy of the step unit length, with $(n + n_+ + n_-)$ being the total number of uncompensated bounds per unit length of an atomically smooth step ($n = 1/a$); and $S_l = k \ln (n!/n_+!n_-!n_0!)$ is the entropy of an ideal one-dimensional "gas" on the step, which consists of n_+ positive kinks, n_- negative kinks, and n_0 sites with no kinks. Indeed, for a step parallel to the close-packing direction we have $\varphi = 0$ and $n_+ = n_-$; it is easy to show that on solving the (1.27–29) for $\varphi = 0$ we have

$$\frac{n_+}{n} = \frac{n_-}{n} = \frac{\eta}{1 + 2\eta}, \quad \frac{n_0}{n} = \frac{1}{1 + 2\eta}; \quad \eta = \exp(-w/kT).$$

Substituting these values into the expression for entropy S_l and using (1.31), we get

$$\alpha_l = nw - nkT \ln(1 + 2\eta) = -nkT \ln \eta(1 + 2\eta). \tag{1.32}$$

It follows that the specific free linear energy of the step, α_l, reduces to zero at $\eta = \eta_R = \frac{1}{2}$, i.e., at $w/kT = \ln 2 \simeq 0.69$. At higher temperatures α_l becomes negative [1.25a], and the step can no longer exist (Sect. 1.3.4).

The above consideration of the roughness of a step at the crystal–vapor interface can also be generalized for the crystal–melt case. Here the so-called lattice model of a melt is used: it is assumed that the atoms of the liquid are packed in the same lattice as in the crystal, but the binding energies in the two phases are different. Let ε_{SS} be the binding energy between the atoms in a crystal (previously denoted ε_1), ε_{MM} that in the melt (medium), and let ε_{SM} be the binding energy of two atoms, one of which belongs to the crystal and one to the melt. Then the excess energy of the boundary per atomic site will be

$$w = \frac{\varepsilon_{SS} + \varepsilon_{MM}}{2} - \varepsilon_{SM} \quad \text{and} \quad \eta = \exp(-w/kT).$$

For the crystal–vapor interface, $\varepsilon_{SM} = \varepsilon_{MM} = 0$, i.e., $w = \varepsilon_{SS}/2 = \varepsilon_1/2$.

1.3.4 Surface Roughness

The vanishing of the linear energy of the step implies a qualitative change in the structure of the entire surface. Indeed, a step is the boundary of an uncompleted layer; it separates a plane net (half plane), which is largely filled with atoms (with the exception of individual vacancies, i.e., positions of type 5 in Fig. 1.7), from a net whose filling is limited to only a few molecules (of type 1 in Fig. 1.7). If, however, the interfacial energy of the two regions becomes negative, such regions can no longer exist separately, and their mutual dissolution and homogenization take place. A well-known example of homogenization is mutual dissolution of a liquid and a vapor at the critical point where the surface energy between these phases reduces to zero. The physical picture of the vanishing of α_l, i.e., of the attainment of equality $U_l = TS_l$, consists in the fact that with an increase in temperature the step fluctuations grow continuously stronger, the step "smears out" over an ever-widening band near its average position, and finally at some critical temperature T_R this width becomes infinite, i.e., homogenization of the two-dimensional gas and the condensate occurs. At above-critical temperatures, the steps no longer exist, and the surface of the F face is no longer smooth on the atomic scale, but consists instead of more or less accidental atomic clusters and separate atoms, as shown schematically in the last of the surfaces in Figs. 1.7, 1.8. The kinks on such a surface are distributed uniformly over the whole area and have a much higher density than on the vicinal faces, where they belong to separate steps of macroscopic length.

The free energy (1.32) was obtained under the assumption that the step has only kinks of monatomic height. If we remove this limitation, then in place of (1.32) we get [1.25a]:

$$a\alpha_l = w - kT \ln \frac{1+\eta}{1-\eta}$$

and

$$\alpha_l = 0 \quad \text{for} \quad \eta_R = \sqrt{2} - 1, \quad w/kT_R = 0.88.$$

Fig.1.8a,b. Atomic structure of a close-packed face according to computer simulation data. ▶ **(a)** Face with steps; **(b)** face without steps. The *numbers* indicate the values of w/kT [1.25]

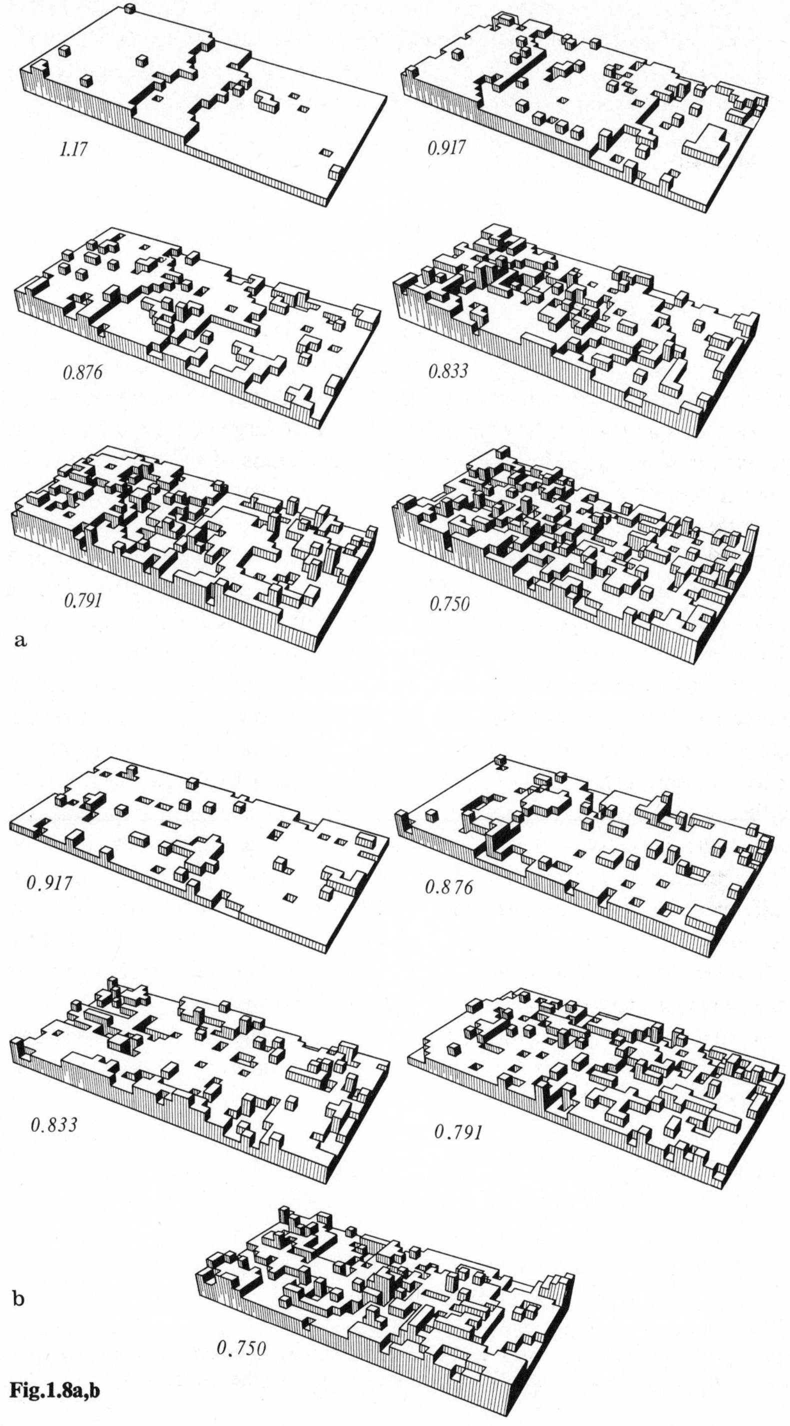

Fig.1.8a,b

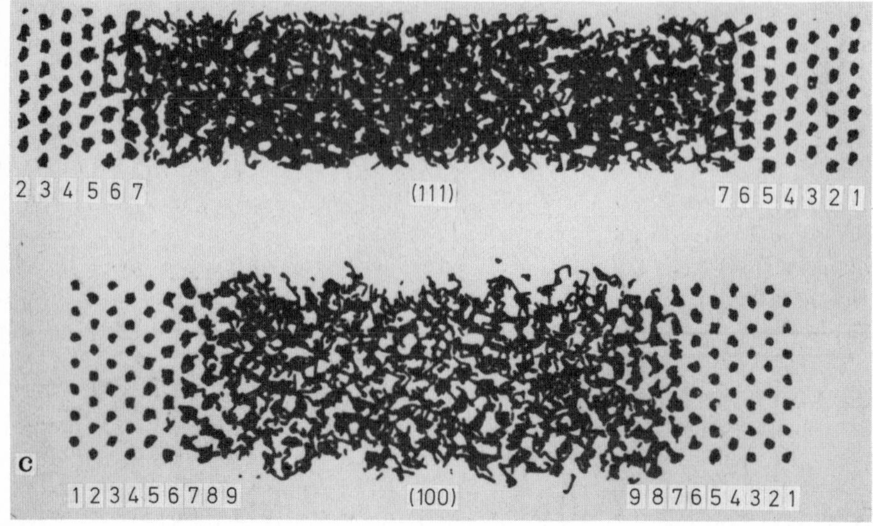

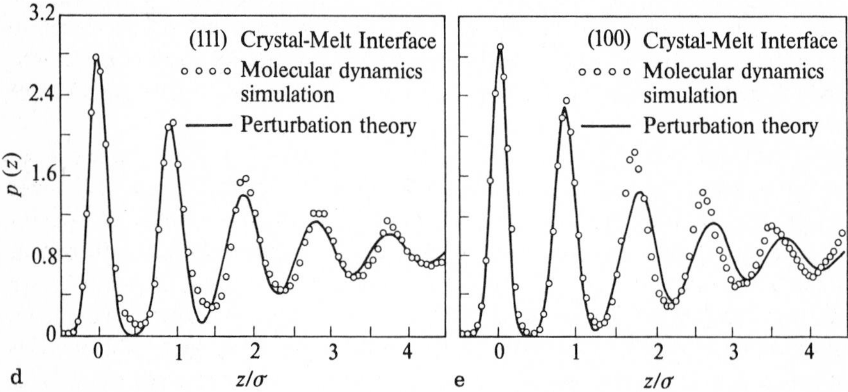

Fig.1.8c–e. (**c**) Trajectories of atoms (molecules) during the simulation in a slice perpendicular to the (111) and (100) faces. Delocalization of atoms close to the interface may be noted [1.26e]; (**d**) density profile of the melt neighboring the (111) face [1.26f]; (**e**) the same as (**d**), for the (100) face

Interestingly, the expression $\eta_R = \sqrt{2} - 1$ coincides with the exact *Onsager's* solution for the critical transition temperature in the two-dimensional Ising model (see below). Monte Carlo computer simulation gives $w/kT_R \simeq 0.7$ for both the $\langle 001 \rangle$ and $\langle 011 \rangle$ steps on the (100) face of a simple cubic lattice [1.25b]. T_R is called roughening temperature.

The disappearance of the step on transition from an atomically smooth to a rough surface is seen in Fig. 1.8a, which reflects the results of computer simulation of the surface structure [1.26]. This simulation was performed as follows:

A "crystal" made up of identical atoms packed in a simple cubic lattice with a surface (100) and a rectangular shape of 20 × 40 atoms was considered. To avoid the effect of the boundaries of this area of the surface (i.e., the edges), cyclic boundary conditions were used. The program was compiled in such a way that the atomic chains forming two long (40-atom) edges of the isolated "face" were assumed to be adjacent to each other. Such joining is equivalent to rolling each atomic surface into a cylinder and "gluing" together the opposite ends of the plane. In joining short edges they were shifted by one interatomic distance along the normal to the "face," so that the whole "crystal" consisted of a continuous rolled-up atomic net with an axis parallel to the short edge. As a result, the step reaching the right edge of the selected area is appeared as the beginning of a new layer on the left edge. The random-number generator chose a pair of an atom and an unoccupied site (vacancy) on the surface. In cases where interchanging the atom with the vacancy would have reduced the system energy (i.e., the number of uncompensated bonds), this interchange was actually accomplished. Where the interchange would have increased the energy by ΔE, however, it was only accomplished with a probability $\exp(-\Delta E/kT)$.

The above-described Monte Carlo simulation method was until recently the most popular technique in theoretical investigations of crystal surface structure and growth. But now molecular dynamics simulation (MD) is being carried out, in which equations of each atom's motion are solved with a preassigned law of the atom's interaction with the other atoms [1.27].

In MD simulation the time interval between the subsequent moments for which the positions and velocities of each atom are calculated is about 100 times shorter than the period of atomic vibrations. Thus, quite a number of calculation steps and considerable amounts of computer time are needed to obtain reliable functions for atomic distribution, energy, and entropy, as well as other thermodynamic functions of phases and interfaces. (The investigations made up until now have been devoted primarily to equilibrium situations.) Despite this difficulity, several two-dimensional [1.27a-c] and three-dimensional [1.27d, e] systems with Lennard-Jones interatomic potentials have been simulated in the last few years in different approximations. Recently, a three-dimensional system consisting of as many as 7 × 7 × 36 = 1764 atoms was analyzed [1.27e]. The longest edge of this 7 × 7 × 36 atomic aggregate was chosen to be perpendicular either to the (111) or the (110) face of the fcc lattice, and the coexistence of solid and liquid phases was investigated by means of "temperature" changes (Fig. 1.8c). The trajectories of the atoms show smoothly increasing delocalization from the solid phase (left and right sides of the aggregate in Fig. 1.8c) to the liquid phase (central part of Fig. 1.8c). The crystalline long-range order transfers to liquid short-range order within 4–5 atomic layers parallel to the interface. This picture of gradual delocalization is more realistic than the one based on the lattice model of the liquid (Fig. 1.8a, b). In the latter model each atom belongs either to the solid or the liquid, and the interface may be diffuse only on the average.

MD simulation shows that the material density decreases in a different way when one crosses the (111) or (100) interface from the crystal to the liquid. For the (111) face this decrease is realized via a decrease in the atomic density in each layer parallel to the interface, with the interlayer spacing remaining constant. For the (100) face on the other hand, the density decreases because the interlayer spacing increases, with the 2-D desntiy of the atoms in each layer remaining constant. This difference in the liquid structure in the vicinity of the (111) and (100) faces can be seen from a comparison of Figs. 1.8d and 1.8e, which depict the average atomic density of the layer parallel to the interface as a function of the distance z from this layer to a fixed crystalline layer ($z = 0$). The MD data are shown as dots. The solid curves were obtained analytically according to perturbation theory [1.27f]. The difference in structure between the (111) and (100) faces gives rise to the slightly higher potential energy of the former as compared to the latter. This potential energy difference is not compensated for by the calculated entropy terms, and thus the close-packed (111) face should be more easily transformed into the rough state than the less dense (100) face. This is in contradiction to what is expected from the lattice model and what is actually observed experimentally (as discussed below in this section). However, the difference in the free surface energies of the two faces is small compared to the energies themselves, and more precise analysis is needed. Except for an attempt made by *Gilmer* to dynamically simulate vapor condensation very little has been done on MD simulation of solidification kinetics.

The conclusion with regard to the transformation of an atomically smooth surface into a rough one at point $\alpha_l = 0$ was not rigorous enough. Indeed, as the critical temperature is approached and laceration develops, configurations of the "overhang" type appear which were not taken into account above (Fig. 1.7b) and which can turn into separate islands in front of the step. But the problem of surface structure can be tackled by other (though also approximate) methods which make allowance for surface roughness from the very outset. In one of them we define the roughness as the ratio $(U - U_0)/U_0$, where U is the total surface energy at a given temperature, and U_0 that at a temperature $T = 0$. In the nearest-neighbor-interaction model, the roughness is simply the ratio between the number of unsaturated bonds "parallel" to the close-packed xy face (see Fig. 1.7) and the total number of bonds "normal" to this face. The uncompensated bonds parallel to the xy face appear if the surface atoms adjacent in the xy plane (i.e., the atoms whose x and y coordinates differ by one lattice spacing) have different z coordinates, i.e., are located at different z levels. In other words, each uncompensated horizontal bond corresponds to a unit jump (a jump one interatomic distance high) of the z level between the above-mentioned neighboring atoms of the surface. Since the excess energy of a real rough surface as compared with an ideally smooth face ($T = 0$), is related to precisely these jumps of the z coordinate, the number of such jumps and the way they are distributed over the surface yields the entropy, the total energy (roughness), the free energy, and the other thermodynamic functions of the surface. At the same time the mutually independent parameters characterizing the surface relief are the z coordinates (levels) of the surface atoms, rather than the magnitudes of the jumps. Indeed, the atomic levels can be assigned independently of each other, but the jumps at various joints of the surface atoms which have neighboring x and y coordinates cannot, because the number of jumps exceeds that of the atoms. For instance, with each N atoms of a square face (100) of a simple cubic lattice $2N$ joints (bonds) between them are associated. Therefore the surface atoms are said to be interrelated by cooperative interaction. Cooperative interaction greatly impedes investigation of the structure of surfaces. Nevertheless, it can be performed for a model with two allowable values for the z coordinates of the surface atoms (a two-level model). Here, the surface is similar to a two-dimensional magnetic lattice, each point of which has a spin oriented in one of two mutually opposite directions. The spin interaction energy is different for parallel and antiparallel neighboring spins. This is the so-called Ising model, which is widely known in phase transition theory. Corresponding to the two directions of the spins in the magnetic lattice are, in our case, two levels of surface atoms, and the difference in their energies is the energy of the "horizontal" bond.

As is known from the theory of magnetism, because of the cooperative nature of spin interaction in a two-dimensional (and also three-dimensional) lattice, an ordering in the direction of the spins, i.e., the resultant magnetic moment of the lattice, appears (or disappears) at a strictly definite temperature. This is the Curie point. An atomically smooth surface is similar to an ordered spin distribution (ferromagnetic state), and a rough one to a disordered spin distribution. Hence a critical parameter $(w/kT)_R$ must exist, which defines, at a given w, the critical transition temperature T_R such that at $T < T_R$ almost all the surface atoms are located at the same level, i.e., the surface is smooth on the atomic scale, and at $T > T_R$ the numbers of atoms at the different levels are commensurate. The approach described provides one more technique for determining T_R.

The simplest method of appraising cooperative interaction is the self-consistent-field approximation. Here we shall describe this method (ascribable to Gorsky, Bragg, and Williams) in its thermodynamic version for a two-level model, although it permits the analysis of systems with three or an infinite number of levels [1.17, 28].

Consider a close-packed atomically smooth face containing N sites to which atoms (or molecules) of only one next layer can be added. Suppose there are N_1 atoms in all in this new layer, and, of course, $N_1 \leq N$. The quantity $\Theta = N_1/N$ is the degree of coverage of the surface. We denote by Z_1 the total number of possible bonds with the first nearest neighbors in a plane parallel to the surface under consideration. For the (100) face of the simple cubic lattice, $Z_1 = 4$, and the (111) for face of the fcc lattice, $Z_1 = 6$ (in the diamond lattice each new layer consists of two atomic planes, and the following reasoning cannot be directly applied to it). The problem is to find the coverages which correspond to equilibrium between the crystal, adsorbed layer, and environment.

To solve the problem thus formulated it will suffice to consider the free energy due to the "horizontal" bonds, i.e., those parallel to the face in question. The above-mentioned free energy

$$\Delta F = U - TS,$$

where U is the total energy of the uncompensated bonds, T is the temperature, and S is the entropy of distribution of N_1 adatoms under review among N sites on the substrate. The self-consistent-field approximation used here assumes that the atoms are distributed over the surface randomly and uncorrelatedly at any coverage density Θ. In this approximation, an atom of the new layer under consideration chosen at random will have $Z_1\Theta$ neighbors in this layer, while $Z_1(1 - \Theta)$ of its horizontal bonds will remain unsaturated. Accordingly,

$$U = N_1 Z_1 (1 - \Theta) w .$$

Since

$$S = k \ln \frac{N!}{N_1!(N - N_1)!} = -kN_1 \ln \Theta - k(N - N_1) \ln (1 - \Theta) ,$$

we get

$$\frac{\Delta F}{NkT} = \frac{Z_1 w}{kT} \Theta(1 - \Theta) + \Theta \ln \Theta + (1 - \Theta) \ln (1 - \Theta) .$$

The dependences of $\Delta F/NkT$ on Θ for different values of $Z_1 w/kT$ are represented in Fig. 1.9a. The curves $\Delta F = \Delta F(\Theta)$ are invariant with respect to the replacement $\Theta \to (1 - \Theta)$ and attain a minimum or a maximum at $\Theta = \frac{1}{2}$, because at $\Theta = \frac{1}{2}$,

$$\frac{\partial}{\partial \Theta} \frac{\Delta F}{NkT} = \frac{Z_1 w}{kT}(1 - 2\Theta) + \ln \frac{\Theta}{1 - \Theta} = 0 .$$

The adsorption layer at hand will obviously have densities Θ corresponding to the minima of $\Delta F(\Theta)$. Corresponding to the larger values of $Z_1 w/kT$ are curves with two minima, as can readily be verified. In this case the extremum at $\Theta = \frac{1}{2}$ is a maximum. Indeed, the second derivative

$$\frac{\partial^2}{\partial \Theta^2} \frac{\Delta F}{NkT} = \frac{-2Z_1 w}{kT} + \frac{1}{\Theta} + \frac{1}{1 - \Theta}$$

at $\Theta = \frac{1}{2}$ is negative if $Z_1 w/kT > 2$.

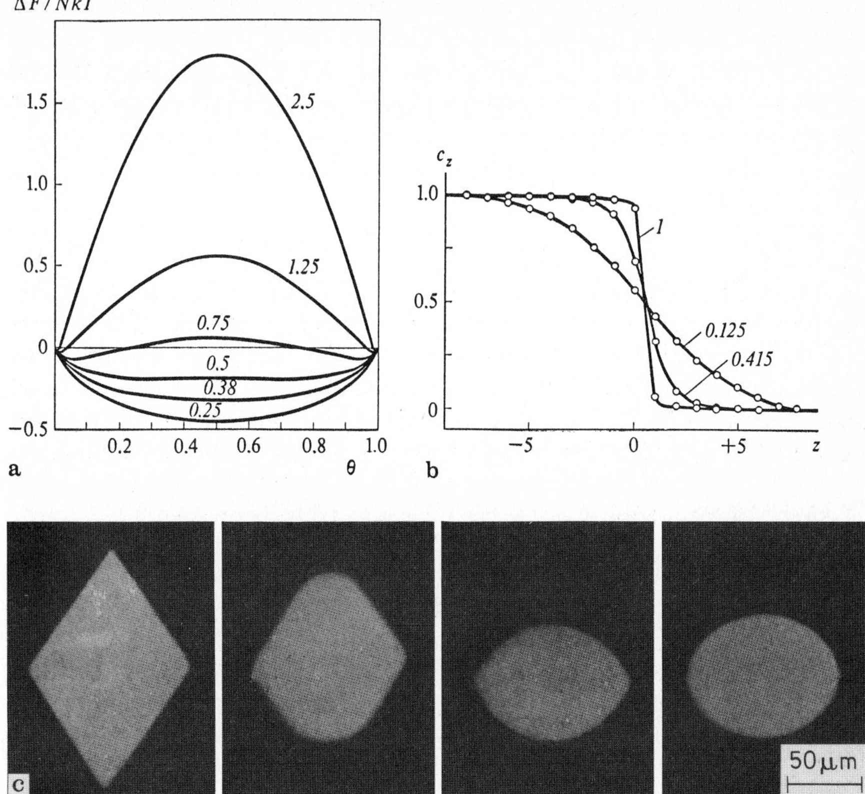

Fig.1.9a–c. Conditions of interface roughening (**a**), calculated interface profile (**b**), and morphology of naphthalene crystals (**c**). The *figures* on the curves in (**a, b**) indicate the values of parameter w/kT. (**a**) Dependence of specific free energy ΔF for the (100) face of a simple cubic lattice on the adsorption layer coverage Θ. The energy is computed from that of the atomically smooth interface [1.28]. (**b**) Probability c_z that an atom located at a distance z from the average position of the interface will belong to the crystal. The semispace $z < 0$ is occupied by the crystal, the semispace $z < 0$ by the medium [1.29a, b]. (**c**) Shapes of naphthalene crystals grown at low supercooling ($\Delta T = 0.05°$C) from naphthalene–pyrene solutions of different compositions (24.5–34.5 mass % of pyrene) corresponding to equilibrium liquidus temperatures of 70.5, 69.7, 67.1, and 65.9°C from left to right [1.30c]

Consequently, the free energy of adatoms has two minima if the energy per surface bond is sufficiently high ($w/kT > 2/Z_1$). But two minima of equal depth on the free-energy curve mean that it is advantageous for the adatom association to reside in one of the two states. The density Θ characterizing the first of them is small ($\Theta < \frac{1}{2}$, and if $w/kT \gg 2/Z_1$, then $\Theta \ll \frac{1}{2}$), while that of the second approaches unity ($\Theta > \frac{1}{2}$). Thus equilibrium takes place for two structures which are equivalent to one another: separate adsorbed atoms on an unfilled face or, conversely, separate vacancies in a nearly completed layer. In both cases the surface is atomically smooth. If, however, $w/kT < 2/Z_1$, the surface is rough ($\Theta = \frac{1}{2}$).

Thus, the criterion for transition between a smooth and a rough surface in terms of w has the form

$$Z_1 w/kT_R = 2\,. \tag{1.33}$$

Equilibrium structures of the surface for different values of parameter $kT/2w$ have been obtained [1.26] by Monte Carlo simulation and are presented in Fig. 1.8b. In contrast to Fig. 1.8a, the surface in Fig. 1.8b contains no steps, i.e., it has an average orientation (100).

Within the framework of the approach described, it is also easy to take the interaction of farther neighbors into account. The result is the same criterion, (1.33), in which, however, $Z_1 w$ must be replaced by the sum $Z_1 w + Z_2 w_2$ (with due regard for the first and second nearest neighbors). This criterion may serve for semiquantitative estimates, provided the relationship between w and w_2 is known. Here, Z_2 is the number of second-order neighbors in a plane parallel to the face under consideration, and w_2 is a value determined via the binding energies of the second nearest neighbors in a crystal and a liquid by the same rule as w is expressed through the binding energies of the first nearest neighbors.

Analysis is simplified if the atomic structure of the surface is considered for a model with a limited number of location levels for the surface atoms—a limitation which does not exist on a real surface. This limitation is particularly substantial at high temperatures. Lifting the limitation imposed on the width of boundary smearing-out is also especially important in the analysis of growth kinetics, because only thus does it become possible to consider not only the equilibrium structure, but also the motion of the interface consistently. *Temkin* was the first to consider the face structure without any limitation on the number of levels [1.28].

The smearing-out of the (100) face near its average position $z = 0$ for a simple cubic lattice is shown in Fig. 1.9b. The ordinate marks the probability c_z of finding an atom belonging to the crystal at a level spaced z interatomic distances from the average position of the boundary; $c_z = 1$ corresponds to the crystal phase and $c_z = 0$ to the liquid or gas phase. It is clear from Fig. 1.9b that the larger the ratio w/kT indicated on the curves, the less smeared out is the boundary. At $w/kT = 0.415$ the boundary occupies a layer of thickness $\sim$ 10 interatomic distances. Assuming by analogy with (1.22) that $w = \Delta H/Z_1$, the roughness criterion (1.33) for the (100) face of a cubic lattice can be expressed by

$$\Delta H/kT < 2\,.$$

An analogous roughening criterion may be obtained also from the equation $w/kT = 0.88$, which gives the condition when the linear free energy of a step on the (100) face vanishes. Again using (1.22) one gets:

$$\Delta H/kT < 5.3 .$$

The discrepancy between the two critical values of $\Delta H/kT$ is caused by inaccuracies associated with the different models. An additional error with respect to a real value may come from the approximate nature of (1.22). To exclude the latter we may replace (1.22) by the following empirical relation found mainly for metals [1.30c]:

$$\alpha\Omega^{2/3} \simeq (0.3\text{–}0.5)\Delta H$$

where Ω is atomic volume in the crystal, and α is the interfacial free energy. The lower values in parentheses are valid for Ge, Sb, Bi, Al, Pb, and H_2O, and the higher ones for Pt, Ni, Pd, Co, Mn, Fe, Cu, Au, Ag, Sn, Ga, and Hg (see also Table 2.2). Assuming $s_1 w \simeq \alpha\Omega^{2/3}$, one gets

$$w \simeq (0.3\text{–}0.5)\Delta H/s_1 ,$$

where s_1 is the number of first nearest neighbors which an adatom has in the substrate. For the (111) face possessing the highest atomic density in an fcc lattice, $s_1 = 3$; for the (110) face of a bcc lattice, $s_1 = 2$. With these values of s_1, the empirical w-versus-ΔH relation gives lower critical values of $\Delta H/kT$ than (1.22) does. For instance, (1.22), substituted into criterion (1.33) rewritten for the corresponding lattices, gives: $\Delta H/kT < 4$ for the (111) face and $\Delta H/kT < 6$ for the (100) face of the fcc lattice, and $\Delta H/kT < 4$ for the (110) face of the bcc lattice. The empirical $w(\Delta H)$ relation gives $\Delta H/kT < 3.3 \div 2$ for the (111) fcc, $\Delta H/kT < 5 \div 3$ for the (100) fcc, and $\Delta H/kT < 3.3 \div 2$ for the (110) bcc lattice, respectively. Experimental investigation of the growth shapes (see below in this section) give the critical values of $\Delta H/kT$ between 2 and 4.

As we see, crystallographically different faces have differently structured plane nets (parallel to different PBC systems) and hence different critical values of the ratio w/kT_R. This difference is clearly defined for F, S, and K faces, but it must also exist, in a weaker form, for different F faces. Therefore in the neighborhood of transition temperature T_R for F faces of one type, F faces of other types may be either smooth or rough.

If the surface is atomically smooth, the kinks are concentrated only on steps generated by different sources (Sect. 3.3). Accordingly, smooth faces grow layerwise (Sects. 3.1,2) and remain macroscopically plane in the course of growth; the growth rates of different faces diverge considerably. In the long run the crystals will grow in the shape of polyhedra. Crystals with atomically rough surfaces, on the other hand, can receive new particles at practically any point on the surface. Therefore the rates of growth in different directions are nearly equal, and the crystals acquire the rounded shapes of crystallization isotherms in the course of growth (Sect. 5.2). Thus crystal macromorphology supplies information on atomic processes on the surface.

The heats of evaporation of crystals are high, so that the $\Delta H/kT$ ratio for the solid–vapor interface usually exceeds ~ 20. This fully agrees with polyhedral crystals growing from vapors. Conversely, for the solid–melt interface in the case of most metals, $0.8 \lesssim \Delta H/kT \lesssim 1.5$, and their crystals grow rounded. Silicon crystals, for which $\Delta H/kT \simeq 3.5$, have both atomically rough and atomically smooth areas (Sect. 5.2). Crystals of many organic substances with entropies of fusion grow as polyhedra, even from melts.

In the crystal–solution system the simple criterion (1.33) has qualitative meaning only, since the interrelation of w with the dissolution heat is not so straightforward. The structure of an interface between a one-component crystal and a two-component solution was first considered by *Voronkov* and *Chernov* [1.31a] and *Kerr* and *Winegard* [1.31b] for eutectic systems. The formers predicted that there exist a critical equilibrium concentration and a corresponding temperature which divide the liquidus curve in the phase diagram into two parts. At a concentration and temperature corresponding to one part, the crystal–solution interface must be atomically rough, whereas it must be smooth for the conditions corresponding to the other. Rough and smooth interfaces may exist both for low- and high-concentration solutions, i.e., at both sides of the diagram. This conclusion may be extended to include general two-component systems.

Direct experimental proof of the existence of a critical point on the liquidus curve at which roughening transition occurs has now been found for several systems. For instance, Fig. 1.9c shows naphthalene crystals grown very slowly, at supercooling $\Delta T = 0.05$ K, from naphthalene–pyrene solutions of different equilibrium concentrations and temperatures, in other words, under conditions corresponding to different points on the liquidus curve. Transition from faceted to nonfaceted shapes occurs in a relatively narrow interval of temperatures from 70.2°C to 66.0°C (24.5 wt. % to 34.5 wt. % of pyrene, respectively) [1.31c]. The width of the interval is evidently determined by the interface anisotropy: the rounding occurs first for orientations corresponding to the corners, and may be observed on the close-packed faces at higher pyrene concentrations (i.e., at lower equilibrium temperatures).

At $T \to 0$ K the entropies of both the crystal and the melt (the only substance which is liquid at $T = 0$ K is superfluid helium) approach zero, and thus $\Delta S \to 0$, i.e., the crystal–melt interface should be rough. However, the entropy of an interface that is rough in the classical sense exceeds zero. This contradiction disappears if proper quantum analysis is used [1.32a]. A kink on the step on the quantum interface is a nonlocalized quasiparticle. The energy of kinks decreases if they form a band ~ 1 K wide (here the energy is expressed in temperature units). Thus, in the ground state the kinks on the bottom of the band should have an energy ~ 0.5 K below the energy of an isolated kink. On the other hand, the excess energy per crystal–melt bond for ^{4}He is only $\sim$ 0.1 K. Therefore the step energy should be negative and the step must disappear, i.e., the interface must be rough. Indeed, rounded shapes of ^{4}He crystals in the ^{4}He melt have been observed [1.32b, c, d].

Quantum growth kinetics are very fast compared to classical kinetics (Sect. 3.1.2).

1.4 Phase Equilibrium with Allowance for Surface Energy. Equilibrium Shape of a Crystal

1.4.1 Phase Equilibrium over a Curved Surface

When considering phase equilibrium in Sect. 1.1 we took only the chemical potentials reflecting the three-dimensional properties of the substance into account, and neglected the additional energy of the phase boundary. This approximation, however, is insufficient for disperse phases, in which the surface energy is comparable with the bulk energy. Assume, for instance, that one of the phases (crystal) with a chemical potential μ_S is a sphere of radius R surrounded by another phase (medium) with a potential μ_M of the substance of the crystal, and that $\mu_M > \mu_S$, i.e., the bulk phase S is thermodynamically more advantageous than phase M.[7] The surface energy in such a system is equal to $4\pi R^2\alpha$, where α is the specific free energy of the interface. The transfer of some number δN of particles to the more advantageous solid phase S will reduce the thermodynamic potential of the system by $(\mu_M - \mu_S) \cdot \delta N$. But then the size of the crystal must increase so that its radius changes by $\delta R = \Omega\delta N/4\pi R^2$, where Ω is the unit volume occupied by an atom, molecule, or aggregate in the crystal, i.e., by the kind of particle that is exchanged by the phases and to which the indicated values of the chemical potentials refer. Of course, the new particles need not necessarily be distributed uniformly over the sphere surface, but may also form a local overgrowth (for the general case see Fig. 1.10). But it is easy to see that a uniform increase of the radius corresponds to a minimum increment of the total surface energy. An increase of solid phase size by δR will increase the surface energy of the system by $8\pi R\alpha\delta R$. If the latter increase is less than the gain $(\mu_M - \mu_S)\delta N$, then the energy accumulated in the system in the form of the difference between the chemical potentials of the phases is sufficient to perform the work of building up new areas of the interface. Then the growth of the sphere will be advantageous and must continue. Otherwise the size of the sphere must reduce. Equilibrium sets in when the gain in bulk energy becomes equal to the loss in surface energy,

[7] Hereafter μ_S and μ_M will stand for the chemical potentials of phases of sufficiently large volume, which are independent of effects associated with the presence of the phase boundary.

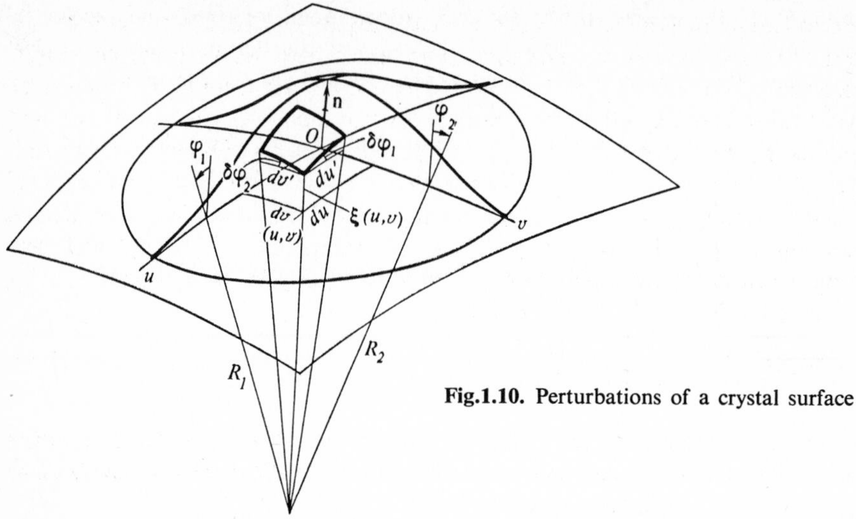

Fig.1.10. Perturbations of a crystal surface

i.e., when

$$\mu_M - \mu_S = 2\Omega\alpha/R\,. \tag{1.34}$$

The equality given in (1.34), which is called the Gibbs-Thomson equation, defines the shift of phase equilibrium at the spherical interface. This shift can easily be expressed in terms of the deviation in the temperature, pressure, or concentration from their equilibrium values according to the equations of Sect. 1.1.

In the general case of a nonspherical interface, phase equilibrium at a given point of the interface will be achieved if the difference in chemical potentials of the phases in the vicinity of this point is

$$\mu_M - \mu_S = \Omega\alpha\left(\frac{1}{R_1} + \frac{1}{R_2}\right), \tag{1.35}$$

where R_1 and R_2 are the main curvature radii of the surface at the point in question. Equations (1.34, 35) neglect, however, the anisotropy of the surface energy.

We shall now obtain the condition for phase equilibrium in the vicinity of a given point O (Fig. 1.10) on a curved surface of a crystal with anisotropic surface energy. Let the main curvature radii at this point be R_1 and R_2 (Fig. 1.10). We construct, at the given point O, two mutually perpendicular planes of main sections and take the traces of intersection of these planes with the surface as the axes u and v of the orthogonal curvilinear system of coordinates on the surface. The orientation of the surface at each point will be defined by angles φ_1 and φ_2, which are indicated by arrows in Fig. 1.10. These angles are in one-to-one correspondence with the angles φ and ϑ used in Sect. 1.2 (Fig. 1.5). Let us now transfer some number of particles δN from the mother medium to the crystal. To find out the condition for local equilibrium near point O we concentrate the new particles in the vicinity of this point, building up a bump perturbation. The shape of this bump is determined by a function $\xi(u, v)$, which represents the distance between the old surface and the new surface, reckoned along the normal to the old surface at each of its points. If, as assumed, $\xi \ll R_1$ and $\xi \ll R_2$, then the orientations of the old and new surface will differ from one another everywhere only by small angles

$$\delta\varphi_1 = \frac{\partial \xi}{\partial u}\,; \qquad \delta\varphi_2 = \frac{\partial \xi}{\partial v}\,. \tag{1.36}$$

Let us now see how the thermodynamic potential of system Φ changes on the transfer of δN particles from the medium to the crystal. Since $\Omega\delta N = \iint \xi(u, v)\, du dv$, the change in potential

$$\delta\Phi = -(\mu_{\mathrm{M}} - \mu_{\mathrm{S}}) \iint \frac{\xi(u, v)}{\Omega}\, dS + \iint (\alpha\delta dS + \delta\alpha dS)\,. \tag{1.37}$$

Both integrals in (1.37) are extended over the whole interface. The first term in the right-hand side of (1.37) is, as in the isotropic case, $(\mu_{\mathrm{M}} - \mu_{\mathrm{S}})\delta N$, i.e., the decrease in the system potential due to the greater advantage of phase S over phase M, or an increase in potential in the opposite case (at $\mu_{\mathrm{S}} > \mu_{\mathrm{M}}$). The second term stands for the change in surface energy. The area element of the interface $dS = dudv$. The specific surface energy α is anisotropic and can be regarded as a function of the two angles φ_1 and φ_2, which determine the orientation of the surface element at each point.

The term $\alpha\delta dS$ in (1.37) is due to the increase (change) in the area of a surface element with the variation in the number of particles in the crystal considered here, and is not related to anisotropy. As can be seen from Fig. 1.10, the length elements du' and dv' on a perturbed surface are related to du and dv on an unperturbed surface with an accuracy to terms of the first order with respect to $\xi(u, v)$ and to derivatives of this function, by the expressions:

$$du' = \left(1 + \frac{\xi}{R_1}\right) du\,, \quad dv' = \left(1 + \frac{\xi}{R_2}\right) dv\,,$$

whence

$$\delta dS = \left(\frac{1}{R_1} + \frac{1}{R_2}\right) \xi dS\,. \tag{1.38}$$

The variation $\delta\alpha\, dS$ in the surface energy of element dS at point (u, v) is associated with the change in orientation of this element as a result of perturbation: $\delta\alpha = (\partial\alpha/\partial\varphi_1)\,\delta\varphi_1 + (\partial\alpha/\partial\varphi_2)\,\delta\varphi_2$. Hence using (1.36) and integrating by parts, we have

$$\iint \delta\alpha dS = \iint\left(\frac{\partial\alpha}{\partial\varphi_1}\frac{\partial\xi}{\partial u} + \frac{\partial\alpha}{\partial\varphi_2}\frac{\partial\xi}{\partial v}\right)dudv$$

$$= \int\xi\left(\frac{\partial\alpha}{\partial\varphi_1}dv + \frac{\partial\alpha}{\partial\varphi_2}du\right) - \iint\left(\frac{\partial}{\partial u}\frac{\partial\alpha}{\partial\varphi_1} + \frac{\partial}{\partial v}\frac{\partial\alpha}{\partial\varphi_2}\right)\xi dudv\,. \quad (1.39)$$

The first of the integrals on the right-hand side extends over the contour bounding the perturbation at hand. On this contour the perturbation disappears, i.e., $\xi = 0$, and hence the contour integral is also zero.

At equilibrium, the thermodynamic potential is minimal with respect to the transfer of particles from one phase to the other. Therefore we must have $\delta\Phi = 0$. Substituting (1.38, 39) into (1.37) and noting that $\delta\Phi$ must vanish for any $\xi(u, v)$, we arrive at the equilibrium condition (the *Herring* equation [1.33]):

$$\mu_M - \mu_S = \frac{\Omega}{R_1}\left(\alpha + \frac{\partial^2\alpha}{\partial\varphi_1^2}\right) + \frac{\Omega}{R_2}\left(\alpha + \frac{\partial^2\alpha}{\partial\varphi_2^2}\right). \quad (1.40)$$

In deriving (1.40) we used the relations following from Fig. 1.10:

$$\frac{\partial}{\partial u}\frac{\partial\alpha}{\partial\varphi_1} = -\frac{1}{R_1}\frac{\partial^2\alpha}{\partial\varphi_1^2}\,; \qquad \frac{\partial}{\partial v}\frac{\partial\alpha}{\partial\varphi_2} = -\frac{1}{R_2}\frac{\partial^2\alpha}{\partial\varphi_2^2}\,.$$

1.4.2 Equilibrium Shape of a Crystal

Thus a melt, solution, or vapor will be in equilibrium with a crystal over a given area of its curved surface if the chemical potentials of these phases exceed that of an infinitely large crystal by the value in the right-hand side of (1.40). Generally speaking, this value changes from point to point on the surface because of the change in the curvature and crystallographic orientation of the surface. Therefore if we immerse, for instance, a crystal of arbitrary shape into a solution which is, on the average, saturated, some areas of its surface will dissolve, while others will grow as a result of the supersaturation due to the surface energy. The crystal will not cease to change its shape until equilibrium is established over the entire surface. The surface shape ensuring the fulfillment of this condition is called the equilibrium shape. Thus the equilibrium shape obeys a second-order nonlinear differential equation which is obtained from (1.40) if we require that the difference $\mu_M - \mu_S$ be constant along the whole interface. It can be shown [Refs. 1.1, 17, Sect. 1.4.3] that the solution of this equation is the envelope of a family of planes

$$\boldsymbol{nr} = 2\Omega\alpha(\boldsymbol{n})/\Delta\mu\,. \quad (1.41)$$

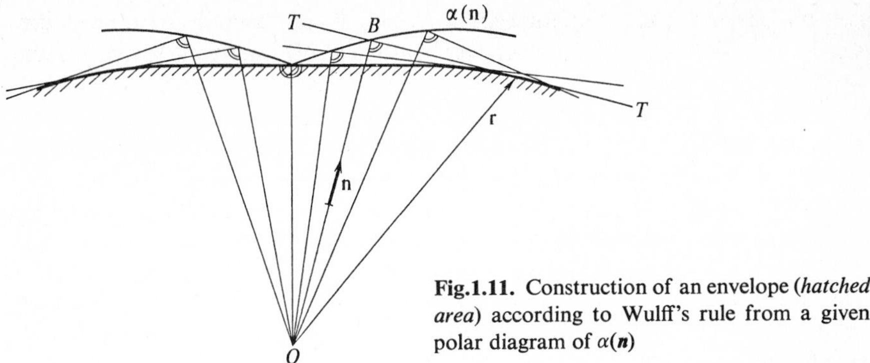

Fig.1.11. Construction of an envelope (*hatched area*) according to Wulff's rule from a given polar diagram of $\alpha(\boldsymbol{n})$

Here, $\boldsymbol{n}$ is the vector of the normal to the envelope, i.e., to the surface of a crystal of equilibrium shape, at a point defined by radius vector $\boldsymbol{r}$ (Fig. 1.11). Expression (1.41) is known as the *Gibbs-Curie-Wulff* rule: a crystal of equilibrium shape is formed by faces such that the distance from them to the center of the crystal is proportional to the surface energies of these faces. From (1.41) follows the rule of construction of the equilibrium shape. We first assign some value of parameter $\Delta\mu/2\Omega$, which will determine the crystal size scale. Then we choose an arbitrary crystallographic orientation $\boldsymbol{n}$ (Fig. 1.11) and find, on the polar diagram $2\Omega\alpha(\boldsymbol{n})/\Delta\mu$, point B corresponding to surface energy $\alpha(\boldsymbol{n})$. Then we draw plane TT perpendicular to $\boldsymbol{n}$ and intersecting segment OB at point B. We repeat this procedure for all the $\boldsymbol{n}$ and obtain a family of planes (1.41). After that the internal convex envelope of this family, which will be the equilibrium shape of an isolated crystal, remains to be constructed. It is clear from Fig. 1.11 that the orientations corresponding to sharp singular minima on polar plot $\alpha(\boldsymbol{n})$ are represented by flat faces on the equilibrium shape. Indeed, for orientations of singular faces the first derivatives $\partial\alpha/\partial\varphi_1$ and $\partial\alpha/\partial\varphi_2$ are discontinuous, and hence the second derivatives appearing in (1.40) are infinite. Therefore the right-hand side of (1.40) will be finite only if the curvatures of surfaces with singular orientations are zero. Atomically rough surfaces for which $\alpha + \partial^2\alpha/\partial\varphi_i^2$ $(i = 1, 2)$ are positive and finite will be represented by rounded areas on the equilibrium shape. Finally, the orientation areas where $\alpha + \partial^2\alpha/\partial\varphi_i^2 < 0$ will not be represented at all on the equilibrium shape, which must contain no convex areas (negative R_i). The pairs of values φ_i, where $\alpha + \partial^2\alpha/\partial\varphi_i^2 = 0$, will correspond to the crystallographic orientations of the surfaces intersecting at edges and vertices. Here, $R_i = 0$, but the quotient $(\alpha + \partial^2\alpha/\partial\varphi_i^2)/R$ is again finite. If orientations for which $\alpha + \partial^2\alpha/\partial\varphi_i^2 \neq 0$ converged at the vertex or edge $(R_i = 0)$, the surface addition (1.40) to the difference in chemical potentials would be infinite and would consequently cause a rapid reconstruction of the surface near the vertex or edge. Therefore the immediate vicinities of the vertices and edges must always have an equilibrium configuration [1.18].

Suppose the polar diagram $\alpha(\boldsymbol{n})$ is made up of spheres, or, in the plane section, of circles (Fig. 1.6b). Then in the first quadrant, $\alpha(\vartheta)$ is given by (1.18) without the modulus signs, so that $\alpha + \partial^2\alpha/\partial\vartheta^2 = 0$ and hence $R_{1,2} = 0$ for all $0 < \vartheta < \pi/2$. Geometrically, this means that all the planes similar to TT in Fig. 1.11 intersect at the same point, precisely where vector $\boldsymbol{A}$ (diameter of the circle in Fig. 1.6b) terminates.

At low temperatures the surface energy has sharp minima for many faces parallel to different PBC systems. The minima for faces with simple crystallographic indices parallel to PBCs of the nearest neighbors are especially prominent (Sect. 1.2). Therefore the equilibrium shape will be formed by flat faces, and the faces with the simplest indices will be the largest. The others will be represented as planes truncating the edges and vertices between the main faces. With an increase in temperature (more precisely, in the ratio kT/w of the temperature to the excess energy of the surface bond; see Sect. 1.3), faces with complex indices become atomically rough, and the higher the temperature, the fewer singular faces remain. Accordingly, more and more rounded areas are found on the equilibrium shape. Finally, if only the main F faces remain singular, and if, moreover, this singularity is weak, the equilibrium shape will be nearly spherical, with the exception of individual small flat areas. As noted in Sect. 1.3, the roughness is the more pronounced, the lower the ratio of the heat of crystallization to the equilibrium temperature. Therefore the equilibrium shapes in the crystal–gas system at low temperatures must be faceted, and the equilibrium shapes in the crystal–melt system with $\Delta H/kT_0 < 2$ must be rounded. In crystal–solution systems, different cases are possible, but usually the value w/kT is sufficiently large and the shapes are faceted.[8]

1.4.3 Average Detachment Work. Finding Faces of Equilibrium Shape

In analyzing concrete types of faces represented on a faceted equilibrium shape at low temperatures, it is sometimes convenient to use *Stransky* and *Kaishev's* method of average detachment work. To explain the essence of the method we

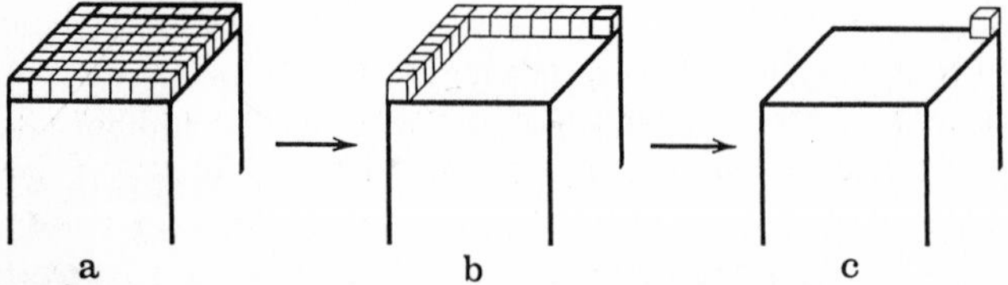

Fig. 1.12a–c. The removal of one plane net from the crystal: the difference between the work required to detach particles from the filled layer **(a)**, from a row **(b)** and to detach the last isolated atom **(c)**

[8] In the crystal–solution system, criterion (1.33) is not straightforward, since w has to be expressed via the solution concentration and a more general approach must be used ([1.3.1a] Sect. 1.3.4).

consider the (100) face of a simple cubic crystal and take into account interaction between the first nearest neighbors only (Fig. 1.12) [1.10]. Let the face length be equal to N interatomic distances. We remove all the atoms of the surface layer consecutively, beginning, say, at one of the vertices. Let us first separate $(N - 1)^2$ atoms, retaining only the two extreme rows as shown in Fig. 1.12b. This requires work $3\varepsilon_1(N - 1)^2$, because each of the atoms is removed from the kink, expending work $3\varepsilon_1$. The detachment of each of $2(N - 1)$ atoms in the extreme rows shown in Fig. 1.12b calls for work $2\varepsilon_1$, and finally, the detachment of the last corner atom (Fig. 1.12c) requires the breaking of only one bond. Hence the work of detachment per atom of the removed layer, i.e., the average detachment work

$$\bar{\varepsilon} = \frac{3\varepsilon_1(N-1)^2 + 2\varepsilon_1 \cdot 2(N-1) + \varepsilon_1}{N^2} = 3\varepsilon_1 - \frac{2\varepsilon_1}{N}. \tag{1.42}$$

By definition, the chemical potential of a crystal is the work needed to increase the number of particles in the crystal by unity at a constant temperature and pressure. Consequently, the chemical potential of an infinitely large crystal at $T = 0$ is equal to $-\varepsilon_{1/2}$, and the value $-\bar{\varepsilon}$ can be regarded as the chemical potential of the finite crystal under consideration. To achieve equilibrium with the medium, it is necessary that the potential of the latter, μ_M, be equal to $-\bar{\varepsilon}$. Hence (1.42) is equivalent to the condition

$$\mu_M - \mu_S = 2\varepsilon_1/N, \tag{1.43}$$

which is, in turn, similar to the macroscopic expressions (1.34, 40) and also expresses Wulff's rule for the particular case discussed. The numerator in the right-hand side of (1.43) is different for the various faces and lattice types.

At equilibrium, the chemical potentials of the medium over all the crystal faces must be equal. Therefore the works of detachment from different positions on faces of equilibrium shape must also be the same. By calculating these works for crystals of different structures one finds the faces present on their equilibrium shapes at $T = 0$.

Suppose, for instance, we have to find the faces of equilibrium shape for a simple cubic lattice with due regard for interaction between the first and second nearest neighbors. We begin with the shape containing cube faces only (Fig. 1.13). The particles located at the cube vertices have an energy of binding with the crystal $3\varepsilon_1 + 3\varepsilon_2$, which is less than the binding energy in a kink, $3\varepsilon_1 + 6\varepsilon_2$. Therefore such vertices cannot be present on an equilibrium shape. The corner particles remaining after the vertex atoms have been removed are bonded with the crystal with an energy of $3\varepsilon_1 + 4\varepsilon_2$, which is also weaker than the bond in a kink. Therefore all the particles along the edges must also be removed, and this will produce the shape pictured in Fig. 1.13b. The energy of the bonds of the corner particles on this shape is equal to $3\varepsilon_1 + 5\varepsilon_2$, and therefore they

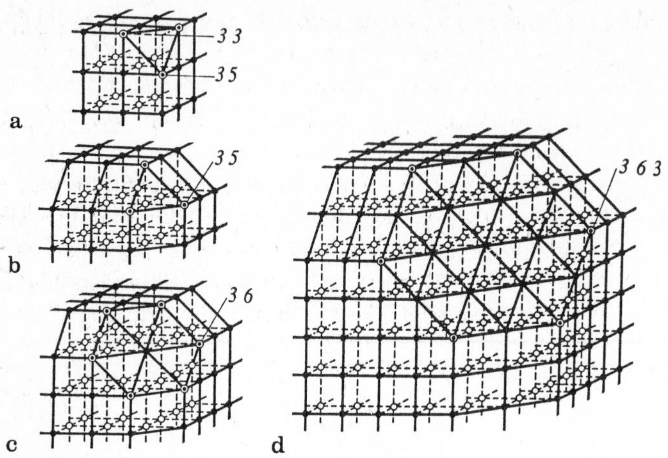

Fig.1.13a–d. Determination of the faces belonging to the equilibrium form of a crystal by the detachment-work method. The detachment of particles and the calculation of the energies necessary for the detachment begin with the particle located at the apex of the simplest form **(a)**, continue to form **(c)**, and end with **(d)**. For each apical particle of **(d)** the detachement work is equal to that of detachment from the kink (in the first- and second-nearest-neighbor approximation) [1.9]. The *numbers* give the formulae for the binding energy of the indicated atoms (see Sect. 1.3.1)

cannot exist on the equilibrium shape either. By removing them we arrive at the shape shown in Figs. 1.13c, and containing the (100), (110), and (111) faces. This is the predicted set of faces of equilibrium shape in the approximation to the second nearest neighbors inclusive. In the approximation of the first nearest neighbors alone, the equilibrium shape has only cube faces. The size of the faces can be determined by separating whole plane nets from the indicated faces until equality of the average work of detachment for all the faces is achieved in conformity with (1.43).

The method described for determining the equilibrium shape is efficient for one-component crystals. If a crystal is formed from atoms of different kinds, prediction of the equilibrium habit becomes more difficult and sometimes even ambiguous at the present level of knowledge on the binding energies in crystals. Some examples of the determination of PBCs in crystals of complex composition and of the habit of these crystals were discussed by *Hartman* [1.11].

The equilibrium shape can be determined not only from the condition that the equilibrium chemical potential of the medium must remain constant over different surfaces, but also from the condition for the minimum of the total surface energy of a crystal $\iint \alpha dS$ at its constant volume. It can be shown that this definition is equivalent to the one used above.

As we shall see in Sect. 5.2, the main faces belonging to the equilibrium shape are also the main ones on the growth form. Therefore the methods for predicting the faces of the equilibrium shape are also those for predicting the habit of growing crystals.

1.4.4 Experimental Observation of an Equilibrium Shape

Let us estimate the time within which a nonequilibrium shape would be transformed to an equilibrium shape under strictly isothermal conditions in a solution with a diffusion coefficient D if this time depended on the diffusion in the solution bulk, rather than on the surface processes. Here, the rate of variation in crystal size in some direction

$$\frac{dR}{dt} \simeq \Omega D \frac{\partial C}{\partial n} \simeq \Omega^2 D C_0 \alpha / kTR^2 , \tag{1.44}$$

where $\partial C/\partial n$ is the gradient of the solution concentration along the normal to the surface, $D\partial C/\partial n$ is the mass flow, C_0 is the equilibrium concentration of the solution, and R is the characteristic nonequilibrium curvature radius. The last estimate in (1.44) follows from (1.40) in the isotropic approximation and the expression for supersaturation in solution, $\Delta\mu \simeq kT(C - C_0)/C_0$. Integrating (1.44), we obtain the characteristic time of transformation $\tau \simeq kTR^3/3\Omega^2 DC_0\alpha$. At $\alpha = 50$ erg/cm^2, $\Omega C_0 \sim 10^{-1}$, $D = 10^{-5}$ cm^2/s, $T \simeq 300$ K, $\Omega \simeq 3 \times 10^{-23}$ cm^3, $R \simeq 10^{-3}$ cm, we have $\tau \simeq 2.5$ h. For $R \simeq 10^{-2}$ cm an appreciable change in shape under the effect of the surface energy will not take place within less than $\sim$100 days, etc. Therefore an equilibrium shape can be obtained experimentally only for very small (10^{-3}–10^{-4} cm) crystals.

Experiments for obtaining the equilibrium shape of small crystals and inclusions were discussed by *Lemmlein* [1.34]. In a corresponding experiment staged by *Kliya* [1.35], a drop of an ammonium chloride solution in water, saturated at $\sim$ 40°C, was inserted into an organic substance, polycyclohexenylethyl, which is extremely hydrophobic. This precluded an increase in solution concentration and crystal growth due to loss of water. The drops were $\sim$30 μm in diameter. Sandwiched between two cover glasses, the preparation was placed on the microscope stage and cooled to room temperature. As a result of the cooling, dendritic crystals (Fig. 1.14a) appeared in the drops and then spontaneously transformed at constant room temperature. The latter process can be seen from a series of microframes in Figs. 1.14b–f. The entire process took $\sim$10 h. The final shape (Fig. 1.14f), which does not change on further storage for 23 h, is very close to spherical. Careful inspection reveals quite a number of rounded "faces" on the surface. This should be precisely the case for atomically rough surfaces showing nonsingular (rounded) minima in the polar diagram of the specific surface energy (Sects. 1.3, 4). The process of spontaneous change of shape by a faceted crystal is highly sensitive to temperature fluctuations: the larger the amplitude of the temperature fluctuations or regular oscillations, the more rapid is the transformation in shape. This is evidently due to the nonequivalence of growth and dissolution on the faceted and rounded areas of the surface, and must be taken into account in the quantitative treatment of experiments on the kinetics of the formation of equilibrium shape. Temperature fluctuations are also important during the process of coarsening in an ensemble of crystals (Sect. 7.4).

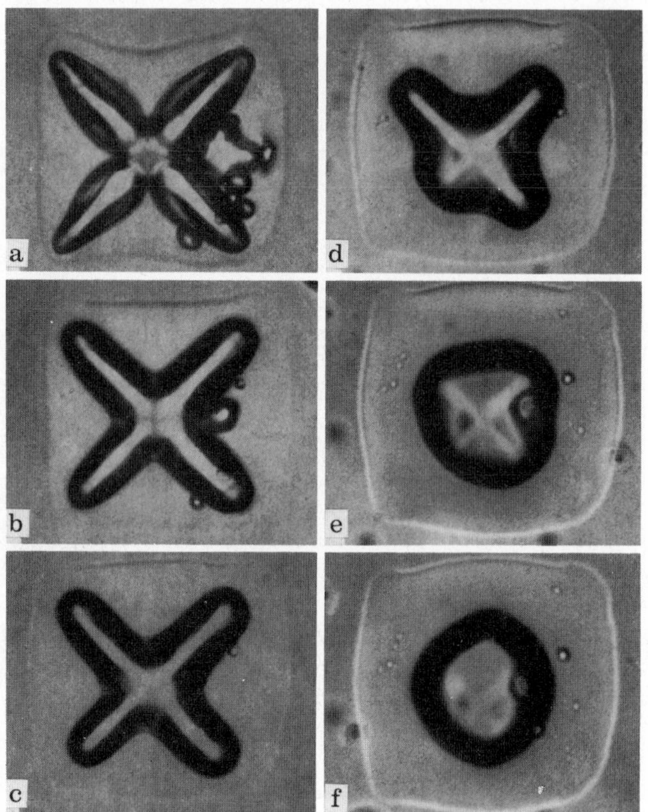

Fig.1.14a–f. Transformation of the dendritic shape **(a)** of a NH_4Cl crystal in a sealed droplet of solution into an equilibrium shape at a constant temperature. Intervals between frames: **(a)**–**(b)** 10 min, **(a)**–**(c)** 1 h, **(a)**–**(d)** 2 h 20 min, **(a)**–**(e)** 4 h 30 min, and **(a)**–**(f)** 36 h. Magnification $\times 525$ [1.35]

Equilibrium shapes of Au [1.36a] and Pb [1.36b] crystals several microns in size have been obtained on graphite substrates starting from UHV-deposited and annealed Au and Pb thin ($\simeq 3000$ Å) films, Au frozen droplets, and Pb tabular crystallites. The samples were annealed at 1000°C (Au) and 250°C(Pb) in a UHV closed cell in order to preclude evaporation and growth processes. The equilibrium shapes were spheres truncated with facets, the faceting being less pronounced on Au than on Pb. Extended rounded parts have also been observed on many metal tips in field emission microscopes. Statistical analysis

[1.36c] shows that the relative anisotropy of the gold surface energy, $[\bar{\alpha} - \alpha_{(111)}]/\alpha_{(111)}$, equals 0.034 ($\bar{\alpha}$ is the energy of the curved surface), whereas $[\alpha_{(001)} - \alpha_{(111)}]/\alpha_{(111)} = 0.019$.

Equilibrium shapes that are numerically calculated by means of Morse and Mie (Lennard-Jones) interatomic potentials may be fitted with the observed rounded equilibrium shapes of metals by suitable choice of potential parameters [1.36d].

2. Nucleation and Epitaxy

The creation of a new phase in the body of the mother phase, be it gas, liquid, or solid, is one of the most fundamental aspects of phase transitions in general and of crystal growth in particular. The potential barrier which a system must overcome in order to create a (crystalline) nucleus in the ideally homogeneous mother phase and which determines the rate of nucleation is defined, in homogeneous nucleation, by the interface energy. In heterogeneous nucleation on solid or liquid surfaces, microclusters, and ions, the properties of these foreign bodies are an additional factor upon which this barrier and rate depend. In heterogeneous nucleation on solid surfaces the role of surface inhomogeneities (steps, point defects) is also very important. Homogeneous and heterogeneous nucleation are discussed in Sect. 2.1,2 from both the macroscopic and microscopic points of view.

Epitaxy, misfit dislocations, and specific features characterizing the appearance of a new phase on crystalline substrates are discussed in Sect. 2.3.

2.1 Homogeneous Nucleation

2.1.1 Work and Rate of Nucleation. Size and Shape of Nuclei

Experience shows that in suprcooled liquids and gases crystals may not appear for a long time. Glasses, which are supercooled melts, retain their amorphous state for thousands of years, although the difference in the chemical potentials of a glass and a crystal greatly exceeds the energy of the thermal vibrations of the atoms. Molten specimens of many metals can be supercooled by several hundred degrees, down to temperatures of 0.7–0.8 T_0, T_0 being the melting point. Gallium droplets have been supercooled to 0.5 T_0 (−123°C) [2.1], and bismuth droplets to 0.6 T_0 (44°C) [2.2]. The water vapor pressure may exceed the equilibrium pressure more than fivefold without the formation of mist drops.

The cause of such stability in metastable systems lies in the difficulties involved in the nucleation of a new phase, in particular of crystals, in supercooled or supersaturated media. Let us now establish the factors determining the rate of nucleation. We first consider a supersaturated vapor, the chemical potential of whose particles, μ_V, exceeds that of the crystal, μ_S. The constituent atoms or molecules of the vapor may, on collision, join into groups of two, three, four, or more particles, forming dimers, trimers, tetramers, etc. On the other hand, some of these polymers disintegrate as a result of fluctuations in the oscillatory energy of their constituent atoms and molecules. As a result, a metastable size distribution of the aggregates is established in the vapor. Similar processes occur in solutions and melts. Let us estimate, with a given supercooling, the number of aggregates consisting of a certain number of "elementary" particles — atoms or molecules — and forming spheres of radius R. If the number of particles in the aggregate is not very small, so that the aggregate contains atoms of at least 2–4 coordination spheres, the macroscopic concepts of surface energy and chemical potential can be applied to it. Recently [2.3] it was shown that the equations of macroscopic theory also describe the rate of nucleation on the whole correctly when the nucleus consists of only a few atoms. For the atomistic approach, see Sect. 2.2.2.

The process by which crystal nuclei are formed is, in principle, similar to that for the formation of liquid drops in a supercooled vapor, of drops of one liquid in another during liquation, or of bubbles in a stretched liquid. At the same time, the principal regularities of the phenomenon are best understood in the isotropic approximation, with which we shall now begin.

Repeating the reasoning given at the beginning of Sect. 1.4, we obtain the change in the thermodynamic potential of a system upon the formation in it of a spherical crystalline aggregate of radius R:

$$\delta\Phi = -\frac{4\pi R^3 \Delta\mu}{3\Omega} + 4\alpha\pi R^2 . \tag{2.1}$$

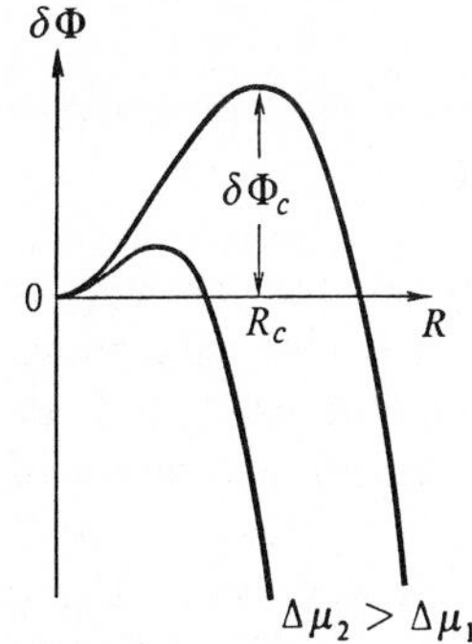

Fig.2.1. Thermodynamic potential of a system containing an aggregate of a new phase with radius R

Function $\delta\Phi(R)$ has the form shown schematically in Fig. 2.1. At small R, the second, positive term in the right-hand side of (2.1) which is associated with the formation of the aggregate's surface, prevails. At large R (relatively large aggregates) the decisive role is played by the first term, i.e., the reduction in the system potential due to the joining of the particles into a phase with a lower chemical potential. The maximum of $\delta\Phi$ is achieved (as follows from the condition $\partial\delta\Phi/\partial R = 0$) at the critical value

$$R_c = 2\Omega\alpha/\Delta\mu . \tag{2.2}$$

This relation is identical to (1.34) and expresses the condition for equilibrium of the aggregate with the environment. But this equilibrium is unstable. Indeed, a reduction in the size of an aggregate with $R < R_c$ increases the system potential, and therefore aggregates less than R_c will tend to decompose into monomers. Conversely, if the aggregate has already achieved a size exceeding the critical one, its further increase will reduce the system energy, and therefore such aggregates must grow like a snowball rolling down the right-hand branch of the potential hill in Fig. 2.1. Hence it is they that become the nuclei of the new phase. An aggregate with $R = R_c$ is called a critical nucleus. To form it, the system must overcome the potential barrier which is depicted in Fig. 2.1 and has a height

$$\delta\Phi(R_c) = \delta\Phi_c = \frac{16\pi}{3}\frac{\Omega^2\alpha^3}{(\Delta\mu)^2} \,. \tag{2.3a}$$

In the process described this may result from fluctuation, where by sheer chance the number of particles joining the aggregate with $R < R_c$ for a certain period will exceed the number of monomers leaving it. According to fluctuation theory, the probability of the event at which the potential differs from the mean by $\delta\Phi_c$ is proportional to $\exp(-\delta\Phi_c/kT)$. Therefore the density of aggregates with a size R_c in the gas bulk is also proportional to this exponent. The pre-exponential factor can be assumed to be equal in order of magnitude to the gas density n[cm^{-3}], although its actual value is slightly lower (by approximately the number of particles in the aggregates). If we presume the gas in which the nuclei are formed to be ideal, then $4\pi R_c^2 P/\sqrt{2\pi mkT}$ molecules (atoms) of the gas, where m is the mass of the molecule (atom), and P is the vapor pressure, arrive at the surface of a spherical aggregate of raduis R per unit time. Hence the number of nuclei formed in unit volume per unit of time, i.e., the rate of nucleation, is

$$J \simeq \frac{4\pi R_c^2 Pn}{\sqrt{2\pi mkT}} \exp\left(-\frac{\delta\Phi_c}{kT}\right) \equiv B\exp\left(-\frac{\delta\Phi_c}{kT}\right). \tag{2.4}$$

A more accurate calculation of the kinetics involved in the aggregate's overcoming the critical barrier $\delta\Phi_c$ results [2.4a, b] in an additional factor Z, the so-called *Zeldovich* factor, in the expression for the rate of nucleation (2.4):

$$Z = \sqrt{-\frac{1}{\pi kT}\frac{\partial^2\Phi}{\partial N^2}\bigg|_{N=Nc}} = \sqrt{\frac{\delta\Phi_c}{3\pi kTN_c^2}} \,,$$

where $N = 4\pi R^3/3\Omega$ is the number of particles in an aggregate of radius R, and N_c the number in an aggregate of critical radius R_c [2.4]. For typical numerical values during nucleation in the melt, $Z \simeq 10^{-2}$ [2.5].

The preexponential factor B in (2.4) can be transformed as before, assuming the saturated vapor to be the ideal gas, i.e., taking $P = nkT$, and also representing the equilibrium vapor pressure P_0 as

$$P_0 = (\nu \sqrt{2\pi mkT}/\kappa a^2)\exp(-\Delta H/kT)\,.$$

This last formulation reflects the equality of the flows from the kink to the vapor and back. Here, κ is the effective "condensation coefficient" on the kink. It can be estimated by comparing the calculated and tabulated values of P_0. Putting $\nu = 10^{13}\ \mathrm{s}^{-1}$, we find that for simple substances $\kappa \simeq 10^{-1}$–10^{-3}. For a simple cubic lattice we can assume, within the order of magnitude, that $\alpha \simeq \Delta H/6a^2$, $\Omega \simeq a^3$. Taking $P \simeq P_0$, using (2.2) for R_c, and expressing $\Delta\mu$ in it in terms of vapor supercooling ΔT, we obtain the following simplified expression for the preexponential factor, B, as determined by (2.4):

$$\ln B = \ln 16\pi(2\pi m/kT)^{1/2}(\nu T/6\kappa a\Delta T)^2 - 2\Delta H/kT\,.$$

A crystalline nucleus does not, generally speaking, have a spherical shape, and therefore (2.1–3a) are applicable to it only qualitatively. If, for instance, a nucleus is a cube of side L, then $\delta\Phi = -L^3\,\Delta\mu/\Omega + 6\alpha L^2$ where α is now the specific free surface energy of the cube face. The work $\delta\Phi$ of formation of the aggregate has, as does that given by (2.1), a maximum, which is attained for a cube of size $L_c = 4\Omega\alpha/\Delta\mu$(following from $\partial\delta\Phi/\partial L = 0$ at $L = L_c$). The height of this maximum, i.e., of the potential barrier to nucleation,

$$\delta\Phi_c = 32\Omega^2\alpha^3/(\Delta\mu)^2\,. \tag{2.3b}$$

If the surface energy of the cube face is equal to that of the sphere, the indicated nucleation barrier for a cube is about twice that for a sphere, cf. (2.3a). This reflects the fact that in the isotropic approximation the equilibrium shape is a sphere, while a cube does not correspond to the energy minimum at a given volume. Therefore in calculating the barrier to the formation of nuclei it is the equilibrium shape of the crystal nucleus that must be used.

The preexponential factor for a cubic or any other crystal nucleus has the same structure as that in (2.4), the only difference being that the cube surface $6L_c^2$ must be substituted for the sphere surface $4\pi R_c^2$.

The above-considered fluctuational process of nucleation does not presuppose the participation of any foreign particles or sufaces, and is called homogeneous. Otherwise one speaks of heterogeneous nucleation (Sect. 2.2).

2.1.2 Critical Supersaturation and Metastability Boundary in Vapors

The number J of nuclei appearing in a unit of volume per unit of time depends very strongly on supersaturation $\Delta\mu$, and there exists a critical supersaturation

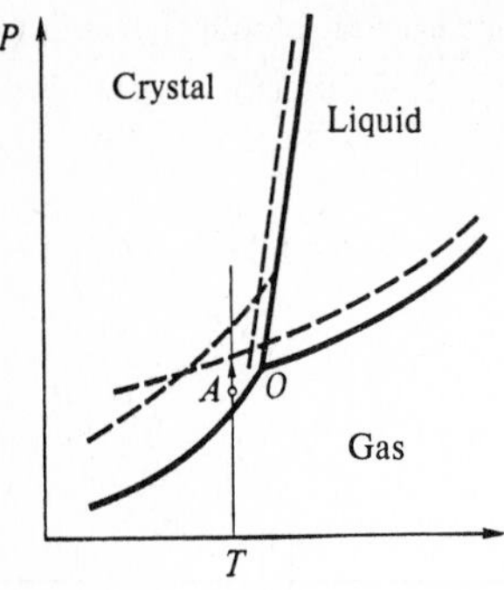

Fig.2.2. Phase diagram of vapor, solid, and liquid (*solid lines*) with metastability boundaries (*dashed lines*) (cf. Fig.1.1). With an increase in the vapor pressure at a given temperature T (indicated by the *arrow*), a melt may arise first and a crystal only afterwards

below which there is practically no nucleation, and above which nucleation proceeds vigorously enough to be detected experimentally. This critical deviation from equilibrium defines the metastability boundary. The metastability boundaries are shown schematically by the dashed lines in Fig. 2.2.

The widths of the metastable zones are different for different phases; therefore, for instance, near the triple point O (Fig. 2.2) an increase in gas pressure at a temperature T, which corresponds to crystal stability, may at first result in the appearance of the liquid phase, This is because point A, which reflects the state of the system, will cross the metastability boundary of the liquid first and that of the crystal only afterwards.

Quantitatively, the critical supersaturation $\Delta\mu_M$ can be defined as that at which $J \simeq 1\ \mathrm{cm^{-3}\ s^{-1}}$. Thus, taking the logarithm of (2.4) for spherical nuclei,

$$\ln P_c/P_0 = \frac{1}{kT}\left(\frac{16\pi}{3}\,\frac{\Omega^2\alpha^3}{kT\ln B}\right)^{1/2}. \tag{2.5}$$

From (2.5) it follows, first, that the critical supersaturation depends only slightly on preexponential factor B, which partly justifies the insufficient accuracy of its calculation. At $\Omega \simeq 3\times10^{-23}\ \mathrm{cm^3}$, $m \simeq 20\times1.7\times10^{-24}\mathrm{g}$, $T\simeq 300$ K, $\alpha\simeq 10^2\ \mathrm{erg/cm^2}$, $P_c = 10$ torr, $P/P_0 \simeq 3$, we have $B\simeq 10^{25}\ \mathrm{cm^{-3}\ s^{-1}}$, i.e., $\ln B \simeq 57$. To estimate the critical supersaturation, it is necessary to operate with more accurate values for the parameters. According to *Volmer* and *Flood* [2.6] for water droplets formed from a water vapor where $T = 275.2$ K, $\alpha = 75.23\ \mathrm{erg/cm^2}$, $m = 18\times1.7\times10^{-24}\mathrm{g}$, $\Omega = m/\rho$, where $\rho = 1\ \mathrm{g/cm^3}$ is the density of water, we have (using the adjusted value of B), according to (2.5)$P_c/P_0 = 4.16$; for $T = 261.0$ K and $\alpha = 77.28\ \mathrm{erg/cm^2}$, $P_c/P_0 = 4.96$. The agreement between the calculated values of P_c/P_0 and exepriment is excellent, as the measured values are 4.21 and 5.03, respectively. The calculated and measured values of P_M/P_0 for the formation of droplets in other experiments by *Volmer* and *Flood* are given in Table 2.1, which indicates good agreement between theory and experiment. But *Pound* and *Lothe* [2.5, 7.8] noted that in determining J no allowance was made for the translatory and rotary motion of

Table 2.1. Nucleation conditions for liquid droplets in supersaturated vapors[a] [2.6]

Substance	Molecular weight	Density	Surface energy [erg/cm^2]	P[torr]	T_0[K]	T_C[K]	(P_C/P_0) exp	(P_C/P_0) theor
Methyl alcohol	32	0.81	24.8		270.0	295	3.2 ±0.1	1.8
Ethyl alcohol	46	0.81	24.0		273.2	289.5	2.34±0.05	2.28
Propyl alcohol	60.1	0.82	25.4	2.8	270.4	289	3.05±0.05	3.22
Isopropyl alcohol	60.1	0.81	23.1	3.4	264.7	283.2	2.80±0.07	2.89
Butyl alcohol	74.1	0.83	26.1	1.12	270.2	291	4.60±0.13	4.53
Nitromethane	61	1.2	40.6	2.39	252.2	291.5	6.05±0.15	6.22
Ethyl acetate	88.1	0.94	30.6		240 244	290	12.3 8.6	10.37

[a] The supersaturation was caused by a temperature decrease from the equilibrium value T_0. Here, $P_0=P(T_0)$, $P_C=P(T_C)$, T_C is the droplet nucleation temperature

the nuclei in the gas. If the contribution of these motions is taken into account, the result is an increase in velocity J by a factor of $\sim 10^{17}$, which impairs the would-be favorable situation: the calculated values of $\ln(P_c/P_0)$ reduce by about 30%. The discussion on adequate use of the standard state in [2.7] has lasted up until now [2.9a, b]. The present status of the theoretical problem is discussed in [Ref. 2.9c, pp. 1–102, 205–279].

The critical supersaturation is proportional to $\alpha^{3/2}$, in other words, it is highly sensitive to changes in surface energy (by which it is actually determined). Therefore the introduction of very small amounts of surface-active impurity may facilitate nucleation and narrow down the metastability boundary. If, however, there is so much impurity that it covers most of the surface of the nuclei or of the aggregates of subscritical size, the exchange of particles between aggregates and gas is impeded, and the rate of nucleation drops off.

If the impurity atoms (molecules) are bonded more strongly with vapor atoms (molecules) than among themselves, then stable aggregates, which facilitate nucleation, may arise on impurity atoms.

Nucleation is also facilitated if the vapor contains ions. Indeed, the energy of the electrostatic field around the ion decreases when vapor molecules (atoms) stick to the ion. The aggregates formed on the ions are stable at much lower supersaturations than the critical one for a clean vapor, and serve as centers of condensation. The cloud-chamber principle is based on this effect.

Equations (2. 4, 5) define the nature of changes with temperature in the width of the metastable zone in a clean system.

The width of the metastable zone can be expressed either in terms of relative supersaturation $(P_c - P_0)/P_0$ or relative supercooling $\Delta T_c/T_0 \equiv (T_0 - T_c)/T_0$, where T_c is the temperature at the boundary of the metastability zone at a given pressure. At relative supercoolings which are not too high, $\Delta\mu \simeq \Delta H \Delta T/T_0$ (Sect. 1.1), and (2.4) at $J = 1\ \mathrm{cm^{-3}\,s^{-1}}$ gives the following analog of (2.5):

$$\frac{\Delta T_c}{T_0} \simeq \left(\frac{16\pi\Omega^2\alpha^3}{3kT\Delta H^2 \ln B}\right)^{1/2}.$$

Assuming again, for the sake of estimation, that $\alpha \simeq \Delta H/6a^2$, and putting $\Omega \simeq a^3$, we obtain

$$\Delta T_c/T_0 \simeq 0,3(\ln B)^{-1/2}(\Delta H/kT_0)^{1/2}. \tag{2.6a}$$

The estimate (2.6a) can, in principle, also be applied in determining the critical supercooling in melts and solutions. Using the empirical relationship of the crystal–melt surface energy to the heat of fusion ΔH in the form $\alpha \simeq (0.3\text{–}0.5)\ \Delta H\Omega^{-2/3}$[2.10a] (see final part of Sect. 1.3.4), we arrive at the following estimate in place of (2.6a):

$$\Delta T_c/T_0 \simeq (0.7 - 1.4)(\ln B)^{-1/2}(\Delta H/kT_0)^{1/2}. \tag{2.6b}$$

By substituting the typical values for B and $\Delta H/kT_0$ into (2.6), we find that $\Delta T_c/T_0$ for different substances and changes in the state of aggregation lies between 0.05 and 0.3, i.e, the critical supercooling in clean systems may reach tens of per cent of the absolute value of the equilibrium temperature.

2.1.3 Nucleation in Condensed Phases

The preexponential factor in the expression for the rate of nucleation in solution is proportional to the density of the dissolved substance, $n[\mathrm{cm^{-3}}]$, and to the particle flux towards the surface of the crystal nucleus, whose area is $\simeq 4\pi R_c^2$. In solution this flux depends on the rates of diffusion and addition of particles to the nucleus. The addition of particles requires the breaking of several of their bonds with the solvent, i.e., overcoming the potential barrier. This process has been very poorly studied. The available data permit only estimates of the activation energy for the complete process whereby the particles are delivered to the lattice of the macroscopic crystal. For the growth of the (0001) face of quartz in hydrothermal solutions the activation energy is about 20 kcal/mol. Thus the preexponential factor

$$B \simeq 4\pi R_c^2 n^2 \nu a \exp(-E/kT), \tag{2.7}$$

where $\nu \simeq 3 \times 10^{12}\ \mathrm{s^{-1}}$, $a \simeq 3 \times 10^{-8}$ cm, and E is the activation energy for the addition of particles to the nucleus. At $R_c \simeq 3 \times 10^{-7}$cm, $n \simeq 3 \times 10^{20}\ \mathrm{cm^{-3}}$ ($\sim$1 particle of solute per 10^3 particles of solvent), we have $B \simeq 10^{34}$

$\cdot \exp(-E/kT)$. If $E \simeq 20$ kcal/mol and $T \simeq 650$ K, then $E/kT \simeq 15.4$ and $B \simeq 2 \times 10^{27}\,\mathrm{cm}^{-3}$.

In a melt, the problem of the delivery of molecules to the nucleus does not exist, and all that remains is the barrier associated with the rearrangement of the short-range order which occur when the particles cross to the nucleus. It is believed that this rearrangement is characterized by an activation energy, and thus the preexponential factor in melts also has the form (2.7). It is often assumed that the activation energy for melt growth is similar to that of a viscous flow in which relative displacement of molecules also occurs. However, detailed experiments on the melt growth kinetics of some organic crystals (cryclohexanol, succinonitrile; see [2.10b]) show that the growth rate cannot be adequately described on the basis of similarity between the atomic processes of crystallization and those of the viscous flow. Thus it is more realistic to expect the activation energy E to be proportional to the heat of fusion ΔH, the proportionality factor being $\lesssim 1$. For melts consisting of simple species (atoms, small molecules) it should even be assumed that $E \ll \Delta H$ (Sect. 3.1.2). For $n \simeq 3 \times 10^{23}\,\mathrm{cm}^{-3}$, $R_c \simeq 1 \times 10^{-7}$ cm, $\nu \simeq 3 \times 10^{12}\,\mathrm{s}^{-1}$, $a = 3 \times 10^{-8}$ cm, we obtain from (2.7) $B \simeq 10^{39} \exp(-E/kT) \simeq 10^{35}$–$10^{38}\,\mathrm{cm}^{-3}\,\mathrm{s}^{-1}$.

The width of the metastable zone can be roughly estimated with the aid of (2.6b). For $B \simeq 10^{37}$ and $\Delta H/kT \simeq 3$ we have $\Delta T_c/T \simeq 0.2$. From Table 2.2, which is discussed below, it follows that the maximum experimental values of $\Delta T_0/T_c$ attained so far are $\simeq 0.2$–0.5.

The rate of nucleation $J(T, T_0)$ is proportional to $\exp(-\delta\Phi_c/kT)\exp(-E/kT)$, which results in the fact that the rate of nucleation first increases with a decrease in temperature, because the supercooling increases and the work of nucleation $\delta\Phi_c$ reduces, and then drops together with the mobilities of the particles in the liquid phase, which are proportional to $\exp(-E/kT)$. If the mobility is low, no nucleation takes place despite the high supercooling: glass is formed. The curve in Fig. 2.3 possesses maxima which precisely correspond to the picture described above. The experiment is described on p. 57.

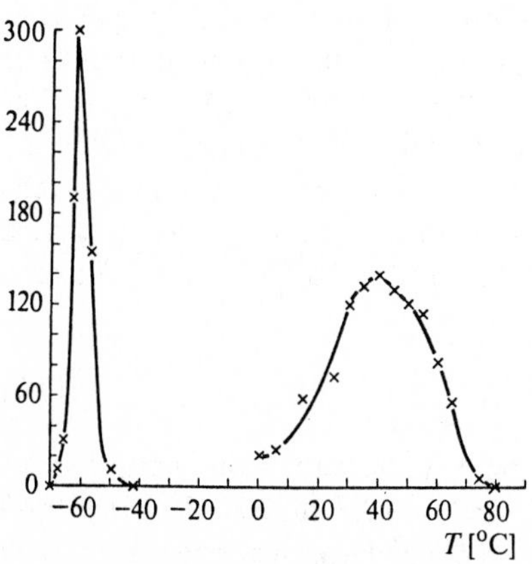

Fig.2.3. The number of nuclei in glycerin crystals in 1 cm³ of melt (*left-hand curve*) and piperine in 1.2 cm³ of melt (*right-hand curve*) vs temperature [2.11]

Crystal nuclei also form in glasses, though sometimes at a very low rate; so-called devitrification is observed, which results in glass turbidity. Of late, compositions of glasses and techniques for their preparation (heating followed by quenching) have been found which produce an extremely large number of very small crystals separated by a glasslike mass. These materials, among them Pyroceram, show very high strength compared wtih ordinary glasses.

Nucleation is a necessary stage of any first-order phase transition, including the transition from one solid phase (amorphous or crystalline) into another (crystalline). Since the new phase has a different specific volume and structure, the appearance of a nucleus in the initial matrix causes stresses [2.12a]. Let u_{ik}^0 be the deformation of the initial lattice due to the phase transition, so that $u_{ii}^0 = u_{xx}^0 + u_{yy}^0 + u_{zz}^0$ is the relative change in volume, and let u_{ik}^0 with $i \neq k$ characterize the shear, i.e., the change in the angles between the corresponding crystallographic planes on phase transition. Then, if N is the number of particles in the nucleus, Ω is the specific volume per particle, and G is the shear modulus, then the elastic energy of the matrix and nucleus is, to an order of magnitude, $G(u_{ik}^0)^2 N\Omega$, i.e., it is proportional to the volume of the nucleus. The surface energy of the system is equal, also to an order of magnitude, to $\alpha\,(N\Omega)^{2/3}$. These expressions hold true if the factors containing the parameters of the precipitate's shape and the Poisson coefficients are assumed to be unity, while the shear and dilatation moduli of the old and new phase are the same. For typical values of $G \simeq 10^{12}$ erg/cm^3, $\alpha \simeq 10$–10^2 erg/cm^2, $u_{ii}^0 \simeq 10^{-2}$–10^{-1}, $u_{ik}^0 \simeq 10^{-2}$–3×10^{-1} $(i \neq k)$, we find that the total elastic and surface energy are comparable when the linear size of the nucleus $(N\Omega)^{1/3} \simeq \alpha/G(u_{ik}^0)^2$, i.e., is of the order of several interatomic distances. It follows from the foregoing that the elastic energy plays the leading role in transformation in the solid phase.

The change in the thermodynamic potential of the system on nucleation is, to an order of magnitude,

$$\delta\Phi \simeq -\,N[\Delta\mu - G(u_{ik}^0)^2\Omega] + \alpha(N\Omega)^{2/3}\,.$$

Hence nucleation is only possible if $\Delta\mu \gtrsim G(u_{ik}^0)\Omega$, i.e., at $u_{ik}^0 \simeq 3 \times 10^{-2}$, $\Omega \simeq 2 \times 10^{-23}$ cm^3, if $\Delta\mu \gtrsim 300$ cal/mol. Recalling (see Sect. 1.1) that $\Delta\mu \simeq \Delta H \Delta T/T_0$ where ΔH and T_0 are the transition heat and temperature (usually $\Delta H \lesssim 1$ kcal/mol), we find that the new phase can appear in the bulk of the solid matrix if the relative supercooling $\Delta T/T_0 \gtrsim 0.3$. At smaller supercoolings the appearance of the new phase in a rigid matrix, which does not permit stress relaxation, is thermodynamically disadvantageous.

If the stresses can relax within times comparable to that necessary for nucleation, the formation of nuclei in the solid phase will proceed as in the liquid phase.

In the case of a nonrelaxing matrix, the elastic energy of the system depends on the form of the precipitate. This means that the expression for the elastic energy of a matrix containing precipitate, $G(u_{ik}^0)N\Omega$, must also include a factor which depends on the shape of the nucleus [2.12b].

With this factor included, the elastic energy introduced by a platy precipitate or a precipitate in the shape of a disc with diameter d and thickness h, $d \gg h$, is markedly lower than the elastic energy introduced by a relatively isometric cylindrical precipitate with $d \sim h$ or by a spherical precipitate whose shape would be energetically most favorable if it depended on the surface energy alone. Hence the deviation from equilibrium sufficient for the nucleation of disc-shaped and strip-shaped precipitates in the solid matrix must be less than for the nucleation of spherical precipitates of the same volume.

With an increase in deviation from equilibrium and a decrease in the number of particles in the nuleus, the surface energy (the second term in the expression for $\delta\Phi$) plays an ever-increasing role, and accordingly, the appearance of increasingly isometric critical nuclei becomes advantageous.

In all analysis above the heat and mass transport was supposed to be infinitely fast. The influence of finite rates is considered in [2.12c,d].

A quantitative experimental investigation of nucleation is extremely complicated. This is due primarily to the small sizes of the nuclei, which as yet makes it impossible to observe them directly and measure the condensation parameters. Field emission microscopy allows one to observe atoms and atomic clusters [2.13a, b], find phase diagrams for two-dimensional condensation [2.13c], and study nucleation process phenomenologically [2.13d, e]. A combination of these approaches might be useful in future analysis of heterogeneous nucleation. Recently, clusters containing an arbitrary number N of atoms ($1 \leq N \lesssim 6000$) have been successfully produced in the vapors of about 30 substances (Sect. 2.2.2). This technique offers very good prospects for the direct study of homogeneous nucleation. At the moment, however, use is generally made of the "developing" method proposed by *Tamman* [2.11]: the initial phase (a melt, a liquid or solid solution, a glass, etc.) is held for some time in the supercooling state; i.e., it is "exposed" at a temperature for which the rate of nucleation is to be found. Then the liquid (glass, etc.) is heated rapidly to the "developing" temperature, at which (as established experimentally beforehand) new nuclei are not formed (the metastability region), and the nuclei appearing during exposure grow to a visible size. The number of small crystals thus obtained, divided by the exposure time and the vessel volume, yields the rate of formation of nuclei per unit of volume. The number of glycerine and piperine nuclei counted in sepecimens with size of 1 cm^3 and 1.2 cm^3, respectively, is given in Fig. 2.3.

The second difficulty in nucleation studies is connected with the contamination of the crystallizing phase. Nucleation can proceed much more easily on the surfaces of several foreign particles (dust, microparticles of oxides, etc.) than in a homogeneous phase. Nuclei also form more easily on vessel walls. This activity of foreign surfaces is treated in Sect. 2.2 and depends, in the long run, on how good the adhesion of the crystals onto these surfaces is. The number of foreign particles in melts can be reduced, for instance, by filtering off [2.14a]. With metals, where insoluble oxides of these metals with high melting points constitute the greatest danger, the work is done in a reducing atmosphere, and the specimen under investigation is surrounded with a layer of liquid glass, which prevents oxidation and the ingress of particles from the outside, and also excludes and contact with the solid walls. The small-drop method proposed by *Turnbull* [2.10a, 15, 16] has proved to be highly efficient. The specimen is dispersed into a multitude of drops of diameter ≈ 100 μm, which are placed in a nonoxidizing liquid matrix. With such a division, impurity particles (macroscopic ones, of course) occur only in individual drops. Upon cooling, these drops crystallize before the others, i.e., at smaller supercoolings, but the fact that they solidify does not now mean that the whole specimen solidifies. The

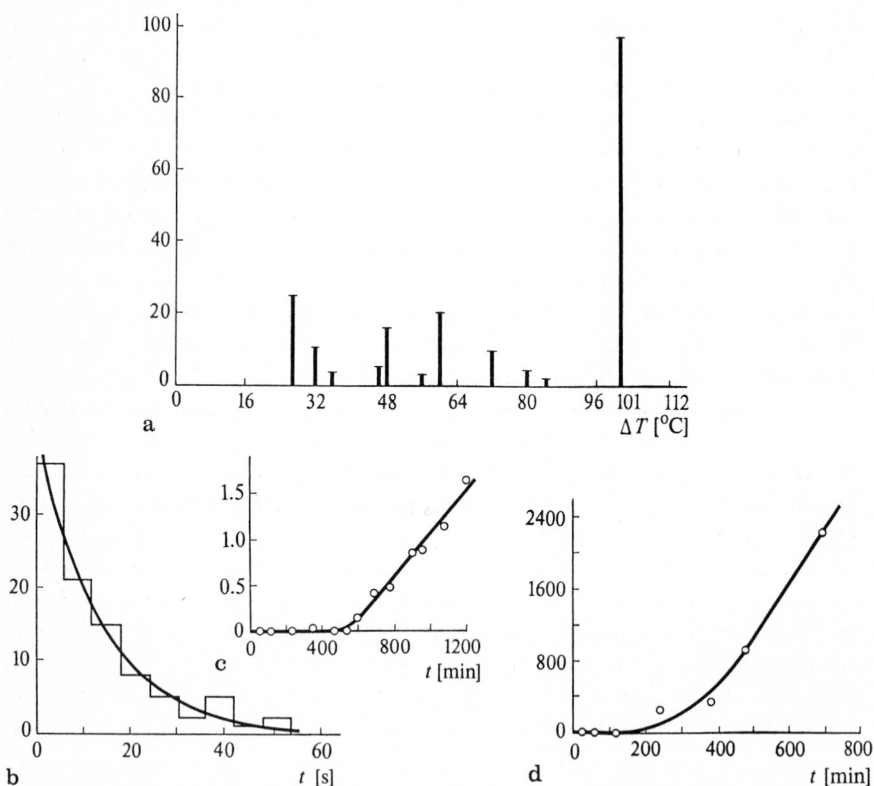

Fig.2.4a–d. The number of nuclei appearing at different supercoolings **(a)** and times **(b–d)**. **(a)** The number of droplets of isopropylene crystallized at different supercoolings ΔT. At $\Delta T < 101°C$ heterogeneous nucleation takes place; at $\Delta T = 101°C$ the nucleation is homogeneous [2.17]; **(b)** the number of tin droplets which had not frozen by moment t. The supercooling is constant and equal to 110°C [2.18]. **(c, d)** the number density of nuclei (in units of 10^6 cm^{-3}) formed by moment t in a Graham glass containing particles of gold **(c)** and iridium **(d)** at constant temperatures of 332°C **(c)** and 302°C **(d)** [2.19]

bulk of the drops undergoes the greatest supercooling and, it is assumed, crystallizes as a result of homogeneous nucleation. In contrast to supercoolings at which heterogeneous solidification of the drops takes place, maximum supercooling is, in the case of homogeneous nucleation, highly reproducible, which helps in judging the nature of the nucleation. The percentages of isopropylene drops frozen at different temperatures are shown as vertical segments in Fig. 2.4a.

From the kinetic point of view, maximum supercooling is not a quantity which enables one to find the nucleation rate, since the act of nucleus formation results from fluctuation and is therefore accidental. The probability of its occurrence within a time dt is $JVdt$, where V is the drop volume. Therefore the fraction of the drops not frozen by moment t is equal to $\exp(-JVt)$, and the fraction of the frozen ones to $1-\exp(-JVt)$. The experimental dependence of the fraction of the tin drops not frozen by moment t is shown in Fig. 2.4b, and indeed decreases exponentially with time, which confirms the randomness of nucleation. Knowing the volume of each drop V, we find the desired rate J from the shape of curves of the type given in Fig. 2.4b. Work with an aggregate of drops is, of course, statistically equivalent to repeated experiments with a single drop; Fig. 2.4b was obtained in precisely the latter way. In practice, it is more convenient to investigate nucleation not at a constant temperature, but at a decreasing one, when the number of nuclei arising within the given temperature range passes a maximum [2.20, 21].

Deep supercoolings of metal specimens weighing up to 500 g (e.g., 470°C for Ni; see Tables 2.2, 3) have been achieved recently by cleaning surfaces thoroughly and protecting them from oxidation.

The nucleation rate J depends strongly on supercooling ΔT, and therefore the practically achievable range of supercoolings does not usually exceed a few tens of degrees; this is not always sufficient for checking the $J(\Delta T)$ dependences, which are substantially different in the homogeneous and heterogeneous cases.

The use of the small-drop technique and of statistical methods to process the lifetimes of supercooled drops has produced the values of solid–liquid interfacial energies (chiefly for metals) arranged in Table 2.2 in order of increasing melting points. The energies determined by measuring dihedral angles at grain boundary grooves (DA) and the depression of melting points for small ($\lesssim$ 500Å) particles (DMP) are also presented for comparison. The experimental values of the preexponential factors in experiments with drops vary between 10^{33} and 10^{42} cm^{-3} s^{-1}. These values correspond qualitatively to the theoretical values and favor the homogeneous formation of nuclei: in heterogeneous nucleation the preexponential factor must be proportional not to the density of molecules, but to that of the impurity centers which provoke nucleation. The density of such centers, each of them being a single foreign macroparticle, is many orders of magnitude lower. Based on the conviction that the observed $J(\Delta T)$ dependence is determined by a stochastic homogeneous nucleation

Table 2.2. Supercoolings ΔT_C experimentally achieved for substances with different melting temperatures T_0, entropies of fusion ΔS, and crystal-melt interface energies α

Substance	T_0[K]	$\Delta S = \Delta H/T_0$ [cal /mol deg]	ΔT_C [deg]	α [erg/cm]
Hg	234	2.32	79[a], 51.2[b]	31.2[a], 23.0[b]
H_2O	273	5.28	39[c], 36.4[b]	32.1[c], 28.3[b]
Ca	310	4.31	76[c], 99[b], 106[d], 153[k]	55.9[c] 42.8[d], 40.4[b]
In	429	1.81	81[b]	30.8[a]
Sn	505	3.36	105[c], 122[b]	54.5[c], 59.0[b] (62, DMP)[m]
Bi	544	4.83	90[c], 100.4[b], 115[d],230[l]	54.4[c], 61.2[d] (74, DA; 55–80, DMP)[m]
Pb	610.5	1.9	80[e]	33.3[c], 71[m], 55[m] (40, DMP)[m]
Sb	903	5.28	135[c]	101[c]
Al	933	2.80	130[c]	93[c] 122[m] (158, DA)[m]
Ge	1231	4.94	227[c],316[b],200[b]	181[c], 251[b]
Ag	1234	2.19	227[c], 292[b]	126[c], 143[b], 172[m]
Au	1336	2.27	230[c], 190[f]	132[c], 191[m] (270, DMP)[m]
Cu	1356	2.29	236[c], 277[b], 180[f], 218[g]	177[c], 200[b], 254[m] (237, DA)[m]
Mn	1517	2.31	308[c]	206[c]
Ni	1726	2.40	319[c], 480[h], 290[i], 400[j]	255[c], 322[m]
Co	1765	2.1	330[c], 470[h], 310[f]	234[b]
Fe	1803	1.97	295[c]	204[c]
Pd	1825	2.0	332[c],310[f]	209[c]
Pt	2042	2.5	370[c]	240[c]

Note. [a][2.15], [b][2.18], [c][2.10$_a$], [d][2.22], [e][2.16], [f][2.23], [g][2.24], [h][2.25], [i][2.26], [j][2.27], [k][2.1], [l][2.2], [m][1.116]. DMP—depression of melting point; DA—dihedral angle method.

Table 2.3. Solid–liquid interface energies for alloys [1.12b]

Solid A	Liquid AB	T_0 [K]	X_{AL} molar fraction of A in liquid	α [erg/cm^2]
Al	Al-Sn	673	0.093	243
		823	0.440	197
	Al-In	673	0.018	290
		823	0.060	260
	Al-Bi	823	0.070	254
	Al-Pb	893	0.010	370
Cu	Cu-Pb	1093	0.115	390
Zn	Zn-Sn	600	0.473	124
	Zn-Pb	600	0.022	175
Ni[a]	Ni-Pb	100	0.0085	190
		1100	0.051	258
Fe[b]	Fe-Cu	1373	0.032	430
Nb[c]	Nb-Cu	1773	0.093	430
Cr	Cr-Ag	1673	0.022	540
Mo	Mo-Sn	1873	0.0015	735
W	W-Sn	2273	10^{-5}	1000

[a] $X^S_{Pb} = 6.10^{-3}$ [b] $X^S_{Cu} = 0.2$ [c] $X^S_{Cu} = 0.2$

process governed by the solid–liquid interface energy, one can find the latter quantities for the materials under investigation.

Using the tabulated values of surface energies and supercoolings, it is easy, with the help of (2.2), to estimate the size of the critical nucleus, R_c. For nickel, $R_c \simeq 6$Å, for gallium 4 Å, and for mercury 13 Å; in other words, for the supercoolings indicated, the nuclei contain several hundred atoms each, which justifies the application of the macroscopic approach even at the indicated high supercoolings. This approach fails, however, at supercoolings of $\sim 0.5\ T_0$ achieved for gallium [2.1]. At this supercooling the nuclei must consist of several atoms only. Even more importantly, the nucleation rate will be so high at these supercoolings that the droplets must solidify at lower ones. However, this is not the case in experiment. One of the possible explanations is based on the fact that atomic configurations and vibration modes in tiny clusters and, hence, their chemical potentials, do not reach the values corresponding to the solid macrophase. This is consistent with the appearance of at least five metastable crystalline modifications of Ga at high supercoolings. Metastable modifications of water crystallized at high supercoolings have also been reported.

Much evidence has been found for noncrystallographic atomic arrangements in small clusters in the gas phase (Sect. 2.2.2). Metastable phases have also

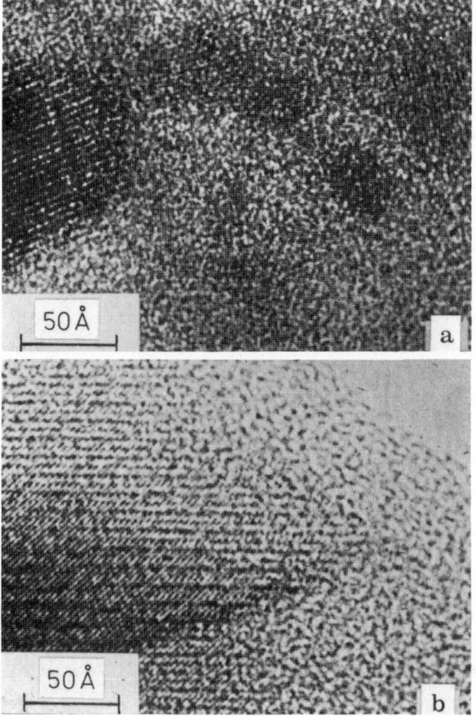

Fig.2.5a,b. Metastable structure of nuclei. Crystallites appearing in 3Mn: 1Mg silicate glass after initial (**a**) and subsequent (**b**) annealing have different lattice spacings. Lattice imaging in a transmission electron microscope. Courtesy of D.A. Jefferson

been observed in the initial stages of devitrification of 3Mn : 1 Mg silicate glass by direct lattice imaging in an electron microscope [2.27b]. After 30 min annealing at 800°C islands of pyroxene structure appear in this glass (Fig. 2.5a). After further annealing pyroxenoids with longer periods (Fig. 2.5b) and, finally, the stable pyroxmangite chain repeat can be observed.

Rasmussen [2.2] has found some features of spinodal decomposition in nucleation phenomena occurring at very high supercoolings in melts and solutions. However, as crystallization is a first-order phase transition, it should not be spinodal decomposition, though nucleation of metastable phases may produce such effects.

2.1.4 Transient Nucleation Processes

The above-discussed steady state nucleation rate is achieved only after a steady-state size distribution of the aggregates corresponding to a given supercooling is established in the crystallizing medium (vapor, melt, or solution). Since at the moment the supercooling is "turned on," the distribution corresponds to the initial temperature, the attainment of a new steady-state distribution requires some time. This transient period should not be confused with the simple expectation time for a nucleus to appear at a given steady-state nucleation rate. Let us estimate the transient time. We denote the aggregate by a point on the axis of sizes, which are expressed by the number N of atoms or molecules in the nucleus. The addition of a single atom to the aggregate will make this point jump by unity to the right, and the detachment of a single atom will cause it to jump by unity to the left; i.e., the point representing the aggregate will execute random walks on the size axis. It is in the course of such walks along the x axis that the aggregate rises to the crest of the potential barrier in Fig. 2.1 and, having overcome it, rolls down the right-hand branch, turning into a macroscopic crystal. The coefficient of diffusion along the size axis in the course of the walks described is equal in order of magnitude to the frequency $w_+ S_N$ of the addition of an atom to an aggregate of N particles with a surface area S_N, if w_+ is the frequency of addition per atomic position on the surface. Since there are few large aggregates in the system before the supercooling is "turned on," the time required to attain steady-state distribution, τ, is equal in order of magnitude to the time necessary for the diffusion of the nucleus along the size axis from point $N = 1$ to point $N = N_c$, where N_c is the number of atoms in the critical nucleus. This time $\tau \simeq N_c^2/4w_+ S_{N_c} \propto N_c^{4/3}/w_+$. In one-component condensed phases the frequency $w_+ \simeq \nu \exp(-E/kT)$, where E is the above-mentioned activation energy for the addition of one particle to the nucleus.

If the nucleus is formed on a foreign surface "wetted" by it (Sect. 2.2), its surface and volume are smaller, and the transient period will also be shorter. But a change in the order of magnitude of τ is possible only when the substrate is almost completely wetted by the crystal.

In the gas phase, in the absence of chemical reactions, $w_+ \simeq Pa^2/\sqrt{2\pi mkT}$, and at $P \simeq 1$ mm Hg, $w_+ \simeq 10^6\ \mathrm{s}^{-1}$. Therefore in the gas phase for $N_c \simeq 10^2$, τ will be $\simeq 10^{-4}$ s. In melts at $N_c \simeq 10^3$, $\tau \simeq 10^{-7}$ s, which is also rather short. In glasses, however, where the viscosity may be as high as 10^{10}–10^{13} P, while $E/kT \simeq 30$–40, the frequency $w_+ \simeq 10^{-2}\ \mathrm{s}^{-1}$, and at $N_c \simeq 10$, $\tau \simeq 10^5$ s, i.e., a period of the order of 24 h at $E/kT \simeq 33$. As the temperature increases, τ drops abruptly. The experimentally observed time τ for silicate glass $Li_2O \cdot 2SiO_2$ at temperatures of 430°C is $\simeq 9 \times 10^4$ s [2.28]. Experiments with nucleation of Cd and Ag on Pt electrodes in electrolytic solution showed $\tau \simeq 10^{-4}$–10^{-1} s [2.19].

A macroscopic description of the atomistic picture of non-steady-state nucleation yields the time-dependent nucleation rate [2.4b]:

$$\tilde{J}(t) = J[1 + 2\sum_{s=1}^{\infty}(-1)^s \exp(s^2 t/\tau)]\,,$$

where J is the steady-state nucleation rate (2.4) and τ is the induction period qualitatively discussed above. Direct calculations [2.4b] give

$$\tau = 8kT/\pi D^+(-\partial^2\delta\Phi/\partial N^2)_{N_c}\,.$$

Here $\delta\Phi$ is the change (2.1) in the Gibbs thermodynamic potential due to the formation of a cluster, and D^+ is the diffusivity along the cluster size axis ($\simeq w_+ S_{N_c}$) taken for critical nucleus $N = N_c$.

The number of crystallites experimentally observed to appear per unit volume after exposure during time t at a given temperature is

$$\int_0^t \tilde{J}(t)dt = J\left[t - \frac{\pi^2}{6}\tau - 2\tau\sum_{s=1}^{\infty}\frac{(-1)^s}{s^2}\exp\left(-\frac{s^2 t}{\tau}\right)\right] \simeq J\left(t - \frac{\pi^2}{6}\tau\right),$$

where the last equality is valid at $t \gtrsim 5\tau$.

Analogous equations hold for heterogeneous nucleation on foreign particles. The corresponding experiments have been staged by *Gutzov* and *Toshev* [2.19] on Graham glass with specially introduced gold and iridium particles $\lesssim 1$ μm in size since in real systems, most nuclei may form not homogeneously, but on the surfaces of foreign particles. The numbers of $Na_3P_3O_9$ crystals grown in these glasses after different exposure periods are presented in Figs. 2.4c, d, respectively. Although the activities of iridium and gold are different, as is evident from a comparison of these figures, an induction period is observed in both cases.

2.2 Heterogeneous Nucleation

2.2.1 Work and Rate of Nucleation. Size and Shape of Nuclei

Heterogeneous nucleation is the name given to the formation of nuclei at the interface: on foreign particles, vessel walls, the surfaces of existing crystals, etc. Nucleation is not equally probable at all points on the surface (Sect. 2.2.2). Nucleation around individual ions, as in the bubble and Wilson chambers, can also be called heterogeneous. The size of nuclei at supercoolings which ensure an experimentally detectable rate of nuclei formation does not exceed 10^{-6} cm (Sect. 2.1.3). Therefore surfaces with a curvature radius of $\gtrsim 10^{-5}$ cm can be regarded as flat during the formation of nuclei on them. Such surfaces may include vessel walls, the surfaces of foreign particles, or the faces of existing crystals.

Let us consider nucleation on a flat surface in more detail. We begin with the isotropic phenomenological model[1] and approximate the nucleus shape

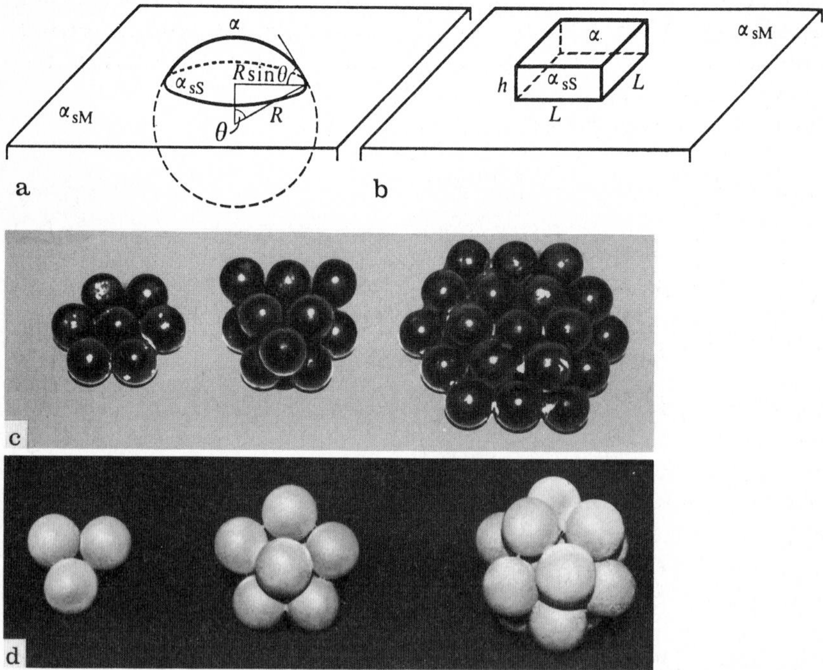

Fig.2.6a–d. Shapes of nuclei on a substrate. Macroscopic model: spherical segment **(a)** and parallelepiped **(b)**. Atomistic model: closest packing **(c)** and noncrystallographic packing of pentagonal symmetry with 3, 7, and 13 atoms **(d)** [2.29]

[1] The atomistic picture of nucleation is treated in Sect. 2.2.2.

by a sphere segment, whose surface forms a wetting angle θ with the substrate (Fig. 2.6a):

$$\alpha \cos\theta = \alpha_{SM} - \alpha_{sS} , \tag{2.8}$$

where subscript s denotes the substrate, and α_{sM} and α_{sS} stand for the substrate–medium and substrate–crystal interface, respectively.

The wetting angle determines the shape of the segment, i.e., the ratio between its height and the base diameter. The work of formation of the segment, as can be shown by reasoning similar to that used in Sect. 2.1 and with the help of trigonometric equations, is

$$\delta\Phi = \frac{\pi R^3}{3\Phi}(1 - \cos\theta)^2(2 + \cos\theta)\,\Delta\mu + \pi(R\sin\theta)^2(\alpha_{sS} - \alpha_{sM}) + 2\pi R^2(1 - \cos\theta)\alpha , \tag{2.9}$$

where R is the curvature radius of the spherical surface of the nucleus. The first term of (2.9) reflects the gain in bulk energy, the second arises because of the replacement of the substrate–medium interface by the substrate–crystal interface, and the third yields the energy of the spherical boundary of the nucleus with the medium.

The work of nucleation reaches a maximum [2.6]

$$\delta\Phi_c = \frac{16\pi\Omega^2\alpha^3}{3(\Delta\mu)^2}\,\frac{(1 - \cos\theta)^2(2 + \cos\theta)}{4} \tag{2.10}$$

at

$$R_c = \frac{2\Omega\alpha}{\Delta\mu} . \tag{2.11}$$

A comparison of (2.11) and (2.2) shows that the curvatures of the free nucleus and the nucleus on the substrate are equal at a given supercooling. This is understandable, as the equilibrium between nucleus and medium must exist over all points of its surface, [see (1.34, 35)]. Therefore at a given $\Delta\mu$ the equilibrium over any area of the surface depends exclusively on the curvature of this area, and not on the shape of the remaining surface, whether it is a complete sphere or is truncated as in Fig. 2.6a. The above reasoning is valid, of course, only if there is equilibrium along the line of contact between the three phases, i.e., if the equilibrium wetting condition (2.8) is satisfied.

The work (2.10) of nucleation on the substrate is less than that of homogeneous nucleation in the bulk of the medium, and is given by (2.3a). If the precipitating substance wets the substrate completely, i.e., $\theta = 0$ and $\cos\theta = 1$, crystallization or condensation on the substrate is not connected with the potential barrier, and occurs at zero or even negative supercooling. Even with moderate wetting, when $\theta \sim 45°$, the height of the potential barrier (2.10) for nucleation on the surface is one order of magnitude less than for nucleation in the bulk, i.e., the formation of nuclei on foreign surfaces is much more advantageous than homogeneous nucleation in the bulk. It follows from the foregoing that in order to achieve considerable supercoolings, a material which is wetted as poorly as possible by the crystallizing substance must be found for the vessels in which the supercooled solutions, melts, or vapors are to be placed. Graphite, which is used as crucible material, meets this condition. The same conclusion naturally holds good for solid foreign particles suspended in the supercooled liquid, which serve as nucleation "catalysts." The values of angles θ for boundaries involving two solid phases are as a rule, poorly known, but estimates can be made with the aid of contact angles characterizing the corresponding melts, because the surface energies of the melts differ from those of the crystals approximately as the densities of these phases, i.e., by no more than $\sim 10\%$.

The above discussion should be expanded in several directions. Firstly, one must take into account the anisotropy of the interfacial energy of the precipitating crystal, the substrate, and the medium. This anisotropy affects the crystal—substrate orientation (so-called epitaxy, see Sect. 2.3) as well as the shape of the nucleus and the values of the numerical coefficients in expressions like (2.10, 11) for the size of the critical nucleus and the height of the potential barrier. Secondly, at large supersaturations $\Delta\mu$, when the size of the nucleus becomes comparable with the interatomic spacing, the behavior of the system acquires a number of peculiarities (Sect. 2.2.2).

Let us consider some aspects of both of these questions within the framework of the phenomenological model of a nucleus having the shape of a rectangular parallelepiped of side L and height h (Fig. 2.6b). When such a nucleus appears on the substrate, the thermodynamic potential of the system changes to

$$\delta\Phi = -(L^2h/\Omega)\Delta\mu + L^2\Delta\alpha + 4Lh\alpha\,, \tag{2.12}$$

where $\Delta\alpha = \alpha + \alpha_{sS} - \alpha_{sM}$ is the variation of the free energy per unit area of the substrate interface associated with the replacement of the substrate–medium interface possessing specific free energy α_{sM} by the substrate–crystal interface (α_{sS}) plus crystal–medium interface (α). The value of $\Delta\alpha$ can be expressed via the specific free energy of adhesion α_s. This magnitude represents the work (per unit of interface area) which has to be performed in order to achieve reversible isothermal separation of the crystal from the substrate. Using the "cut and glue" procedure to build up a crystal–substrate interface (Sect. 1.2.1), we easily obtain

$$\alpha_{sS} = \alpha + \alpha_{sM} - \alpha_s\,.$$

Hence

$$\Delta\alpha = 2\alpha - \alpha_s\,, \tag{2.13}$$

i.e., $\Delta\alpha$ is the measure of how much more difficult it is to split the crystal than to separate it from the substrate. When the adhesion is strong, $\Delta\alpha < 0$, and one observes complete wetting of the substrate by the crystal, whereas $\Delta\alpha > 0$ corresponds to weaker adhesion and to poor or moderate wetting. If substrate and crystal are identical, then $\alpha_s = 2\alpha$ and $\Delta\alpha = 0$.

The work of formation of an aggregate depends both on its absolute dimensions L and h and on their ratio, i.e., on the shape of the aggregate. For an aggregate of a given volume L^2h, the work required to form a nucleus, $\delta\Phi$, is minimal when the sum of the last two terms in (2.12) is minimal, i.e., when the nucleus has an equilibrium shape. Minimization of the interfacial energy $L^2\Delta\alpha - 4Lh\alpha$ at $L^2h = \text{const}$ yields the equilibrium-shape condition

$$h/L = \Delta\alpha/2\alpha\,, \tag{2.14}$$

while minimization of total work $\delta\Phi$ under this condition yields the dimensions of the critical nucleus, L_c and h_c and the work of its formation, $\delta\Phi_c$:

$$L_c = 4\Omega\alpha/\Delta\mu, \quad h_c = 2\Omega\Delta\alpha/\Delta\mu, \quad \delta\Phi_c = 16\Omega\alpha^2\Delta\alpha/\Delta\mu^2\,. \tag{2.15}$$

These equations, which are analogous to (2.10, 11) for the anisotropic case, are meaningful as long as $L_c, h_c > a$, but inapplicable if formal substitution of $\Delta\alpha$ and $\Delta\mu$ into (2.13) gives $L_c, h < a$. Naturally, the closer the dimensions of the nucleus are to being atomic, the less exact (2.12, 15) become, since they use macroscopic values of the chemical potential of the nucleus and of its surface energy, which depend on the dimensions of the nucleus (Sect. 2.2.2). Remaining at the level of qualitative concepts, however, we shall ignore these dependences; we are concerned in particular with the formation of plane nuclei, $L \gg h$, when L may be considerable on the atomic scale even at $h = a$. Such a situation exists at $2\alpha/|\Delta\alpha| > 1$ [see (2.13)]. For instance, the last in equality always holds if $\Delta\alpha > 0$ and $\alpha_s > 0$, so that $L = 2\alpha h/\Delta\alpha > h$ [see (2.14)].

We now find (always bearing in mind the approximate nature of the approach) the region of $\Delta\alpha$ and $\Delta\mu$ where the equations in (2.15) are valid at least qualitatively, and the relations to which they change in the opposite case. To do this, we imagine the shape of the surface $\delta\Phi = \delta\Phi(L, h)$, using first a set of intersections of this surface with the planes $h = \text{const}$, see (2.12). Each such intersection is a parabola passing through the origin and reaching, at

$$L = L_c^*(h) = 2\Omega\alpha/(\Delta\mu - \Omega\Delta\alpha/h) \tag{2.16}$$

a maximum or a minimum with the value

$$\delta\Phi = \delta\Phi_c^*(h) = 4\Omega\alpha^2h/(\Delta\mu - \Omega\Delta\alpha/h)\,. \tag{2.17}$$

The maximum of $\delta\Phi$, which is the top of the potential barrier at a given h, is achieved when

$$\partial^2\delta\Phi/\partial L^2|_{h=\text{const}} = (2/\Omega)(\Omega\Delta\alpha - h\Delta\mu) < 0\,, \tag{2.18}$$

i.e., when $\delta\Phi_c^* > 0$, and the minimum is achieved in the opposite case. Of all the barriers $\delta\Phi_c^*(h)$, the lowest is the one for which $\partial\delta\Phi_c^*/\partial h = 0$ and $\partial^2\delta\Phi_c^*/\partial h^2 < 0$. Where $\Delta\alpha > 0$ and $\Delta\mu > 0$ such a situation is realized at a saddle point with coordinates $h = h_c$, $L = L_c^*(h_c) = L_c$, $\delta\Phi = \delta\Phi_c^*(h_c)$, which naturally coincide with (2.15).

Intersections of the surface $\delta\Phi = \delta\Phi(L, h)$ with the planes $L = \text{const}$ are straight lines with a slope $\partial\delta\varphi/\partial h|_{L=\text{const}} = -L^2\Delta\mu/\Omega + 4L\alpha$ with respect to the plane (L, h). This slope is positive when the size of the aggregate is less than the critical value $L_c < 4\Omega\alpha/\Delta\mu$ given by (2.13), and negative in the opposite case. A straight line passing through the saddle point is parallel to the plane (L, h). At $\Delta\alpha > 0$ and $\Delta\mu > 0$ a nucleus is formed when the system surmounts the saddle and moves from the initial state $L = 0$, $h = 0$, $\delta\Phi = 0$ into the region of large L and h, where $\delta\Phi < 0$.

When the supersaturation $\Delta\mu > 0$ increases and/or the value of $\Delta\alpha > 0$ decreases, the saddle point shifts and reaches the plane $h = a$ (since $L = 2\alpha h/\Delta\alpha$ and $2\alpha/\Delta\alpha > 1$, we always have $L > h > a$ in this case). Penetration by the saddle point beyond this plane, into the region $h < a$, is physically meaningless, and therefore in the region of parameters $\Delta\mu$ and $\Delta\alpha$, where $h_c = 2\Omega\Delta\alpha/\Delta\mu < a$, the system potential will change only along the curve $\delta\Phi = \delta\Phi(L, a)$ in the plane $h = a$. The maximum on this curve $\delta\Phi = \delta\Phi_c^*(a)$ is reached at $L = L_c^*(a)$, i.e., it has coordinates

$$\begin{aligned} &h = a, \qquad L = L_c^*(a) = 2\Omega\alpha/(\Delta\mu - \Omega\Delta\alpha/a)\,, \\ &\delta\Phi = \delta\Phi_c^*(a) = 4\Omega\alpha^2 a/(\Delta\mu - \Omega\Delta\alpha/a)\,. \end{aligned} \tag{2.20}$$

This point [which is, generally speaking, not a saddle point for the surface $\delta\Phi = \delta\Phi(L, h)$] is precisely the top of the potential barrier to condensation in the region $\Delta\mu > 2\Omega\Delta\alpha/a(\Delta\alpha > 0,\ \Delta\mu > 0)$, where the estimates given by (2.15) are no longer applicable.

At $\Delta\mu < 0$ (and $\Delta\alpha < 0$), the slope of the straight lines in the sections $L = \text{const}$ is always positive, i.e., an increase in h requires an increase in energy. On the other hand, in the case at hand

$$\delta\Phi = (L^2/\Omega)(h|\Delta\mu| - \Omega|\Delta\alpha|) + 4Lh\alpha \tag{2.21}$$

and at small h, when $a \le h < \Omega|\Delta\alpha|/|\Delta\mu|$ and $L > L_c^*(h)$, the work required to form the aggregate decreases to values $\delta\Phi < 0$ with an increase in its length L in the substrate plane. Thus at $\Delta\alpha < 0$ condensation is possible even from undersaturated vapor ($\Delta\mu < 0$); it occurs by means of the formation of flat nuclei and generally requires overcoming the potential barrier (2.20). If, however, $|\Delta\mu| > \Omega|\Delta\alpha|/a$, i.e., if undersaturation is high enough, $\delta\Phi$ increases with increasing L even at $h = a$, and no condensation is possible.

At $\Delta\mu > 0$ (and $\Delta\alpha < 0$), the slope of the lines in sections $L = \text{const}$ is positive for $L < 4\Omega\alpha/\Delta\mu$, that is, throughout the region $0 < L < 2\Omega\alpha/\Delta\mu$, where the branch of the hyperbola $L = L_c^*(h)$ passes, while $\delta\Phi = (L^2/\Omega)(-h\Delta\mu - \Omega/|\Delta\alpha|) + 4Lh\alpha$ and decreases with increasing L at $L > L_c^*(h)$ for any h and $\Delta\mu$. Therefore at $\Delta\alpha < 0$ and $\Delta\mu < 0$ the barrier is also described by (2.20), but is naturally minimal in relation to all the other cases; the condensation of the supersaturated vapor proceeds readily on a completely wetted substrate.

Indeed, the greatest barrier height in the latter case, reached at $\Delta\mu = 0$, is $\delta\Phi_c^*(a) = 4\alpha^2 a^2/|\Delta\alpha| \simeq 2\alpha\varepsilon_1/|\Delta\alpha|$, where $\varepsilon_1 = 2\alpha a^2$ is the binding energy of the nearest neighbors. In other words, if $\Delta\alpha$ is not too low, the barrier is comparable to the desorption energy of a single atom. A similar estimate is also true for the smallest barrier height in the preceding case where $\Delta\alpha < 0$ and $\Delta\mu < 0$, so that here, too, the barrier may be comparatively small.

The process of nucleation on a substrate at different values of parameters $\Delta\alpha$ and $\Delta\mu$ is described by the relations given in Table 2.4.

Of special interest is condensation on an intrinsic substrate when $\Delta\alpha = 0$, i.e., crystal growth. It proceeds at $\Delta\mu > 0$ and is described by (2.20). If the critical nucleus has the shape of a disc of thickness a, its radius r_c and work of formation $\delta\Phi_c$ are equal to

$$r_c = \Omega\alpha/\Delta\mu\,, \qquad \delta\Phi_c = \pi\Omega\alpha^2 a/\Delta\mu\,. \tag{2.22}$$

The expression for the rate of nucleation on the surface has the same structure as (2.4). The preexponential factor B, however, now has the dimensionality $\text{cm}^{-2}\,\text{s}^{-1}$ and a slightly different form. If nucleation is possible around any one of n_s adsorbed atoms (molecules, ions) on a unit surface area, and the new particles join the nucleus from the bulk of the medium, then in order to obtain the value of B in the heterogeneous case one must replace $n \to n_s$, $4\pi R_c^2 \to 2\pi R_c^2(1 - \cos\theta)$ (the surface area of the spherical segment) in the

Table. 2.4. Supersaturations and wetting conditions at which nucleation is described by different relationships

$\Delta\alpha>0,\ \Delta\mu<0$		Condensation impossible
$\Delta\alpha>0,\ \Delta\mu>0$	$\Delta\mu<2\Omega\Delta\alpha/a$	(2.13)
	$\Delta\mu>2\Omega\Delta\alpha/a$	(2.14)
$\Delta\alpha<0,\ \Delta\mu<0$	$\Delta\mu<\Omega\Delta\alpha/a$	Condensation impossible
	$\Delta\mu>\Omega\Delta\alpha/a$	(2.14)
$\Delta\alpha<0,\ \Delta\mu>0$		(2.14)

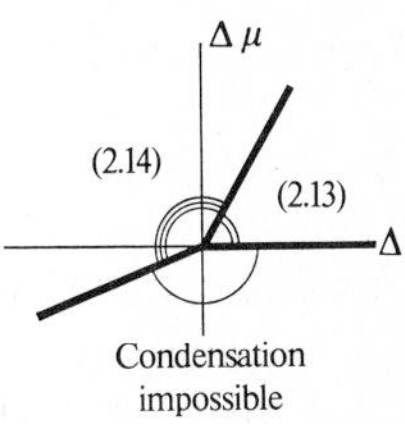

expression for B in (2.4). For nucleation in a condensed medium (melt, solution) one must, in addition, replace the flow $P/\sqrt{2\pi mkT}$ with $na\nu\exp(-E/kT)$, where all the values have the same meaning as in (2.7). In the formation of nuclei from atoms of the adsorption layer, the product $4\pi R_c^2 P/\sqrt{2\pi mkT}$ must be replaced by $2\pi R_c \sin(\theta)\, n_s a\nu \exp(-E/kT)$.

The work of nucleation on steps, scratches, crevasses, and other irregularities (defects) of the surface (Sect. 2.2.3) may be greatly reduced in comparison with regular sites of the surface. If the density of such nucleation-active defects on the surface is n^*, then in the formula for B in (2.4), n must be replaced by n^*, and $\delta\Phi_c$ by the work of nucleation on the defect.

In the above-mentioned cases the product of the preexponential factor for heterogeneous nucleation multiplied by the total surface area of the vessel walls and foreign particles may be much less than that of the preexponential factor of homogeneous nucleation multiplied by the total volume of the system, but the reduction in the work of nucleating on sufficiently strongly adsorbing active surfaces indicates a higher probability of heterogeneous nucleation.

The nucleation on substrates described occurs in the same fluctuational way as in the bulk, the crystal aggregate "climbs" up onto the crest of the potential barrier (which is lower than that in the homogeneous case) as a result of accidental attachments and detachments of particles, that is, of random walks along the size axis. Therefore heterogeneous nucleation will also be characterized by a transient induction period. This period will be shorter in the heterogeneous case than in the homogeneous case by as many times as the square of the number of particles in the critical nucleus in the first case is less than it is in the second. In the isotropic approximation this ratio is $(1\text{-}\cos\theta)^4\,(2+\cos\theta)^2/16$.

2.2.2 Atomistic Picture of Nucleation. Clusters

We have just described the phenomenological approach to nucleation, which holds true in the range of not-too-high supersaturations (supercoolings) when the critical nucleus contains many tens of atoms and can be considered a macroscopic formation. Accordingly, we attributed a definite macroscopic shape (sphere, cube, parallelepiped, etc.) to the nucleus and used the surface energy concept. But at very high supersaturations, when the size (2.2) of the critical nucleus is near atomic, the applied approach is unjustified. Here the rate at which nucles are formed and their sizes and shape must be determined from atomistic rather than macroscopic considerations. The atomistic approach is necessary in particular when analyzing the condensation of a molecular beam on a substrate which is so cold that the reverse flow of thermal (or chemical) evaporation of the condensing substance is many orders of magnitude less than the incident one. Very high supersaturations are also attained in electrolytic deposition from solutions.

We shall outline the general features of the atomistic approach, focusing primarily on condensation on the substrate, although the same features characterize the analysis of nucleation in the bulk. In the following we shall make use of the study conducted by *Stoyanov* [2.3]. A comparison of theory and experiment is reviewed in [2.29].

Suppose that N atoms of the gas phase or of the adsorption layer join together into an aggregate (cluster, complex) located at some site of the substrate. The change in the thermodynamic potential of the system will then be

$$\delta\Phi(N) = \mu^0(N) - N\mu_V \equiv - N(\mu_V - \mu_S) + [\mu^0(N) - N\mu_S]\,. \tag{2.23}$$

Here, μ_V is the chemical potential of the atom in the vapor phase (molecular beam). When the atoms of the adlayer join together into a complex, μ_V must be replaced by their chemical potential μ_s. If the vapor is in equilibrium with the adsorption layer, $\mu_V = \mu_s$. On condensation from a molecular beam, μ_s may be less than μ_V. The quantity $\mu^0(N)$ is the chemical potential of one separate aggregate of N atoms without regard for its translatory, rotary, and vibratory motions as a whole. Hereafter we restrict ourselves to the analysis of nucleation on a substrate, where these motions can justifiably be neglected. Since μ^0 refers to one aggregate, it does not contain a term related to the configurational entropy of the distribution of the aggregates over the substrate. The second of the above equalities for $\delta\Phi(N)$ is quite similar to (2.1); the first terms of both expressions give a reduction in system potential due to the formation of an island of a thermodynamically more advantageous phase ($\Delta\mu = \mu_V - \mu_S > 0$), while the second term characterizes the difference between the thermodynamic potential of N atoms in the aggregate, $\mu^0(N)$, and the potential of these N atoms in a bulk crystal. For a fixed macroscopic aggregate this difference is, by definition, its free surface energy, and for an atomic-size aggregate, the difference $\mu^0(N)$-$N\mu_S$ is due to the unsaturated bonds of the outer atoms of the aggregate and to the differences in the mutual arrangement, interatomic distances, and vibration states ("phonon spectrum") of the atoms of the aggregate as compared with the same number of atoms in the bulk crystal.

Statistical computations of the thermodynamic values of small aggregates, including computer simulation, are reported in [2.30 29]. We shall restrict ourselves here to a reference to Sect. 1.1, where it was shown that the energy term in the chemical potential of the macrophase, (1.1), plays the predominant role, and note that qualitatively, the same situation obtains for the difference $\mu^0(N) - N\mu_S$ for the microphase. Indeed, the difference between the average potential energies per atom in a small aggregate (where most atoms are surface atoms) and in a bulk crystal is close to binding energy ε_1. For gold, for instance (heat of evaporation $\Delta H = 84$ kcal/mol, fcc lattice), the value of $\varepsilon_1 \simeq$ 14 kcal/mol, and for silicon ($\Delta H = 111$ kcal/mol, diamond lattice) $\varepsilon_1 \simeq$ 55.5 kcal/mol. Hence for gold at $T = 700$ K, $\varepsilon_1 \simeq 10\,kT$, and for silicon at 900 K, $\varepsilon_1 \simeq 30\,kT$. At the same time, the entropy term of the difference $\mu^0(N) - N\mu_S$ per particle evidently does not exceed that in the chemical potential of a bulk crystal, i.e., a few kT. Therefore in estimating $\mu^0(N) - N\mu_S$ we can restrict ourselves to the energy term.

The number of unsaturated bonds of surface atoms belonging to an aggregate with a given total number N of atoms essentially depends on the aggregate's shape. For instance, three atoms on the substrate may form a chain or a triangle. It is obvious that the number of unsaturated bonds for a triangle is minimal, and hence the value of $\delta\Phi(N)$ at $N = 3$ for this shape of aggregate is also minimal. The work needed to form an aggregate of four atoms is minimal for a tetrahedron, but not for a flat rhombus, a chain, etc. In atomistic calculations of the rate of nucleation, use is made, as in the macroscopic approach, of the minimum work $\delta\Phi(N)$ at a given N, i.e., of the equilibrium shape, or, more precisely, the equilibrium atomic configuration in the aggregate.

The atoms of elements (e.g., metals) in small aggregates may either be stacked according to the laws of crystallographic packings or form complexes of noncrystallographic pentagonal symmetry. Figure 2.6c shows models of complexes of the first type consisting of black balls, and in Fig. 2.6d, complexes of the second type consisting of white balls can be seen. The existence of pentagonal aggregates (Fig. 2.6d) is reasonable, because the interatomic distances in the first one or two coordination spheres of such a packing deviate only slightly from crystallographic ones, and the number of saturated bonds in these spheres is greater than for crystallographic closest packings. As the number of atoms in the complex increases, the law of the impossibility of filling the space with polyhedra having fivefold symmetry comes into force. This can be seen from the fact that the interatomic distances increases with the number of atoms in the pentagonal complex, and such packing becomes disadvantageous.

An increase by unity in the number of particles N in the cluster always reduces the first term in $\delta\Phi(N)$ by $\Delta\mu = \mu_V - \mu_S$. The second term, $\mu^0(N) - N\mu_S$, then increase as a rule, because the number of atoms at the boundary of the cluster increases, as does its "surface area." But in contrast to the macroscopic approximation, in which the shape of the aggregate is assumed to remain unaltered upon the addition of particles, the second term in the atomistic model increases nonmonotonically with N. The point is that the addition to the aggregate of an atom which completes some coordination shell gives fewer new unsaturated bonds than the addition of an atom which initiates the filling of the next coordination shell. Deviations from monotonicity with an increase in the number of particles in a pentagonal cluster occur, in particular, for $N = 7$ and $N = 13$, as can be seen from Fig. 2.7a which shows the dependences $\delta\Phi(N)$ for two values of supersaturation $\Delta\mu$ (only the values of $\delta\Phi$ for integral values of N have meaning in Fig. 2.7a, of course). The dependence given in Fig. 2.7a was obtained by *Stoyanov* [2.3] for the case where the binding energy of each of the condensing atoms with the substrate is equal to the energy of one bond ε_1 of these atoms with each other. The general shape of the dependence $\delta\Phi(N)$ in Fig. 2.7a is the same as in the macroscopic approximation (Fig. 2.1): an increase at small $N (N < 7)$ and a decrease at large N. The maximum of $\delta\Phi$ is reached at $N = N_c$, the critical size of the complex nucleus (in Fig. 2.7a $N_c = 7$). The linear dependence $\delta\Phi(N)$ within intervals of several units corresponds

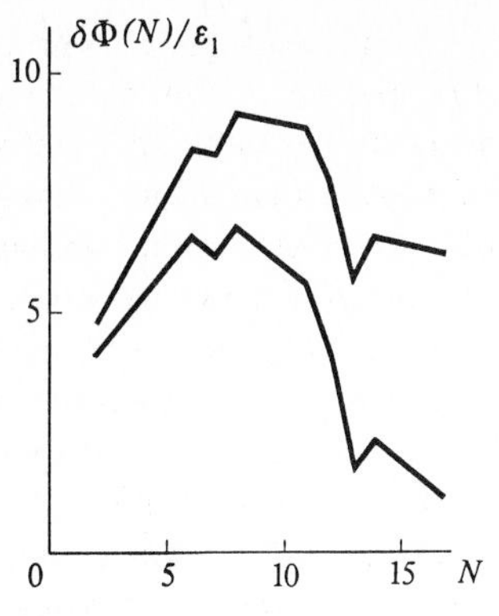

Fig.2.7a–c. Work of formation (**a**) and mass spectra of atomic clusters (**b, c**). (**a**) Change in the thermodynamic potential of a system during the formation of a pentagonal aggregate of N atoms (Fig. 2.6d) on a structureless substrate where the binding energy of each atom with the substrate is the same as that between two atoms in the aggregate. The *upper curve* is calculated for supersaturation $\Delta\mu = 2.4\ \varepsilon_1$, the *lower* for $\Delta\mu = 2.1\ \varepsilon_1$ [2.29]; (**b, c**) mass spectra of lead (**b**) and xenon (**c**). The "magic numbers" of atoms in clusters are indicated at the corresponding peaks. The number $N = 7$ is singular in both the calculation (**a**) and experiment (**b, c**)

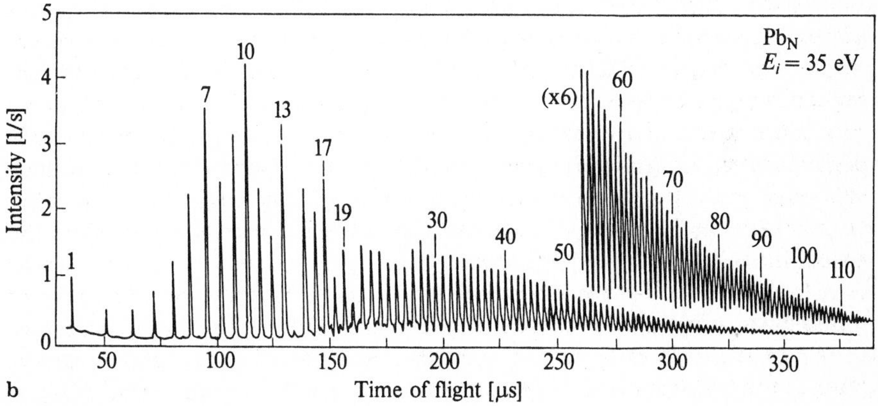

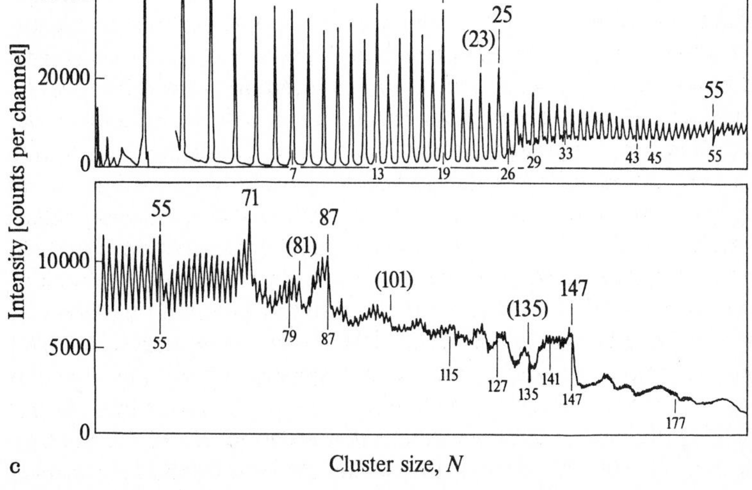

to the addition of new particles into kink positions on the cluster surface, i.e., to the completion of coordination shells.

The nonmonotonic variation of the cluster energy was recently confirmed when *Sattler* and co-workers discovered "magic numbers" of atoms in clusters [2.31]. These authors succeeded in producing for the first time atomic clusters in the entire range of mass numbers, from a single atom to thousands of atoms, thus covering the span between atomic vapor and microparticles $\simeq 10^{-6}$ cm in size.

Two different techniques of producing clusters have been developed, depending on whether the substance being investigated has a low or high vapor pressure at room temperature. For substances with low pressures, the vapor, produced in an oven, effuses into a cooled chamber filled with helium [2.31c]. In this chamber the vapor atoms collide and form clusters. At helium pressures above 10 mbar the gaseous mixture of clusters, vapor atoms, and helium effuses continuously into a vacuum. There are no more collisions in the vacuum and the aggregates cease to grow. They flow to a specially constructed time-of-flight mass spectrometer, where they are ionized by a pulsed electron beam crossing the gas flux and are accelerated by the electric field to different velocities depending on their masses. By measuring the time of flight of the ions through the known distance (1.7 m), the authors measureed the masses of the ions corresponding to the masses of the clusters.

Substances which are in a gaseous state at room temperatures were effused into a vacuum through a nozzle 2 cm long and 200 μm in diameter under a pressure of about 10 atm. Adiabatic expansion of the gas caused it to cool and cluster. The cluster masses were again measured by a time-of-flight mass spectrometer. [2.31d].

The spectra of lead obtained in this way are presented in Fig. 2.7b. [2.31e]. The different times of flight plotted along the abscissa correspond to different masses, some of which are indicated at the peaks. Intensity peaks (counts/s) are present for all the masses, from a single atom (first peak on the left) to more than 100 atoms. Figure 2.7b shows that the intensity of the peaks decreases very slowly with the mass number as compared to conventional spectra, in which the intensity falls by 3–4 orders of magnitude as the mass numbers decrease from 1 to 4–5 [2.32]. The most important feature of the spectrum in Fig. 2.7b is the nonmonotonic decrease in the peak intensity. Indeed, starting with clusters of $N = 3$ atoms, the intensity rises to clusters with $N = 7$ and then suddenly drops by at least 30% for $N = 8$. The peak intensity reaches new local maxima at $N = 10$ and again falls by about half for $N = 11$. The next maxima correspond to $N = 13$, 17, and 19. This structure of spectra is quite reproducible with respect to evaporation temperature, helium pressure, ionization voltage, and other experimental conditions. Analogous spectra have been obtained for Bi [2.31c].

The observed "magic numbers" (analogous to the numbers in nuclear physics) may be explained by noncrystallographic packing of spheres: $N = 7$

corresponds to a pentagonal bipyramid, $N = 13$ to an icosahedron (Fig. 2.6d), and $N = 19$ to an icosahedron with an added pentagonal cap of 6 atoms [2.31 f]. The absence of $N = 10$ and 17 among the calculated clusters with minimal energy might be due to the discrepancy between the Lennard-Jones potential used in calculations and the real interatomic potential in lead clusters.

The mass spectrum of Sb exhibits magic numbers $N = 4, 8, 12, 16, 20, 36, 52, 84, \ldots$, because Sb_4 tetramers seem to be a major species in the gas. This sequence may be attributed to subsequent crystallographic packing of the Sb_4 tetrahedra [2.31 f]: by adding a Sb_4 group to each face of the initial Sb_4 tetrahedron, one obtains $N = 4, 8, 12, 16, 20$, thus completing the first coordination shell of the initial Sb_4 tetrahedron. Adding one Sb_4 group onto each face of the tetrahedron consisting of 20 atoms gives $20 + 4 \times 4 = 36$ atoms. If, instead, two groups are added on top of each face, one obtains the next shell and $N = 52$. The addition of four Sb_4 groups to each face complete the third shell and yields a cluster of $20 + 4 \times 16 = 84$ atoms. However, this simple geometric model does not give cluster energies and does not sufficiently explain other features of the spectrum.

The xenon spectrum presented in Fig. 2.7c shows magic clusters of 13, 19, 55, 71, 87, and 147 atoms [2.31d]. Again, geometrical analysis and computer simulation [2.33a] show that noncrystallographic packing of spheres is energetically most favorable, especially for icosahedrons of 13, 55, 147, 309, and 561 atoms [2.33b].

Sodium halide clusters have the chemical formula Na_nX_{n-1}, probably due to neutralization and subsequent loss of halogen atoms X during ionization. The structure of the observed magic ionized clusters $N = 5, 14, 23, 38$ may be understood to have simple cubic packings ($3 \times 3 \times 1$ atoms for $N = 5$; $3 \times 3 \times 3$ atoms for $N^+ = 14$; $3 \times 3 \times 5$ for $N = 23$, etc.) [2.31a]. Calculations for neutral clusters show that the most favorable structures consist of 3-molecule rings stuck together. This model gives $N = 3, 6, 9, 12, 15, \ldots$. The charged clusters of 5, 8, 11, 14, ... atoms might represent ionized fragments of the neutral ones with ring structures.

Electron diffraction studies of silver clusters obtained in the gas phase by conventional techniques [2.32 a] show that the smaller the clusters are, the stronger are the deviations from the fcc structure [2.34 a]. However, icosahedral structures in ihe clusters have not yet been unambiguously demonstrated by diffraction data. Numerous electron microscopic observations of pentagonal microparticles [2.34b] may also be considered indirect evidence of the noncrystallographic cluster structure. Extensive information on recent cluster studies is given in [2.32a, 2.33a].

Summarizing the experiments on magic numbers described above, we conclude that the energetically most favorable clusters are indeed those with completed shells, and that the noncrystallographic packing in small aggregates like the one shown in Fig. 2.6d and used in the calculations for Fig. 2.7a is often more favorable than the packing typical of the corresponding bulk crystal. Thus a transformation of the noncrystallographic aggregate structure or its matching with the bulk structure in the course of cluster growth may cause the twins in microparticles which are observable in electron microscopes and which have kinetic consequences in nucleation.

Let us now return to the $\delta\Phi(N)$ dependence presented in Fig. 2.7a. An important feature of this dependence at high supersaturations is that the critical size of the nucleus is independent of supersaturation within a certain range of supersaturations $\Delta\mu$. This is revealed in Fig. 2.7a in the coincidence of the maxima of the two curves, which are plotted for different supersaturations. Since $\Delta\mu = kT\ln(P/P_0)$, the vapor pressure P_1 corresponding to $\Delta\mu_1$ exceeds the pressure P_2 corresponding to $\Delta\mu_2$ by a factor of $\exp(\Delta\mu_2 - \Delta\mu_1)/kT$. At $\Delta\mu_1 = 2.4\ \varepsilon_1$, $\Delta\mu_2 = 2.1\ \varepsilon_1$, $\varepsilon_1 = 14$ kcal/mol, and $T = 700$ K, we have $P_1/P_2 \simeq 20$. Thus in the range of such high supersaturations $P/P_0 = \exp(\Delta\mu_2/kT) \simeq 1.5 \times 10^9$, the critical size remains unaltered, upon a 20-fold change in supersaturation. At first glance, the independence of N_c from $\Delta\mu$ is obtained at variance with the increase in the size of the critical nucleus with a reduction in supersaturation which follows from (2.11). The point is, however, that the number of particles in the nucleus can change only discretely. Therefore function $N_c(\Delta\mu)$ is, in coordinates N_c *vs* $\Delta\mu$, a staircase descending towards large $\Delta\mu$ and consisting of steps of unit height. The length of the steps, i.e., the interval $\Delta\mu$ within which N_c^* is constant, increases with supersaturation $\Delta\mu$. The classical continuous dependence $N_c = 32\pi\Omega^2\alpha^3/2\Delta\mu^3$, which is equivalent to (2.11), since $N_c\ \Omega = 4\pi R^3/3$, approximates fairly well the described "stepped" dependence at large N (i.e., at small $\Delta\mu$), but is too imprecise for small N_c.

The function $N_c(\Delta\mu)$ for nucleation on a substrate at small N_c depends on the structure of the complex and is not known in the general case. This function is readily obtainable, however, in the following simple, essentially macroscopic model. Suppose that atoms in the crystal phase form a simple cubic lattice with an energy ε_1 for a single bond between the nearest neighbors. Assume that the binding energy of the adatom with the substrate is ε_s, and the nucleus is a rectangular parallelepiped made up of N particles of size $m \times m \times n$ interatomic distances, so that $m^2n = N$. It is clear from symmetry considerations that the interfacial energy in the complex–substrate system is minimal if the square "face" of the nucleus is parallel to the substrate. Then

$$\begin{aligned}\delta\Phi(N) &= -\ N\Delta\mu + [\mu^0(N) - N\mu_S] \\ &= -\ N\Delta\mu + 2m^2(\varepsilon_1/2) + 4mn(\varepsilon_1/2) - m^2\varepsilon_s \\ &= -\ N\Delta\mu + m^2(\varepsilon_1 - \varepsilon_s) + 2mn\varepsilon_1\,,\end{aligned} \tag{2.24}$$

where $m^2\varepsilon_s$ is the energy released on adhesion of the nucleus to the substrate, and the remaining three terms in the middle expression give the work required to form an aggregate in the gas phase. By minimizing $\delta\Phi(N)$ with respect to m and n, subject to the additional condition $m^2n = N = \text{const}$, we obtain

$$m^3 = N/(1 - \varepsilon_s/\varepsilon_1), \qquad n^3 = N(1 - \varepsilon_s/\varepsilon_1)^2\,. \tag{2.25}$$

Substituting these expressions into the equation (2.24) and minimizing the result with respect to N, we find the following analog of relation (2.10):

$$N_c = 8\varepsilon_1^3(1 - \varepsilon_s/\varepsilon_1)/\Delta\mu^3 ; \tag{2.26}$$

$$\mu^0(N_c) - N_c\mu_s = 12\varepsilon_1^3(1 - \varepsilon_s/\varepsilon_1)/\Delta\mu^2 , \tag{2.27}$$

$$\delta\Phi(N_c) = 4\varepsilon_1^3(1 - \varepsilon_s/\varepsilon_1)/\Delta\mu^2 . \tag{2.28}$$

(2.26) is the above-described stepped dependence $N_c(\Delta\mu)$; the width of the step, which has an ordinate N_c, equals

$$2\varepsilon_1(1 - \varepsilon\ /\varepsilon_1)^{1/3}[(N_c - 1)^{-1/3} - N_c^{-1/3}] .$$

If $\Delta\mu > 2\varepsilon_1(1 - \varepsilon_s/\varepsilon_1)^{1/3}$, the nucleus consists of a single atom ($N_c = 1$), and the step width is infinite. At $\varepsilon_1 \simeq 14$ kcal/mol, $T = 700$ K, and $\varepsilon_s/\varepsilon_1 = 0.5$, this must take place at supersaturations $P/P_0 \simeq 10^7$. With a decrease in supersaturation, the number of particles in the critical nucleus increases, while the length of the intervals $\Delta\mu$ within which N_c remains constant drops off as $\Delta\mu^4/24\varepsilon_1^3(1 - \varepsilon_s/\varepsilon_1)$. The above-discussed model correctly reflects the principal dependences, but is too simplified to obtain absolute values, whose computation requires analyzing concrete atomic configurations of the type shown in Figs. 2.6c, d.

Let us now consider the rate of nucleation. Suppose there are, per unit surface, $n(N_s)$ critical complexes of N_c atoms each. A complex (cluster) is critical if the addition of one atom to it turns it into a nucleus, whose further growth can only proceed with a decrease in the thermodynamic potential of the system. We assume that addition of the next atom to the complex proceeds only with the overcoming of the barrier of surface diffusion U_D, and that only atoms adsorbed at sites located at a distance of one diffusion jump, i.e., of $\sim a$, from the edge of the complex are added. If ζ is the fraction of such sites along the perimeter, then

$$J = \zeta(n_s/n_0)\nu n(N_c)\exp(-U_D/kT) \tag{2.29}$$

aggregates of above-critical size arise on a unit surface area per unit time. Here, n_s is the density of the number of single adatoms on the surface, and $n_0 \simeq a^{-2}$ is the number of adsorption sites per unit surface area [so that $\nu n_0^{-1}\exp(-U_D/kT) \simeq D_s$, i.e., the coefficient of surface diffusion]. The density $n(N_c)$ of the number of complexes of critical size on the surface is obtained from the mass action law as applied to the formation reaction of a complex consisting of N adsorbed atoms:

$$\mu^0(N_c) + kT\ln[n(N_c)/n_0] = N_c\mu_s . \tag{2.30}$$

Hence $n(N_c) = n_0\exp[-\delta\Phi(N_c)/kT]$. The same approach can naturally be applied in determining the number of aggregates of a given size in a vapor, solution, melt, etc. (Sect. 2.1).

If we express the steady-state density of the adatoms in terms of their lifetime on the surface, τ_s and the intensity of the incident flux I[cf. (1.24)] $n_s = I\tau_s = (I/\nu)\exp(\varepsilon_s/kT)$, and substitute $n(N_c)$ and n_s into the equation (2.29), we get the general expression for the rate of nucleation

$$J = \zeta I \exp[(\varepsilon_s - U_D)/kT] \exp(N_c\Delta\mu/kT) \exp\{-[\mu^0(N_c) - N_c\mu_S]/kT\} . \quad (2.31)$$

It is worth noting that $\exp[(\varepsilon_1 - U_D)/kT] \simeq n_0\lambda_s^2$. The expression obtained describes the nucleation rate J on the surface both at low and high supersaturations. At high supersaturations, N_c is constant in a wide interval of $\Delta\mu$ values and changes jumpwise to $N_c + 1$ or $N_c - 1$ at the ends of this interval. Within each interval J must increase linearly with $\Delta\mu/kT$, with the slope of the lines giving the value of N_c. At low supersaturations, N_c may be considered a continuous function of $\Delta\mu$, and we return to the classical dependence, which has the following form for the above-discussed model:

$$J = \zeta I \exp[(\varepsilon_s - U_D)/kT] \exp[-4\varepsilon_1^3(1 - \varepsilon_s/\varepsilon_1)/\Delta\mu^2] . \quad (2.32)$$

At high supersaturations, the general expression for nucleation rate J can be transformed. Note that

$$N\Delta\mu/kT = N \ln P/P_0 = \ln (I/I_0)^N , \quad (2.33)$$

where $I_0 = \nu n \exp(\varepsilon_{1/2})$ is the equilibrium flux of the crystallizing substance towards (or from) the surface at the substrate temperature. It has been shown above that the entropy terms in the difference $\mu^0(N) - N\mu_S$ are small compared to the potential energy terms. Thus, having in mind that $\mu_S \simeq -\varepsilon_{1/2}$

$$\mu^0(N_c) - N_c\,\mu_S = -U(N_c) + N_c\varepsilon_{1/2} , \quad (2.34)$$

where $-U(N_c)$ is the potential energy of an aggregate of critical size on the substrate, computed from the energy of isolated atoms in the gas phase [$U(N_c) > 0$]. For a two-dimensional nucleus,

$$\mu^0(N_c) - N_c\mu_S = -E(N_c) - N_c\varepsilon_s + N_c\varepsilon_{1/2} , \quad (2.35)$$

where $-E(N_c)$ is the energy of formation of a flat nucleus from atoms in the gas phase [$E(N_c) > 0$], and $N_c\varepsilon_s$ is the energy released on adhesion of this nucleus to the substrate. Substituting the expressions (2.33), (2.35) and that for I_0 into (2.31), we get

$$J = \zeta I(I/\nu n_0)^{N_c} \exp\{[E(N_c) + (N_c + 1)\varepsilon_s - U_D]/kT\} . \quad (2.36)$$

This equation was first obtained in a somewhat different way by *Walton* [2.35, 36] in connection with the problem of nucleation on substrates at high supersaturations upon condensation of molecular beams. It is also called the *Walton-Rhodin* equation [2.37].

2.2.3 Decoration. Initial Stages of Growth

The approach described in the preceding section can be applied not only to the formation of nuclei at arbitrary, regular points on the surface, but also at singular points where the potential barrier which must be overcome in order for a critical-size aggregate to form is lower than at regular, "ideal" sites. These singular points on the surface may be of different kinds. For instance, impurity adatoms, whose bond with the atoms of the crystallizing substance is stronger than that of the latter among themselves, may serve as nucleation centers. Surface vacancies and associations of vacancies may also serve as singular points; deposited atoms may form stronger aggregates around them than around adatoms. For instance, during gold precipitation, a gold atom in an anionic (Cl^-) vacancy on the (100) face of a KCl crystal leads to the formation of an aggregate with a strong metallic bond [2.38]. Tightly bonded aggregates may also arise around other point defects, at steps, on grain boundaries, domains, segregations of the second phase, etc. The nuclei in the reentrant angles of steps, near and around impurity aggregates, and on vacancy depressions are shown schematically in Fig. 2.8. In the macroscopic representation, the advantage of nucleation at the sites listed follows from the fact that, for instance, the peripheral energy of the nucleus and the step (Fig. 2.8a) is lower by $Lh\alpha_s$ than that of a separated nucleus and step, α_s being the specific adhesion energy. In precipitation from the gas phase or from a molecular beam in vacuum, $\alpha_s > 0$ simply because of the difference in densities between the gas and the solid.

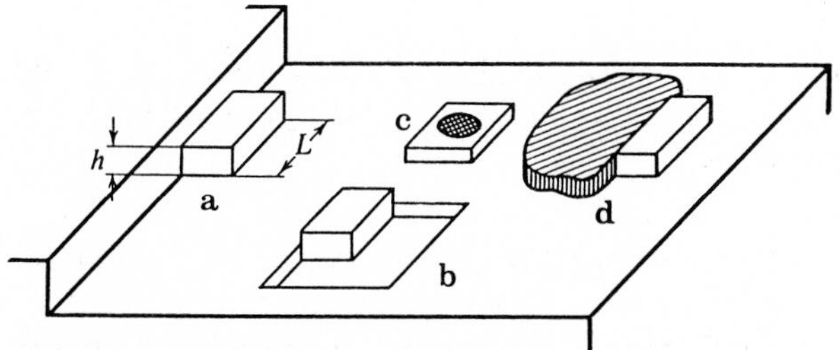

Fig.2.8a–d. Different possibilities of preferential nucleation on the surface. **(a)** At a macrostep; **(b)** in the reentrant angle of a unit-depth pit; **(c)** around or above an impurity particle; **(d)** at an impurity particle

Similar reasoning is possible at the atomic level for nuclei containing only a few atoms. If nucleation on certain irregularities (defects) is associated with overcoming a lower potential barrier, it will occur predominantly on these defects. The activity of the isolated surface sites depends on the gain $U(N_c)$ in the energy of formation of the aggregate at these sites. The greater $U(N_c)$ is, the higher the probability that nuclei will appear on the indicated irregularities (defects).

The rate of nucleation on active centers of a given type differs from the general expression (2.29) by the factor n^*/n_0, where n^* is the population density of centers on the surface [cm^{-2}]. In this case quantities N_c and $\mu^0(N_c)$ refer to the nucleus on the center.

If the supersaturation in the adsorption layer exceeds the critical value for nucleation on a group of defects characterized by sufficiently large $U(N_c)$, but is below the critical value for the other (among them regular) sites of the surface, nuclei will appear on the defects of the indicated type only; this is called decorating. With further growth these nuclei become visible in a microscope and can supply information on the distribution of the isolated high-activity sites. The electron-microscopic picture of the decoration of the (100) surface of a NaCl crystal with small crystals of gold is given in Fig. 2.9. The sample was obtained by condensing a molecular beam of gold in a vacuum. The decorating technique is briefly described in Sect. 3.4.2. The sites of preferential formation of nuclei are elementary steps of height 2.82 and 5.64 Å. The decorating particles between the steps are also seen in Fig. 2.9.

The important role of defects as nucleation centers follows in particular from experiments on the irradiation of substrates (NaCl, KCl) with neutrons, electrons, and x-rays prior to condensation of molecular beams on these substrates [2.38, 40–43]. Irradiation produces point defects in the substrates, i.e., vacancies and interstitials and, in large doses, a rather rough surface. As a result, the population density of small crystals arising on the irradiated substrates increases by several times. The density of the decorating particles also increases when an impurity is introduced into the crystal [2.44]. The density also depends on the pressure of the residual gases at which the condensation is carried out.

It would, however, be a mistake to set up a one-to-one correspondence between all the small decorating crystals and point defects formed on the surface and always equate the surface density of defects to that of the number of decorating particles. To clarify this, let us consider what actually determines the observed density of the crystals formed on the surface.

In the first moments following the beginning of condensation of the molecular beam on a substrate which is cooler than the source, the population density of adatoms is low, and $\mu_s \ll \mu_V$. As adatom accumulation proceeds on the substrate, the adatom density and chemical potential increase, causing nucleation — first at the most active sites, which are characterized by higher values of $U(N_c)$. In this case nucleation may also begin at $\mu_s < \mu_V$, i.e., at $\Delta\mu =$

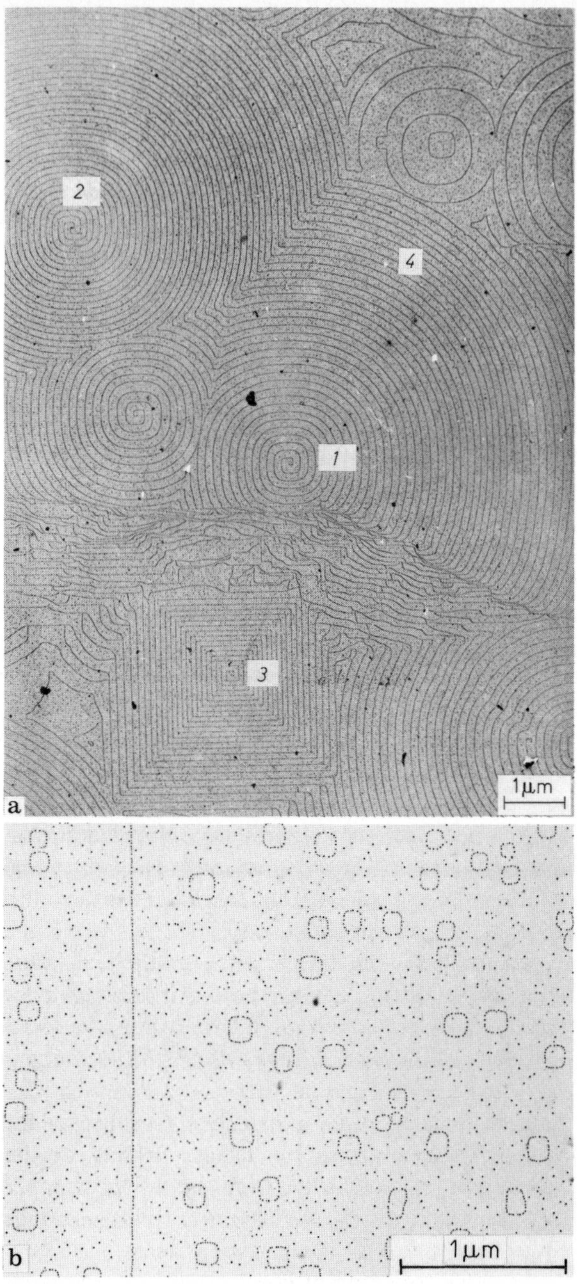

Fig.2.9a,b. Morphology of the gold-decorated evaporation surface of NaCl crystals. **(a)** Different shapes of steps: (*1*) single-step spiral, step height h = 2.82 Å; (*2*) double-step spiral, h = 2.81 Å; (*3*) single-step spiral, h = 5.64 Å; (*4*) concentric layers, h = 2.81 Å [2.39]. **(b)** Circular evaporation steps soon after nucleation. No such evaporation steps are observed in the vicinity of the straight step on the left (Courtesy of M. Krohn. See also [2.39], which contains a similar picture)

$\mu_s - \mu_S < \mu_V - \mu_S$. The crystals formed and growing from the adsorption layer reach a macroscopic size and then form adatom-depleted zones around themselves where the supersaturation $\Delta\mu \ll \mu_V - \mu_S$ is insufficient to produce new nuclei. As a result of the process described, the number of nuclei on the surface at the beginning of condensation increases with time, and then, after the formation of the depleted zones, reaches a constant value[2]. Under these conditions, no new crystals arise, and the impinging material is consumed by the crystals (or droplets) already existing on the surface. The corresponding growth kinetics were calculated by *Sigsbee* [2.45, 46].

The mutual effect of decorating particles is seen from Fig. 2.9a. Indeed, the probability of finding a decorating particle in the immediate vicinity of the decorated steps is lower than in the central zones of atomically smooth terraces between the steps. This can be interpreted as follows: After the molecular beam of gold begins to fall on the NaCl surface that is to be decorated, the gold atoms form aggregates on steps, which are the most active sites of the surface. As a result each step turns into a linear sink for gold adatoms, and the supersaturation in the gold adlayer near the step falls off. Therefore the probability that viable aggregates will appear is lower near the step than away from it. This is, however, a nonzero probability.

Indeed, just after the start of condensation the steps are not sufficiently effective as sinks, and the aggregates on atomically smooth surface areas may prove viable in the competition for the building material; thus they will grow to a size visible in a microscope.

Interaction between potential nucleation centers via the adsorption layer is possible, of course, not only during decorating, but also during the growth and evaporation of the crystalline surface itself. This effect is evident in Fig. 2.9b, which is also an electron micrograph of the gold-decorated surface of a NaCl crystal, taken soon after the start of vacuum evaporation of the crystal. A straight step passes near the left edge of the photo and parallel to it, while the rest of the field is covered with a multitude of nearly circular loops. Similar photos showing earlier and later stages of evaporation reveal that the loops increase in diameter with time; they represent elementary steps 2.81Å in height. Loops visible in an electron microscope originate from evaporation nuclei, i.e., from flat-bottomed pits of elementary depth and a critical radius. In the course of evaporation a straight step emits NaCl molecules, which transfer into the adsorption layer near the step. Therefore the undersaturation near the step is insufficient for two-dimensional nucleation even in the first moments after the beginning of evaporation and remains insufficient later on. Hence the "dead zone" surrounding the step in Fig. 2.9b on both sides arises. The width of the "dead zone" on the NaCl crystals is close in order of magnitude

---

[2] If the active centers are filled irreversibly, after the adatom first settles there the characteristic time for the filling of the centers with adatoms diffusing over the surface is $\tau_s/(1 + 4D_s\tau_s n^*)$, provided $n^* \ll n$.

to the diffusion length of the adsorbed NaCl molecules ($\lambda_s \simeq 2 \times 10^{-5}$ cm at $T = 400°C$ and $\lambda \simeq 10^{-4}$ cm at $T = 300°C$, as measured by *Krohn* [2.47]).

If the step absorption (or emission) capacities with respect to the two terraces separated by this step are not equal (the adatom exchange with the "lower" terrace is easier than with the "upper"), then the "dead zones" near the step on these terraces may be asymmetric. Asymmetry of this kind in growth and evaporation processes has not yet been studied. It has, however, been observed during the diffusion of atoms of one substance over the surface of another. For instance, *Ehrlich* and co-workers [2.48] directly observed migration of separate atoms over the surface of some metals in a field emission ionic microscope, and noted the "reflection" of adatoms moving over the upper terrace of the projector tip from the elementary step surrounding this terrace. More recent data are reviewed in [2.13 b, 49].

Step asymmetry with respect to the addition of atoms of a decorating substance (gold) "from below" and "from above" was observed by *Klaua* [2.50] on the (111) face of a silver crystal. He established that the difference in the potential barriers for the addition of atoms to the step rise "from above" and "from below" is 0.67 eV $\simeq$ 15 kcal/mol for gold on silver.

If there are growing crystallites of perceptible size on the surface of the substrate, supersaturation $\Delta\mu = \mu_s - \mu_S$ is not constant along the surface and is rather low in the depleted zones around these crystallities from which the crystallites have absorbed atoms. The radius of each zone is equal in order of magnitude to the diffusion length λ_s. With an increase in substrate temperature the ratios $[\mu^0(N_c) - N_c\mu_S]/kT \infty U(N_c)/kT$ and $(\varepsilon_s - U_D)/kT$ reduce, as does the supersaturation $\Delta\mu$, and the radius of the feeding zones $\sim\lambda_s$. Therefore according to the eqns (2.31) and (2.36) for nucleation rate J, it would be natural to expect a decrease in the population density of nucleated crystals with increasing temperature. Indeed, on condensation of gold from an atomic beam $I = 10^{13}$ cm^{-2} s^{-1} impinging on a (100) NaCl substrate, 3×10^{11} crystallites of gold are formed per cm^2 of the substrate at a temperature $T = 125°C$, whereas only 9×10^{10} crystallites per cm^2 appear on such a substrate at a temperature $T = 300°C$ [2.51]. The indicated population densities of crystallites were achieved within times of the order of hundreds of seconds. The rates measured by *Robinson* and *Robins* of the increase with time in the number of crystallites visible in an electron microscope were interpreted in line with the above-described nucleation process after the beginning of condensation. The resultant nucleation rate was $J = 8 \times 10^8$ cm^{-2} s^{-1} at $T = 300°C$ and $I = 10^{13}$ cm^{-2} s^{-1}. For fluxes $I \simeq 2 \times 10^{13}$ cm^{-2} s^{-1} in the temperature range of 100°–250°C, the rate of appearance of gold crystals on the (100) NaCl face was $J \propto \exp[(1.10 \pm 0.15\,\text{eV})/kT]$, while the maximum population surface density of crystallites was $\propto \exp[(0.09 \pm 0.015\ \text{eV})/kT]$. At higher temperature (250°–350°C) the maximum density of the number of crystallites was $\propto \exp[(0.34 \pm 0.1\ \text{eV})/kT]$. To obtain the adsorption energy ε_s and the activation energy of surface diffusion U_D of gold atoms on (100) NaCl from these experimental data, *Robinson* and *Robins* used *Walton's* model and equation [2.35, 36]. Bearing in mind that $J \propto I^2$, they concluded that the critical nucleus consists of one atom, and therefore 1.10 eV = 2 $\varepsilon_s - U_D$. Moreover, they showed that the energy characterizing the increase in the maximum density of the crystallites must be $U_D/3$, i.e., $U_D/3 = 0.09$ eV. Hence $\varepsilon_s = 0.68$ eV and $U_D = 0.27$ eV.

Stowell [2.52] summarized the data on condensation of Au on (100) NaCl and obtained the following dependences for the lifetime and the surface diffusion coefficient: $\tau_s = 9.2 \times 10^{-13} \exp[(0.69 \text{ eV})/kT]$s, $D_s = 3.4 \times 10^{-4} \cdot \exp[(-0.31 \text{ eV}/kT]$ cm^2 s^{-1}. These dependences are characterized by energies close to those found by *Robinson* and *Robins*.

The increase with time in the number of crystals visible in an electron microscope can also be interpreted in another way [2.53, 29]. The crystals, nucleated just after the beginning of condensation [within times of the order of $\tau_s(1 + 4D_s\tau_s n^*)^{-1}$], are distributed over the surface randomly enough so that the distances to the nearest neighbors vary from one crystal to another. Therefore the nutrition of these crystals from the adsoption layer must be unequal, and they will attain microscopically resolvable size at different times. If the mechanism described actually plays the decisive role, then the observed "nucleation rate" in fact reflects the growth rate of the crystallites and the nature of their distribution on the surface, but not the dynamics of nucleation. In this case the temperature parameters and the parameters of the other dependences of the observed "nucleation rate" have a different physical meaning. The theory of nuclei formation on active centers is also treated by *Markov* and *Kashchiev* [2.54, 55].

It is evident that the foregoing considerations and the (2.29, 31, 36) are applicable not only in describing the condensation of vapors or molecular beams on foreign substrates, but also in describing condensation on atomically smooth substrates of the condensing material, and the evaporation of atomically smooth faces. The formation of evaporation nuclei must be easier around surface vacancies and at all other points where the chemical potential of the crystal is higher than at regular sites. With vacancies this increase is due to additional uncompensated bonds, while at the edge dislocation outcrops it is due to the enhanced elastic energy, and around some impurity atoms, to both factors simultaneously,

2.2.4 Activity of Solid Surfaces in Melts

Heterogeneous nucleation occurs in practice in all systems whose preparation does not include thorough purification. *Danilov's* investigations [2.14a] showed that the value of critical supercooling in insufficiently clean systems depends on the superheating at which the melt was held before the supercooling was attained (Fig. 2.10): the higher the superheating temperature, the higher is the critical supercooling. This effect increases only up to superheatings of 15–20°C, after which the critical supercooling ΔT_c remains unaltered. The phenomenon

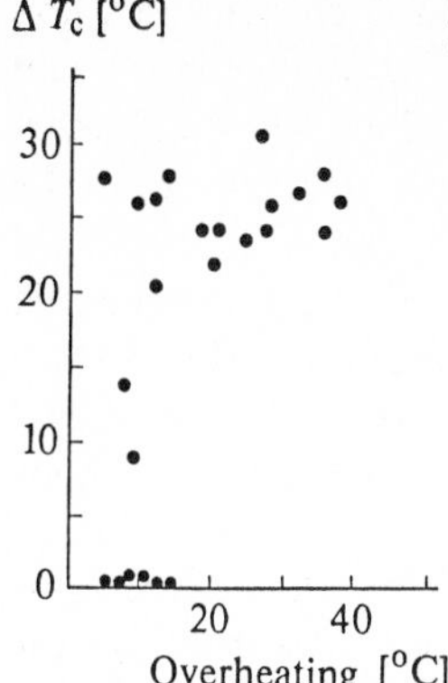

Fig.2.10. Dependence of the maximum critical supercooling of a bismuth melt on preliminary overheating [2.14]

was investigated in experiments with bismuth, mercury, tin, and water. Filtration of the melt removed the described effect, and the supercooling became equal to the maximum value in experiments with superheating of a contaminated melt. The introduction of solid foreign particles (e.g., PbO and WO_3 into tin), on the other hand, restored the dependence of the critical supercooling on the superheating. Consequently, the dependence of the maximum supercooling ΔT_c on the preliminary superheating of the melt is associated with insoluble impurity particles, which are deactivated as potential centers of crystallization in the course of superheating. If the melt with impurities is first superheated to the minimum deactivation temperature or higher, and then held at low superheating (where no deactivation occurs), the critical supercooling still corresponds to the deactivated state of the impurity. The activity of the foreign particles is restored only after the melt is crystallized. These results can be interpreted as an indication that on the surface of the particles or in their microcaverns, there exists a layer with a crystalline or a similar structure which disintegrates on superheating and then reappears after the liquid is crystallized. Such ready crystallization centers must begin to grow immediately upon attaining the corresponding supercooling; i.e., the process will not be fluctuational or characterized by a definite time lag for the appearance of nuclei.

The effect of deactivating solutions by superheating is also known [2.14b].

2.3 Epitaxy

2.3.1 Principal Manifestations

Epitaxy is the oriented growth of one crystal over another. The term derives from the Greek *επι* (over) and *ταξισ* (orderliness). The consistency of this overgrowing lies in the coincidence of distinct, usually crystallographically simple planes and directions in the growing crystal and the substrate crystal. Figure 2.11 shows, for instance, an overgrowing of saltpeter ($NaNO_3$) crystals

Fig.2.11. Small crystals of saltpeter ($NaNO_3$) grown from an equeous solution and located along the macroscopic steps on the rhombohedron face of a calcite ($CaCO_3$) crystal. The orientation of the saltpeter crystals is determined by the rhombohedron face, rather than by the direction of the steps. Courtesy of M.O. Kliya

from an aqueous solution on the rhombohedron face of calcite $CaCO_3$ crystals. One can see the identical orientation of the overgrowing crystals with respect to the cleavage steps on the calcite surface and hence with respect to one another. Here, the saltpeter and calcite crystals contact along the $(10\bar{1}1)$ rhombohedron faces, so that the simple crystallographic directions in these faces of the two crystals coincide. The saltpeter and calcite cystals have identical structures (are isostructural). However, epitaxial overgrowing also takes place with pairs of crystals of whatever structure or bond type, though with a different degree of facility [2.56]. The phenomenon of epitaxy from a vapor and solution is used extensively in growing thin single-crystal films, which are a working element in electronic equipment, magnetic storage devices, integral optics, etc. Epitaxy is the focus of much attention, as can be seen from monographs by *Palatnik* et al. and *Palatnik* and *Papirov* [2.44, 57], and from reviews [2.58]. An extensive review of basic phenomena in the early stages of epitaxy was given by *Kern* et al. [2.59a].

Numerous important phenomena in the initial stages of epitaxial nucleation and growth have been observed in situ in electron microscopes by *Honjo* and co-workers. They are reviewed in [2.59b].

Consistent orientation is possible not only with contacts along a single surface, but also during the formation of one crystal inside another, for instance during the decomposition of supersaturated solid solutions and in polymorphous transformations. With such three-dimensional epitaxy, crystals join along several nonparallel surfaces. Here, the orientation is also characterized by parallelism of crystallographically simple planes and directions. Hereafter we shall consider two-dimensional epitaxy on substrates only.

Crystallites of several epitaxial orientations may exist on one and the same substrate surface with altogether unoriented crystallites. The fractions of crystallites of different orientations with respect to the total number characterize the degree of epitaxy. If the crystallite sizes are approximately equal, the degree of epitaxy can be estimated from the electron diffraction pattern of the crystallite system on the substrate. Strict epitaxy is characterized by point electron diffraction patterns corresponding to one orientation of the crystallites, and partial epitaxy by the superposition of patterns from different orientations; in the absence of epitaxy, Debye rings arise. The formation of a continuous crystal layer on the substrate often begins with the appearance of separate nuclei and aggregates, which then grow and merge into a continuous layer. Therefore they must be obtained under conditions of complete epitaxy in order to exclude defects on the forming films.

The advantage of a particular orientation with respect to the substrate may be manifested both in the stage in which nuclei are formed and in the course of their further growth. *Kern* and co-workers showed [2.59, 60] that small crystals of gold (several tens of angstroms in diameter) on the (100) NaCl face are not oriented with respect to this face in the initial stage of formation. But on annealing for several tens of minutes at temperatures of ~100–150°C, the gold coating produces point electron diffraction patterns instead of Debye rings in the initial stage. The absence of orientation in the initial stages of epitaxy had been noted earlier by *Distler* and *Vlasov* [2.61], who discussed the possibility of epitaxy in the stage of crystallite coalescence.

In the course of detailed investigations of gold crystallites grown by vacuum deposition on (100) KCl, *Kern* and co-workers [2.62a–c] found evidence not only of rotational, but also of translational motion of 20–30 Å crystallites as a whole and of interaction among them which was probably due to elastic and electrostatic forces.

Many attempts have been made to observe displacement of a particular crystallite either by comparing the same piece of substrate with deposited crystallites before and after annealing in an electron microscope or by in situ electron microscopy. Several Au clusters 10–30 Å in diameter have been noted to move by ~20 Å on the (111) MgO face while the other clusters remained stagnant [2.59b]. In other experiments, displacement rates of up to 150 Å/s have been noted in situ [2.63]. Other results are reported and reviewed in [2.64].

Another interesting phenomenon was found in a series of epitaxial pairs used in microelectronics, e.g., Si and GaAs on sapphire. It turns out that the correspondence between the close-packed planes and directions of the overgrowth and the substrate is not exact: misorientations of up to 2–4 degrees always exist between them. The inclination depends on the substrate surface orientation with respect to the substrate lattice, and on growth conditions [2.65].

2.3.2 Thermodynamics

The mutual orientation of crystal and substrate depends primarily on their interface energy. With a given size and state of the growing crystal the system energy should be close to its minimum, provided the surface energy of the system is minimal as well. The latter consists of the crystal–medium, substrate–medium, and crystal–substrate energies.

These values appear in macroscopic form as $\Delta\alpha$ in the work (2.10, 15, 20) of formation of an aggregate on a foreign surface (Sect. 2.2.1) and in microscopic form as $\varepsilon_1 - \varepsilon_s$ in the relations of Sect. 2.2.2. The specific free surface energy α_{sS} of the substrate–crystal (solid) interface and the adhesion energy α_s, which appears in the quantity $\Delta\alpha$, are anisotropic even when the substrate is amorphous. If we repeat the reasoning used in Sect. 1.2 in discussing the nature of the anisotropy of a crystal's surface energy, we realize at once that α_{sS} has singular minima for certain mutual crystal–substrate orientations. If the substrate is crystalline, α_{sS} depends on the orientations of the contacting surfaces of both crystals with respect to their lattices, as well as on the mutual orientation of these surfaces. From the considerations of PBC systems given in Sect. 1.2 it follows that the energy of the interface between two crystals is at a minimum when there is coincidence not only between the close-packed planes of the substrate and the crystal, but als between the close-packed atomic rows within these planes. The fullest saturation of the surface bonds of crystal and substrate will be achieved if the two plane nets of their surface atoms, which form the strongest bonds with each other, coincide. Exact coincidence between the structure and parameters of the plane nets of different crystals is improbable, but similarity between their parameters is rather common, because due to the atomic or ionic radii, the interatomic distances in many crystals differ by less than a factor of two. Indeed, epitaxy occurs most often at comparatively small misfits ($\lesssim$ 10%–12%) in lattice parameters. Oriented overgrowing, however, also takes place at parameter differences as large as tens of percent. Precisely this kind of situation exists, for example, at an epitaxy of fcc metals at the (100) faces of alkali halide crystals (AHCs); here, contrariwise, the degree of epitaxy is higher for those pairs for which the misfit is still greater (see below). At a large parameter misfit, however, the correspondence of close-packed planes and directions in them is always retained [2.56].

The value of $\Delta\alpha$, which is important in epitaxy, may also be expressed via the binding energy. Suppose, for instance, that crystal and substrate have simple cubic lattices with the same atomic spacing a and that the binding energy of two atoms, one belonging to the substrate and one to the precipitating crystal, is equal to the adsorption energy ε_s. Then, neglecting the structure rearrangements in the contact area and the entropy terms in the expression for the free energy of adhesion α_s, we have $\alpha_s = \varepsilon_s/a^2$. In the same simplest approximation $\alpha \simeq \varepsilon_1/2a^2$, and hence

$$\Delta\alpha = 2\alpha - \alpha_s = (\varepsilon_1 - \varepsilon_s)/a^2 \, .$$

The difference $\varepsilon_1 - \varepsilon_s$ indeed determines all the main characteristics of the nucleation process in atomistic theory (Sect. 2.2.2).

The stronger the adhesion of the crystal on the substrate is, the smaller are $\Delta\alpha$ and the work $\delta\Phi_c$ of formation of the nucleus according to (2.15, 20), and the smaller is the deviation from equilibrium, $\Delta\mu$, required for it to appear (see Sects. 2.2.1, 2). Therefore the magnitude $\Delta\alpha$ serves as a basis for contemporary classification of the various mechanisms of heterogeneous nucleation and epitaxy. These mechanisms are listed in Table 2.5, which is closely related to Table 2.4. Its main content and the several examples quoted below are taken from the survey by *Le Lay* and *Kern* [2.66], the original classification being due to *Bauer*.

We first consider the two extreme cases in Table 2.5, i.e., the *Frank-van-der-Merve* ($2\alpha < \alpha_s$) and *Volmer-Weber* ($2\alpha > \alpha_s$) mechanisms, and then the intermediate, *Stransky-Krastanov* ($2\alpha \simeq \alpha_s$) mechanism.

The Frank–van-der-Merve mechanism is typical of condensate–substrate pairs, which are characterized by such strong adhesion that the crystal–substrate binding energy exceeds that between the crystal atoms; $\varepsilon_s > \varepsilon_1$. In this case of complete wetting $\Delta\alpha < 0$, and according to the results in Sect. 2.2.1, the formation of one or several condensate layers on the substrate is thermodynamically profitable even if the vapor is undersaturated or, more precisely, when $-\Omega\Delta\alpha/a \simeq \varepsilon_s < \Delta\mu < 0$[3]. According to (2.20), the work $\delta\Phi_c$ of nucleation on the substrate is low.

Experimentally, the complete-wetting mechanism is observable for pairs of isostructural metals (Au/Ag, Fe/Au) and semiconductors (Sect. 2.3.4) with very close lattice parameters (relative misfit < 1%), and also for noble gases (Xe, Kr) on the basal (0001) face of graphite. In the Frank–van-der-Merve overgrowth the atomic spacings in the condensate and substrate atomic nets parallel to the interface are found to be precisely identical. In other words, the condensate film is either expanded or contracted in the substrate plane. Accordingly, in a direction perpendicular to this plane the film will also be contracted or expanded; this deformation is determined, to a first approximation, by the value of the Poisson ratio in the elasticity equations. If we abstract ourselves from this deformation, we may say that the condensate atoms fit themselves into the substrate lattice, building it up and forming a so-called pseudomorphous layer (Sect. 2.3.4).

The epitaxial relationships in the case at hand are trivial, and amount to a coinciding of all the similar crystallographic planes and directions of substrate and condensate.

If one studies the condensation of Xe and Kr on the (0001) face of graphite by ellipsometry or by measuring the amount of the deposit, one observes a jumpwise change in the amount of condensate with increasing vapor pressure

[3] All of the reasoning here and in Sect. 2.2 applies, of course, to epitaxy from solutions and melts as well.

Table 2.5. Epitaxy mechanisms

Mechanism and its successive stages	Energy condition	Minimum supersaturation $\Delta\mu$ for beginning of condensation	Examples of condensate–substrate pairs	Investigation technique[a]
Frank–van-der-Merve (strong adhesion, complete wetting) 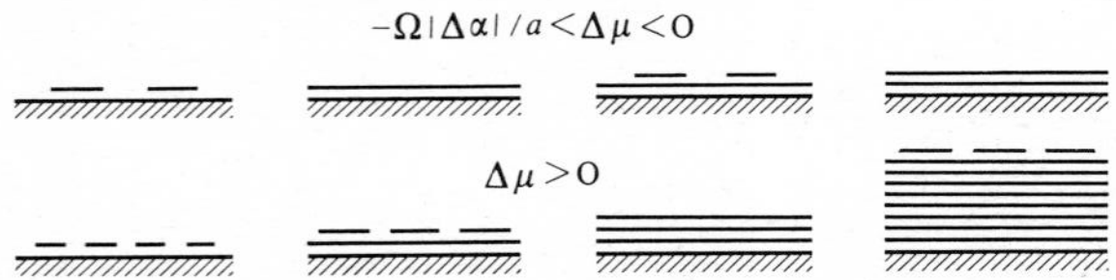 $-\Omega\lvert\Delta\alpha\rvert/a<\Delta\mu<0$ $\Delta\mu>0$	$\Delta\alpha<0$ or $2\alpha<a_S$	$\varepsilon_1-\varepsilon_s<\Delta\mu<0$ undersaturation relative to macrophase	Au/Ag, Ag/Au, Fe/Au, Au/Pd, Xe, Kr/graphite	LEED, RHEED, AES, TEM, ellipsometry, measuring of adsorbate amount
Stransky-Krastanov 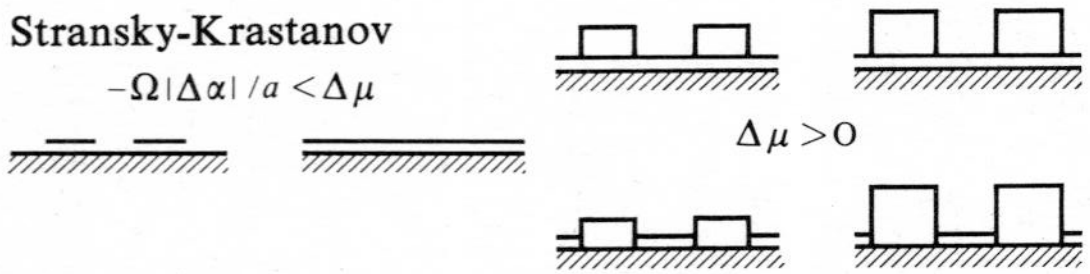 $-\Omega\lvert\Delta\alpha\rvert/a<\Delta\mu$ $\Delta\mu>0$	First layers: $\Delta\alpha<0$ Three-dimensional crystals: $\Delta\alpha>0$	First layers: $\Delta\mu<0$ Three-dimensional crystals: $\Delta\mu>0$	Ag/Si, Au/Si	LEED, RHEED, AES, TEM, TDS, isothermal TDS
Volmer-Weber (weak adhesion) 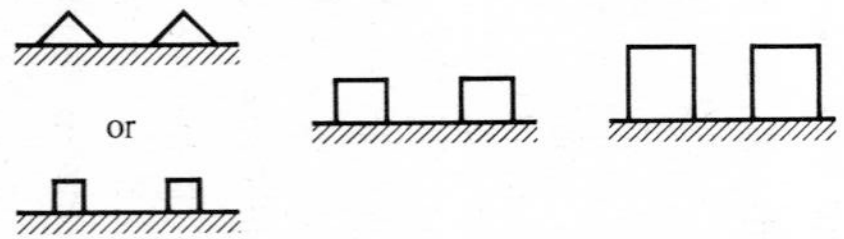 or	$\Delta\alpha>0$ or $2\alpha>a_S$	$\Delta\mu>0$	Au/NaCl Ag/KCl	RHEED, AES, TEM

[a] LEED: low-energy electron diffraction. RHEED: reflection high-energy electron diffraction. AES: Auger electron spectroscopy. TEM: transmission electron microscopy. TDS: thermodesorption

over the substrate; the value of each jump in the degree of coverage corresponds to one monolayer. This dependence corresponds to the predictions of the model discussed in Sect. 2.2.1 for the case where $\Delta\alpha < 0$ and $\Delta\mu < 0$, and means that an increase in pressure and hence in the chemical potential of the Xe vapor over pure graphite leads first to the formation of a rarefied two-dimensional gas (surface coverage $\Theta < 10^{-2}$–10^{-3}). This gas is not detectable by the methods used, and later, at a pressure of 1.1×10^{-4} torr ($T = 97$ K), is condensed with the formation of a crystalline epitaxial monolayer (according to LEED data). The formation of the second monolayer requires a pressure of 1.22×10^{-1} torr, and that of the third, fourth, and fifth 2.0×10^{-1}, 2.3×10^{-1}, and 2.4×10^{-1} torr, respectively. A normal condensate of macroscopic thickness is formed at pressures exceeding 2.5×10^{-1} torr, which is the pressure of the saturated vapor at $T = 97$ K. At $T > 100$ K the fine structure of the first of the above-described jumps appears: it decomposes into two jumps of smaller height along the coverage axis. The first jump is interpreted as a phase transition between a two-dimensional gas and a two-dimensional liquid, and the second as one between a two-dimensional liquid and a two-dimensional crystal.

The jumpwise change in Θ described above indicates that the chemical potential of the atom in the thin crystalline (and liquid) film is lower than that in the bulk crystal, and becomes equal to it only at thicknesses of the order of 5 to 6 atomic spacings (for Xe). At the same time, according to (2.15, 20), the condensation of the two-dimensional gas on the substrate into a crystalline (or liquid) layer occurs as a first-order phase transition, and hence requires the formation of two-dimensional nuclei and possibly of a certain critical supersaturation with respect to equilibrium pressures, which are evidently close to those indicated above for layers of different thickness. The process of such two-dimensional nucleation, which is shown schematically in the first line of the first column in Table 2.5, is not yet understood. The role of substrate defects probably amounts to a contracting of the metastability regions of the two-dimensional phases, and is also vague.

Let us now consider the *Volmer-Weber mechanism* (the third part of Table 2.5), which is typical for weak adhesion of a crystal on the substrate. Here, the work required to form a nucleus on the substrate is larger than in all the other cases, and may even approach the work of homogeneous nucleation at zero adhesion ($\alpha_s \ll 2\alpha$) when $\Delta\alpha \simeq 2\alpha$. Accordingly, the role of inhomogeneities in the substrate (Sect. 2.2.2), which lower the nucleation barrier, is greatest in the Volmer-Weber mechanism. Weak adhesion also leads to a less rigorous interreleation between the orientations of the crystalline condensate and the substrate. This fact manifests itself in the existence, for one and the same crystal–substrate pair, of several epitaxial relationships which are realized simultaneously under different deposition conditions.

The mechanism of weak adhesion is strikingly manifested in the condensation of noble metals (Au, Ag, Pt, Pd) on alkali-halide crystals (NaCl, KCl, LiF), oxides (MgO), salts (MoS_2), and graphite.

The epitaxy of Au and Ag on NaCl and KCl has been studied most thoroughly. For these substances, the characteristic features of the Volmer-Weber mechanism have been traced by the modern methods listed in the last column of Table 2.5. In particular, gold crystallites of ~ 10 Å or more in size can be detected by transmission electron microscopy after gold is evaporated on (100)-cleavage faces of NaCl and KCl made in a high (10^{-8}–10^{-10} torr) vacuum. The typical number density of crystallites is 10^{10}–10^{11} cm^{-2} and depends on the evaporation conditions (Sect. 2.2.2). The intensity of the Auger spectrum of gold due to such a coat is much lower than the one that should have been observed if the gold were distributed uniformly over the surface. Hence it follows that the Auger signal is obtained only from crystallites, while the coverage Θ of the substrate by the gold atoms is below the attained sensitivity limit of the method ($\Theta \simeq 10^{-3}$). In other words, three-dimensional gold crystals appear when the gold adlayer is extremely rarefied, as should be the case for a weak-adhesion mechanism.

In accordance with the foregoing assumption with regard to the possibility of different crystal–substrate Au/NaCl orientations, the following two principal epitaxial relationships are observed, depending on the deposition conditions:

$$(100)\langle 110\rangle \mathrm{Me} \| (100)\langle 110\rangle \mathrm{AHC}\,,$$

$$(111)\langle 110\rangle \mathrm{Me} \| (100)\langle 110\rangle \text{ or } (100)\langle 100\rangle \mathrm{AHC}\,.^4$$

Cations or anions on the (100) face of the AHC form simple square nets with axes along $\langle 110\rangle$ and with internodal spacings equal, e.g., (for NaCl) to $a_1 = 3.99$ Å (the NaCl lattice parameter is 5.64 Å). The metal atoms in the (100) plane of an fcc lattice also form a simple square net. For gold the atomic spacings, i.e., the period of this square net, are, in the $\langle 100\rangle$ direction, $a_2 = 2.88$ Å (lattice parameter: 4.07 Å). Thus, for the first of the epitaxial relations written above, the relative misfit of the parameters of contacting nets is $\Delta a/a_1 = (a_2 - a_1)/a_1 = (3.99–2.88)/3.99 = 0.28$, i.e., 28%. At the same time, for an epitaxial relationship $(100)\langle 110\rangle$Au $\|$ $(100)\langle 100\rangle$NaCl, when the metal atoms come into contact alternately with substrate cations and anions spaced crystal. at $5.64/2 = 2.82$ Å, the relative misfit $\Delta a/a_1 = (2.88 - 2.82)/2.88 = 0.02$, i.e., only 2%. Nevertheless, such an epitaxial ratio is not realized, as a rule. This attests to a substantial difference between the interaction energies of metal atoms with cations and anions.

The adhesion energy of gold on NaCl and other AHCs is unknown. Only estimates of the adsorption energy $\varepsilon_s \simeq 0.68$ eV (Sect. 2.2.2) of individual Au atoms on the (100) NaCl surface are available; here, these atoms are probably arranged over the cations [2.67, 68]. If we relate this energy to the surface area

---

[4] In this notation the matching plane is indicated first, and one of the mutually corresponding matching directions lying in this plane second. Me: metal. AHC: alkalihalide crystal.

$a^2 = (2.82)^2 \times 10^{-16}\text{cm}^2$ per cation (or anion) on the (100) NaCl surface, we obtain the adhesion energy $\varepsilon_s/a^2 = 1.4 \times 10^3\text{erg/cm}^2$. In actuality, in adhesion with a poor correspondence (28%) of the Au and NaCl nets, the number of gold atoms located over salt ions must be four to five times lower. Moreover, the bond strength in adhesion is less than in adsorption, because the redistribution of electron density between gold condensate and salt is not so well pronounced in the first case. On the other hand, within the same approximation (Sect. 1.2.1) for the (111) face of Au, $\alpha \simeq 4 \times 10^3$ erg/cm^2 (the binding energy of the nearest-neighbor atoms, ε_1, is $\simeq$ 14 kcal/mol, the spacing between them 2.88 Å). Thus, for the Au/NaCl pair one should confidently expect the fulfillment of the inequality $2\alpha > \alpha_s$ and, indeed, operation of the *Volmer-Weber mechanism*.

An analysis of the degree of epitaxy of metals (Au, Ag, Al, Cu, Ni, Pd, Pt) on crystals of Na, K, and Rb halides shows that epitaxy improves not with a reduction in the misfit, which lies between 10% and 50% for the pairs indicated, but with an increase. More precisely, the epitaxy improves with a reduction in the value $1 + \Delta a/a_1$, which indicates how many parameters of the contacting nets are needed before a sufficiently exact confrontation of substrate and condensate atoms is achieved. For (100)⟨110⟩Au‖(100)⟨110⟩NaCl, $1 + \Delta a/a_1 = 4.6 \simeq 5$ [2.69a]. Thus the closeness of the lattice parameters determines the ideal epitaxy in the initial stages of Frank–van-der-Merve growth, but is not directly responsible for the epitaxial relations in the Volmer-Weber mechanism. In the latter, the epitaxy is determined by the correspondence of the close-packed planes and the directions in them.

The arrangement of the closest-packed (111) plane of an fcc metal parallel to (100) AHC may be interpreted as a consequence of the fact that the bond of metal atoms with the substrate is weaker than that among themselves. This results in a prevailing trend towards two-dimensional close packing of the adatoms, which yields a plane cluster representing an islet of the (111) net. In the case of (111) Me ‖ (100) AHC, even the symmetry groups of the contacting nets fail to coincide. Accordingly, the coincidence of any one direction of the close packing in one of the nets with the same direction in the other does not lead to coincidence of the remaining directions of the close packing in the two matching nets. The arbitrary azimuthal orientations of the condensate result in a texture which is also characteristic of condensation on amorphous substrates.

The formation of the (111)-net cluster probably leads to the attainment of only a partial minimum of the system energy in condensation, and Ostwald's step rule[5] is realized; the role of the intermediate phase is played by crystallites with orientation (111) Me‖(100)AHC. The final, more profitable state is (100) Me‖(100)AHC.

---

[5] This rule states that in a system deviated from thermodynamic equilibrium the intermediate, metastable phases may precipitate and only later transform into final, stable phases.

Under conditions of a comparatively weak condensate–substrate bond, the deposition conditions — temperature, supersaturation (incident beam intensity), and degree and nature of substrate defects, which are discussed at the end of this section — become particularly important in the manifestation of epitaxy.

The *Stransky-Krastanov mechanism*, whose features are listed in the middle Part of Table 2.5, is intermediate between the two mechanisms already discussed. It is characterized by the fact that the condition of good adhesion, $2\alpha < \alpha_s$, holds for the first monolayer, while the adhesion of three-dimensional aggregates, i.e., of bulk cyrstals, is weaker, and subject to the opposite condition, $2\alpha > \alpha_s$. Evidence of the validity of this concept was obtained when Ag and Au were deposited on (111) Si. The use of Auger spectroscopy and LEED shows that in the initial stages of silver condensation on (111) Si, ordered and sufficiently dense layers (two-dimensional surface phases) of silver appear — first (111) (3 × 1) Ag[6] with a degree of filling $\Theta = \frac{1}{3}$, and then (111) $(\sqrt{3} \times \sqrt{3})$Ag with $\Theta = \frac{2}{3}$. The subsequent increase in the amount of silver deposited leads to the appearance of three-dimensional crystallites with orientation $(111)\langle 110\rangle \text{Ag} \| (111)\langle 110\rangle$ Si, rather than to the thickening of the uniform coat, as in the Frank–van-der-Merve mechanism. It is not clear as yet whether there is direct contact between the three-dimensional crystallite and the silicon or whether the first adlayer survives as an intermediate layer between them. Accordingly, the first column in Table 2.5 shows the two possible successive development patterns of the epitaxial layer.

2.3.3 Kinetics

Epitaxial deposition usually takes place under conditions strongly diverging from thermodynamic equilibrium. This is particularly typical of deposition from molecular beams; the effective pressure in the beam exceeds the condensate equilibrium pressure at the substrate temperature by many orders. Under these conditions the joining of the atoms into nuclei may be virtually irreversible, so that along with thermodynamic factors, kinetic factors also begin to play a substantial role. The kinetic factors are manifested in the dependence of the degree and nature of the process of epitaxy upon the substrate temperature, the supersaturation (incident beam intensity), and the degree and nature of the substrate defects, including the quantity and composition of the impurities adsorbed on the surface or embedded in the surface layer of the crystal.

Inclination from equilibrium shifts the temperature regions where the three epitaxy modes may be observed (Sect. 2.3.2). For instance [2.69b], monomolecular islands of gold appear on the (110) Mo face after deposition of a 0.1 monolayer of Au, thus demonstrating the Frank–van-der-Merve mode. These islands are transformed, however, into three-dimensional crystallites after 2 min

---

[6] A designation such as (3 × 1) denotes the parameters of a two-dimensional Ag net expressed via the elementary parameters of the (111) Si net.

annealing in the same UHV at $T = 500$–$700°C$, showing that the Stransky-Krastanov mode is thermodynamically more favorable. This mode was, indeed, observed directly after deposition at 500–700°C. In the intermediate temperature range $20°C < T < 500°C$, an incomplete second layer appears above the first complete monolayer. Analogous phenomena have been reported for Au and Ag on (110) and (100)W [2.69c], Cu/(110)W [2.69d], Fe/(100)Cu [2.69e], Co/Cu(100) [2.69f], Cu/Ag(111) and Fe/(111)Cu [2.69h]. An explanation of this is given in the following [2.69i].

An adsorption layer appearing on the substrate during deposition at low temperatures is supersaturated even with respect to the less favorable monomolecular islands. At higher temperatures and lower impinging beam intensities, three-dimensional crystallites form so fast that monomolecular islands have no chance to appear. In other words, at high temperatures and surface mobilities the thermodynamically stable Stransky-Krastanov deposit structure may no longer be suppressed by the kinetically more effective Frank–van-der-Merve decay of the supersaturated adsorption layer.

The effect of crystallization conditions on the number density of the condensate crystallites and on their orientation increases with the weakening of condensate adhesion to the substrate, i.e., from top to bottom in Table 2.5 This trend finds expression in particular in the several possible orientations in the Au/NaCl system.

At a given pressure in the system or a given intensity of the molecular beam incident on the substrate, the degree of epitaxy increases with the substrate temperature. Conversely, by lowering this temperature one can obtain entirely unoriented and even amorphous films. The temperature above which the condensate film becomes monocrystalline is called the epitaxial temperature. This is not a rigorous concept, as the degree of epitaxy changes continuously, rather than jumpwise, with temperature (see also Sect. 8.2.3). The epitaxial temperature increases with increasing supersaturation and depends on the presence of impurities in the system. For instance, if the pressure of residual gases is 10^{-5}–10^{-6} torr, then ZnSe is deposited on Ge as an epitaxial film at $T \gtrsim 420°C$, while in a vacuum of 10^{-7}–10^{-8} torr it is deposited at as little as $T \gtrsim 300°C$. Epitaxial temperatures of this order of magnitude are typical of many substances.

The effect of the substrate temperature on the epitaxy is due first of all to the fact that it, together with the temperature of the vapor source, sets the supersaturation in the system. Secondly, and more importantly, it determines the mobility of condensing adatoms on the surface, i.e., surface diffusivity (1.25), and also the probability of attachment of new atoms to the cluster and their detachment from it, i.e., the possibility of selection by trial and error in the course of nucleation and growth. As the temperature increases, rotations (or even migration) of the atomic cluster as a whole become possible, which also promotes the attainment of a unified epitaxial orientation corresponding to the minimum energy (Sect. 2.2.2). It is natural to expect higher chances of cluster mobility in systems with lower adhesion energy α_s, i.e., those at the bottom of Table 2.5.

The Ostwald rule mentioned above is also manifested in condensation of the liquid rather than the solid phase at substrate temperatures below the melting point of macroscopic amounts of condensate. This phenomenon was discovered and investigated by *Palatnik* and *Komnik* [2.57] when depositing bismuth on glass. They showed that the deposit islets are crystalline only at $T \lesssim 95°C$ whereas bismuth melts at 271°C. The temperature above which a substance is condensed on the substrate in the form of liquid drops rather than crystals depends as does the melting point of bulk samples, only a slightly on the vapor pressure. Therefore a pressure–temperature phase diagram of a two-dimensional system consisting of crystalline islets, liquid islets, and the gas of adatoms was found to be identical with the conventional phase diagram of a crystal–melt–vapor system, but shifted by 176° towards the lower temperatures.

With an increase in supersaturation (incident beam intensity), the degree of epitaxy reduces somewhat, whereas the number of nuclei increases (Sect. 2.2.2).

When speaking of supersaturation in the course of epitaxy or of precipitation in general, one should bear in mind the difference between supersaturation in the bulk of the condensing gas or molecular beam and that in the adsorption layer from which at sufficient adatom mobility, nucleation actually proceeds (cf. Sect. 3.2). The latter supersaturation is much lower than the former. Indeed, the condensate islands and steps existing on the surface are sinks for adatoms or admolecules. Therefore when they are absorbed by the steps at a sufficiently high rate (a large kinetic coefficient β_{st}; see Sect. 3.2), the supersaturation near the steps is near zero and increases with the distance from the step (island) on atomically smooth portions of the surface. Far away from the steps, where the supersaturation is high enough, new two-dimensional nuclei may be formed, and hence new steps will appear, which in turn reduces the supersaturation [2.70].

The estimate (corresponding to the above model) of the average spacing λ between steps on the (111) surface of silicon condensing from an atomic beam yields $\lambda/2\lambda_s \simeq 0.1$ for a condensing-beam intensity of $I = 10^{16}\mathrm{cm}^{-2}\mathrm{s}^{-1}$ and for a substrate temperature of 1000 K, i.e., for a relative supersaturation of 10^9 and a layer growth rate of 3.5×10^{-4}cm/s [2.70b]. Then the maximum concentration of adatoms on atomically smooth portions of the interface is $\sim 10^{-2} I\tau_s$, and hence the supersaturation is two orders of magnitude less than it would have been in the absence of steps. In the case under review, however, it is still very high ($\sim 10^7$).

If the supersaturation is so high that the time required to fill one monolayer, $(Ia^2)^{-1}$, is much shorter than the time of one diffusion jump, $(Ia^2)^{-1} \ll (1/\nu) \exp(U_D/kT)$, the arranging of atoms into a regular lattice may prove impossible, and an amorphous condensate is formed.

Condensation may not, however, occur on the substrate if the lifetime and hence the concentration of the adatoms on the substrate is so small that collisions between them are not frequent enough. For instance, during the evaporation of gold from a source with a temperature of $\sim 1500°C$ upon the (100) NaCl face with a temperature of ~ 700 K, the respective equilibrium gold vapor pressures are 10^{-1} and 10^{-9} torr, i.e., the supersaturation is $\sim 10^8$. Nevertheless, a molecular beam of Au is fully reflected (reevaporated) from this surface at a temperature of $\gtrsim 350°C$; i.e., the concentration of Au in the adsorption layer is insufficient for its condensation. This is due to the comparatively low energy of gold adsorption on rock salt.

A second possible cause for slow condensation even under conditions of high supersaturation is the inefficiency of collisions of adatoms with respect to the formation of aggregates. Indeed, the formation of crystal aggregates from adatoms requires that bonds be formed between adatoms. The formation of such bonds, however, may require the overcoming of the potential barrier even in the course of physical condensation, to say nothing of chemical deposition. Let us consider the condensation of a metal on the surface of an AHC. The metal adatoms are arranged on the salt surface over ions of like sign (probably cations), and therefore the spacing between adatoms is equal to that between the metal ions in the plane net corresponding to the surface structure. Owing to the presence of anions, this distance is 1.5 to 2 times as long as the equilibrium distance between atoms in the metal crystal. Therefore the exchange of valence electrons between adatoms will be weak, thus predetermining a weak attraction between them. The adatoms cannot move closer together because of their interaction with the substrate. Another reason for the appearance of a barrier to the bonding of two adatoms is the interaction between the valence electrons of adatom and substrate: in adsorption the electron density is redistributed and the adatom becomes charged positively or negatively. Together with the compensating charge in the substrate, this charge forms a dipole [2.71, 72]. Investigations of Na adsorption on tungsten with the aid of ion and electron projectors have established that the adsorbed Na atoms are charged negatively, and that a dipole moment of $\sim$3–5 debyes[7] is related to each adatom [2.73]. There have as yet been no direct measurements of the dipole moment of metal atoms at the surface of salt crystals, but theoretical estimates point to a "drawing" of adatom electrons onto the surface ions of substrate metals; i.e., in this case the dipole moment must have the opposite sign. The absolute value of the moment was found in estimates to be less, but of the same order of magnitude. Thus the adatoms must repulse each other with an energy of $\sim s^2/R^3$, where s is the dipole moment and R is the spacing between adatoms. The adatom repulsion energy is comparable to the binding energy of the atoms in the crystal; hence dipole–dipole repulsion will strongly inhibit condensation, thereby increasing the barrier to nucleation.

In the condensation of metals on salt crystals, the barrier to the formation of condensation nuclei must be lowered on surface layer defects. A surface vacancy of chlorine on (100) KCl, when filled with an atom of the condensing metal, forms a metal islet with the neighboring potassium atoms. A metal

[7] 1 debye $= 10^{-8}$ cgs. The dipole moment of an electron–position pair spaced at 1 Å is $\simeq$5 debyes.

crystallite is then formed more easily from adatoms over this islet, since the adatoms can accommodate themselves over the cluster at a distance dictated by the metal structure, rather than by the substrate parameter. With this mechanism the orientation of the crystallite [2.42] is determined by that of the islet, and corresponds to the relationship (100)⟨110⟩ Au‖(100)⟨110⟩ KCl. This agrees with experiments on the effect that the irradiation of a KCl substrate with electrons (energy 100 eV, flux density 10^{16} electrons/cm^2 s) before and during deposition has on the epitaxy of Au, Ag, and Cu [2.42]. In these experiments the substrate irradiation lowered the epitaxial temperature in gold deposition from 250°C to 200°C. On the other hand, if a gold film of average thickness ~1 Å was first applied to an irradiated substrate with a temperature of −195°C to +80°C, and deposition was then continued without irradiation, single-crystal epitaxial films formed in the substrate temperature range of $T > 100$°C. In the procedure described, low-temperature ($T < 80$°C) deposition under irradiation gave rise to individual, uniformly oriented single crystallites, which, on further high-temperature ($T > 100$°C) condensation, grew over the substrate and coalesced into a single-crystalline epitaxial film.

2.3.4 Misfit Dislocations and the Conditions of Pseudomorphism

We now return to systems with strong adhesion between an overgrowing crystal and the substrate, and consider in more detail the contact between two crystals (e.g., metal and metal) with similar bond types and with lattices having interplanar distances a_1 and a_2. Let us match the ends of two arbitrarily chosen atomic nets of the contacting lattices, with the nets perpendicular to matching surface C and thus continuing into one another (Fig. 2.12a). The other atomic planes form a nonius: a distance l enclose l/a_1 nets of the lower crystal and l/a_2 nets of the upper crystal, so that at $a_1 < a_2$ the distance l between the closest

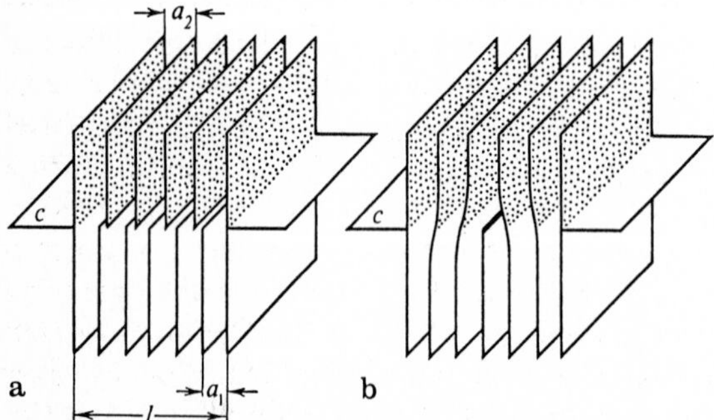

Fig.2.12a,b. Matching of atomic planes at the boundary C of two crystals. **(a)** All but the extreme atomic planes are broken but not curved; **(b)** all but one of the planes are continuous; the edge of the only frayed plane forms a misfit dislocation

pairs of nets fitting one another meets the condition:

$$\frac{l}{a_1} = \frac{l}{a_2} + 1 .$$

Compensation of the bonds at boundary plane C will be more complete, and hence the energy will be minimal when the atomic planes on the crystal and substrate continue into one another (Fig. 2.12b), whereas one plane of the lower crystal per l/a_1 planes is broken off.

The configuration along the edge of this excess plane is actually the edge dislocation in the lattice, which is now, however, composed of two halves with different atoms. Such a dislocation is called a misfit dislocation. From the foregoing it follows that the density of misfit dislocations is

$$\frac{a_2 - a_1}{a_1 a_2} \simeq \frac{\Delta a}{a^2} ,$$

where $\Delta a = a_2 - a_1 \ll a_1$, $a_1 \simeq a_2 \simeq a$. For the Au/Ag contact (in this notation the growing crystal comes first and the crystal substrate second), $\Delta a/a \simeq 2 \times 10^{-3}$, $a \simeq 2 \times 10^{-3}$ cm, and the linear density of misfit dislocations is $\simeq 10^5$ cm^{-1}, i.e., the distance between the dislocations is $\simeq 1000$ Å. Generally speaking, when crystals are joined together at least two systems of nonparallel dislocations must arise in order to compensate the misfit of the lattice parameters along two nonparallel directions in the matching plane. The misfit dislocation net at the Au/Ag interface is presented in Fig. 2.13.

In the preceding reasoning we presumed that the two contacting lattices were not deformable as a whole. In actual fact, if the overgrowing crystal is very thin, each of its atomic planes may continue those of the substrate; this gives rise, for instance, to a pseudomorphous crystal of gold with interatomic distances of silver. If the thickness of the pseudomorphous layer is h, the density of its elastic energy per unit surface is $\sim Gh(\Delta a/a)^2/2$. This elastic energy will disappear upon the formation of $\sim 2\Delta a/a^2$ misfit dislocations of unit length in unit interface. Their total energy per unit interface will be $\simeq 2\alpha_d \Delta a/a^2$, where the specific linear dislocation energy $\alpha_d \simeq (Gb^2/4\pi)\ln(ka/\Delta a)$ [Ref. 2.75]. The logarithmic factor in the expression for α_d takes the interaction of dislocations into account; here, $b \simeq a$ is the Burgers vector, G is the shear modulus, and k is a numerical coefficient of about unity. In order of magnitude, one can adopt $\alpha_d \simeq Gb^2/8$, taking into account the weak logarithmic dependence of α_d on $\Delta a/a$. The growing layer remains pseudomorphous until it reaches a thickness h_c, at which the two above-mentioned homogeneous elastic and dislocation energies become equal. Consequently, there is a critical thickness of the pseudomorphous layer $h_c \simeq a^2/2\Delta a$. For the Au/Ag pair, when $\Delta a/a = 1.8 \times 10^{-3}$, we have $h_c \simeq 500$ Å. Indeed, as Au/Ag grows, a pseudomorphous Au film of a thickness of up to 600 Å is observed. When this

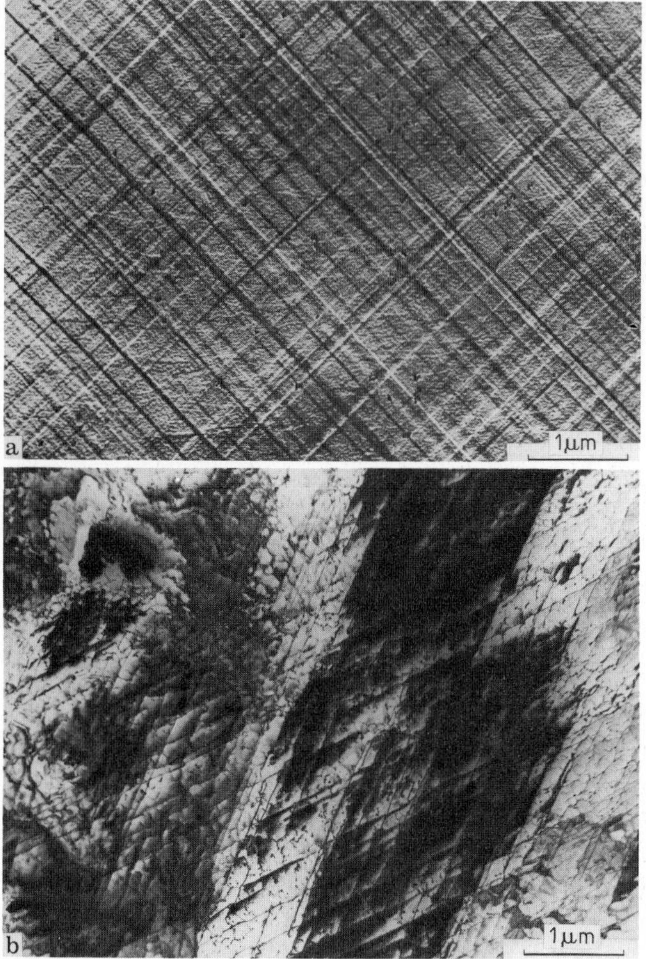

Fig.2.13a,b. Electron micrograph of the net of misfit dislocations at (**a**) the (100) Au/(100)Ag and (**b**) the (110) Au/(110)Ag interfaces [2.74]

value is exceeded, a misfit dislocation network is formed at the matching interface.

The density of misfit dislocations can be gradually changed by changing the lattice spacing of the overgrowing crystal, for instance by introducing impurities into it, in accordance with Vegard's law [2.76]. This effect is observed during the formation of the heterojunction epitaxial pair of semiconductors $GaAs_{1-x}P_x$/GaAs. The density of misfit dislocations at the heterojunction can also be reduced by gradually changing the impurity concentration normal to the epitaxy plane. Then the ultimate difference between the parameters of the substrate lattice and the growing layer will "smear out" over the region where the impurity concentration gradient is set up. The dislocation density in each of the atomic planes in this region drops off accordingly. The dislocation density dependence on the concentration gradient of phosphorus (in % per μm) in films of identical thickness can be seen from Fig. 2.14.

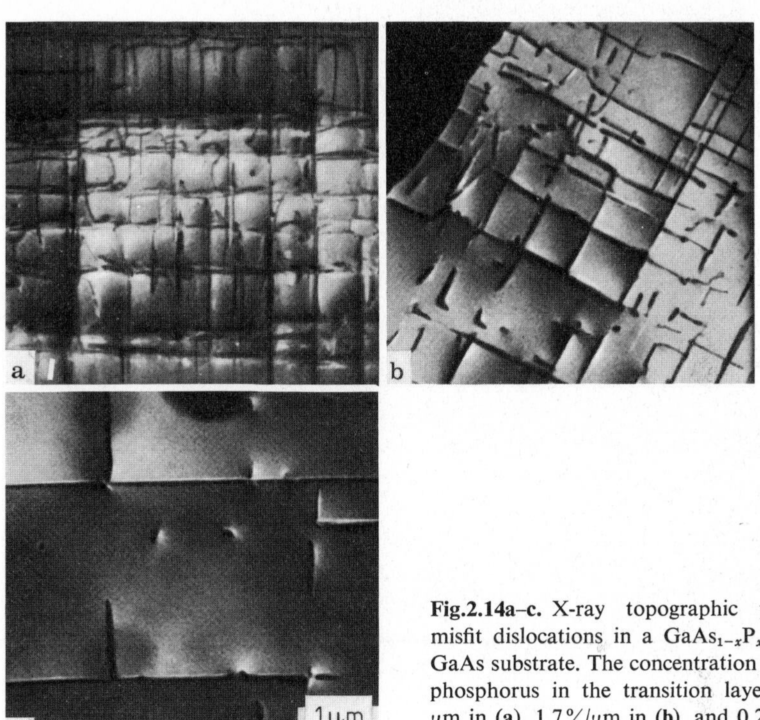

Fig.2.14a–c. X-ray topographic pattern of misfit dislocations in a $GaAs_{1-x}P_x$ film on a GaAs substrate. The concentration gradient of phosphorus in the transition layer is 5.0%/μm in (**a**), 1.7%/μm in (**b**), and 0.21%/μm in (**c**). The *scale* is the same in (**a–c**) [2.77a]

During the pseudomorphous growth of a film, the dislocations in the substrate emerging at the surface of the substrate penetrate into the overgrowing layer until the latter reaches the critical thickness. After that the dislocations which have Burgers vectors parallel to the substrate bend, form misfit dislocations, and only then penetrate into the overgrowing layer (Fig. 2.15a). Subsequently, only separate dislocation segments will be observed, rather than the misfit dislocation network or dislocation segments together with it (Fig. 2.15a). Finally, misfit dislocations may close up the emergences of conventional dislocations at the substrate surface (Fig. 2.15b). If one adjusts the parameters of the matching lattices by changing the concentration of the impurity (e.g., of phosphorus in GaAs), and thus selects the desired density of the misfit dislocations, one can obtain dislocation-free epitaxial films with high mobilities and low carrier densities.

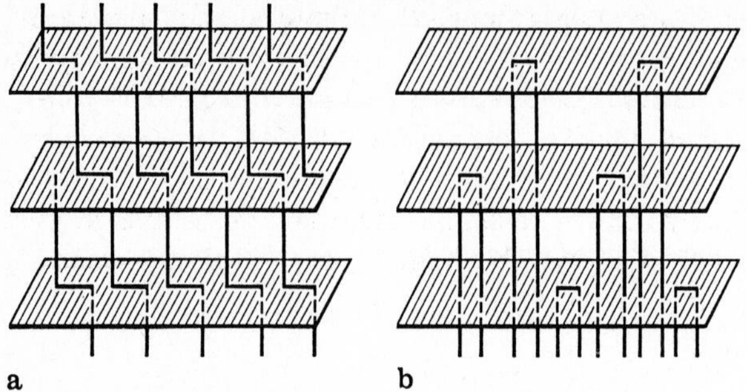

Fig.2.15a,b. Dislocations at the interface between the substrate (*below*) and the built-up layer. **(a)** Transfer of dislocations to the new layer with the formation of misfit dislocation segments; **(b)** closing up of substrate dislocations (*vertical lines*) by misfit dislocations (*horizontal segments*) [2.77a]

The creation of partial dislocation caused by misfit strain in Au/(001)Pd has also been reported [2.77b]. The problems of misfit strain and the formation of misfit dislocations are reviewed in [2.77c].

Epitaxial growth does not necessarily require proximity of lattices. Failing such proximity, the distance between the misfit dislocations must be comparable to the lattice parameter, so that the minimum of the interface energy α_{sS} is attained not by setting up a one-to-one correspondence between the atoms of the contacting atomic networks, but by correspondence between the deepest valleys and ridges of the potential relief, which are disposed along simple crystallographic directions and define the mutual orientations of crystal and substrate.

In 1825, *Wackernagel* found that epitaxy is possible not only in the growth of one crystal directly on the surface of another, but also in growth on a substrate coated beforehand with a thin film of plastic material [2.78–80]. Subsequent investigations on this effect, including those carried out recently with the use of electron microscopy [2.81, 82], showed that the "long-range interaction" effect probably has nothing to do with the pores in the intermediate layers, which may be amorphous or crystalline, dielectric or conducting (metallic), and of thickness $\sim$100–1000 Å. The cause of orientation through an intermediate layer has not yet been established. It is possible that the orientation is due to elastic interaction [2.83, 59a] between the forming crystallite and the substrate, which manifests itself when the crystallite size exceeds the thickness of the intermediate layer [2.83]. The "long-range interaction" effect is highly sensitive to the coating method, the purity of the atmosphere, and the temperature, and is not always reproducible. The cause of this nonreproducibility may be sought in the weak adhesion of the film to the substrate, which hinders the transmission of elastic interaction from substrate to film.

Other unsolved problems in epitaxy concern the following: the mechanism governing the effect of impurities, whose presence often improves epitaxy in the sense that several orientations of the growing crystal are replaced by one; epitaxy in solutions where desolvation[8] of the surface is necessary; the formulation of rules for theoretical prediction of epitaxial orientations under different growth conditions; the mechanisms of crystallite migration and rotation; and the role of kinetics and defects in the three modes of epitaxy described in Table 2.5.

---

[8] Desolvation means cleaning the interface from the adsolved molecules and/or ions of solvent or cleaning the crystallizing species from the solvation shell.

3. Growth Mechanisms

Normal growth of atomically rough interfaces and layer-by-layer growth of atomically smooth surfaces are the two mechanisms by which the crystalline lattice is built. The conditions, general kinetic features, and other characteristics of these two modes are discussed in Sect. 3.1. The specific features of layer growth from vapor, solution, and melt are the subject of Sect. 3.2. The sources of growth steps — two-dimensional nucleation and screw dislocations — often determine the growth rate (Sect. 3.3), and are responsible for the morphology of the growing interface (Sect. 3.4). This morphology also reveals many other distinctive features of growth, such as the participation of the liquid phase in vapor growth, the shock waves of steps, etc.

The general characteristics of optical techniques used to study surface morphologies are also described in Sect. 3.4.

3.1 Normal and Layer Growth of Crystals

3.1.1 Conditions of Normal and Layer Growth

The growth of a crystal results from the addition of new atoms, molecules, or more complex aggregates. The last alternative is possible if the substance in the initial gas or liquid phase tends to polymerization (sulfur vapors, for instance, contain an appreciable amount of S_2, S_4, S_6, and S_8; selenium atoms in the liquid phase form polymer chains). Growth resulting from the addition of aggregates, however, has as yet been only very poorly investigated and in what follows we discuss growth by the addition of uniform units, in particular single atoms or molecules.

The addition of a new particle of the same substance is not necessarily the same thing as crystal growth. For instance, adsorption on an atomically smooth surface is not sufficient to be regarded as growth, because it may cease once a certain concentration of adatoms is achieved; this happens when the chemical potential of the adsorbed atoms or molecules becomes equal to that of identical atoms or molecules in the environment (vapor, solution). Atoms adsorbed on steps may also have a chemical potential differing from that of the particles in the crystal, because when such adsorbed particles are detached, the number of free bonds on the surface, and hence the surface energy, changes. Removing an atom from or adding one to a kink, on the other hand, does not affect the surface energy (Fig. 17). Therefore the chemical potential of a particle in the kink can be identified with that of the crystal.

Thus the addition of new particles to the kink means growth of the crystal. As indicated in Sect. 1.3, thermal fluctuations ensure a certain (often quite high) density of kinks on steps and atomically rough surfaces. On atomically smooth surfaces, kinks are possible because of the individual vacancies in the surface layer and at steps having a length of several interatomic distances, these steps making up the boundaries of vacancies or adatom micro-aggregates. The filling of vacancies and their aggregates in the surface layer leads, sooner or later, to their disappearance. Adatom aggregates are built up by two-dimensional nucleation, which requires that a potential barrier having a height determined by (2.22) be overcome. Thus growth on atomically rough surfaces and steps requires merely that the potential barriers to the incorporation of separate atoms or molecules be overcome, while the growth of atomically smooth surfaces requires the formation of steps as well.

From the macroscopic standpoint, the addition of new particles to atomically rough (diffuse) interfaces can occur anywhere, so that in the course of growth the surface shifts along the normal towards itself at each of its points. Such growth is termed normal. Atomically smooth surfaces, on the other hand, grow by consecutive deposition of layers, i.e., by tangential motion of steps. This is called tangential or layerwise growth. Let us first consider the kinetics of normal growth.

3.1.2 Kinetic Coefficients in Normal Growth

The changes in the average energy of a particle (atom, molecule, or ion) as it moves to or from the interface (kink) is given in Fig. 3.1. Here, ε_S and ε_M are the average energies of the particles occupying equilibrium positions in the crystal and the medium, and the wavy lines on both sides of the boundary represent the changes in energy which take place when particles are displaced from equilibrium positions in medium and crystal (the amplitudes may actually be much larger than those in the figure).

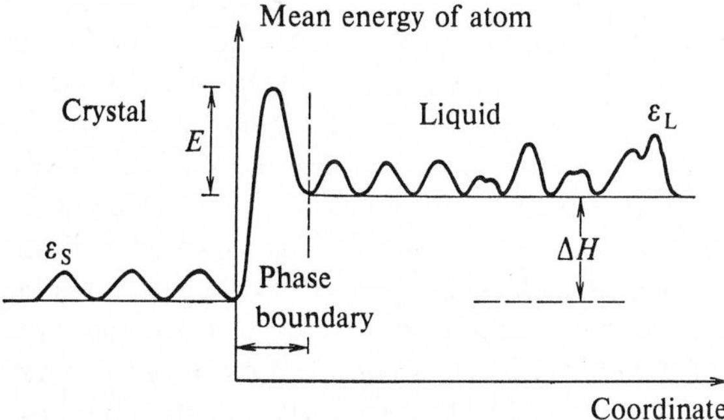

Fig.3.1. Mean energy of an atom in the vicinity of a crystal–liquid boundary

As shown in Sect. 1.1, when a particle moves from the crystal into the medium under equilibrium conditions, it must change its energy by

$$\varepsilon_M - \varepsilon_S = T_0(s_M - s_S) \equiv \Delta H .$$

Moreover, it must generally overcome a potential barrier E (Fig. 3.1). The activation energy E depends on the configuration of the activated complex in the liquid, i.e., on the disposition of the nearest neighbors of the particle shifting from the liquid to the solid phase. The real role is played by the value of E which is minimal with respect to the various configurations of the activation complex.

At the crystal–melt interface these configurations depend on the structures of both the liquid and the solid phases, and the probability that the complex most suitable for the transition will appear is determined by the ease with which the short-range order in the liquid changes to that in the solid. The atomic configurations also change in a viscous flow of the liquid. Therefore for the crystal–melt interface, E was expected to have values of the order of the activation energy for a viscous flow. However, recent developments [2.10b] show that the growth rate exceeds that predicted on the basis of viscous flow activation energy (Sect. 2.1.3). Normal growth from melts has been discussed extensively by *Jackson* and co-workers [3.1].

At the crystal–solution interface the solvate shell forms or disintegrates, which requires much higher activation energies (for aqueous solutions of various substances the experiments carried out give values in the range of 10–25 kcal/mol).

During the growth of chemically simple substances (elements or substances with high-symmetry molecules) from the gas phase without chemical reactions, the activation energy E for the addition of particles at kinks is close to zero ($E \lesssim kT$). Complex molecules, however, cannot join a crystal in an arbitrary orientation, and therefore an appreciable effective barrier must exist for them. This entropy-type barrier must exist even when there is no necessity for desolvation, chemical reactions, or rearrangement of the short-range order (for instance, during growth from vapors).

Let us now estimate the velocity of motion of an atomically rough boundary during *growth from the melt*. The number j_+ of atoms going from the melt over to the crystal per unit time at a single kink, and the opposite flow j_- from crystal to melt can, in accordance with Fig. 3.1, be written as

$$j_+ = \nu \exp(-\Delta s/k) \exp(-E/kT); \quad j_- = \nu \exp[-(E + \Delta H)/kT] . \tag{3.1}$$

Here, ν is the frequency of thermal vibrations of an atom in the crystal and the liquid (we assume the two frequencies to be equal), and $\exp(-\Delta s/k)$ is the probability of finding an atom of the liquid in the immediate vicinity of the kink in the most advantageous activation complex corresponding to barrier E. It is clear intuitively that the configuration of such a complex must be close to solid phase. According to the principle of detailed equilibrium, the flows are equal ($j_+ = j_-$) at $T = T_0$. Therefore we must have $\Delta s = s_M - s_S$.

If the average distance between the kinks is λ_0, the probability of finding a kink on the surface is equal to $(a/\lambda_0)^2$. Then the velocity of the phase boundary is

$$\begin{aligned} V &= \left(\frac{a}{\lambda_0}\right)^2 a(j_+ - j_-) = \left(\frac{a}{\lambda_0}\right)^2 a\nu \exp\left(-\frac{E}{kT}\right) \\ &\quad \times \exp\left(-\frac{\Delta s}{k}\right)\left\{1 - \exp\left[-\frac{\Delta H}{k}\left(\frac{1}{T} - \frac{1}{T_0}\right)\right]\right\} \\ &\simeq \beta^T \Delta T, \end{aligned} \tag{3.2}$$

$$\beta^T \simeq \left(\frac{a}{\lambda_0}\right)^2 a\nu \frac{\Delta s}{kT} \exp\left(-\frac{\Delta s}{k}\right) \exp\left(-\frac{E}{kT}\right);$$

β^T(cm/s K) is called the kinetic coefficient of growth from the melt. In obtaining the last of the equations (3.2), use was made of condition $\Delta s \Delta T/kT \ll 1$, i.e., the condition of low supercooling ΔT at the growth front.

At $\Delta T = 0$, the flow $j_- = j_+$ is called the exchange flow at the kink. The exchange flow can be determined similarly for a step and for a surface.

The obtained linear dependence (3.2) of the growth rate on the supercooling at low ΔT is the most characteristic feature of normal growth, and follows from the statistical independence of the acts of attachment and detachment of atoms at kinks. On a singular face, which contains no steps and hence no kinks, the only statistically independent addition of atoms possible is their entry into the adsorption layer, but not incorporation into the crystal lattice. An elementary step arises on such a face in the course of two-dimensional nucleation, during which the probabilities of detachment of particles are not independent, but vary with the atomic cluster configurations formed on the surface as a result of previous acts of attachment and detachment. In this sense nucleation is a cooperative process, and therefore the dependence of its rate on the supersaturation or supercooling is nonlinear. Accordingly, the dependence of the growth rate on supersaturation (supercooling), which is determined by nucleation processes, is also nonlinear.

To estimate the value of β^T from (3.2) we shall take $\Delta H/kT_0 = \Delta s/k \simeq 3.5$, $T_0 = 1685$ K (the data for silicon). Assuming that in order of magnitude, $a \simeq 3 \times 10^{-8}$ cm, $\nu = 10^{13}\,\mathrm{s}^{-1}$, and $\lambda \simeq 3a$, and taking E to be small compared with ΔH, we obtain $\beta^T \simeq 2$ cm/s K. Experimental investigations of normally growing metal crystals yield estimates of 1–50 cm/s K for β^T.

If we replace $(a/\lambda_0)^2$ with a/λ_0, expression (3.2) describes the rate of growth of an elementary step, whose rise can essentially be regarded as a strip of rough surface. Hence the kinetic coefficient for a step is $\beta^T_{st} \simeq (\lambda_0/a)\beta^T$, and in the above-discussed case of silicon, $\beta^T_{st} \simeq 6$ cm/s K. The experimental value of $\beta^T_{st} \simeq 50$ cm/s K, i.e., the above rough theoretical estimate, yields a value understated by one order of magnitude (see Sect. 5.2.5, [5.16c]).

Let us now turn to a rarer case of *normal growth from solutions*. Suppose that C [cm^{-3}] is the average concentration of solution near the surface and at kinks; a^3 are the volumes per particle of solute or solvent in solution, equal in order of magnitude; λ_0 is the average distance between kinks; and E is the potential barrier at the kink (Fig. 3.1). Then the probability of finding a particle of the crystallizing substance near a kink is $\sim Ca^3$, and the flow of the crystallizing substance towards the solid phase at one kink is $j_+ = \nu Ca^3 \cdot \exp(-E/kT)$. The opposite flow j_- of the solute particles is proportional to the probability that the space at the kink in question contains free solvent ions (molecules), which can form the solvate shell of the particle going into solution. This probability can be equated to the probability $(1 - Ca^3)$ that a particle of the crystallizing substance will be absent at the kink. Consequently, the flow from crystal to solution at one kink is $j_- = \nu(1 - Ca^3)\exp[-(E + \Delta H)/kT]$, where ΔH is now the heat of dissolution. The normal growth rate of a face from solution is defined with the aid of these new expressions for j_+ and j_- by an equation similar to (3.2):

$$V = a(a/\lambda_0)^2(j_+ - j_-)\,. \tag{3.3}$$

At equilibrium, $V = 0$, $j_+ = j_-$, whence follows the expression for the equilibrium concentration C_0 at a temperature T:

$$C_0 a^3 = \frac{\exp(-\Delta H/kT)}{1 + \exp(-\Delta H/kT)} \tag{3.4}$$

The flux $(a/\lambda_0)^2 j_+$, which the crystal and solution exchange in equilibrium, is called the exchange flux. The exchange flux at a step $(a/\lambda_0)j_+ = (a/\lambda_0)j_-$ has so far been found experimentally only for the elementary step on the (100) face of a silver crystal in a 6n aqueous solution of $AgNO_3$, and is equal to $\sim 3 \times 10^6\,\mathrm{s}^{-1}$ [3.2] (Sect. 3.3).

If the solution is supersaturated, then $j_+ > j_-$ and $V > 0$. Using expression (3.4), it is easy to show that $j_+ - j_- = \sigma\nu\exp[-(E + \Delta H/kT]$, where $\sigma = (C - C_0)/C_0$ is the relative supersaturation on the surface of the growing crystal. Hence

$$V = a\nu\sigma(a/\lambda_0)^2\exp[-(E + \Delta H)/kT] \equiv \beta\Omega C_0\sigma = \beta\Omega(C - C_0);$$

$$\beta = a\nu(a/\lambda_0)^2(\Omega C_0)^{-1}\exp[-(E + \Delta H)/kT] \simeq a\nu(a/\lambda_0)^2\exp(-E/kT)\,. \tag{3.5}$$

Here β[cm/s] is called the kinetic coefficient of growth from solution, and $\Omega \simeq a^3$ is the average volume occupied by a particle in its own crystal. Proceeding again from the statistical independence of the acts of attachment and detachment of particles at kinks, the linear growth rate versus supersaturation dependence was obtained.

Experimentally, the normal mechanism of growth from solutions is observed, for instance, in crystals of NH_4Cl grown in aqueous solutions at room temperatures and in the basal face (0001) of quartz crystals (α-SiO_2) grown under hydrothermal conditions. The typical linear dependence $V(\sigma)$ for the (0001)

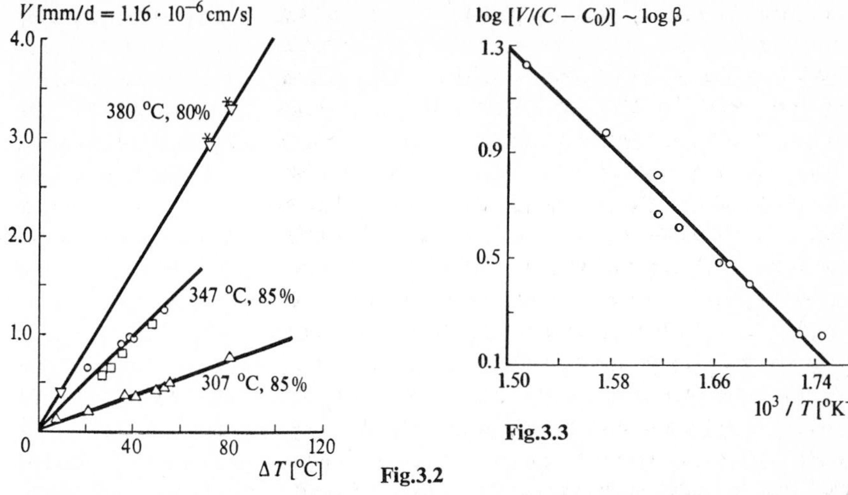

Fig.3.2. Dependence of the normal growth rate of the basal face of quartz growing from hydrothermal solutions at supersaturation ΔC. The measure of supersaturation is ΔT, i.e., the difference between the temperatures of the dissolution and growth zones; $\Delta T \propto \Delta C$ (Sect. 9.3.5). The *figures* on the curves are the saturation temperatures and degrees to which the autoclaves are filled with solution at room temperature. The linearity of $V(\Delta T)$ shows that the kinetic coefficient is constant [3.3]

Fig.3.3. Temperature dependence of the kinetic coefficient of growth of the basal face of quartz growing in a hydrothermal solution [3.3]

face of a quartz crystal is given in Fig. 3.2. In accordance with (3.5) one should expect a linear dependence of $\ln(\beta)$ on $1/T$, which is indeed observed for the (0001) quartz face and is presented in Fig. 3.3. The slope of the line in Fig. 3.3 corresponds to the observed activation energy of 20 ± 1 kcal/mol. According to (3.5) it would be natural to identify it with quantity E. Knowing E and putting $\lambda_0 \simeq 3a$, $\nu \simeq 3 \times 10^{12}$ s^{-1}, $a = 3 \times 10^{-8}$ cm, $T = 658$ K, we have $\beta \simeq 2.9 \times 10^{-3}$ cm/s. For a 3% solution of SiO_2 in a mixture H_2O–NaOH, which is used in growing quartz, $\Omega C_0 \simeq 3 \times 10^{-2}$. One can, knowing the temperature dependence of the solubility and the difference in temperature between the growth and dissolution zones, roughly estimate the supersaturation in the solution bulk. For a solution of SiO_2 in a 0.5m aqueous solution of NaOH at 385°C and at a temperature difference of 60°C between zones (as measured outside the autoclave) the supersaturation is 7%. The actual supersaturation at the front must be lower, partly because the temperature drop is smaller inside the autoclave than it is outside, and partly because of the depletion of the mother liquor near the growing surface. Assuming, for the sake of estimation, that $\sigma \simeq 3 \times 10^{-2}$, we find that the normal growth rate (3.5) must be of the order of 2×10^{-6} cm/s, which is in qualitative agreement with experiment. An exact comparison with experiment during growth from solution is difficult in this and many other cases, because of the lack of data on supersaturation directly on the surface of the growing crystal (and not in the bulk of the mother liquor; see Sect. 5.1) and also on the activation energies E.

Estimates of the kinetic coefficient for a step in electro-crystallization of the (100) face of the above-mentioned silver crystal made with the use of data on the exchange flux give $\beta_{st} = 0.2$ cm/s, which is two orders higher than the kinetic coefficient for the basal face of quartz. The divergence indicates widely differing resistances of growth at the step and on the face as a whole, as well as a difference in the crystallization behavior of the Ag^+ ion and the SiO_2 molecule.

Quantum crystallization of the ^{4}He melt (superfluid) at low temperatures ($T < 1$K) is a completely different process from the classical growth discussed above. The driving force for the growth of ^{4}He crystals is a pressure excess above the equilibrium value of 25 atm. The helium atoms joining the crystal do not overcome potential barriers at the kinks. Being quantum particles, these atoms are transferred to the solid under the potential barrier by quantum tunneling. Since the steps and kinks on the quantum interface are delocalized (Sect. 1.3.4), the motion of the interface is coherent and is not associated with a dissipation of energy at $T = 0$K. This corresponds to infinitely fast growth kinetics. At $T > 0$K, however, phonons and rotons exist in the system. They should be scattered by the moving interface, at which the density changes discontinuously. If phonons are the main excitations in the system, this interaction gives rise to energy dissipation $\simeq E_{ph} V^2/c$, where $E_{ph} \simeq kT^4/\Theta^3\Omega$ is the phonon energy, Ω is the specific volume per atom, Θ is the debye temperature, c is the velocity of sound, and V is the growth rate. On the other hand, the same dissipation must be equal to $V\Delta\mu/\Omega$, and thus one obtains [1.32a]:

$$V \simeq (c\Theta^3/kT^4)\,\Delta\mu = \beta^\mu \Delta\mu\,.$$

At $T = 0.6$K, $\Theta = 32$K, $c = 2.4 \cdot 10^4$ cm/s, the kinetic coefficient $\beta^\mu = 6.10^{25}$ cm/s·erg. A considerably lower experimental value, $\beta^\mu = 3.10^{23}$ cm/s·erg at $T = 0.6$K, shows that rotons and, probably, other factors (such as ^{3}He impurities) are important in the deceleration of the phase boundary. Still, the quantum β^μ coefficient greatly exceeds the classical value $\beta^\mu = (\Delta T/\Delta\mu)\beta^T \simeq \beta^T/\Delta S$. At $\beta^T = 10$ cm/s·K, one has $\beta^\mu = 6.10^{14}$ cm/s·erg.

The low resistance for quantum growth makes it possible to observe crystallization waves, which include periodic growth and melting at any point of the interface [1.32b]. The mechanical inertia forces needed for these waves result from the density change occurring during growth and melting. The crystallization waves look like conventional capillary waves on a liquid–gas surface and obey the corresponding dispersion law:

$$\omega^2 = (\alpha + \alpha'')\,\frac{\rho_L k^3}{(\rho_L - \rho_S)^2}\,.$$

Here ω is the frequency, k is the wave number, ρ_L and ρ_S are the densities of liquid and solid ^{4}He, and $\alpha + \alpha''$ is the effective interfacial energy (Sect. 1.4.1).

Experiments with capillary waves [1.32b] give $\alpha + \alpha'' = 0.21$ erg/cm^2 at $T =$ 0.6K. The liquidlike behavior of the solid–liquid interface of helium made it possible to measure α by the conventional technique of capillary rise [1.32c]. For hcp ^{4}He, $\alpha = 1$ erg/cm^2 at $T = 0.4$K and $\alpha = 0.11$ erg/cm^2 at $T = 1.31$K. For bcc ^{4}He, $\alpha = 0.1$erg/cm^2 at $T = 1.67$K.

The crystal is round-shaped at equilibrium at $T = 0.5$K and becomes polygonized during growth [1.32 b, c, d].

3.1.3 Layer Growth and the Anisotropy of the Surface Growth Rate

Normal growth occurs at practically any place on the surface, whereas in layer-by-layer growth the kinks are concentrated at steps only. Elementary steps (of a height of or less one lattice parameter) are separated by atomically smooth areas (Fig. 3.4), for whose growth nuclei of new steps must be formed. Therefore the growth of a stepped surface at supersaturations insufficient for nucleation on atomically smooth areas is achieved only by the motion of the existing steps. The problem of step sources is discussed in Sect. 3.3.

The most comprehensive information on the kinetics and processes of layer growth is obtained by observing this growth directly in a microscope, but it is much simpler and hence more common to study the surface morphology of grown crystals. The photomicrograph presented in Fig. 3.5a gives evidence of layer growth on a dislocation (Sect. 3.3) on the (100) face of a man-made diamond crystal. The crystal was obtained at high temperatures and pressures, which made observation of the growth impossible. A cubic face of the diamond lattice is, in the first-neighbor approximation, atomically rough in the structural sense; i.e., it belongs to the category of K faces according to the *Hartman* and *Perdok* (see Sect. 1.3.4) The presence of step on it points to the important role of the second nearest neighbors and of the interaction between the crystal surface and the solution (diamond crystals of this type are obtained from a carbon solution in a metal–catalyst melt; see Sect. 9.4).

Figure 3.5b gives an idea of the stepped surface structure of a crystal of the organic substance paratoluidine ($CH_3C_6H_4NH_2$) growing from vapor. Paratoluidine has a large lattice spacing (23 Å) in the direction of the c axis perpendicular to the plane of Fig. 3.5b and a high birefringence (the difference between

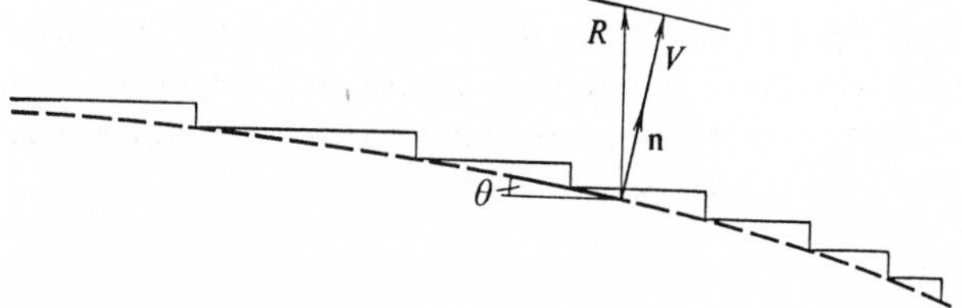

Fig.3.4. Stepped surface of a crystal. Angle θ describes the local deviation of the orientation from the singular face; R is the growth rate of the interface along the normal to this sigular face; V is the velocity along the normal $\boldsymbol{n}$ to the stepped surface

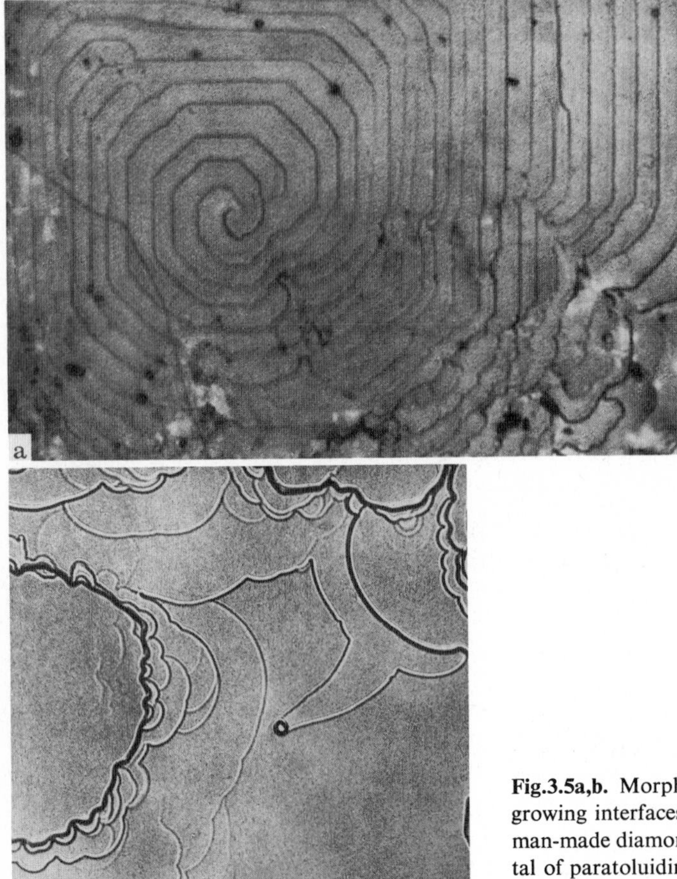

Fig.3.5a,b. Morphology of layerwise-growing interfaces. (**a**) (100) face of a man-made diamond [3.4]; (**b**) thin crystal of paratoluidine in polarized light between crossed polaroids [3.5]

the refractive indices of two beams propagating along the c axis is 0.27). This permits use of polarized light technique in investigating growing crystals. The step heights can be determined from interference colors of the type shown in Fig. 3.5b. The principles of some optical methods used to study surface morphology are briefly discussed in Sect. 3.4. The picture in Fig. 3.5b is a frame from a cinefilm taken by *Lemmlein* and *Dukova* during growth [3.70]. Such films are generally made from model substances (paratoluidine, diphenyl, β-methylnaphthalene, cadmium iodide) in special chambers of simple design. A diagram of one such chamber, used to study growth from vapors, is given in Fig. 3.6. A crystallizing substance *2* is placed in a glass chamber *1* which is either pumped out or communicates with the atmosphere. The chamber has double walls; a thermostated liquid (water) is circulated through the lower and upper parts of the chamber from different thermostats, which maintain temperatures T_2 and T_1, respectively. If $T_2 > T_1$, substance *2* is evaporated

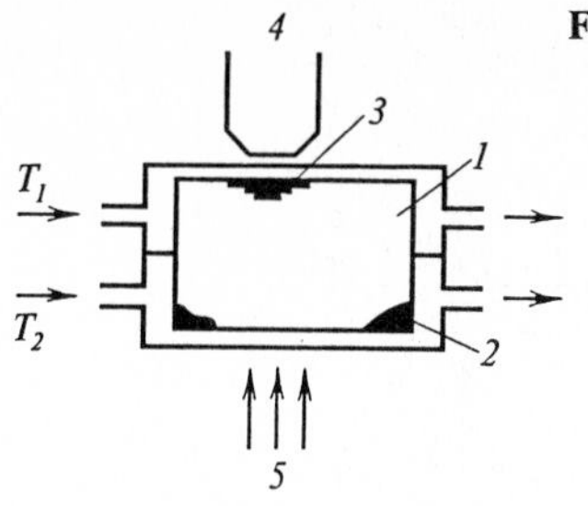

Fig.3.6. Setup for investigating growth from vapors. (*1*) chamber, (*2*) substance evaporated, (*3*) growing crystal, (*4*) microscope, (*5*) polarized light

from the lower and condensed on the upper part of the chamber in the form of drops (when T_1 exceeds the melting point T_0 or is even a little below it) or small crystals (when $T_1 < T_0$). In the simpler version the chamber walls are single, there are no thermostats, and the vapor source is a substance in a crucible placed inside the chamber and heated by a miniature resistance furnace. Then condensation proceeds on the upper lid of the chamber and is observed through a microscope.

The photo in Fig. 3.5b shows that the growth layers propagate from the place at which the crystal is fastened, and there are no dislocation sources, in contrast to Fig. 3.5a. The photo also shows the splitting of thick layers into thin ones, and the reverse process, by which the layers merge. These phenomena result from growth due to surface diffusion (Sect. 3.2.1). On crystals with dislocation layer sources one can observe spiral steps containing tens of turns. Away from the center of the spiral, the stepped relief represents a staircase of near-parallel steps having equal heights (elementary heights included). At supersaturations of up to $\sim 100\%$, the step velocities usually do not exceed 10^{-3} cm/s, and therefore can be measured simply with the aid of a micrometer eyepiece and a stopwatch. The dependences of step growth rates on the height of the steps, which are presented in Fig. 3.10 and discussed in Sect. 3.2.1, were obtained in this manner.

It follows from the foregoing that on each microarea of the surface, growth proceeds by consecutive deposition of layers. An echelon of parallel identical elementary (or thicker) steps may serve as the simplest model. Suppose each step moves along the face at a velocity v. Then, on the average, the stepped surface as a whole moves parallel to itself along the normal to the crystallographic orientation of the face at a velocity

$$R = \frac{a}{\lambda} v = |p| v \,, \tag{3.6}$$

where λ is the distance between the steps, a is the height of each of them, $p = \tan\theta$ is the slope of the surface to the singular face under consideration (Fig. 3.4), and velocity v is a function of p. Let us reflect the stepped surface depicted in Fig. 3.4 in a plane perpendicular to the singular face and parallel to the step lines. As a result we obtain a vicinal face with step rises facing the opposite side. If these steps of the opposite sign have the same velocity as the initial

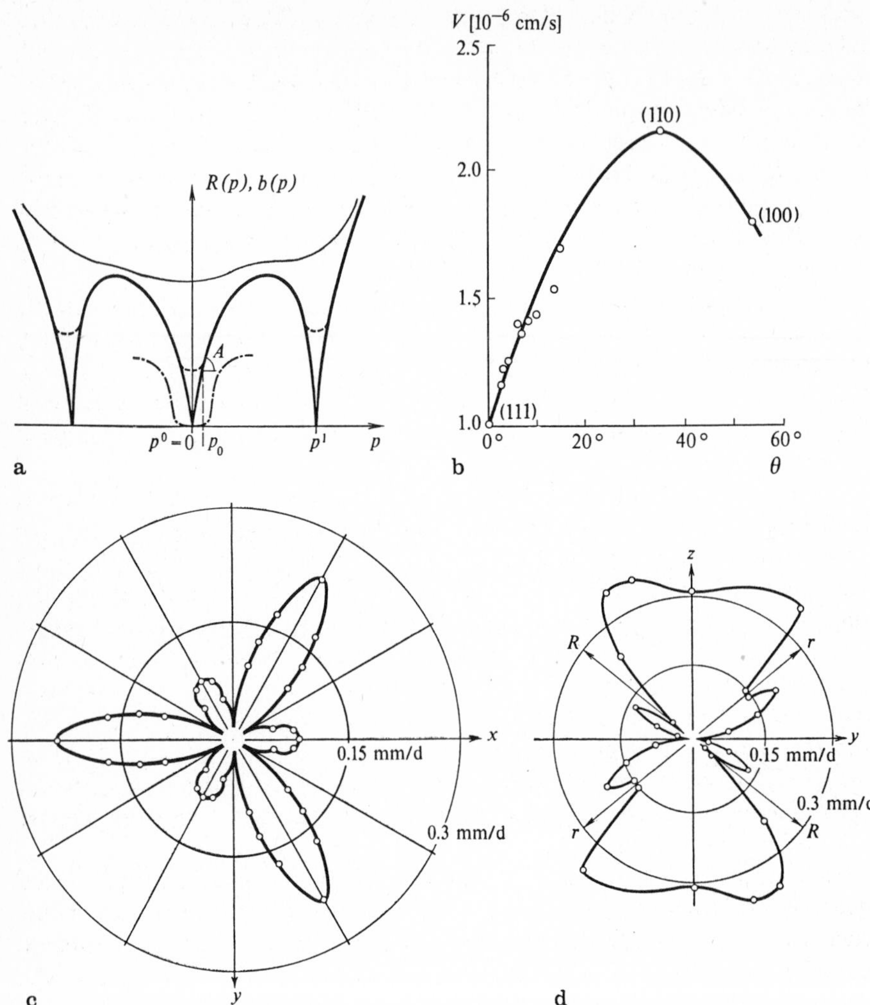

Fig.3.7a–d. Growth rate anisotropy. (**a**) Schematized dependence of growth rate R and kinetic coefficient b (Sect. 3.2.2) on the orientation of surface $p = \tan\theta$, where θ is the angle of deviation of the surface from the singular face (see Fig. 3.4). The growth rate anisotropy Θ is characterized by relative change in the growth rate R with change in the slope p, i.e., $\tan A = \Theta R(p_0)$. The *solid line* indicates an ideal crystal at low supersaturations; the *dot-dashed line* the same, but in the presence of an impurity; and the *dashed line* a crystal with dislocation. The slope on the generatrix of the dislocational vicinal growth hillock is indicated by p_0, and the *thin solid line* shows growth by mass formation of nuclei or near-normal growth for any orientation. (**b**) Growth rates of a silicon crystal in the chloride gas process (see Sect. 8.4.2) with different substrate orientations relative to the (111) face [3.6]. (**c**) Polar diagram of the growth rate of quartz in the plane perpendicular to its optic axis [3.7]. (**d**) Polar diagram of the growth of quartz in the plane of one of the prism faces, $(1\bar{2}10)$. The directions of the normal to the faces of the major and minor rhombohera are given by R and r, respectively; z is direction of the optic axis [3.7]

ones, i.e., if the reflection plane is the symmetry plane of the crystal, then velocity R will still be equal to $(a/\lambda)v$. This means that in the case at hand, growth rate R depends merely on the absolute value of slope p, which is shown by the modulus sign in (3.6). On the other hand, if the indicated reflection plane is not the symmetry plane of the crystal, the slopes of function $R(p)$ at $p > 0$ and $p < 0$ in the vicinity of the singular face ($p = 0$) will be different. The symmetric dependence $R(p)$ is denoted by the continuous line in Fig. 3.7a. This dependence is characterized by anisotropy index Θ, which will be determined in Sect. 3.2.2 by equation (3.21).

If the surface grows layerwise, the corresponding function $R(p)$ necessarily has a sharp, singular minimum at $p = p^0 = 0$. In this case, according to (3.6), the growth rate $R = 0$ at $p = p^0$ and, in general, at any values of $p = p^i$ ($i = 0, 1, 2, \ldots$) corresponding to other singular faces (continuous curve in Fig. 3.7a). For orientations close to p^i, one can use all the reasoning given above for $p = 0$.

It should be noted, however, that the actual *local* slope of the surface is implied by p here, so that $p = 0$ corresponds to the zero local density of the steps. However, if p denotes average slope $\bar{p}$, the local step density will not be zero even for a surface whose orientation coincides, on the average, with the singular one, $\bar{p} = p^i$. Such a situation is realized, for instance, for a singular face covered with vicinal hillocks. Its growth rate is naturally nonzero, because the step density at the slopes of the vicinal hillocks is by no means zero. Also, the singular face with orientation $p = p^i$ has a nonzero growth rate if the supersaturation is so high that a multitude of two-dimensional nuclei form on the face. Such cases are discussed in Sect. 3.3.3. The dependences $v(p)$, which determine the behavior of $R(p)$ at comparatively large deviations from singular orientations are treated in Sect. 3.2.

The normal growth rates V found experimentally for differently oriented quartz crystals under hydrothermal conditions and for silicon crystals in the chemical vapor deposition process are given in Figs. 3.7b–d. From geometric considerations (Fig. 3.4) it is clear that the velocity V of displacement of the crystal surface parallel to itself along normal n[1] is equal to

$$V = R \cos \theta \,. \tag{3.7}$$

At small θ, i.e., at $p < 1$, $V = R$.

It is obvious that kinematic relations (3.6, 7) describe not only growth, but also layer-by-layer decrystallization (evaporation, melting, dissolution, etching, etc.).

---

[1] This velocity is called the normal growth velocity of a surface, as distinct from the tangential velocity of steps v. The normal-growth rate concept is equally applicable to both layer-by-layer and normal (diffuse) growth mechanisms. In contrast to V, the velocity R characterizes the displacement of the growth front along the normal to a selected, usually simple, crystallographic plane.

If the face is atomically rough, i.e., if it contains many kinks and grows according to the normal mechanism, then surfaces deflected from it will not have a substantially higher kink density. Furthermore, the potential barrier to the addition of a particle to the kink in solution and melt growth depends primarily on the atomic configurations in the liquid, whose probabilities $\propto \exp(-\Delta s/k)$ should be nearly equal for all faces. The structure of the liquid boundary layers adjusted to crystallographically different singular faces may vary essentially (especially for chemically complex solutions), whereas it is natural to expect low sensitivity of this structure to changes in the orientation of rough surfaces. For these reasons a change in orientation in the vicinity of atomically rough surfaces must not (and does not) lead to an appreciable change in growth rate. In other words, the nature of growth rate anisotropy in layer growth differs qualitatively from normal growth, just as the surface energy anisotropy of singular faces differs from nonsingular (atomically rough) surfaces. The difference in growth rate anisotropy manifests itself in the formation of faceted and rounded growth shapes, respectively (Sect. 5.2).

Generally speaking, the surface carries not only elementary steps of a height equal to one lattice spacing, but also steps of height h reaching hundreds and thousands of lattice spacings. These steps are easily detectable in a microscope or even by the naked eye (Fig. 3.6). In this case (3.6) must be generalized:

$$R = \sum_{h=a,2a,\ldots}^{\infty} \frac{h\, v(h)}{\lambda(h)},$$

where $v(h)$ is the velocity of tangential motion of steps of height h, and $1/\lambda(h)$ is the density of such steps (the number of steps per cm). Although macrosteps are clearly visible, they do not always make a large contribution to the growth rate, because of their low velocity and density (Sect. 3.2). For instance, during the growth of epsomite ($MgSO_4 \cdot 7H_2O$) from solution, the contribution of microscopically detectable macrosteps does not exceed 5% [3.8].

3.2 Layer Growth in Different Phases

3.2.1 Growth from Vapor

As indicated in Sect. 1.3, the crystal–vapor interface is atomically smooth, and hence it has to grow and decrystallize layerwise. Elementary building blocks (atoms, molecules or their aggregates) form, on a smooth and step-free surface, an adsorption gas (or liquid), which is in equilibrium with the vapor. The equilibrium density of a two-dimensional adsorption gas n_{sV} is defined by relation (1.24), which also follows from the equality of the chemical potentials of the particles in an ordinary three-dimensional and an adsorbed two-dimensional gas.

On the other hand, if there is a step on the surface, the kinks and the adsorption layer exchange particles, and a crystal–adlayer equilibrium is achieved when the chemical potentials of the adlayer and the crystal are equal. The corresponding adlayer concentration is n_{sS}. If the vapor is supersaturated, then $n_{sV} > n_{sS}$, and the relative supersaturation in the adlayer on atomically smooth areas away from the step is the same as in the vapor:

$$\sigma = (n_{sV} - n_{sS})/n_{sS} = (P - P_0)/P_0 \,.$$

Since $n_{sV} > n_{sS}$, surface diffusion of adatoms towards the steps begins, the concentration near the steps becomes higher than equilibrium concentration, and the flux into the crystal phase exceeds the opposite flux (evaporation flux). The steps move, and the crystal grows. If the vapor is undersaturated, the steps move in the opposite direction, i.e., evaporation takes place. Particle exchange between crystal and vapor on steps is possible not only through the adsorption layer, but also directly. This flux is, however, considerably less intensive, because of the low vapor density, the small surface area of the step rise, and the high energy of three-dimensional as compared with two-dimensional evaporation. Indeed, three bonds have to be broken in order to transfer a particle of a simple cubic lattice from the kink to the three-dimensional vapor, whereas only two bonds must be ruptured to transfer this particle to the two-dimensional adsorption gas on the (100) face. Accordingly, the detachment frequencies are higher in the latter case.

The leading role of surface diffusion in crystal growth from vapors was first proved experimentally by *Volmer* and *Estermann* [3.9]. Their experimental set-up can be seen in Fig. 3.8. The small crystals K, which were growing from the mercury vapors filling the chamber, were observed in a microscope. They had the shape of thin platelets. The linear velocity of the end faces of a platelet, v, was calculated from the equation $v = \Omega(P - P_0)/\sqrt{2\pi mkT}$ and found to be $\sim 10^3$ times less than the observed one. This served as a basis for the hypothesis that rapidly growing end faces of the platelet are fed not only (and not to so

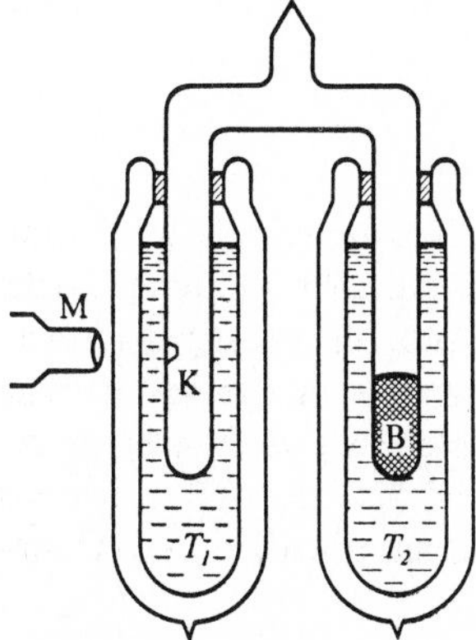

Fig.3.8. Schematic representation of *Volmer* and *Estermann*'s setup for investigating the growth of mercury crystals K from vapors. B is the source of mercury vapors. Temperature $T_2 > T_1$ in the Dewar flash [3.10]

great an extent) from particles arriving there directly from the vapor, but consume the substance from the basal plane areas (strips) adjacent to the end face of the platelet. It can be assumed that the width of these strips framing both parallel basal faces and "cropped" by the end face is at least a thousand times greater than that of the end face (i.e., the thickness of the platelet). The substance must be delivered to the end face by surface diffusion.

Let us estimate the velocity of motion of an isolated step of hight h in the course of growth or evaporation [1.17]. We restrict ourselves to the case where the average distance between the kinks at the step rise, λ_0, is much less than the mean free path of the adsorbed particles on the surface, λ_s. Then the step can be regarded as a continuous linear sink or source for the adsorbed particles. In the course of crystal growth this sink absorbs the particles adsorbed on the surface in strips of width $\sim\lambda_s$ on both sides of the step. Particles farther away from the step are with overwhelming probability reevaporated. Therefore at a distance from the step exceeding λ_s, the adsorbed particles have a concentration ensuring adlayer–vapor equilibrium, i.e., n_{sV} (Fig. 3.9a). Near the step (provided there is intensive particle exchange at the step between adsorption layer and crystal) the concentration in the adsorption layer is equal to n_{sS}, i.e., to the equilibrium value at the crystal temperature. Hence, the surface flux of particles per unit length of the step, j_s, is $\simeq(2D_s/\lambda_s)(n_{sV} - n_{sS}) \simeq 2\lambda_s(P - P_0)/\sqrt{2\pi mkT}$, because $n_{sS}/\tau_s \simeq P_0/\sqrt{2\pi mkT}$. The direct flux from the bulk vapor per unit length of the step rise, j, is $\simeq(P - P_0)\,h/\sqrt{2\pi mkT}$. Thus the step velocity

$$v \simeq \left(1 + \frac{2\lambda_s}{h}\right)\frac{\Omega(P - P_0)}{\sqrt{2\pi mkT}}\,. \tag{3.8}$$

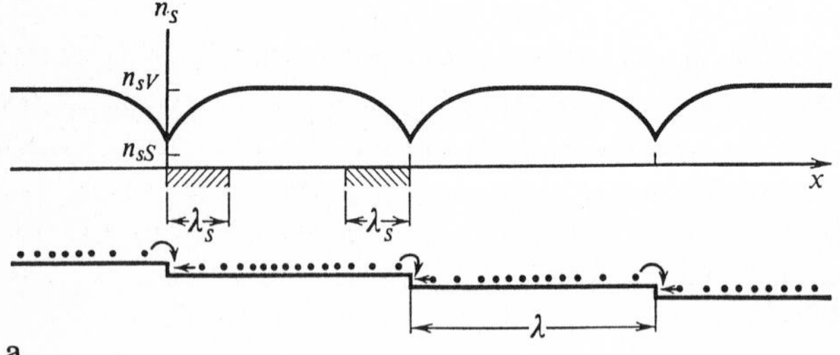

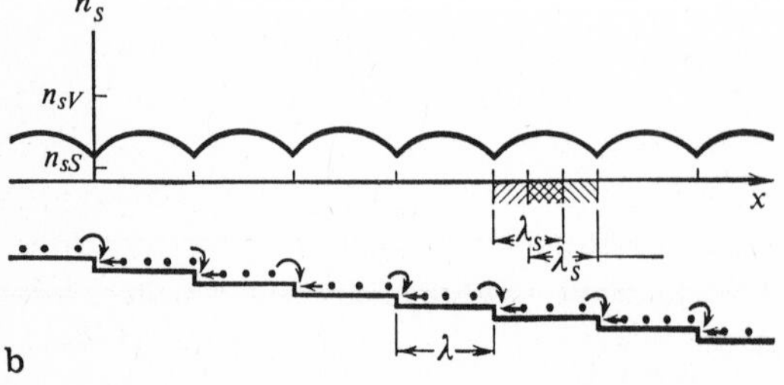

Fig.3.9a,b. Adsorbed atoms on growing stepped surfaces and the corresponding curves of their distribution over the surface. **(a)** Diffusion regions near the steps do not overlap ($\lambda > 2\lambda_s$); **(b)** diffusion regions overlap ($\lambda < 2\,\lambda_s$)

If the step height is much less than diffusion length λ_s, its velocity reduces with increasing height: the surface diffusion is the main source of nutrition. High, macroscopic steps are actually small faces, and if $h \gg \lambda_s$, their velocity is height independent.

Similar reasoning is qualitatively true for solution growth by surface diffusion. If the solution growth rate is determined (Sect. 3.2.2) by the bulk diffusion, the diffusion flux towards a step of height h changes only slightly with an increase in height h at small h ($h \ll D/\beta_{st}$), so that the step velocity reduces by about $(1 + \text{const.}\ \beta_{st}\ h/D)^{-1}$ [1.18]. This is also true of growth from the melt, when the step velocity is limited by the removal of heat (Sect. 3.2.3).

The experimental dependences of the velocity of the steps on their height for crystals of paratoluidine and β-methyl-naphthalene during growth from different media are presented in Fig. 3.10.

When deriving (3.8) it was implied that the particle exchange flux between the step and the adsorption layer proportional to kinetic coefficient β_{st} is intensive enough for the crystallization rate directly at the step to be considered high compared with the characteristic rate of surface diffusion D_s/λ_s. Analogous to (3.5) in Sect. 3.1, the kinetic coefficient for the step

$$\begin{aligned}\beta_{st} &\simeq a\nu(a/\lambda_0)\,(n_{sS}a^2)^{-1} \exp[-(E + \Delta H - \varepsilon_s)/kT] \\ &\simeq a\nu(a/\lambda_0) \exp(-E/kT)\,. \end{aligned} \tag{3.9}$$

Here, $\Delta H - \varepsilon_s$ is the heat of transition from the step to the adlayer. It determines the step growth rate at a relative supersaturation $(n_s - n_{sS})/n_{sS}$ in the immediate vicinity of the step via the expression

$$v = 2\beta_{st}a^2(n_s - n_{sS})\,. \tag{3.10}$$

Here, the factor 2 takes into account the possibility that particles can arrive at the step both "from above" and "from below" (Fig. 3.9a). If the resistance for crystallization at the step is comparable to the diffusional resistance for delivery of the crystallizing substance to the step, i.e., if $\beta_{st} \simeq D_s/\lambda_s$, the growth rate will be defined not only by the surface diffusion, but also by the processes on the step. In this case the right-hand side of (3.8) must acquire a factor approximately equal to $(\beta_{st}\lambda_s/D_s)(1 + \beta_{st}\lambda_s/D_s)$ [see below, equation (3.16)].

We have just estimated the velocity of a single isolated step on the surface. In actuality, there is usually a multitude of steps on the surface, and they form whole staircases, or step echelons. When interstep spacings $\lambda \gg \lambda_s$, their diffusion fields do not overlap, and the velocity of each is given by (3.8). At $\lambda \lesssim \lambda_s$, the neighboring steps draw the material arriving from the gas away from each other, and the velocity of each step is lower than that without overlapping. In order to take consistent account of this circumstance, the problem of surface diffusion towards the steps in the echelon must be solved. The schematic distribution of concentration in the adsorption layer of a step echelon is evident

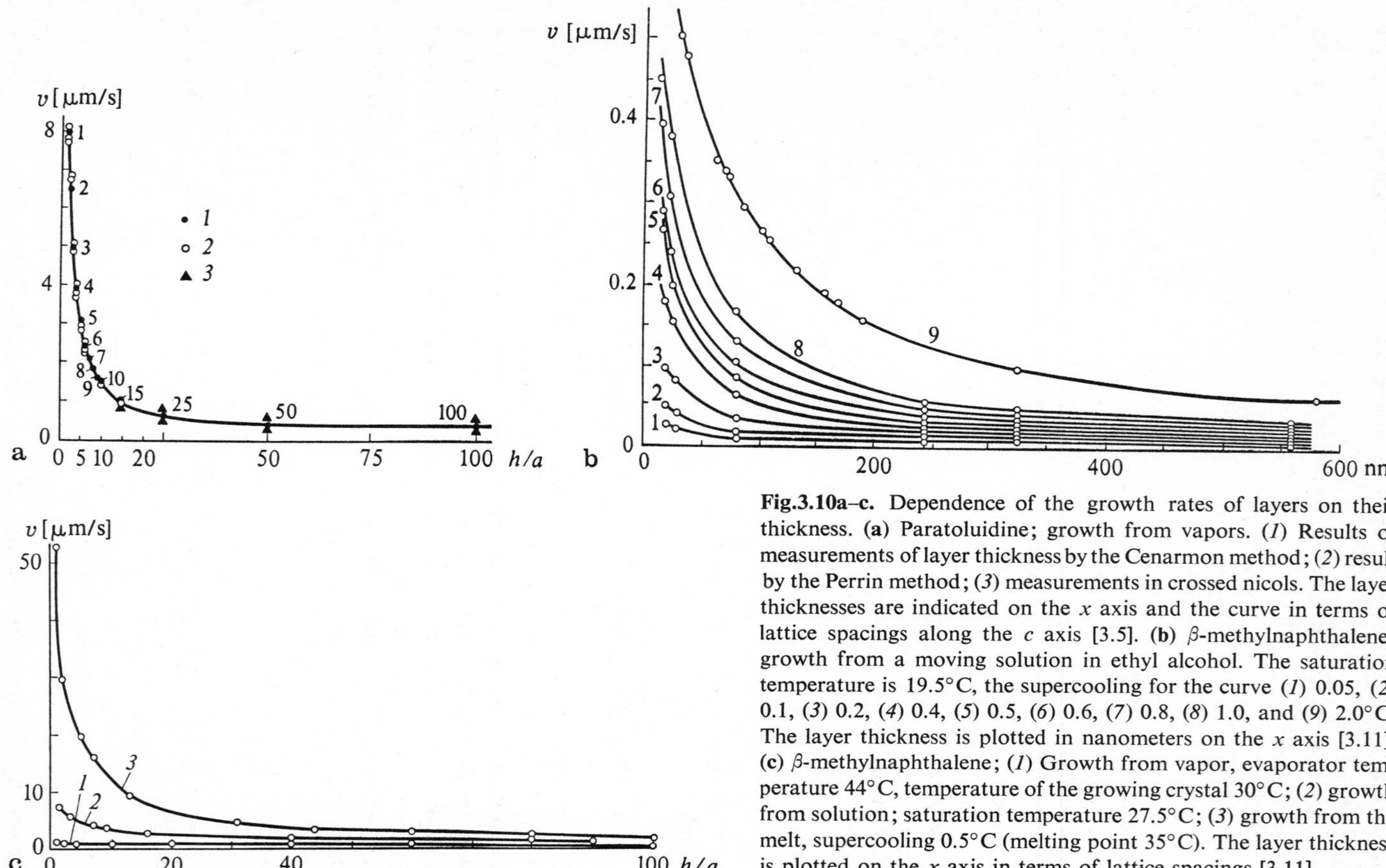

Fig.3.10a–c. Dependence of the growth rates of layers on their thickness. **(a)** Paratoluidine; growth from vapors. (*1*) Results of measurements of layer thickness by the Cenarmon method; (*2*) result by the Perrin method; (*3*) measurements in crossed nicols. The layer thicknesses are indicated on the x axis and the curve in terms of lattice spacings along the c axis [3.5]. **(b)** β-methylnaphthalene; growth from a moving solution in ethyl alcohol. The saturation temperature is 19.5°C, the supercooling for the curve (*1*) 0.05, (*2*) 0.1, (*3*) 0.2, (*4*) 0.4, (*5*) 0.5, (*6*) 0.6, (*7*) 0.8, (*8*) 1.0, and (*9*) 2.0°C. The layer thickness is plotted in nanometers on the x axis [3.11]. **(c)** β-methylnaphthalene; (*1*) Growth from vapor, evaporator temperature 44°C, temperature of the growing crystal 30°C; (*2*) growth from solution; saturation temperature 27.5°C; (*3*) growth from the melt, supercooling 0.5°C (melting point 35°C). The layer thickness is plotted on the x axis in terms of lattice spacings [3.11]

and is given in Fig. 3.9b. Here, in contrast with Fig. 3.9a, the interstep distance $\lambda < 2\lambda_s$.

The concentration in the adsorption layer satisfies the steady-state equation of surface diffusion, which can be obtained as follows: The diffusion flux on the surface

$$j_s = -D_s \operatorname{grad} n_s . \quad (3.11)$$

If the adsorption layer did not exchange particles with the vapor phase, concentration n_s would satisfy the usual diffusion equation div $j_s = 0$, i.e., the two-dimensional Laplace equation:

$$\Delta n_s = 0_s . \quad (3.12)$$

In actual fact, n_s/τ_s particles transfer to the vapor from a unit surface per unit time, while $P/\sqrt{2\pi\, mkT} = n_{sV}/\tau_s$ particles arrive at the surface. Therefore the total flux from the vapor to the surface

$$j = \frac{n_{sV} - n_s}{\tau_s} . \quad (3.13)$$

Consequently, the diffusion equation div $j_s + (n_{sV} - n_s)/\tau_s = 0$ takes the form:

$$(\lambda_s^2/4)\Delta n_s - (n_s - n_{sV}) = 0 . \quad (3.14)$$

In deriving (3.14) use was made of expression (1.26) $\lambda_s = 2\sqrt{D_s \tau_s}$. In addition to (3.14), surface concentration $n_s(x, y)$, where x and y are the coordinates on an atomically smooth surface, must satisfy the boundary conditions at the step. Neglecting the direct flux from the gas phase to the step rise (which can be done for steps of height $h \ll \lambda_s$) and assuming for the sake of simplicity that the substance arrives at the step with identical intensity from the upper and lower terraces as shown by the arrows in Fig. 3.9a, we can write, for a step parallel to the y axis:

$$2D_s \left| \frac{\partial n_s}{\partial x} \right| = \frac{h}{\Omega} v ,$$

where h is the step height. From this, using expression (3.10), we obtain, at steps at $x = 0, \pm\lambda, \pm 2\lambda, \ldots$:

$$D_s \left| \frac{\partial n_s}{\partial x} \right| = \frac{h}{a} \beta_{st}(n_s - n_{sS}) . \quad (3.15)$$

By solving (3.14) in the one-dimensional case ($\Delta n_s \equiv \partial^2 n_s/\partial x^2$) with boundary condition (3.15), it is easy to obtain the velocity of the steps in the echelon [1.17]:

$$v = \frac{\sigma\lambda_s \nu \exp(-\Delta H/kT)}{1 + \dfrac{D_s}{\beta_{st}\lambda_s}\tanh\dfrac{\lambda}{\lambda_s}} \tanh\frac{\lambda}{\lambda_s} = \frac{a^2\lambda_s P_0 \sigma}{\sqrt{2\pi mkT}} \frac{\tanh\dfrac{\lambda}{\lambda_s}}{1 + \dfrac{D_s}{\beta_{st}\lambda_s}\tanh\dfrac{\lambda}{\lambda_s}} \,. \tag{3.16}$$

The first fraction in the last expression is the velocity of an isolated step at $\beta_{st}\lambda_s/D \to \infty$, while $\sigma = (P - P_0)/P_0 = (n_{sV} - n_{sS})/n_{sS}$. The smaller the interstep distance λ is as compared with diffusion length λ_s, the lower the velocity of the steps in the echelon.

Dependence (3.16) for different values of $\beta_{st}\lambda_s/D_s$ is shown in Fig. 3.11. If the fact is taken into account that particles may reach the step rise not only from the adsorption layer by surface diffusion, but also directly from the gas phase, the step velocity will also be nonzero at $\lambda \to 0$.

The face growth rate R follows from (3.6, 16). Its dependence on face orientation $p = a/\lambda$ is schematized in Fig. 3.12. As mentioned in Sect. 3.1.3, the velocity R has a singular minimum at $p = 0$. Thus the actually observed growth rate of a singular surface depends both upon the processes on the steps and atomically smooth areas and upon the step density. This last value depends on the power of the step sources (Sect. 3.3).

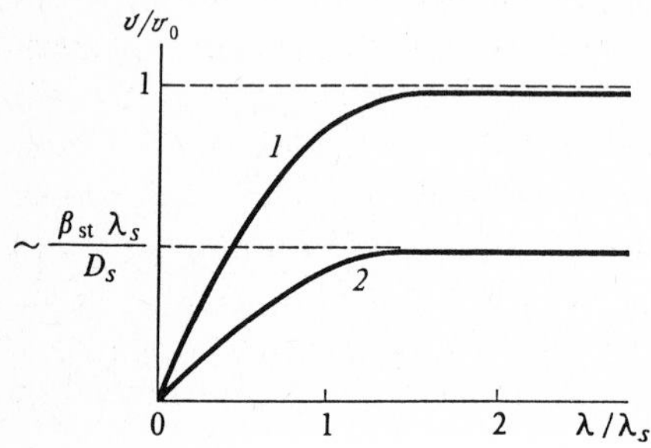

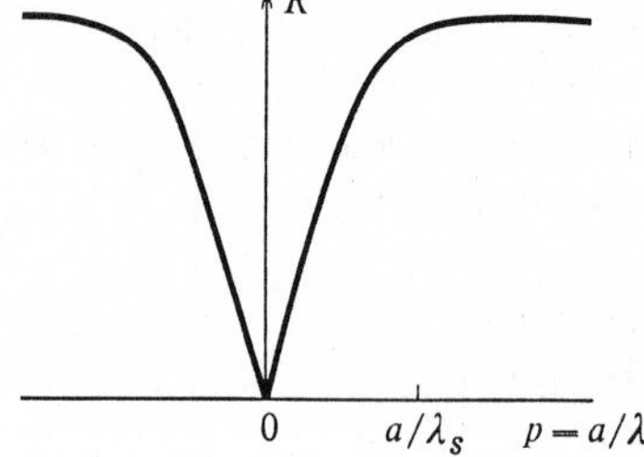

Fig.3.11 Fig.3.12

Fig.3.11. Dependence of the relative velocity of the motion of parallel equidistant steps, v/v_0, on the distance λ between them at different rates β_{st} of crystallization processes at the step. Here, $v_0 = a^2\lambda_s P_0\sigma/2\pi mkT$ [see (3.16)]. (*1*) $\beta_{st}\lambda_s/D_s \gg 1$; (*2*) $\beta_{st}\lambda_s/D_s \ll 1$

Fig.3.12. Face growth rate R as a function of the face's orientation p for growth via an adsorption layer with an adsorption path length λ_s

3.2.2 Growth from Solution

Solution growth usually occurs layerwise (Sect. 3). The crystallizing substance is delivered to the steps by either bulk or surface diffusion. Their relative role has not yet been established.

In many cases of growth from solution, the solution is mixed, either by stirring or as a result of convection which arises due to the depletion of the solution near the growing crystal and the corresponding change in density (so-called concentration flows). However, because the liquid has finite viscosity and sticks to the solid interface, an unmixed boundary layer is formed near the solid surface, in which mass transfer is achieved by ordinary diffusion. As the liquid flows around a thin plate parallel to the oncoming flow, the thickness of the so-called diffusion boundary layer, δ, increases with the distance x from the ledge, where the flow meets the plate. At a distance x from the ledge of encounter [3.12],

$$\delta = 4,5(D/\nu)^{1/3}(\nu x/u)^{1/2}, \tag{3.17}$$

where D and ν are the diffusivity and kinematic viscosity in solution, and u is the velocity of the oncoming flow. For aqueous solutions at room temperatures, $D \simeq 3 \times 10^{-6}$ cm/s, $\nu \simeq 10^{-2}$ cm^2/s; and at $u \simeq 30$ cm/s, $x \simeq 1$ mm we obtain $\delta \simeq 10^{-3}$ cm. Equation (3.17) can, of course, only give a qualitative idea of the thickness of the diffusion layer on the crystal under real mixing conditions, because the flow around a bulk crystal differs from that for a plate. Rigorous hydrodynamical approaches were reviewed by *Rosenbergerr* [1.2b].

Within the boundary diffusion layer, the solution concentration C approximately satisfies the diffusion equation, which can be replaced by the Laplace equation if step velocity $v \ll D/h$ (D/h being the characteristic diffusion rate towards the step), i.e., when the diffusion field in solution manages to "adapt itself" to the moving steps at each instant. If $D \simeq 3 \times 10^{-6}$ cm^2/s, $h \simeq 10^{-7}$ cm, it is sufficient to have $v \ll 30$ cm/s. This inequality is fulfilled with a large margin. Indeed, if step velocity v is limited only by the processes at the kinks, then according to (3.5), $v = \beta_{st}\Omega C_0\sigma$, and at $\Omega C_0 \simeq 10^{-1}$($\sim$10% solution), $\sigma \simeq 10^{-2}$ (i.e. 1%), $\beta_{st} \simeq 2 \times 10^{-1}$ cm/s, we have $v \simeq 2 \times 10^{-4}$ cm/s. If the resistance on the kinks is supplemented by the diffusional resistance, then v will be still lower, so that a diffusion field around each moving step is practically the same as if the step were fixed.

A step plays the role of a linear sink for the solute. Approximating the step rise by a half cylinder with a radius of h/π and hence the same absorbing surface as a step of height h, we get the boundary condition on each step in the form

$$D\frac{\partial C}{\partial r} = \beta_{st}(C - C_0), \qquad r = h/\pi,$$

where r is the radius computed from the center of the cylindrical sink. In the intervals between steps there is no solute flow towards the solid phase:

$$\partial C/\partial n = 0,$$

where $\partial/\partial n$ is the derivative with respect to the normal to the singular surface.

In the bulk of the agitated liquid, particularly if the motion is turbulent, the concentration is virtually constant, although it fluctuates in time and space. Ignoring these fluctuations, we can assume that at a distance δ from the crystal surface the concentration $C = C_\infty$, where C_∞ is the value in the bulk of the mother liquid.[2] The solution of the problem thus stated is easy to obtain for the case of parallel and equidistant steps. Figure 3.13 is a schematic illustration of the equal-concentration lines (dashed) and the current lines (solid) in the plane perpendicular to the steps and the crystal surface (hatched). By calculating the diffusion flux towards the steps we obtain their velocity [1.18]:

$$v = \beta_{st}\,\Omega(C - C_0) = \frac{\beta_{st}\,\Omega\,C_0\sigma_\infty}{1 + \dfrac{\beta_{st}h}{\pi D}\ln\dfrac{\lambda}{h}\sinh\dfrac{\pi\delta}{\lambda}}\,. \tag{3.18}$$

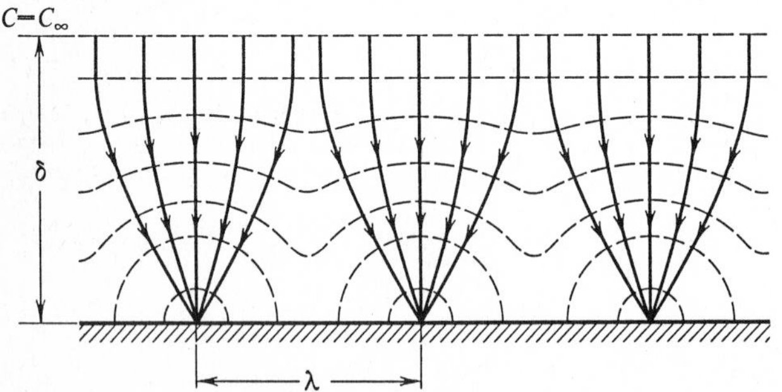

Fig.3.13. Distribution of solution concentration in the boundary layer over a stepped growing surface. The steps are separated by distance λ and are perpendicular to the plane of drawing

According to (3.18), the velocity of the steps reduces with the spacing λ between them. Indeed, the smaller λ is (usually $\lambda \ll \delta$), the more the cylindrical diffusion fields surrounding each step overlap, and the lower is the supersaturation on each step. Conversely, if the steps are so far from each other that the second term in the denominator of (3.18) is small compared to unity, the steps are fed independently of each other, and their velocity is independent of λ. The same result may be obtained if $\lambda \gg \delta$, in which case (3.18) is valid only qualitatively.

In addition to the obvious geometric parameters δ/λ and λ/h, (3.18) includes the dimensionless parameter $\beta_{st}h/D$. It represents the ratio of the rate of the growth processes at the step to the rate of diffusion delivery of the crystallizing substance to it. With β_{st} varying from 10^{-1} cm/s to 10^{-3} cm/s, $D \simeq 3 \times 10^{-6}$ cm²/s, and $h \simeq 10^{-7}$ cm, the above ratio is $3(10^{-3} - 10^{-5})$.

[2] A rigorous solution [1.1, 3.12] takes laminar flow in the whole liquid into account without assuming that the liquid is stagnant within the layer of thickness δ only.

From (3.18) it follows that the higher the step, the lower the velocity of the step. If $\lambda \ll \delta$ and $\beta_{st}h\delta/D\lambda \gg 1$, then $v \simeq D\lambda(C - C_0)/h\delta$, i.e., the velocity is inversely proportional to the step height. In this case v is independent of the kinetic coefficient on the step, i.e., the growth-limiting stage is the delivery of the substance to the steps, rather than its incorporation into the crystal.

In the model described account is taken only of direct incorporation of the crystallizing substance into steps. This is consistent with experiments on electrodeposition [3.32b] (see Sect. 3.3.4). Comparison with conventional less precise solution growth, however, gives reason to believe that the surface diffusion may also be substantial [3.13].

Knowing the step velocity v at a given step density, one can, with the aid of (3.6, 18), find the face growth rate as well:

$$R = b(p)\Omega C_0\sigma_\infty = B(p)\sigma_\infty , \tag{3.19}$$

where $b(p)$ and $B(p) \equiv \Omega C_0 b(p)$ are the kinetic coefficients of crystallization for the face with orientation p. If the diffusion fields of the steps do not overlap, i.e., for sufficiently small p,

$$b(p) = \beta_{st}|p| . \tag{3.20}$$

The nature of the face growth rate and the corresponding kinetic coefficient anisotropy are shown in Figs. 3.7, 12.

The previously used kinetic coefficient β for growth rate V [see (3.5)] is related to b by the geometric expression $b(p) = \beta(p)\sqrt{1 + p^2}$ [see (3.7)]. The dependence $b(p)$ at a given supersaturation has the same character as $R(p)$ and is clear from Fig. 3.7a, where the continuous line corresponds to the growth of an ideal crystal at low supersaturations. Sharp singular minima of $b(p)$ alternate with rounded (or flat) maxima, where $\partial b/\partial p = 0$. The latter correspond to a high step density and hence high kink density on the surface. In the range of orientations p far from that of the singular face, the surface of a growing crystal is usually divided into macroareas (macrosteps) consisting of singular faces. The growth mechanism, kinetic coefficients, and relative dimensions of these faces define the concrete form of $b(p)$ for average orientations p lying outside the vicinity of singular orientations.

The measure of the anisotropy of kinetic coefficient $b(p)$ is its logarithmic derivative

$$\Theta = \partial \ln b(p)/\partial p = (1/b)\partial b/\partial p . \tag{3.21}$$

It is easy to see that $\Theta = \partial \ln R(p)/\partial p$. At small p, i.e. at slight deviations of the growing surface from the singular face, $\Theta \simeq \partial \ln V/\partial p$. The anisotropy parame ter Θ is maximal for orientations close to those of the singular minima, where, according to (3.20), $\Theta \simeq 1/p$. As we recede from the orientations of the singular minima, Θ decreases. For orientations where $b(p)$ reaches its maxima we have $\Theta = 0$.

3.2.3 Growth from the Melt

Growth from a pure one-component melt and growth from solution have many features in common. In the former, however, the problem of delivering crystallizing substance to the surface is replaced by the problem of removing crystallization heat from it. The latent heat of solidification can be removed either through the melt, or what is much more often the case, through the crystal, or both. Each step is a cylindrical heat source, and the temperature field in the crystal near the layerwise growing surface is similar to the concentration field in solution, which is depicted in Fig. 3.13. The dependence of the step velocity on the supercooling and the spacing between the steps is described by a relation similar to (3.18). In the case of a melt, however, the role of quantity $\beta_{st}\Omega$ $(C - C_0)$ is played by $\beta_{st}^T(T_0 - T)$, and parameter $\beta_{st}h/D$ must be replaced everywhere by parameter $\beta_{st}^T T_Q h/a_{S,L}$, where a_S or a_L are the heat diffusivities of crystal and melt, and T_Q is the ratio of the latent heat of solidification to the heat capacity of the solid. T_Q approximately shows how much the temperature of the supercooled melt would have increased if it had crystallized without heat removal through the ambient space. Typical values of T_Q for different simple substances are $\simeq 10^2$ K.

The heat diffusivities of crystals and liquids (10^{-2}–10^{-3} cm^2/s) are about 1000 times higher than the diffusion coefficient, i.e., than the mass diffusivity. Therefore the removal of the crystallization heat proceeds much more rapidly than the diffusion of the crystallizing substance. Moreover, the fact that the potential barriers are much lower for the addition of particles to the crystal from the melt than they are for addition from solution leads in the case of melt growth to kinetic coefficients (Sect. 3.1) such that at the commonly used supercoolings the melt growth rates greatly exceed those for solutions. For instance, in a silicon melt, at $\Delta T \simeq 1$K, i.e., at $\Delta\mu = \Delta S \Delta T \simeq 4.8 \times 10^{-16}$ erg, or at a dimensionless driving force $\Delta\mu/kT \simeq 2 \times 10^{-3}$, the step velocity is tens of cm/s (Sect. 3.1). At the same time, for solution growth at room temperature and identical driving force $\Delta\mu/kT \equiv \sigma = 2 \times 10^{-3}$, even in relatively concentrated solutions ($\Omega C_0 \simeq 10^{-1}$) the step velocity $v = \beta_{st}\Omega C_0 \simeq 2(10^{-5}$–$10^{-6})$ cm/s for typical values $\beta_{st} \simeq 10^{-1} - 10^{-2}$ cm/s. The smallness of the value v, in addition to the high barrier to crystallization, is due to the low density of the crystallizing substance in solution compared with the crystal (usually $\Omega C_0 \lesssim 2 \times 10^{-1}$). Thus, regardless of whether the growth rate depends on the crystallization processes proper or on the mass or heat transfer, the steps move many orders more rapidly during growth from the melt than they do in solution. Later we shall be able to draw the same conclusion with regard to the growth rates of macroscopic faces as a whole.

3.3 Layer Sources and Face Growth Rates

The normal rates R and V of tangentially growing crystalline faces ($R \simeq V$) are proportional to the step density on the surface and the velocity of these steps, (3.6). When discussing the values of the tangential growth rates in the preceding section, we assumed the spacing between the steps, λ, and thus the step density to be given. Let us now see what the step density depends on.

3.3.1 Nuclei

The sources of the growth layers on a singular surface of a dislocation-free crystal are two-dimensional nuclei (Sect. 2.2) whose work of formation depends on the value of the specific free linear energy of the steps, $\simeq \alpha a$, see (2.22). Suppose that nucleation on the surface occurs at the rate of $J[\mathrm{cm}^{-2}\ \mathrm{s}^{-1}]$, L being the size of the face under consideration, and v the step propagation rate, as before. The time needed for the propagation of a new layer from the nucleus over the entire surface with a linear dimension L is $\simeq L/v$. The number of nuclei arising on the surface within this time is $\simeq JL^3/v$. If this value is below unity, i.e., if $L < < (v/J)^{1/3}$, then each consecutive nucleus arises, as a rule, only after the layer begun by the preceding nucleus has propagated over the whole face. Then

$$R = V = JL^2 h\,, \tag{3.22}$$

where h is the height of the nucleus; on a proper substrate h equals the lattice spacing a, or a part of it.

If the supersaturation is high and the size face large enough, new nuclei may arise on the surface before the preceding layer completely covers the face in question and (3.22) ceases to apply. Then the layer formed by the new nucleus at any point on the surface has enough time to grow to a radius $\lambda \simeq (v/J\pi)^{1/3}$ before meeting the layers initiated by other nuclei, the reason for this is that the time needed for the appearance of the next nucleus inside the contour of the previous closed step is of the order of $\lambda/v \simeq 1/\pi J\lambda^2$. Therefore the growth rate is [3.14b–d]:

$$R \simeq \pi\lambda^2 Jh \simeq h(\pi v^2 J)^{1/3} \tag{3.23}$$

If the growth is initiated on the essentially step-free interface, the first generations of nuclei are synchronized in time. This first generation of steps is nucleated just after the moment when the under or supersaturation is switched on. In the beginning, the steps have the form of isolated islands. The islands grow, spreading over the surface, and their total perimeter (total step length) increases in time. When the steps of limiting neighboring islands encounter each other, they are annihilated and the total step length decreases. At the same time, the nuclei of the next layer (next generation) appear and the total length of all steps increases again. As a result, the overall step length and hence the flux of molecules joining the crystal during this first period of growth oscillate in time before reaching a steady-state constant value corresponding to a given driving force. The oscillation period is, obviously, $\simeq(\pi v^2 J)^{-1/3}$. The number of oscillations for vacuum evaporation and deposition of KCl and LiF [3.14e, f] and electrodeposition of Ag [3.15–17] is $\simeq 3$, which agrees with computer simulation data [3.18].

Since $J \sim \sqrt{\Delta\mu/kT}\exp(-\delta\Phi_c/kT)$ and $v \sim \Delta\mu$, it follows from (3.23) that

$$V \sim (\Delta\mu/kT)^{5/6}\exp(-\delta\Phi_c/3kT)\,. \tag{3.24}$$

Thus, noticeable growth may occur only when the supercooling exceeds a certain critical value which is determined by the work of nucleation according to an equation of the type of (2.5, 6a). It is worth noting that (3.24) contains only 1/3 of the work $\delta\Phi_c$. At high supercoolings, when the exponent is close to unity, a nearly linear ($\sim\Delta\mu^{5/6}$) dependence $V(\Delta\mu)$ follows from (3.24), and its approximate extrapolation to zero rates from the range of abrupt increase in $V(\Delta\mu)$ intersects the supercooling axis to the right of the origin.

Equation (3.24), which gives the normal growth rate of a face in terms of v and J, implies that nuclei may arise infinitely close to the step. In actual fact, however, the supersaturation (supercooling) near the step is less than the supersaturation in the bulk of the medium: if nuclei at a given deviation from equilibrium can arise no closer than a distance Λ from the step, then at high supersaturations, when $\lambda \simeq (v/J\pi)^{1/3} < \Lambda$, the distance between the consecutively nucleated steps must be Λ. Therefore the normal growth rate at such high supersaturations is

$$V \simeq hv/\Lambda\,. \tag{3.25}$$

During growth from the gas phase, $\Lambda \lesssim \lambda_s$, and generally speaking, it reduces with increasing supersaturation and decreasing step kinetic coefficient. In solution and melt growth the actual supersaturation on the surface near the step is determined, in accordance with (3.18), by the ratio between three characteristic lengths: δ, $\lambda \simeq (v/J)^{1/3}$, and D/β_{st}(for a solution) or $a_{S,L}/\beta_{st}^T T_Q$(for a melt). The model implying the appearance of a new two-dimensional nucleus in the diffusion field of the steps previously nucleated gives $V\sim\Delta\mu$ at sufficiently high supersaturations [3.19].

If the crystal surface carries adsorbed particles or impurity atoms, which reduce the nucleation threshold, critical supercoolings for the beginning of growth may decrease appreciably. Fundamentally, however, the picture remains unaltered unless the reduction is so great that the critical supercooling becomes less than the minimal supercooling obtainable in a given experiment. In the latter case the growth rate depends linearly on the supercooling in its whole range.

"Normal growth" of this kind is a result of so-called kinetic roughness, rather than the thermodynamic roughness discussed in Sect. 1.3.4. The kinetic roughness is the appearance of such a high density of nuclei on the surface that the average spacing between steps $(v/J\pi)^{1/3}$ becomes experimentally undetectable or even comparable to the interatomic distance. The kinetic roughness concept was introduced by the author [3.20, 21] and reintroduced by *Miller* [3.22] who used computer simulation data. For the attainment of kinetic roughness a high nucleation rate J is necessary. This, in turn, is achieved at high supersaturations and/or in the presence of a large number of defects, which facilitate nucleation. Nucleation is also promoted by low values of the step linear energy, α_l, i.e., under conditions where parameter ε/kT is close to the critical value characterizing the thermodynamical transition from an atomically smooth to an atomically rough surface (Sect. 1.3.4).

3.3.2 Dislocations

The situation may be quite different when a crystal contains dislocations. As first noted by *Frank* in 1949, a step issuing from the point where a screw dislocation crosses the surface cannot disappear in the course of growth, because a crystal with a screw dislocation actually consists of a single atomic plane rolled into a helicoid. Hence, a dislocation outcrop is a continuously acting layer source which eliminates the need for two-dimensional nucleation in order to continue the growth of a singular face. Let us consider in more detail the effect of a screw dislocation with a unit Burgers vector in the course of crystal growth. Suppose that step OA (Fig. 3.14a), which is formed by a screw dislocation crossing the face at point O, is straight at the initial instant. In the course of growth the step starts moving to the right; since the linear velocities of its various parts are the same, their angular velocities decrease with the distance from O. Therefore the step shape will change until the step turns into a spiral which ensures constancy in the angular velocity of the "rotation" of all the parts of the step about point O (Figs. 3.14b–d). The spiral step on the (0001) face of silicon carbide, detected by *Lemmlein* in 1945 [3.14f, p. 278; 3.14g], is presented in Fig. 3.15. *Lemmlein* also indicated at that time that the spiral

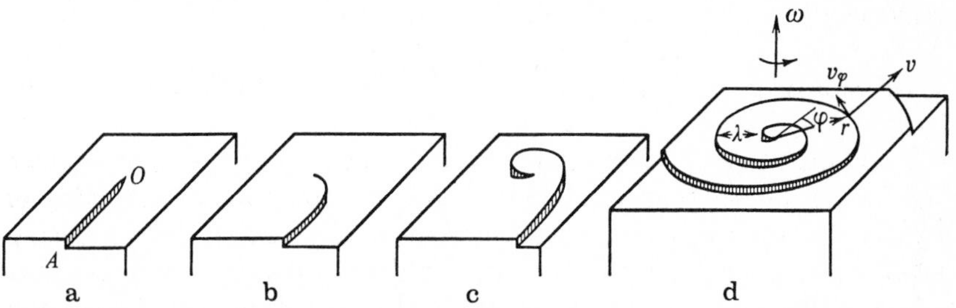

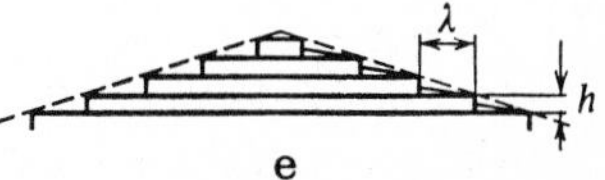

Fig.3.14a–e. Consecutive stages in the formation of a spiral step around the point of emergence of a screw dislocation at the crystal surface

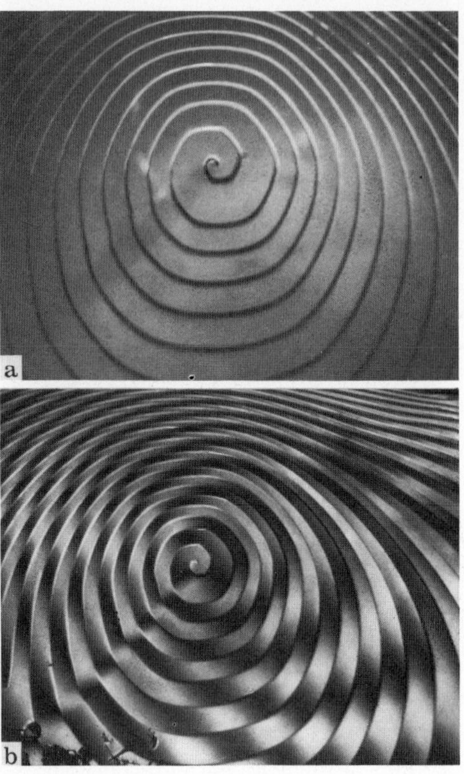

Fig.3.15a,b. Spiral step on the surface of a silicon carbide crystal grown from vapor. Courtesy of G.G. Lemmlein. Magnification × ca.100 (**a**) Surface view in a microscope; (**b**) a silver-coated plate has been placed over the same surface, and interference bands, experiencing jumps on the step, have formed in the gap (see Sect. 3.4.1)

step does not disappear in the course of growth, and drew an animated cartoon depicting this phenomenon, although he failed to establish the relationship between the centers of the spirals and screw dislocations.

Let us now find the normal growth rate dependence on the supersaturation, $R(\sigma)$, which is determined by the screw dislocation. The condition for the steady state "rotation" of the spiral is

$$v_{\varphi}/r = \omega = \text{const} . \tag{3.26}$$

The symbols here are clear from Fig. 3.14d. Taking advantage of the fact that the step velocity is proportional to the supersaturation and of the relationship between the supersaturation on the step and its curvature according to Thomson's equation (1.35) at $R_2 = \infty$, $R_1 = \rho$, and $r_c = \Omega\alpha/(\mu_M - \mu_S)$, we find the dependence of the velocity of the curved step on its curvature radius ρ:

$$v(\rho) = v(1 - r_c/\rho) , \tag{3.27}$$

where v is the velocity of a straight step, and r_c the radius of a two-dimensional critical nucleus. At the point of emergence of the dislocation, where $v(\rho) = 0$, we must have $\rho = r_c$. Hence it is clear that in order of magnitude, $r \simeq r_c\varphi$, so that the spacing between two neighboring turns is

$$\lambda \simeq 2\pi r_c . \tag{3.28}$$

A more accurate calculation gives $\lambda = 19r_c$ [3.23a] so that the normal growth rate is

$$R = (h/\lambda)v = (h/19r_c)v = h\Delta\mu v/19\Omega\alpha . \tag{3.29}$$

At low supersaturations, when r_c is large and the diffusion fields of the steps do not overlap, $v \propto \Delta\mu$ and $R \propto (\Delta\mu)^2$. With increasing supersaturation, the radius r_c and consequently the spacing between the neighboring spiral turns reduces, the diffusion or thermal fields of the steps overlap, and the result is $R \propto \Delta\mu$. The above reasoning with respect to growth also holds true for decrystallization.

At a low ratio of temperature and bond strength the density of kinks on a step must be low, and the velocity of the step is anisotropic with respect to the step orientation on the surface. The anisotropy of the step energy also increases with decreasing temperature, though this change is less pronounced. Both factors make the spiral anisotropic. Analysis of a polygonized spiral with fourfold symmetry [3.23b, c, d] shows that $\lambda = 8r_c$ as opposed to $19r_c$ for a rounded spiral [3.23d]. The anisotropy of the spiral step shape (i.e., the degree of polygonization) is determined by the anisotropy of the step line energy α_l and of the step kinetic coefficient β_{st}. The α_l former is dominant in the center of a spiral, while the latter prevails on its periphery. Thus, if α_l is less anisotropic than β_{st}, the spiral should be more rounded in the center and more polygonized far from the center [3.23c]. Methods of quantitatively estimating crystal growth parameters for the surface diffusion model are given in [3.23e]. Applications of these methods to spiral patterns on SiC [3.23e] and growth hillocks on yttrium–iron garnet [3.23f] have also been reported.

An emerging edge dislocation does not produce a step on a surface to which the Burgers vector is parallel. However, at the point of the dislocation outcrop, impurities or macroparticles may be adsorbed. This may facilitate two-dimensional nucleation and lead to the generation of concentric closed steps. In decrystallization however, the predominant nucleation on the edge dislocation is of a general nature and is associated with the increased elastic energy density and, hence, an increased chemical potential of the crystal around the dislocation line at distances of the order of several b, the Burgers vector of the dislocation. This effect is considered to be the main reason for selective etching on dislocations and the creation of hollow channels along the dislocation lines. Since the density of elastic energy is $\propto Gb^2$, the rate of selective etching of the material around dislocations and the radius of the hollow core increase with increasing Burgers vector b of the dislocation. A detailed analysis of dissolution and etching is given by *Heimann* [3.23k, l].

Some consideration might be given to the question of how the stress field around an edge dislocation is active in growth. A two-dimensional nucleus gives rise to a stress field in the substrate, because the creation of a nucleus means the creation of a step. This nucleus-induced stress must interact with the stress of the edge dislocation, thus leading to a possible decrease in the nucleation barrier and to preferential nucleation either in the compressed or the stretched sectors around the dislocation. A slightly different line of reasoning [3.23g] is based on the stressed state of the surface layer and its relaxation around the dislocation outcrop. The effects suggested above should, however, be suppressed by the increase in chemical potential responsible for preferential etching on dislocations. Some experimental data on the growth activity of dislocations are presented in Sect. 3.3.4.

The stress around a screw dislocation increases the r_c value, thus decreasing both the spiral rotation rate ω and the normal growth rate R [3.23h].

Steps generated by dislocations during growth form conical hillocks (Figs. 3.14d, e; 3.15) and those generated during decrystallization form pits. For a screw dislocation giving rise to a step of height h, the slope of the lateral surface of the hillock equals h/λ (Fig. 3.14e). Since according to (3.29), the spacing between the turns of the spiral step decreases with increasing deviation from equilibrium ($r_c \propto 1/\Delta\mu$), i.e., the strength of the step generator increases, the slope of the growth hillock or the decrystallization pit increases. If the step velocity is isotropic, the spiral is rounded (Fig. 3.15), but if the propagation rate of the step has singular minima for individual directions, then the spiral consists of separate straight segments, i.e., it is polygonized (Figs. 2.9a, 3.5a). Accordingly, the vicinal growth hillocks may have rounded or polygonal intersections by the face plane.

The supersaturation (supercooling) at each point of the growing surface is less than the supersaturation in the mother medium because of the absorption of the crystallizing substance by the steps (or by kinks in general) on this surface (Sect. 3.2). The higher the density of steps and kinks and the higher the rate of attachment of particles onto them (i.e. kinetic coefficient β_{st}), the lower is the actual average supersaturation at the growth front. The point of emergence of a screw dislocation at the surface is no exception. The part of the spiral step adjacent to this point is in the diffusion field of the entire remaining spiral and above all, of its first turn. Therefore the supersaturation or supercooling determining radius r_c in (3.28) and thus the average step density is less than the supersaturation in the mother medium, and is itself dependent on the step density. Consequently, at low supersaturations, the average slope of the vicinal hillock increases linearly with the supersaturation, and this dependence then becomes less steep. Finally, at very high supersaturations, when the spacing between the turns is much smaller than the distance at which the diffusion fields of the steps overlap, the hillock slope ceases to depend on the supersaturation (unless, of course, two-dimensional nuclei begin to form everywhere and the hillock disappears, transforming into a plane; i.e., kinetic roughness of the surface is achieved, Sect. 3.3.1). This so-called back-stress effect [3.24] is particularly pronounced during the evaporation of crystals into a vacuum, when the relative undersaturation in the bulk, $\ln(P/P_0)$, is infinite, $(P = 0)$, but evaporation proceeds according to the spiral-layer mechanism (Fig. 2.9a). The spacing λ between the turns of the spiral in such a case can be estimated as follows: We replace the first turn, which affects the supersaturation at the point of emergence of the dislocation most strongly, by a circle of radius λ. Calculating the supersaturation at the center of such a circle, we find λ from the transcendental equation $\lambda = 19r_c$, which then takes the form

$$\frac{2\lambda}{\lambda_s} \ln I_0\left(\frac{2\lambda}{\lambda_s}\right) = \frac{19\Omega\alpha}{kT\lambda_s}, \tag{3.30}$$

where I_0 is a zero-order Bessel function of the imaginary argument, and λ_s is defined according to (1.26). For the numerical values of function on the left-hand side of (3.30), see [3.25]. Thus the spacing between the steps and hence the slope of the vicinal hillock will, at sufficiently high supersaturations or undersaturations, depend only on the temperature, primarily via the diffusion length λ_s.

During growth from solutions by means of bulk diffusion, the step diffusion fields are of a long-range nature. Here the superpositions of the diffusion fields of all the steps are taken into account, as was done in (3.18) when determining the supersaturation at the point of dislocation emergence and, hence, the λ value.

It is important to emphasize that the above-described dependence of the face growth rate on the degree of deviation from equilibrium, $R(\Delta\mu)$, is basically similar for vapor, solution, and melt growth. This is because the factors governing the activity of the dislocation and the mutual retardation of the steps are similar in all three cases.

The dependence $R(\Delta\mu)$, which is parabolic at small deviations from equilibrium and linear at large ones, presupposes that the spiral step is elementary, i.e., it cannot break up into an echelon of lower steps. If, however, the screw dislocation Burgers vector is equal to several lattice spacings, then the macroscopic step created by it can decompose into lower or even elementary steps: a multiple-step spiral arises (Fig. 3.16). The spacing between the turns in a multiple-step spiral is smaller and the average kink density on the surface larger than in a single-step spiral. The number of branches in a spiral often increases with supersaturation, as can be seen from the series of photos in Fig. 3.17. In this case the large the Burgers vector is, the lower are the supersaturations at which the transition to the linear section of $R(\Delta\mu)$ occurs. A sufficiently large surface usually has dislocations with large Burgers vectors, and therefore the measurement of normal velocities on large crystals may produce linear dependences at practically all supersaturations despite the presence of spiral layers and the vicinal conical growth hillocks formed by them.

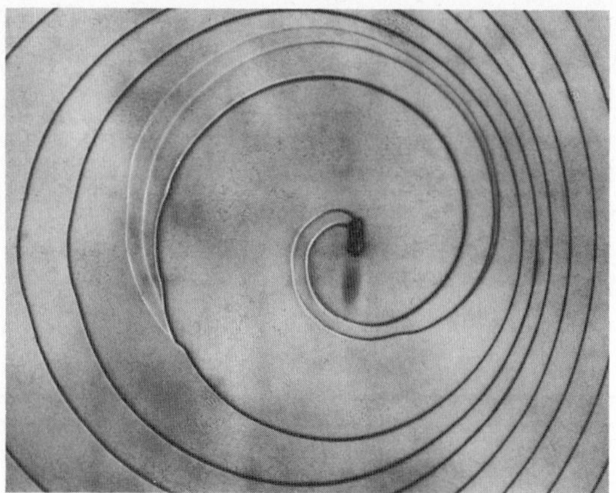

Fig.3.16. Double-step spiral on the basal face of a silicon carbide crystal grown from vapor. Courtesy of G.G. Lemmlein. Reflected light, magnification × ca.200

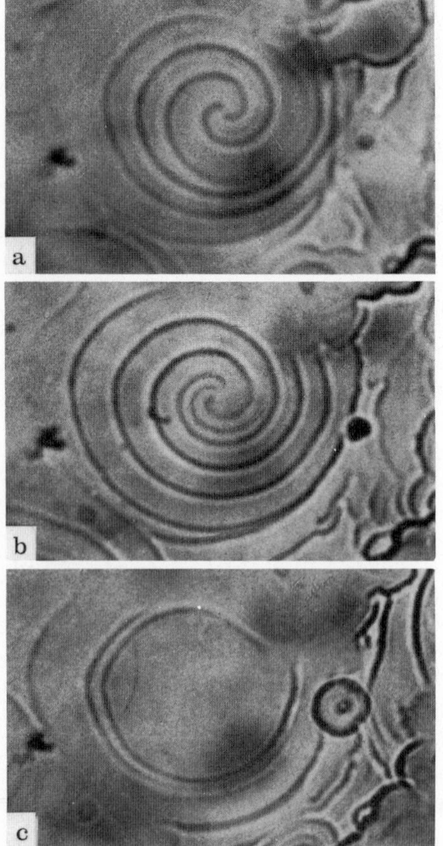

Fig.3.17a–c. Splitting of a macrostep (**a**) into several steps of lower heights with increasing supersaturation, resulting in the formation of a triple-step spiral (**b**) and a flat-topped growth hillock (**c**). Paratoluidine crystals, growth from vapor [1.18]. Magnification × 200

During growth from vapor near the triple point where the vapor, crystal, and melt coexist, and at high supersaturations, crystals showing absolutely uniform colors in polarized light are observed [3.21] (for observation procedure see Fig. 3.5). At low supersaturations these crystals exhibit vicinal hillocks and growth layers. The absence of visible layers at high supersaturations can be interpreted as evidence of growth by multiple production of nuclei (a kinetically rough surface, Sect. 3.3.1) or possibly as evidence of the normal growth of a surface coated with a film of melt or solution (Sect. 3.4.4). The surface energy of the crystal–liquid interface is lower than that of the crystal–vapor interface, which may lead to thermodynamic (Sect. 1.3.4) or kinetic (Sect. 3.3.1) roughness and normal growth.

3.3.3 Kinetic Coefficient and Anisotropy of Face Growth

The above analysis of nucleation and dislocation growth mechanisms shows that both are characterized by nonlinear dependences of the normal growth rates of faces on the supersaturation (supercooling) at low supersaturations, and linear dependences at high supersaturations. By using the linear approximation of $R(\Delta\mu)$ within a supersaturation range that is feasible and not too wide ($\sim 50\%$–100% for vapor growth), it can be ascertained that the extension of the linear section of function $R(\Delta\mu)$ (or of a small section of this dependence containing the inflection point) into the low-supersaturation region does not intersect the origin, but makes a positive intercept $\Delta\mu_0$ on the x axis. Thus dependence $R(\Delta\mu)$ can be approximated by the expression

$$R(\Delta\mu) = b(p)\Omega C_0(\Delta\mu - \Delta\mu_0)/kT \tag{3.31}$$

or, for growth from solution, when $\Delta\mu = kT\sigma$,

$$R = b(p)\Omega C_0(\sigma - \sigma_0) \,. \tag{3.32}$$

Supersaturation σ_0 is termed the threshould or critical supersaturation. Its meaning is clear from Fig. 3.21, which is discussed in Sect. 3.3.4. In the nucleation growth mechanism it is close, both in meaning and magnitude, to the critical supersaturation for two-dimensional nucleation [3.26]. In the dislocation growth mechanism it is defined by the linear energy of the step and the characteristic size of the region supplying the step with the nutrient — that is, by the surface diffusion length [in accordance with (3.16)] or by the thickness of the boundary layer [in accordance with (3.18)]. The critical supersaturation may also be due to the adsorption of an impurity (Sect. 4.2) or of the solvent on the growing face. Its nature has not been conclusively established in all cases.

The kinetic coefficient $b(p)$ in (3.31) depends, according to (3.16, 18–20), on the rates of the processes of incorporation at the steps. These processes explain the difference in the growth rate of different faces on the linear sections of $R(\sigma)$. During solution growth the real step sources ensure the existence, on the faces, of vicinal hillocks with slopes $p \simeq 10^{-2}$–10^{-1}, which differ for different faces. Together with the above estimate $\beta_{st} \simeq 10^{-1}$ cm/s, expression (3.20) gives the typical value $b\Omega C_0 \simeq (10^{-3}$–$10^{-2})\Omega C_0 \simeq 10^{-4}$–$10^{-3}$ cm/s, which is in qualitative agreement with experiment [3.21, 27, 28].

The kinetic coefficient b changes with temperature in conformity with (3.5), often obeying the Arrhenius law.

The work of nucleation depends on the step free energy and therefore differs from face to face. The dislocation activity is different on different faces for the same reason. Finally, it is significant that at a given concentration in the medium, the adatom density is also different on different faces (Sect. 1.3), and therefore the step velocities and normal growth rates are different for different faces as well. At high supersaturations, when the density of the steps and kinks is already high enough [$R(\Delta\mu)$ is linear], the decisive contribution to the anisotropy must be made by the different adatom densities, and also by the incorporation processes at kinks on the steps. Therefore the normal growth rates from chemically simple melts and from the gas phase at high supercoolings are only slightly anisotropic. In condensation from vapors or molecular beams this means that the condensation coefficient[3] is close to unity for surfaces of any orientation.

Let us now return to the question raised at the end of Sect. 3.1 concerning the anisotropy of the normal growth rate over the entire range of face orientations, and not merely for separate singular orientations. Equation (3.6) leads to the conclusion that the growth rate $R = 0$ for $p = p^0 = 0$ and for $p = p^i$, This is valid only if p implies the true local step density in the vicinity of the surface point; it is the growth rate at this point which interests us. However, as we have seen above, a face intersected by screw dislocations may, on the average, have an orientation $p = 0$ but exhibit a finite growth rate corresponding to the local step density on the slopes of the vicinal growth hillocks. If the orientation for these hillocks is $p = p_0$, then expression (3.6) is clearly applicable only at $p > p_0$, while at $p < p_0$ we have $R = |p_0|\, v$. To put it differently, if the average slope $p = 0$, then R is already nonzero. For growth at an elementary dislocation (Sect. 3.3.2), $p_0 = h\Delta\mu/19\Omega\alpha$. This circumstance is reflected qualitatively in the function of the singular minima shown by the dashed line in Fig. 3.7a. Since usually $p_0 \lesssim 10^{-2}$, curve $R(p)$ will be perceived, as a whole, as a curve with singular minima, at which, however, $R \neq 0$.

The simplest method for rough experimental estimation of the anisotropy is to prepare differently oriented cuts of a crystal and grow them. In this way the growth rate anisotropy of quartz in a hydrothermal solution (Figs. 3.7c, d, and Sect. 9.3) and of silicon in the chloride gas phase epitaxy process (Figs.

---

[3] Recall that the condensation coefficient is the fraction of the particle flux remaining on the interface forever, with respect to the amount $(P - P_0)/\sqrt{2\pi mkT}$ of such particles in the case of their ideal absorption by the interface.

3.7b and Sect. 8.3) was studied. The first of the dependences is shown in polar coordinates R, θ, and the second in rectangular coordinates R, $\theta(\theta = \arctan p)$. This method gives only a general picture of anisotropy and is unsuitable for investigating orientations only slightly deflected ($|p| \lesssim |p_0|$) from the singular orientation.

The normal rates R and V given by (3.7) are also different from zero for surfaces which have, on the average, a singular orientation, but are covered with steps from numerous two-dimensional nuclei (kinetically rough surfaces, Sect. 3.3.1). In this case the sharp minima must be rounded out (Fig. 3.7a, dashed line). Finally, for atomically rough surfaces, an isotropic growth rate must be expected (Fig. 3.7a, dot-dashed line), as was indicated at the end of Sect. 3.1.

The anisotropy of the growth rate for slopes $|p| \lesssim |p_0|$ can be found if it is known how the slope of the vicinal hillock p and the face growth rate R, determined by this hillock, depend on the supersaturation. It is these two dependences that define function $R(p)$ [of $b(p)$] in the parametric form. The supersaturation serves as the parameter [3.29]. Another phenomenological procedure [3.30] and shock-wave analysis technique [3.31] were applied to determine silicon growth rate anisotropy in the chemical vapor deposition process at $|p| \lesssim 10^{-2}$.

3.3.4 Experimental Data on Layer Sources

Let us now discuss some quantitative experiments demonstrating growth due to two-dimensional nucleation and dislocations. Investigation of the growth kinetics due to two-dimensional nucleation requires the use of dislocation-free crystals, which is possible, as a rule, in high-purity systems and under specially controlled conditions. Series of quantitative investigations have been conducted by *Kaishev*, *Budevski*, *Bostanov*, and co-workers [3.2, 16, 17, 32a, b] for crystallization of silver in the course of electrolysis of a 6n aqueous solution of $AgNO_3$. A diagram of the electrolytic glass cell is presented in Fig. 3.18. The crystal was obtained during the growing of the seed into a capillary having a diameter ~ 100 μm. The crystal end face approaching the capillary outlet was

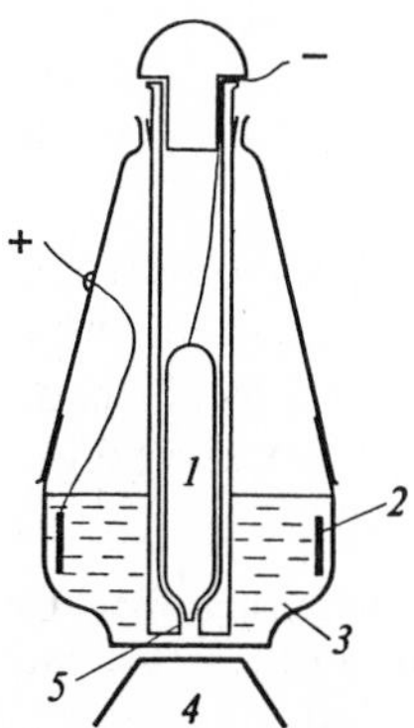

Fig.3.18. Diagram of an electrolytic cell for investigating the growth of silver crystals. (*1*) Crystal; (*2*) ring-shaped anode; (*3*) electrolyte solution; (*4*) microscope; (*5*) capillary

observed in an interference microscope with a phase-contrast device (Sect. 3.4). The seed was oriented by the ⟨100⟩ or ⟨111⟩ axis along the capillary, so that it was possible to investigate the growth of cube or octahedron faces. Dislocations in the fcc silver lattice usually have ⟨110⟩ direction, so that if they do penetrate from the seed into the crystal growing in the capillary they emerge sooner or later at the lateral surface of the crystal and cease to intersect the growing face. Therefore by preventing the production of new dislocations in the course of growth one can obtain cube or octahedron faces free of dislocation outcrop points. When such faces were obtained, it was found that the cube faces did not grow if the overvoltage applied to the crystal was lower than the critical overvoltage, which varied between 8 and 12 mV (this corresponds to a supersaturation of 25%–38%). The formation of two-dimensional nuclei was studied in several experimental runs. In one of them the current through the cell with the disclocation-free crystal was maintained constant. Here it turned out that the overvoltage across the crystal pulsed spontaneously: after the current was switched on, it rapidly reached a value of ~10 mV, at which one or several two-dimensional nuclei were formed on the crystal face. The nucleus grew, the step bounding it extended, and the surface resistance of the crystal fell. The overvoltage across the crystal also fell accordingly. When the layer had covered the entire face and the step had disappeared, the resistance and hence the overvoltage increased again, and the process was repeated. The higher the assigned current through the cell, the shorter the oscillation period was. Calculation of the electric charge passing through the crystal within one period showed that it corresponded precisely to the deposition of one monolayer on the crystal face. In another run, the voltage across the cell, rather than the current, was kept constant. In these experiments the overvoltage was maintained at a level of 6 mV, which was unquestionably below the critical value for spontaneous nucleation on a cube face. Against the background of this overvoltage, 1μs-pulses were fed to the electrolytic cell from a special generator. During these

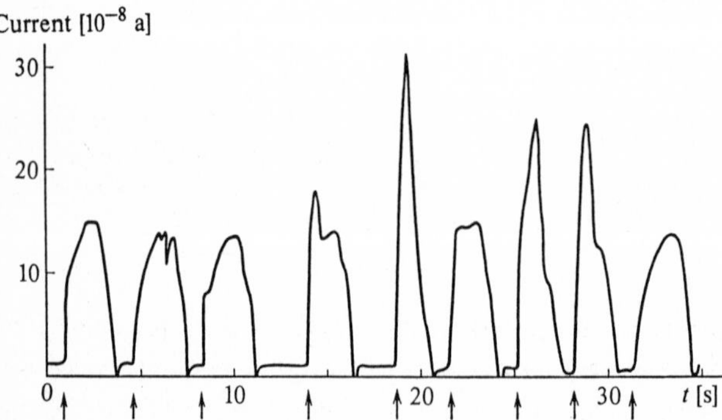

Fig.3.19. Current induced by the progress of individual elementary layers on the (100) face of a silver crystal. The layer nuclei appeared under the influence of short pulses applied at the moments marked by *arrows* [3.32]

pulses the overvoltage reached ~13 mV. The moments of pulse delivery are marked by the arrows in Fig. 3.19, where the current is plotted upwards along the y axis, and the time along the x axis. Each pulse produced a nucleus of the step, whose propagation along the face determined the subsequent current through the cell. When the layer had covered the whole face, the current stopped until the next pulse was delivered. The surges of current with a duration of ~4 s thus obtained are shown in Fig. 3.19. The area under each current surge on this curve is the total quantity of electricity which has passed through the cell during the surge. By dividing it by the Faraday number it is easy to obtain the number of electrons which have passed through the cell and, consequently, the number of univalent silver ions discharged at the step and deposited upon the surface as a new lattice layer. This calculation showed that the layer thickness with the capillary diameter used is precisely 2.04 Å, i.e., the height of one monatomic layer (the parameter of the unit cell of Ag along [100] is equal to 4.08 Å). Thus, growth proceeds by two-dimensional nucleations.

If nucleation is initiated by short pulses of different duration, the fraction of cases where pulse delivery results in nucleation can be found. By thus maintaining the probability of nucleation at the level of 50%, it was possible to obtain the dependence of duration τ of the necessary pulses on overvoltage η. As can be seen from Fig. 3.20. the dependence of $\log \tau$ on $1/\eta$ is linear in accordance with theory [Sect. 2.2, (2.22), where $\Delta\mu \sim \eta$]. The specific free linear energy of the elementary step $\langle 110 \rangle$, determined from the slope of this straight line, was found to be 2.1×10^{-6} erg/cm, which corresponds to an energy of 370 erg/cm^2 for a singular face (100) and 320 erg/cm^2 for the (111) face [3.21].

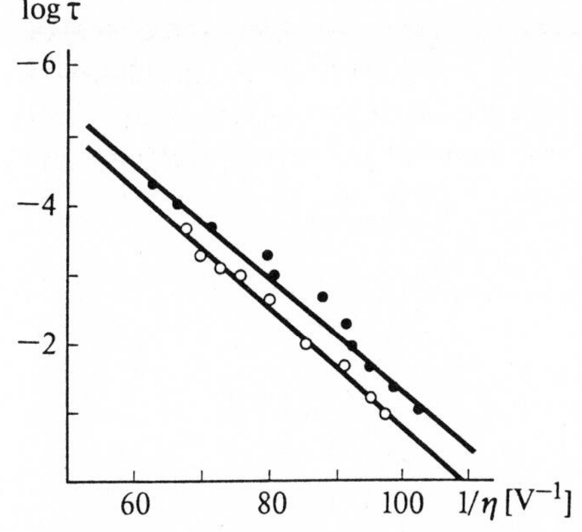

Fig.3.20. Expectation time τ for the appearance of a nucleus with a probability of 50% as a function of the overvoltage in the nuclei-producing impuls [3.32]

When screw dislocations were present in the crystal under investigation, pyramidal growth hillocks appeared on the surface, and the dependence of the overvoltage upon the growth rate underwent a cardinal change: the threshold overvoltage disappeared, while the dependence of the growth rate $R(\eta)$ became parabolic over the entire range of overvoltages used, as should indeed be the case for growth on dislocations. Again in accordance with theory, the slope of the vicinal hillocks covering the growing face in this case increased linearly with the overvoltage.

Measurements of the current caused by the motion of the step from a single two-dimensional nucleus in the capillary of a rectangular section indicated that the steps were polygonal and oriented on the cube face along $\langle 110 \rangle$, and also helped to find the value of the kinetic coefficient at the step and exchange current, which were mentioned in Sect. 3.1.

One of the most important advantages of the procedure used in the above-described experiments on electrocrystallization is the possibility of measuring the amount of crystallized substance accurately and without inertia. Unfortunately, in ordinary growth from vapors, solution or melt, there is as yet no similar method, and measurements of growth rates are made optically. Therefore no direct information on the elementary acts of nucleation and growth has been obtained as yet. Nevertheless, the preparation of highly perfect, dislocation-free crystals makes it possible to measure the threshould supercoolings for the growth of gallium crystal faces from the melt. The gallium single crystal investigated was grown by *Alfintsev* and *Ovsienko* [3.33 a, b] in a flat glass cuvette. The crystal, which was 1–2 mm thick, had the shape of a plate with an end face formed by the (001) or (111) faces. This shape promoted the emergence of dislocations at the lateral surfaces of the crystal that were in contact with the glass, and also the growth of dislocation-free faces on the rises. The growth rate dependences of the (001) and (111) faces are depicted in Fig. 3.21a. The threshold supercoolings typical for these faces disappear and the rate increases abruptly if the growing face is touched lightly with a glass rod. The dependence of the growth rate on the supercooling for both faces thus deformed is given by the points in the left-hand part of Fig. 3.21a. This dependence may be estimated as $R \simeq 2.5 \times 10^{-3} \Delta T$ cm/s if ΔT is expressed in degrees. The curves for undeformed (001) and (111) faces straighten out in coordinates $\ln R - 1/(T\Delta T)$, which also supports the theory of growth by two-dimensional nucleation. The threshold supercooling of $\simeq 1$K was also determined for dislocation-free silicon.

Critical supersaturation (or supercooling equivalent to it) is also necessary to start the growth of a number of crystals from aqueous solutions. For instance, whiskers (see p. 222) of NaCl grow only at supersaturations $\sigma \gtrsim 0.3\%$ (Fig. 3.21b). Critical supercoolings necessary for the growth of various faces of quartz, corundum, and sodalite from a hydrothermal solution are of the order of several degrees [3.35, 36]. Such supersaturations may be evidence of growth by two-dimensional nucleation under actual growing conditions; moreover, recent x-ray topography investigations do not exclude the possibility that some crystals of SiO_2 (quartz) [3.37, 3.8], KH_2PO_4(KDP) [3.39] and NaCl [3.26] had no dislocations during growth.

At the same time a dependence of the type of (3.31) at $\sigma > \sigma_0$ does not always correspond to the complete absence of growth at $\sigma < \sigma_0$. For instance, the faces of the potash alum octahedron also grow at $\sigma < \sigma_0 \simeq 40\%$, but at a low rate.

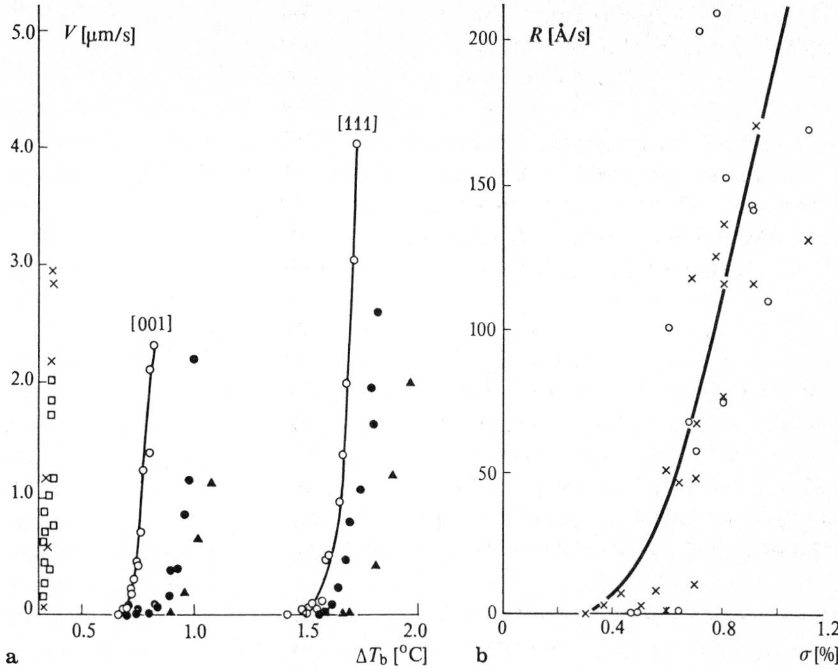

Fig.3.21a,b. Dependence of the growth rate on supercooling (**a**) and supersaturation (**b**). (**a**) Dependence $V(\Delta T)$ for $\{001\}$ and $\{111\}$ faces of gallium and gallium alloy crystals: ○ – pure gallium, ● = Ga + 0.01 wt. % In, ▲ = Ga + 0.1 wt. % In, × = Ga + 0.01 wt. % Ag, □ = deformed crystals of pure gallium [3.33b]; (**b**) dependence of the growth rate of NaCl whiskers from an aqueous solution on the supersaturation [3.26]. The data were obtained for crystals grown initially in air by the Gyulai method [3.34]. The equality of the mean rates of growth along the axis of the whisker (*circles*) and perpendicular to it (*crosses*) attests to the identity of nucleation processes on all faces of the whisker. The *continuous curve* shows the calculation for the multiple-nucleation mechanism [see (3.23)] with specific energy 5.9×10^{-15} erg/ion at the step rise (which is equivalent to ~ 30 erg/cm² if 2×10^{-16} cm² is taken to be the surface per ion) and with potential barrier $E = 18.7$ kcal/mol for the attachment of ions

In such cases the dependence (3.31) in the range of $\sigma < \sigma_0$ can also be due to dislocation sources. *In situ* x-ray topography was successfully used recently to observe directly the growth of dipyramidal $\langle 110 \rangle$ faces of ADP crystals which definitely had no dislocation outcrops. The growth rate of these crystals at supercooling ~ 1.5 K ($\sigma = 2.7\%$) was about 30–100 times lower than that of the crystals with dislocations [3.40].

The existence of a critical supercooling below which the crystal does not grow at all or grows very slowly may also be associated with adsorption of an impurity (Sect. 4.1.) or with the formation of a partly ordered solution layer — or "protector" — preventing the growth. Such a layer (if it actually exists) has the form of a quasicrystalline "epitaxial film" on the surface which, when in the state of maximum ordering, offers a considerable or even infinitely large resistance to face growth — both to step motion and generation. If, however (at high supersaturations) the growth rate of the face exceeds that of the formation of the "coat of mail," the film will have no time to "crystallize" on the surface and will offer only moderate resistance to growth.

The presence of even small amounts of impurity which are strongly adsorbed on the crystal surface and poorly soluble in the crystal decreases its growth rate and results in a stoppage of the growth front if surface coverage with impurity reaches its critical value (Sect. 4.1.3). A nucleus appearing above this impurity adsorbtion layer may give rise to a new layer of crystallizing substance possessing a new fresh crystal–liquid interface. In spreading over the poisoned surface this layer buries the impurity and opens up a new period of growth. This period may also be terminated sooner or later by the next population of adsorbed impurity atoms, and so on. For dislocation-free crystals, the continuation or termination of growth may be the result of a competition between two-dimensional nucleation events which are random in time, and impurity adsorption above each new crystaline layer. Indeed, if the period between two successive nucleation events at a given active surface area turns out to be longer than the time needed to poison this fresh interface with impurity, no further nucleation will be possible there. This mechanism may be responsible for the experimentally observed growth rate fluctuations on dipyramidal faces of dislocation-free ADP crystals [3.40b].

Fig.3.22. Systems of elementary concentric steps on a (100) NaCl surface growing from a NaCl molecular beam. The surface temperature was 385°C, the effective supersaturation $\ln P/P_0 = 1.7$

The activity of edge dislocations during growth is still a subject of discussion, despite the fact that the first unpublished observations of concentric steps on the (100) NaCl surface were reported by *Bethge* and *Keller* during meetings held in 1965 and 1974 (see also [3.23g]). Later, the hillocks formed by elementary steps were observed by *Shimbo* et al. [3.41a] on (111) Si surfaces grown by chemical vapor deposition. They claimed that the hillocks were not associated with dislocations. It was found [3.41b–d] that during NaCl molecular-beam deposition on the NaCl cleavage face, systems of concentric steps (2.81 Å high) gradually replace the spiral steps and cover the whole surface (Fig. 3.22). Concentric steps were also observed on the (100) surface of GaAs grown by liquid phase epitaxy (LPE) [3.41e]. Finally, an edge dislocation was revealed by x-ray topography to emerge at the apex of one of the hillocks on LPE-grown (100)GaAs [3.41f]. The mechanism of such step generation is not yet clear. A possible explanation [3.41 c, d] is that the surface supersaturation reaches a maximum in the middle of the flat surface surrounded by the (moving) first central step (Sect. 3.3.4).

3.4 Morphology of a Surface Growing Layerwise

Investigations into the morphology of the growth surface suggest important conclusions about the growth mechanism. They become particularly valuable if they are accompanied by a study of growth kinetics. Unfortunately, we are often familiar either with only the morphology or only the kinetics.

3.4.1 Optical Methods Used to Investigate Growth Processes and Surfaces

The principal methods for investigating the morphology of the surface of a growing or grown crystal are optical and electron microscopy. The atomic structure and composition of the surface are studied by the methods of low-energy electron diffraction, mass spectrometry (including secondary ion mass spectrometry), Auger electron spectroscopy, ultraviolet and x-ray electron spectroscopy, and field emission microscopy. These methods are described in [2.48,49; 3.42–46]. The elementary technique of electron microscopy is described in [3.47–51]. Here, we shall only dwell briefly on some principles of optical microscopy and ellipsometry.

In addition to the general manuals [3.52], the reader will find information on classical microscopy (including polarization microscopy) in books by *Tatarsky* [3.53] and *Rinne* and *Berek* [3.54].

Polarization-optical Methods. We begin with the utilization of transmitted polarized light. This method is efficient for studying the growth of thin birefringent crystals. A beam of polarized light transmitted through such a crystal splits up into two beams polarized in mutually perpendicular directions. In a uniaxial crystal these beams are called ordinary and extraordinary. The refractive indices in the crystal are, generally speaking, different for the two beams.

Therefore on emergence of the beams from the crystal plate under investigation, a phase difference between them accumulates due to their optical path difference, i.e., the difference between the refractive indices multiplied by the plate thickness (more precisely, by the length of the beam path in the plate). The two beams combine and produce an elliptically polarized light, with the ellipse orientation and shape depending on the phase difference between the beams, i.e., on the plate thickness. By passing this elliptically polarized beam through an analyzer, one obtains the image of the crystal. If the light is monochromatic, the crystal image will consist of dark and light areas. The light intensity in the various areas of the image depends on the optical path difference of the two indicated beams in the crystal, i.e., on the thickness of its different parts. When a white light is used, interference colors appear. Corresponding to each color is a path difference, which is found in the table of interference colors. If one divides the optical path difference thus found by the known difference between the refractive indices for the beams of the two polarizations, one obtains the crystal thickness at a place with a given color. Accordingly, the difference between the thicknesses of neighboring parts yields the height of the steps separating them. This method helps to measure, on paratoluidine, the heights of steps from 20 to 150 lattice spacings along the c axis (this spacing is equal to 23 Å), i.e., the differences in thickness from ~450 Å to ~3500 Å.

The low step heights on birefringent crystals can also be measured by interposing a compensator in the path of the beam between crystal and analyzer (or polarizer). The compensator is a crystal plate which introduces an additional phase difference between beams with mutually perpendicular polarizations. A turn of the compensator changes the interference colors of the crystal under investigation or the brightness of its image, thus making it possible to measure the thickness of different parts of the crystal separated by steps and in this way determine the step heights. For instance, a turn of a Senarmont compensator (a mica plate which introduces a path difference of a quarter wave-length) by 1° enables one to determine an optical path difference of 30Å by using the monochromatic light of a mercury lamp (the green mercury line with a wavelength of 546 nm = 0.546 μm). For paratoluidine, this path difference corresponds to a change in thickness of 136 Å, i.e., of six lattice spacings [3.5].

The use of a monochromatic light reduces determining the stepped structure of the surface to measuring the intensity. This is achieved, for example, with the aid of a photoelectron multiplier [3.55] or a photocell in combination with modulation of the incident light [3.56], which improves the sensitivity of the method.

Interference Methods. The technique based on the interference colors of thin films in reflected light helps to define the heights of steps down to several hundred angstrom units on the surfaces of thin transparent crystal plates. The method is based on the interference between a beam reflected from the film surface and a beam propagated through the film and back-reflected from the second boundary of the film (along the path of the incident beam). The optical path difference of the interfering beams is thus equal the doubled thickness of the plate (the length of the beam path in it) times its average refractive index. With the same crystal thickness this path difference is about one order of magnitude higher than in the birefringence method described. Accordingly, the thin-plate colors method is appreciably more sensitive, and enables one to measure step heights up to 10 lattice spacings (230 Å) on paratoluidine crystals. Areas not distinguishable by color are separated by thinner steps which manifest themselves as thin lines due to the diffraction of the light on them. If we find an area with an echelon having a sufficient number of identical thin steps, the total difference in thickness between the areas separated by the echelon can be measured by the change in color. By dividing the difference in thickness by the number of steps, we readily obtain the height of steps of one or two lattice spacings on paratoluidine (23 or 46 Å) and β-methylnaphthalene (18 or 36 Å) crystals. It should be emphasized that the thin-film colors method is applicable not only to birefringent, but also to all other transparent materials.

When investigating the growing crystal surface it was possible, in principle, to use the method of thin-film colors in transmitted light as well as in nonparallel beams with localization of interference bands on the surface of the crystal [3.52b]

Laser light sources open up new possibilities in interference methods, in particular the use of holography to observe the surface relief [3.57, 58], the distribution of concentration and flow around a growing or dissolving crystal [3.59a] and to measure interferometrically the growth rate [3.59b–d].

Interference methods include interference microscopy, which will be discussed below, as well as methods of multiple-beam interference in the clearance between the reflecting surface under investigation and the overlying translucent plate.

We shall now turn to the operating principles of phase-contrast and interference microscopes [3.60].

The Phase-contrast Microscope. This microscope makes it possible to observe objects which introduce a small phase difference Δ into the light passing through them or reflected by them. Among such objects are nonuniformities on the surface relief and, with transparent crystals, small nonuniformities in the refractive index caused, for instance, by nonuniform impurity distribution or by stresses. A step on the surface of a transparent crystal causes a shift in the phase of the transmitted light by a value Δ equal to the step height times the refractive index of the crystal.

With the phase-contrast principle, beams which have not touched the object (background beams) are shifted in phase by $\frac{\pi}{2}$ by means of a special device, i.e. by a quarter wavelength relative to the beams diffracted by the object. The diffracted beams and background beams interfere with each other in the image plane of the microscope. Thus in the region of the object image the coherent electromagnetic waves combine, with one of them being shifted in phase by $\pm\frac{\pi}{2}+\Delta$

relative to the other. We take advantage of the smallness of Δ and expand the diffracted wave in the powers of this small parameter. By squaring the expression obtained and integrating the result with respect to time, we obtain the observed light intensity at those points of the image plane where the beams from the object converge behind the microscope objective lens. This intensity is, as is easily seen, equal to the *square of the sum* of the amplitudes of the diffracted and background beams. In the other areas of the image plane it is equal to the square of the background amplitude. Consequently, the contrast of the image, which is equal to the ratio of the difference between the intensities of the light from the object and the background to that of the background, is proportional to Δ. The phase shift of $\frac{\pi}{2}$ is carried out in order to increase the contrast, since without the shift the intensity in the object image area is equal to the *sum of the squares* of the amplitudes of the background and diffracted vibrations. Therefore the contrast is proportional to Δ^2, i.e., it is much weaker.

The separation of the light diffracted by a transparent object and the background light, and the shift in the phase of the latter is achieved as follows. A parallel beam is passed through the object and then the objective lens of the microscope, and the rays are collected in its focus. The rays diffracted as a consequence of the non-uniformity of the refractive index in the sample near a step, for instance, constitute, according to the Huygens principle, a beam diverging radially from this nonuniformity. Therefore the microscope lens collects it into the focus in the image plane of the microscope, rather than in the focal plane of the lens, where the rays of the parallel beam converge. Let us place, in the focal plane of the lens, a so-called phase plate consisting of transparent regions with two distinct refractive indices such that the difference between the indices multiplied by the plate thickness equals a quarter wavelength. One region, small in area, coincides with the zone where the rays of the parallel background beam collect behind the lens, while the other occupies the remaining part of the plate. The shape of the zone is defined by the diaphragm, which cuts out the parallel beam falling on the lens. With an annular diaphragm this is a narrow ring; with a circular one, it is a small disc; and with a slit diaphragm, a narrow strip (in the last case the phase contrast appears for one azimuthal orientation of the specimen).

Thus the light of the parallel beam passes only through a narrow zone of the phase plate, while that of the diffracted beam passes through the entire plate. As a result, behind the plate the vibration phase in one beam is shifted by $\frac{\pi}{2}$ with respect to that of the other, which is precisely what is required to obtain the phase contrast in the image. A Nomarsky-type phase plate, which is intended for reflection rather than transmission, is very convenient. It is a glass plate coated with silver everywhere except for a narrow zone, from which the focused rays of the parallel beam are reflected. Diffracted rays are reflected from the whole surface. The phase difference is selected by varying the thickness of the reflecting layer.

A phase contrast can also be obtained from nontransparent specimens.

Then the work is done in reflected light, and the phase difference is introduced by the nonuniformities in the relief. For example, the difference introduced by a step on the surface of a metal crystal in solution is equal to the doubled height multiplied by the refractive index of the solution.

Interference Microscope. This microscope operates as follows: A light beam from a source is split by a translucent plate. One of the two resulting beams traverses the path of the rays in the ordinary microscope, but bypasses the object. The other beam, which is coherent with the first, follows a very similar route, but either passes through the object or is reflected from it. The object shifts the phase in the second beam with respect to the first, as in a conventional interferometer. The two beams interfere with one another in the image plane. Suppose that in the absence of an object, the light of both beams arrives at the image plane in counterphase, which can always be achieved by adjusting the interferometer arms. Then the field of vision (without the object) will be dark. By introducing an object, we change the conditions of interference in the area of its image and see the object as a bright image against a dark field. With a different adjustment of the arms one can obtain a bright-field image. As in the phase microscope, the contrast here is proportional to the path difference Δ rather than to Δ^2.

In some types of interference microscopes the two coherent beams traverse one and the same route, but in different directions (with the exception, of course, of the last segment before the image plane). Each of the beams gives its own image of the object, and one image can be shifted relative to the other by adjusting the mirrors. Then the rays of one beam which have passed through the object (or have been reflected from it) will interfere with those of the other beam which have not touched the object. As a result we obtain a double image of the object under study.

If, in the absence of an object, the image plane does not coincide with the planes of the phase constancy of the interfering coherent beams, but is slightly turned with respect to them, the background intensity will not be uniform, but instead will consist of parallel light and dark interference bands. Following the introduction of the object, the bands at the sites of its image are shifted and/or curved, and the shift of the picture by one interband spacing corresponds to the object-introduced path difference of one wavelength of the light used.

Multiple-beam Interference Microscopy. In these microscopes the initial beam splits not into two beams, as in the above-described double-beam system, but into a multitude of beams. It is based on microscopic observations of the picture obtained in a Fabry-Perot interferometer. This instrument consists, as is well known, of a system of two glass plates facing one another with parallel translucent (silver-plated) surfaces. The beam entering the space between the plates through one of them is multiply reflected from the translucent surfaces, splitting up each time into a reflected and a transmitted beam. All the beams leaving the interferometer through the second plate interfere among themselves, and the light intensity at the outlet is defined by the path difference between the two beams resulting from two consecutive reflections. With a normal incidence of the initial parallel beam, this path difference is equal to the doubled width of the gap multiplied by the refractive index of the substance filling it. If there were only two interfering beams, then two harmonic vibrations would combine. As a result, the intensity of the light passing through the interferometer would be a smooth (and also harmonic) function of the path difference of the rays. In actuality, a multitude of rays interfere with each other, and therefore the resulting intensity is a very steep function of the path difference; it differs appreciably from zero only for narrow ranges of the path difference, which frame the values multiple of the incident monochromatic light wavelength. This sharp dependence is the main peculiarity of multiple-beam interference. It ensures a high sensitivity of the instrument to the introduction of an object with a refractive index different from that of the medium filling the gap between the interferometer plates. Suppose, for instance, that the path difference in the interferometer is chosen at the edge of the range corresponding to the intensity surge. Then even a small path difference introduced by the object will make it visible against a dark background [3.23f].

If the interferometer plates are not strictly parallel, the picture in the image plane of the microscope will consist of very narrow light bands against a dark background. At the site of the object image the bands will be shifted and curved, and their narrowness makes it possible to observe path differences 20–30 times smaller than the wavelength, i.e., down to 150–200 Å.

A multiple-beam interference pattern can also be observed in reflected light between one of the translucent plates and a good reflecting surface (which is sometimes also silver-plated). As in the case of Newton's rings, equal-intensity lines correspond to an equal path difference of two beams — one incident and one reflected from the crystal surface — i.e., to an equal gap width. If there is a large macrostep on the surface, the interference line shifts when crossing it, and experiences a break (Fig. 3.15b). The width of the interference lines in this simple method is comparable to the distance between them, which, in turn, does not usually exceed 3 μm. These factors limit the applicability of the method to investigation of the general surface relief, e.g., the slopes of large vicinal hillocks.

The design and operation of interferometers are treated in *Tolansky*'s book [3.61].

A modification of the double-beam interference microscope is the *polarization microscope*, in which two coherent beams are obtained after a light beam passes through a birefringent crystal plate, where the beam splits up into an ordinary and an extraordinary ray. The object images produced by these two rays are slightly shifted relative to each other, either in directions lying in the image plane or along a normal to it, and the rays interfere with one another.

As a result, illumination with a white light gives a brightly colored picture upon which objects introducing an additional phase difference for one reason or another are seen in the form of areas of different colors.

Ellipsometry. As is well known, elliptically polarized light can be represented as the sum of two linearly polarized coherent vibrations. With time, the terminus of the vector of the electric (and magnetic) field of an elliptic wave rotates with a frequency equal to that of light, in a plane perpendicular to the direction of beam propagation. The ratio of the length of one ellipse axis to that of the other, as well as the orientation of the ellipse are determined by two independent parameters, e.g., by the ratio of the vector lengths of the electric (magnetic) field and the phase difference in the linearly polarized vibrations making up the given elliptical vibration. In describing the reflection it is convenient to choose, as one of these vibrations, that in which the vector of the electric field is perpendicular to the plane of incidence and, hence, parallel to the reflecting plane. As the other vibration, that in which the electric vector lies in the plane of incidence may conveniently be chosen. The former is called the s wave, and the latter the p wave. The angle between the direction of the electric vector of the p wave and the reflecting surface is equal to the angle of incidence of the beam. Because of the different orientation of the electric (and magnetic) vectors relative to the reflecting surface, the amplitudes and phases of the two linearly polarized waves change differently after reflection. Accordingly, the shape and orientation of the ellipse in the reflected wave change as compared with the incident wave. This change is extremely sensitive to the state of the surface, i.e., to the presence on it of foreign substances or phases (even in amounts of the order of one monolayer), systematic nonuniformities of composition and surface relief, and other features changing the conditions of reflection from the surface under investigation. By measuring the difference in the parameters characterizing the ellipticity before and after reflection one gets an idea of the state of the surface. It is customary to measure the phase shift between the s and p waves introduced by reflection, as well as the ratio of their reflection coefficients.

In the ellipsometer, the light from the source passes through a polarizer, reflects from the specimen, passes through an analyzer, and enters a photodetector. A compensator, which changes the phases of the s and p waves and, hence, the parameters of the polarization ellipse, is interposed in the path of the beam between the object and polarizer or analyzer. The theory and practice of ellipsometry has been described in recent books [3.62a,b] and a review [3.63]. The use of ellipsometry helped to establish, for example, the substantial change in the amount of chlorine adsorbed by the (100) face of GaAs occurring with a change in the partial pressure of $AsCl_3$ in chemical precipitation of GaAs from the gas phase [3.64a]. Layer growth of the (0001) Cd surface from the vapor was ellipsometrically studied in [3.64b].

3.4.2 Steps, Vicinal Hillocks, and the Formation of Dislocations During Vapor Growth

Figures 2.9a,b show electron micrographs of a gold-decorated surface of NaCl. The surface to be decorated for investigation purposes is vacuum sputtered for several seconds by a molecular beam emitted by a heated drop of gold (or silver, platinum, or chromium). The temperature of the NaCl surface must not excede $\simeq$ 300°C. After the gold sputtering, a continuous graphite film is sprayed onto the surface (from an electric arc), also in the vacuum. The graphite film with the gold particles retained by it is separated from the crystal (usually by dissolving the latter) and inspected in an electron microscope. In the course of sputtering, the gold forms small crystals, primarily at those sites on the surface where the nucleation barrier is reduced, and thus decorate these nonuniformities (Sect. 2.2.2). As can be seen from Figs. 2.9a, b, the steps are marked by chains of closely spaced gold particles of the order of hundreds of angstrom units in size. An analysis of the interaction of the observed steps with the glide bands makes it possible to measure the height of much thinner steps [3.65]. The minimum height was found to be 2.81 Å, i.e., half of the lattice spacing. In Fig. 2.9a the round-shaped steps (1, 2, 4) have such a height. The polygonal spiral (3) in Fig. 2.9a is built up of steps with a height a = 5.62 Å. These may break up into two steps with heights $a/2$. The islands formed as a result of the growth of evaporation nuclei are seen in Fig. 2.9b.

Prolonged generation of growth or evaporation layers at some points of the surface results in the appearance of growth hillocks or pits of evaporation (dissolution, etching), usually with gentle slopes (0.1°–10°) at these points (Figs. 2.9a, 3.23, 24). It was found that hillocks (or pits) consisting of concentric steps which evidently result from successive acts of a two-dimensional nucleation on their tops (or bottoms) ensure higher rates of growth (or evaporation) than those consisting of spiral steps [3.41b–d]. Consequently, under the conditions described, the mechanism of successive nucleation is more effective than the spiral-dislocation mechanism. This is consistent with computer simulation in which the generation of steps was allowed by both the dislocations and two-dimensional nucleation: the spiral provides a higher growth rate at low supersaturations, whereas two-dimensional nuclei are more effective at high supersaturations [1.26c]. The same conclusion may also be derived from the analytical theory of self-consistent nucleation [3.25, 41c,d].

Figure 3.25 shows frames from a cinefilm of the growth of paratoluidine crystals in the setup depicted in Fig. 3.6. It depicts the successive stages in the formation of a dislocation with a large Burgers vector in the reentrant angle of a skeleton plate-like crystal growing from the gas phase on a glass substrate.

A schematic picture of this phenomenon is presented in Fig. 3.26. Because of the inconstancy of the temperature along the crystal or for some other reason, one of the branches of the skeleton is shifted relative to the other in a direction perpendicular to the plane of the plate (the vector in Fig. 3.26a). The localiza-

Fig.3.23. Vicinal growth hillocks on ZnO crystal formed in a chemical reaction in the gas phase [3.66]

Fig.3.24. Evaporation pits on the surface of a NaCl crystal; photomicrogroph taken in a phase-contrast microscope. The pit slopes form angles of $\sim 10^{-2}$ with the (100) face. Courtesy of M.O. Kliya and Yu. A. Gelman

tion of the stresses and impurities at the vertex of the reentrant angle and the insufficient inflow of the crystallizing substance in this region retard the growth in the vicinity of the vertex and give rise to a void (Figs. 3.25a, 3.26b). When the walls of the void draw together, forming a hollow channel (or one filled with an impurity or mother medium), the atomic nets do not coincide strictly, because of the relative shift of the skeleton branches, and a screw dislocation appears

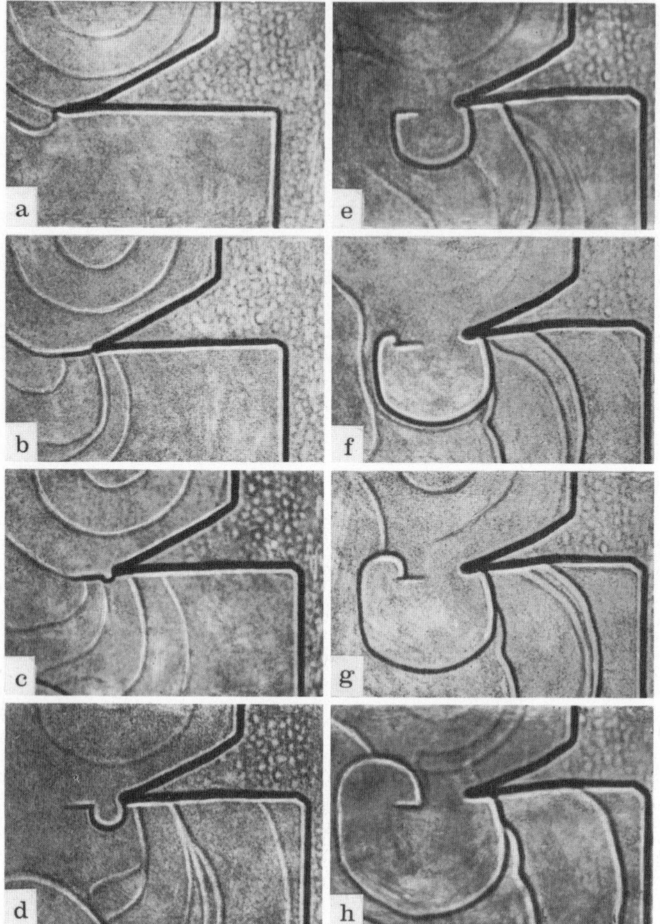

Fig.3.25a–h. Consecutive stages in the formation of a screw dislocation in a paratoluidine crystalline platelet [3.68]

at the meeting of the walls (Figs. 3.25b; 3.26c, d). The corresponding step, which is short at first (Figs. 3.25b, c; 3.26d), elongates in the course of growth within the reentrant angle, and on reaching a length commensurate with the radius of the critical nucleus, begins to grow and acquire a spiral shape (Figs. 3.25d–h; 3.26e, f). By shifting in the course of growth, the above-mentioned reentrant angle can naturally form more than one dislocation. This is evidenced by several dislocations which are usually arranged along the trajectory of the vertex of the reentrant angle. These dislocations have different signs, which attests to the random nature of the deformations in the course of growth within the reentrant angle. A similar process of dislocation formation takes place in the initial stages of growth from solution. Dislocations in plate-like crystals also

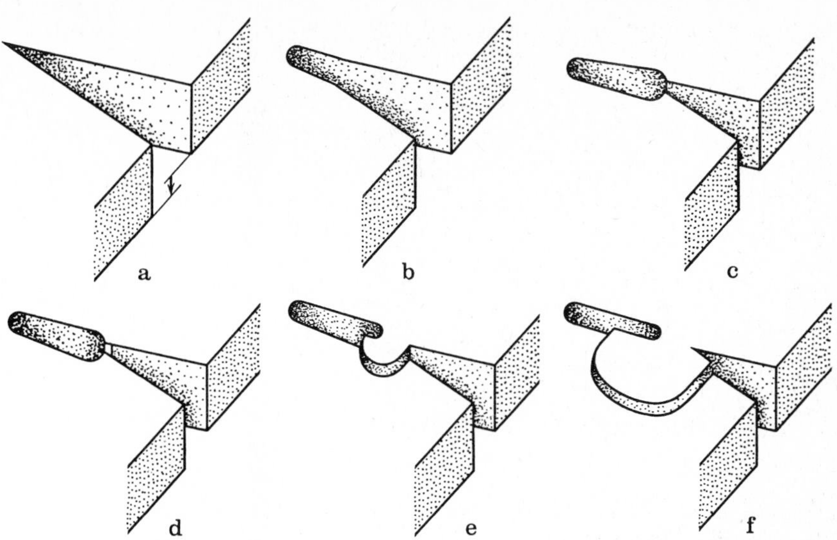

Fig.3.26a–f. Formation of a screw dislocation in the reentrant angle of a dendritic crystal

form when foreign particles are trapped by these crystals [3.69]. The formation of dislocations is also discussed in Sect. 6.1.

3.4.3 Kinematic Waves and Macrosteps

A kinematic wave of steps is a region of increased or decreased step density in a step sequence; it moves at a speed which depends upon the step density in it. Kinematic waves of step density and macrosteps were investigated, for instance, on paratoluidine crystals. The observation procedure consisted in recording the intensity of the monochromatic polarized light passing through the crystal and then through a narrow slit positioned in the image plane of the microscope parallel to the step rises. The change in signal intensity occurring when a step passed the field of vision of the slit was detected by a photoelectron multiplier. In this way it was possible to follow the passage of elementary steps 20–30 Å high [3.55].

Concentric kinematic waves on the surface of a $NaBrO_3$ crystal grown from an aqueous solution can be seen in the photo shown in Fig. 3. 27, which was taken in a conventional microscope. These waves surround the source of the growth layers and were formed on the slopes of the vicinal hillock built by it.

To understand the formation mechanism of kinematic waves, let us inspect the profile of a layerwise growing surface on which a bunch of elementary steps AB was observed at the initial moment (Fig. 3.28a). The cause of this bunch may, for instance, be a temporary increase in supersaturation at the point at which the step emanates. The velocity of the steps depends on the spacing be-

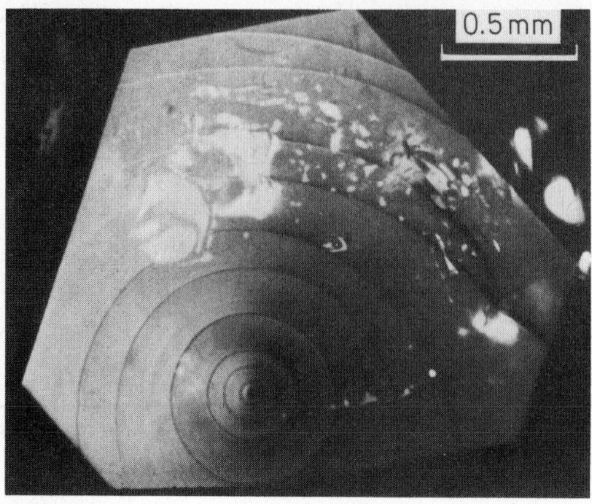

Fig.3.27. Kinematic waves of steps on the surface of a $NaBrO_3$ crystal [3.72]

tween them (Sect. 3.2). For instance, the mutual retardation of steps in accordance with (3.16, 18) leads to a dependence $v(\lambda)$ of the type shown in Fig. 3.11. Therefore the section of the highest step density must move at the lowest velocity: the steps to the left of A (Fig. 3.28a) catch up with the bunch and, combining with it, ultimately from edge C (Fig. 3.28b), on which the slope of the surface, i.e., the step density, changes jumpwise. A bunch of this kind having a discontinuity in step density (i.e., an edge on the surface)is called a shock wave of the step density. The shock wave propagates over the surface with a velocity $(R_1 - R_2)/(p_1 - p_2)$, where p_1 and p_2 are the slopes θ_1 and θ_2 of the surface on both sides of edge C ($p_{1,2} \simeq \theta_{1,2}$, since these angles are small), and R_1 and R_2 are the values of velocity R corresponding to these angles [3.73, 74, 24; 1.18]. With time, the wave velocity and the magnitude of the jump $p_1 - p_2$ decrease, and the wave must ultimately disappear, unless it is stabilized by impurities or other factors. The waves on neighboring growth hillocks usually have no time

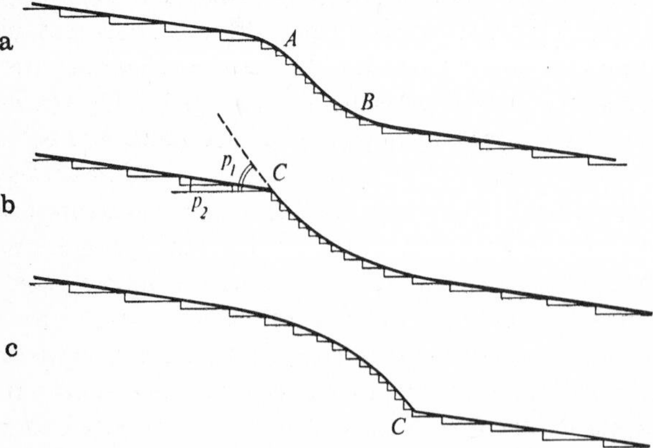

Fig.3.28a–c. Formation of kinematic waves of step density. The tangents of the indicated angles are designated p_1 and p_2

to damp out before annihilation with the steps of the opposite sign on the neighboring growth hillocks. Generally, one and the same sequence of large and small kinematic waves and distances between them is maintained for all the growth hillocks on a given face. This means that the waves were caused by temperature and concentration fluctuations in the bluk of the mother medium due to convection, poor thermostating, etc., which arise virtually simultaneously over various sections of a single face or even over all the faces of the crystal. Sometimes one observes waves [3.55] in which the edge leads the motion (Fig. 3.28c). Such a configuration corresponds to mutual acceleration of the steps.

Mutual acceleration of the steps may in particular result from adsorption, on the growing surface, of an impurity which inhibits growth (Sect. 4.1.2) and is trapped by the crystal. Indeed, the impurity arrives at each given site of the atomically smooth terrace between the neighboring steps in the "exposure period" between the formation of this site by the preceding step and its burial in the crystal after the passage of the next. The higher the step density, the shorter is the above-mentioned "exposure period," the lower is the impurity concentration on the terrace and the step rises, and the higher is the step velocity (Sect. 4.1.2). The described mutual acceleration effect [3.73] leads to dependence $b(p)$, which has a section of positive curvature, where $\partial^2 b/\partial p^2 > 0$. There is such a section on the dot-dashed curve in Fig. 3.7a near $p = 0$. When observed in an optical microscope, kinematic waves are usually perceived as steps of macroscopic height.

Another type of macroscopic step is also possible, however, in which the rises are formed by faces with simple crystallographic indices (true macrosteps). Electron-macroscopic investigations showed that the rises of such macrosteps themselves grow layerwise.

The most general cause for the existence of true macrosteps with singular crystallographic faces on their rises is the anisotropy of the surface energy. Indeed, according to Sect. 1.4, surfaces for which $(\alpha + \partial^2\alpha/\partial\varphi_1^2)$ or $(\alpha + \partial^2\alpha/\partial\varphi_2^2)$ are negative, are unstable, and must disintegrate with the formation of macrosteps. Furthermore, the only configurations of vertices and edges that are stable are those present on the equilibrium form of the crystal [1.18, 32]. Therefore if the equilibrium form contains macroscopically sharp edges and vertices, then in the course of growth of such crystals, macrosteps must form on the singular faces. Conversely, if a crystal of equilibrium form has macroscopically rounded "edges" and "vertices" under such conditions (i.e., with a given composition of the mother medium, temperature, and pressure), its growing surfaces, even if they are singular, cannot have true faceted macrosteps.

When speaking of sharp or rounded vertices and edges, we are not referring to their atomic roughness, which takes place for elementary steps at $T > 0$. Fluctuations causing roughness of the elementary steps must also smear out steps with heights of the order of ten lattice spacings or less, when the cooperative interaction is still moderate. Experiment shows, however, that steps and platy crystals having thicknesses of up to 1000 Å sometimes exhibit rounded growth shapes (Sect. 5.2) and hence rough rises.

It is difficult to distinguish true macrosteps from kinematic waves in an optical microscope. An indirect proof of the absence of faces on a macrostep rise is the rounded shape of the step (see Figs. 3.27, 29) which attest points to a high average density of kinks on the rise.

In the course of growth, macroscopic steps often break down into thinner ones, as can be seen for instance in Figs. 3.17, 25d, 29. Thin steps, on the other hand, which have a high growth rate [see (3.16, 18), Fig. 3.10], catch up with the thick ones and merge with them. Thus a growing surface has a wide gamut of step heights, and the average height depends on the supersaturation, the presence of impurities, and other growth conditions.

3.4.4 Surface Melting

Figure 3.29 demonstrates an interesting phenomenon that occurs during growth from vapors near the triple point in the phase diagram, namely, the formation of drops of a liquid–melt [3.70, 71], or an impurity solution on the surface. Where these drops wet the step rises, the steps no longer grow from the vapor, but from the melt (solution) and therefore exhibit a much higher growth rate. As a result, the drops are followed by the "protuberances" visible in Fig. 3.29. Sometimes the drops disappear as soon as they touch the step. The im-

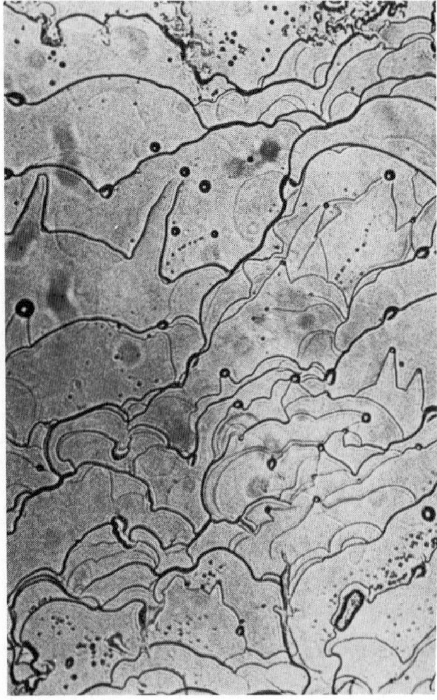

Fig.3.29. Liquid droplets on the surface of a β-methylnaphthalene crystal growing from vapor

proved visibility of the steps near the melting point of the crystal is indirect evidence of the appearance of the liquid phase at the step rises in the course of growth from vapors.

Other evidence for the existence of a liquid or quasi-liquid film on the crystal–vapor interface in the vicinity of the melting temperature is reviewed by *Nenov* et al. [3.75]. The main information has come up until now from optical observations of crystal surfaces under growth and equilibrium conditions. The composition, thickness, and atomic structure of the surface layer cannot be determined in these experiments, and the following three possibilities remain unresolved:

1) The substance under consideration and an impurity present in the system form a solution, the melting point of which is lower than the actual temperature (see VLS growth in Sect. 8.5.1).

2) The vapor–crystal interface becomes atomically rough, forming a diffuse interface layer with a thickness of several interatomic spacings.

3) A melt film considerably thicker than the diffuse (atomically rough) layer appears on the interface.

Nevertheless, morphological observations have provided important information on the surface behavior of organic crystals, ice, and some metals.

Growing diphenyl crystals have the form of polygonal platelets at 55°C and of round discs at 60°C [3.76] (the melting temperature T of diphenyl is 69°C). Flat negative crystals — gaseous inclusions in naphthalene — exhibit an analogous transformation at 70°C (T_0 = 80°C) [3.77]. In both cases, however, the most dense basal planes (F faces, Sect. 1.2.2) remain singular and form the main faces of both the platelets and the flat inclusions. Inclusions in tetrabromomethane are nearly cubic at 40°C and become spherical at 90°C (T_0 = 94°C)

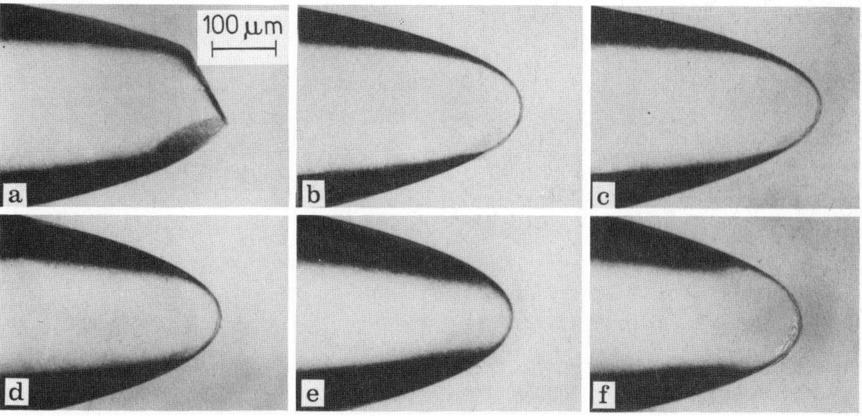

Fig.3.30a–f. Growth shapes of an adamantane crystal at increasing temperatures (°C). **(a)** 225; **(b)** 246; **(c)** 250; **(d)** 256; **(e)** 258.5; **(f)** 259. The disappearance of postive (111) and negative $(1\bar{1}1)$ tetrahedron faces is seen from **(a)** to **(f)** [3.78]

[3.75]. Figs 3.30 shows how singular flat faces disappear on the growing adamantane crystal–vapor interface at $\simeq 260°$ ($T_0 = 272.5°C$) [3.78]. As a rule, the denser the face is, the higher the temperature needed for its surface melting. For instance, the transition temperatures for the (111), (111), and (100) faces of adamantane are 255°C, 251°C, and 177°C, respectively. These temperatures correspond to the transition/melting temperature ratios (0.97, 0.96, and 0.82), which is typical of other organic crystals. For (511) and (100) of platinum this ratio is 0.74 and 0.80 [3.79].

A sample of ice hoarfrost shows at $T < -50°C$ only a wide signal of nuclear magnetic resonance from protons, which corresponds to low proton mobility, i.e., to the solid state. At higher temperatures, however, a narrow peak arises from mobile protons. The integral intensity of this narrow peak is proportional to the amount of liquid in the sample [3.80]. The analogous appearance of a narrow line was observed at ~ 2 K below the melting point of stilbene [3.81] and, in our recent NMP experiments, at $\lesssim 20$ K below the melting point of diphenyl.

An ellipsometric study of the surface of ice in the vicinity of the melting point has also been conducted [3.82]. Quasi-liquid films on ice are discussed in [3.83] and the references therein.

Attempts to observe surface melting on Sn, Bi, and K have been unsuccessful [3.84]. Molecular dynamic simulation of surface melting is discussed in [1.26a–f] (Sect. 1.3.4).

4. Impurities

Like any interfacial process, crystal growth strongly depends on active impurities present in a crystallizing system even in amounts which do not influence the properties of the bulk. On the other hand, many of these properties depend on the amount of impurities or point defects in the grown crystal. Thus the manner in which impurities are trapped by growing crystals is one of the most important characteristics of growth technology. The influence of impurities on growth kinetics is reviewed in Sect. 4.1, and the thermodynamics and kinetics of impurity trapping are analyzed in Sects. 4.2 and 4.3, respectively.

4.1 Effect of Impurities on Growth Processes

4.1.1 Thermodynamics and Structure of Solutions

Impurities are minor additives of foreign substances in the mother medium or the crystal. Thermodynamically, an impurity shifts the equilibrium between crystal and medium (vapor, solution, or melt) in accordance with the phase diagram. Therefore for growth or decrystallization in the presence of an impurity, the supersaturation or supercooling must be determined with due regard for the appropriate phase diagram. For instance, impurities of KCl, KBr, KI, $(NH_4)_2SO_4$, when introduced into an aqueous solution of a potash alum, $KAl(SO_4)_2 \cdot 12H_2O$, reduce the alum solubility in accordance with the mass action law. The addition of chromium increases the solubility of potash alum. It is well known that the melting point of water is depressed by addition of NaCl. The changes in solubility and melting point are usually proportional to the impurity concentration.[1] The proportionality factor naturally differs for different systems. For instance, addition of ~2 wt. % Cr to a solution of potash alum at 25°C increases the alum solubility by about 2 wt. % (without the impurity, its solubility is ~15 wt. %); i.e., the proportionality factor in the dependence of the solubility on the impurity concentration is of the order of unity. The slope of the liquidus line $m = \partial T_0 / \partial C_i$ in the Pb–Bi system at low bismuth concentrations is ~ 3 K per wt. % bismuth. These factors are, of course, by no means absolute, and one must refer to the phase diagrams in each particular case.

[1] In practice, the presence of an impurity results in some smearing-out of the observed melting point of the specimen due to the fluctuation in impurity concentration.

An additive present in solution in a rather large amount (several tens of per cent) generally changes the composition, structure, and population of the complexes in solution, which in turn changes the growth rate for kinetic reasons. If the additive diminishes the solubility of the crystallizing substance, it causes an increase in the kinetic coefficient of proportionality between the growth rate V and the dimensionless driving force $\Delta\mu/kT$ for crystallization in the case of a near-linear $V(\Delta\mu/uT)$ dependence. Such dependences have been found experimentally for nitrates of alkaline metals [4.1,2]. For these nitrates, the above-mentioned kinetic coefficient increases in the series in which the cation hydration heat decreases. These effects are also discussed in terms of exchange fluxes in [4.2].

This generalization of the *Van t'Hoff* and *Bernsted* rules to include solution growth [4.1] appears to be applicable to the pH effect on the solution growth rate.

A third component in a solution may give rise to complexes close to the crystal building units. In this case of course, the kinetic coefficient discussed above may increase. For instance, the addition of HCl to a $FeCl_2 \cdot 4H_2O$–HCl–H_2O solution probably causes one or two water molecules in the octahedral shell in the $[Fe \cdot 6H_2O]^{2+}$ complex to be replaced by Cl^- ions, thus giving rise to a $[FeCl_2 \cdot 4H_2O]$ complex analogous to the crystalline one [4.1]. The influence which the structure of the complex has on the rate of crystallization was splendidly demonstrated by *Troost* [4.3a, b], who a number of times observed different nucleation and growth rates of $Na_5P_3O_{10} \cdot 6H_2O$ crystals from solutions prepared from two different polymorphous modifications of anhydrous substance.

Ionic complexes in a crystal and in its saturated solution give rise to the specific electronic absorption bands in the optical absorption spectra of the crystal and solution. If the wavelengths λ of some absorption bands in both spectra are close to each other, the existence of similar ionic complexes in the crystal and in the solution may be expected. The absorption maxima for the Cs_2CuCl_4 crystal are located at $\lambda = 239$, 297, 410, and 980 nm. The saturated solution of this crystal has absorption bands around $\lambda = 239$, 271, 385, and 870 nm. The 271-nm band may be ascribed to the aquacomplex $[CuCl_m(H_2O)_n]^{-(m-2)}$ and the bands at 239, 385, and 870 nm to the complex $[CuCl_4]^{2-}$ in the crystal structure. It is known that a perfect crystal of an incongruently soluble compound does not grow from its stoichiometric solution [Ref. 1.1a, p.172]. For instance, a perfect Cs_2CuCl_4 crystal does not grow from the aqueous solution which results when such a crystal is dissolved. However, the perfect crystal can be grown using the same solvent. The composition of the mother liquor must correspond to a point on the phase diagram which is located on the solubility curve between the nonvariant points limiting this curve for the desired compound. For instance, to grow Cs_2CuCl_4 one should take an amount of $CuCl_2$ about two times smaller than that in the stoichiometric composition. Analysis of optical spectra gives reason to believe that such a nonstoichiometric solution has more "crystalline" $[CuCl_4]^{2-}$ complexes than the stoichiometric one from which the Cs_2CuCl_4 crystal cannot be grown [4.4a–c].

Investigation of Raman scattering spectra of molten $CaWO_4$ shows that the WO_4^{2-} group still exists in the liquid at temperatures at least close to the melting point of $CaWO_4$ [4.4d].

The assumption that the oxide molecules exist in solution was used to develop a solubility model for rare-earth garnets in a PbO/B_2O_3 solution [4.43].

4.1.2 Adsorption

The adsorption of impurities on the surface substantially changes the average binding forces operating along the surface between the particles of the surface layer. Therefore the addition of an impurity may lead to a transition from an atomically rough surface to a smooth one, and vice versa [1.31]. For instance, crystals of NH_4Cl in an aqueous solution have atomically rough surfaces, which follows from the rounded shape of their equilibrium and growth forms (Fig. 1.14). If bivalent and trivalent cations Cd^{2+}, Fe^{2+}, Ni^{2+}, Co^{2+}, $[Zro]^{2+}$, Fe^{3+}, etc., are added to the solution, crystals of ammonium chloride grow well faceted (Fig. 4.1) [4.5].

Kinetically, the effect of impurities consists in action on the transport processes in the bulk and on the crystallization phenomena on the surface. An impurity generally changes the structure of the liquid phase and therefore affects the diffusion coefficients, but this effect is usually small (at molar concentrations of the impurity $C_i \ll 1$). The effect of an impurity on the thermal conductivity is still smaller at the usual crystallization temperatures. However, change in the structure of the liquid may also affect the association (complexation) of the liquid and the structure of the near-surface layer of the solution, where the impurity concentration is increased as a result of adsorption; it may also change the diffusional resistance more substantially. According to *Mullin* [Ref. 4.6, p. 197], aquo ions $[M \cdot 6H_2O]^{3+}$, where M stands for either Cr, Al,

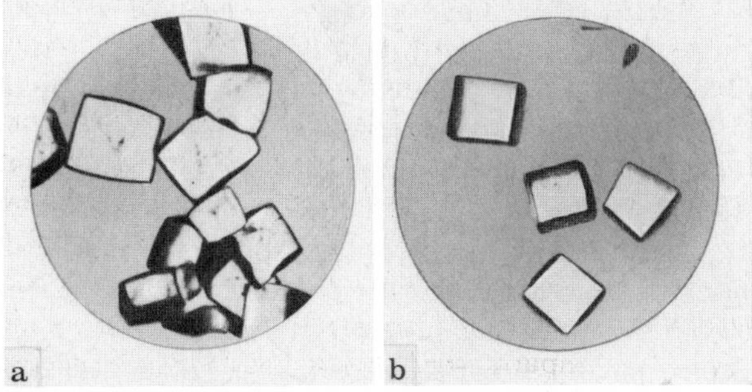

Fig.4.1a,b. NH_4Cl crystals grown in the presence of ions. **(a)** Cd^{2+}, **(b)** $[ZrO]^{2+}$ [4.5] (cf. Fig. 1.14 and Fig.5.8a). Magnification × 37

or more complex formations, have large enough dimensions that when they are attracted by phosphate groups on the surface of crystals of ammonium or potassium dihydrogen phosphate (ADP or KDP), they can screen the surfaces from the delivery of new building blocks. Impurity ions adsorbed on the surface thus retain part of their solvate shell and can screen the growing surface effectively enough.

Transition metals in solution may also be present in the form of different hydrated oxides, depending on the solution pH. For instance, in acid solutions, aluminium and iron are present primarily in the ionic forms Al^{3+} and Fe^{3+}, whereas in alkaline solutions, $Al(OH)^{2-}$, $Al(OH)_2^-$, and $Al(OH)_3$ should be expected in relatively large amounts (the same goes for Fe). Since all these species have different adsorption activities, the solution pH may influence the growth via the state of impurities, and at present, a possible influence of H_3O^+ and OH^- ions adsorbed on the growing faces cannot be excluded either.

In general, an additive forming stable inactive complexes with an impurity which is present in solution and retards growth may cause an increase in the growth rate.

Impurities exert the strongest effect on surface crystallization [4.7]. This effect is again based on the adsorption of an impurity in the form of atoms, molecules, complexes, or even aggregates in different positions on the surface— at kinks, on steps, on atomically smooth areas of a face, etc. The impurity concentration in different positions depends primarily on the binding energy of the impurity particle with the lattice at these sites. In physical adsorption this is equal to a few kcal/mol, and in chemical adsorption, to tens of kcal/mol (and sometimes even over a hundred) [4.8].

Impurity particles differ in size from atoms, ions, or complexes, which they can replace in a crystal, Moreover, an impurity ion, atom, or molecule forms more or less stable aggregates with the particles of the mother medium. this is particularly true of solutions in which the impurity can form stable complexes with solvent ions. The indicated factors often prevent the ingress of the impurity into the crystal.

The role of an impurity can also be played by atoms and ions entering the composition of a chemical compound, or by products of a chemical reaction, provided crystallization is a consequence of this reaction.

If an impurity that has some difficulty entering the crystal is strongly adsorbed by a kink, this kink (or step, or the surface as a whole) loses the opportunity of adding new particles making up the crystal. Therefore an impurity naturally reduces the growth rate to a different extent for faces with different surface structures. A kink with an impurity particle ceases to be a growth site until the impurity is desorbed thermally or chemically, or until it is displaced (or buried) by the building block of the crystal.

Let us estimate the density of impurity-free kinks [1.18]. For a step at a crystal–vapor interface the distance λ_i between free kinks is, in the approximation of the Langmuir adsorption isotherm (i.e., for noninteracting impurity atoms),

$$\lambda_i = \lambda_0 + \xi_V P_i\,, \tag{4.1}$$

where P_i is the impurity vapor pressure, and the quantity

$$\xi_V \simeq \frac{a(kT)^{1/2}}{2(2\pi m_i)^{3/2}\nu_i^3}\exp\left(\frac{\varepsilon_i}{kT}\right). \tag{4.1a}$$

Here, m_i is the mass of the impurity particle, and ν_i is the frequency of its thermal vibrations as a whole. If the adsorption energy at the kink is $\varepsilon_i = 0.5$ eV (i.e., $\simeq 12$ kcal/mol), and if $m_i = 50 \times 1.6 \times 10^{-24}$ g and $T = 300$ K, then $\xi_V \simeq 0.1\ a$ (mm Hg)$^{-1}$. Hence even an impurity with the indicated relatively low adsorption energy will substantially reduce the density of free kinks even at $P_i \simeq 1$ mm Hg. A chemically adsorbed impurity, for which the heat of adsorption may equal many tens of kcal/mol (e.g., $\sim$ 100 kcal/mol for Cl on Si), will, accordingly, poison steps at much lower concentrations in the gas phase.

A very similar consideration for a step at the crystal–solution interface yields a linear dependence of the density of free kinks on the atomic concentration of the impurity in solution near the step. This dependence resembles (4.1):

$$\lambda_i = \lambda_0 + \xi_{sol}\, C_i\,,$$
$$\xi_{sol} = \frac{a}{2\Omega\nu_i^3}\left(\frac{kT}{2\pi m_i}\right)^{3/2}\exp\left(\frac{\varepsilon_i}{kT}\right). \tag{4.2}$$

At $m_i = 50 \times 1.6 \times 10^{-24}$ g, $\nu_i = 10^{12}$ s^{-1}, $\varepsilon_i = 5$ kcal/mol, we obtain $\xi_{sol} \simeq 10^5 a$, i.e., a relative atomic concentration of the impurity $C_i \simeq 10^{-4} = 10^{-2}$ mol. % is sufficient to reduce the growth rate appreciably (if $\lambda_0 \simeq$ 5–10a).

Since the steps are rough, their kinetic coefficients during both growth from vapors and growth from solutions can naturally be assumed to be proportional to the densities of free kinks, see (3.9):

$$\beta_{st}(C_i) = \beta(0)\lambda_0/\lambda_i\,. \tag{4.2a}$$

Using this form of the kinetic coefficient in the relations of Sect. 3.3 yields the normal growth rates of faces with different mechanisms of growth in the presence of an impurity [1.18].

According to (3.16, 18) growth rates reduce with a decreasing kinetic coefficient, and therefore the introduction of an impurity which poisons the kinks results, as a rule, in a drop in the growth rate. The analytical expression of this dependence is extremely cumbersome. Its most essential qualitative feature for dislocation growth is the expansion of the region of quadratic dependence of the growth rate on the supersaturation in the presence of an impurity. The experimental dependences of the growth rate on the impurity concentration are discussed in the next section. They agree in their general nature with the dependence obtained from the above-considered model of kink poisoning. No detailed comparison has been drawn so far because of the lack of quantitative data on the elementary acts of growth and adsorption.

There is another conceivable mechanism of growth retardation under the influence of an impurity. Upon the adsorption of an impurity on atomically smooth areas of the surface, the moving steps must "clean off" this impurity, i.e., do the work of detaching an impurity atom or the complex containing it from the surface. Later on, the atom (or complex) moved up by the step may be adsorbed on the new surface formed by the step. Thus an additional potential barrier to growth arises which is associated with the impurity. The height of this barrier is of the order of the desorption energy of the impurity from the smooth surface. The impurity "coat of mail" (or even a "coat of mail" from the solvent, which is sometimes also regarded as an impurity) is particularly stable if it is bonded by cooperative interaction of the particles on the surface (Sect. 1.3). The impurity or solution adsorption layer naturally interferes both with the motion of the existing steps and with the formation of nuclei, and may cause anomalies in the dependence of the growth rate upon the temperature [4.2].

An impurity exerts an extremely strong effect not only on layer growth, but also on the layer dissolution of crystals. For instance, the addition of 10^{-6} wt. % $FeCl_3$ to water sharply reduces the tangential velocity of steps and changes the nature of the etching of alkali-halide crystals by the water, thus ensuring a transition from the 'polishing mode' to selective etching on dislocations [4.12].

Appreciable impurity concentrations can change the composition and structure of the surface adsorption layer and solution complexes (Sect. 4.11), and cause a reduction or an increase in the growth rate. This may be the mechanism causing the abrupt acceleration (by 30%–60%) in the growth of a crystal of gypsum ($CaSO_4 \cdot 2H_2O$) from an aqueous solution in the presence of an impurity of 0.5 m NaCl [4.13] and also of $BaSO_4$ crystals in the presence $\sim 10^{-6}$ m NH_3 [4.14].

The adsorption of impurities on step rises reduces their specific linear energy, thereby increasing the probability that two-dimensional nuclei will form. In dislocation growth, this adsorption reduces the spacing between the turns of the spiral step. Both these effects also increase the growth rate. But it is not yet clear whether the moderate increase in growth rate occasionally observed upon the introduction of an impurity (see, e.g., [4.3, 15]) is due to these effects or to other factors. Nor is it improbable that the acceleration of growth at low impurity concentrations results from the provocation of two-dimensional nucleation by individual molecules of the impurity on atomically smooth areas.

Acceleration of the surface processes of crystallization and decrystallization is also caused by impurities which ensure the removal of reaction products or play a catalytic role. The addition of fractions of a weight percent of water to glacial acetic acid (etchant) strongly increases the rate of etching on alkali halide crystals, evidently increasing their solubility [4.16].

The charge of the ion and its size and "compressibility" are considered to be the parameters defining the degree of influence of an impurity on growth in solution. This influence increases with an increase in the charge of the cation and a decrease in its radius [4.6]. This is in agreement with the correlation between the kinetic coefficient and the cation hydration heat (Sect. 4.1.1).

In view of what has been said above, the effect of an impurity must depend upon the energy of formation of chemical compounds consisting of an impurity and a crystal or an impurity and a solvent, as well as upon the solubility of these compounds (Sect. 4.1.1), the coefficient of distribution (Sect. 4.2.2), the solution pH, and the crystallization temperature.

The average kink density is much higher on an atomically rough surface than on a layerwise growing surface. Therefore during normal growth, the amount of impurity required to poison the kinks is 3–5 orders higher than in layer growth. Indeed, experiments show that for substances with low melting entropies growing from the melt, the impurity causes only a shift in the melting point. If, however, the amount of impurity is sufficient to reduce the melting temperature or increase the surface energy appreciably, the surface may turn from rough to smooth. Accordingly, the growth rate at the same supercooling will be much lower than in a pure melt.

A transition from normal to layer growth in bismuth crystals grown from Bi–Sn and Bi–Pb solutions was observed in a narrow range of concentrations of the second component: tin in the range of 3.5–4 wt. %, and lead in the range of 16–17 wt. % [4.17].

For the naphthalene-paradibromobenzene system, a roughening transition such as the one predicted earlier [1.31] was actually found in narrow concentration and temperature intervals by *Podolinsky* and *Drykin* [4.18] (see the end of Sect. 1.3.4 and Fig. 1.9c).

4.1.3 Dependences of Growth and Morphology on the Concentration of Impurities

Let us consider in more detail the quantitative data pertaining to the effect of impurities on the growth rate. A typical dependence of the growth rate of a (131) face of a $KClO_4$ crystal on the impurity concentration of the dye Ponso 3R at different crystallization temperatures is presented in Fig. 4.2. The curves clearly show the reduction in growth rate with increasing concentration of the added impurity, and are well described by the empirical equation [4.19]:

$$V = V_0 - (V_0 - V_\infty)\, n_i\,, \tag{4.3}$$

Here, V_0 is the growth rate in the absence of an impurity ($C_i = 0$) and V_∞ that at a sufficiently high impurity concentration. The quantity n_i can be interpreted as the ratio of impurity-covered kinks to the number of them at the crystal–

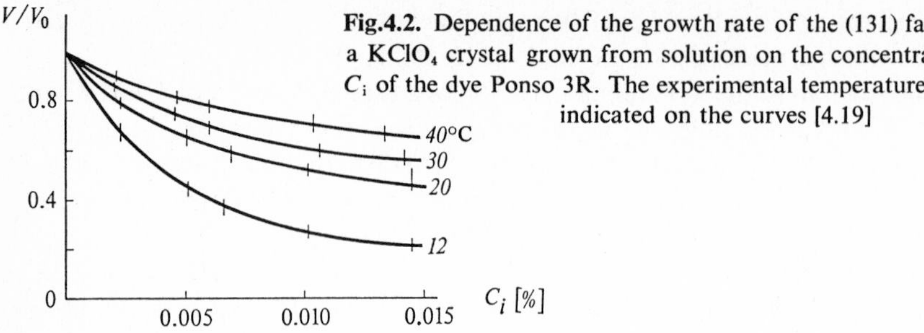

Fig.4.2. Dependence of the growth rate of the (131) face of a $KClO_4$ crystal grown from solution on the concentration C_i of the dye Ponso 3R. The experimental temperatures are indicated on the curves [4.19]

pure-solution interface. At $V_\infty > 0$, expression (4.3) corresponds also to the finite probability of growth at kinks where impurity particles are located. Growth is possible on such poisoned kinks if the impurity particle is either buried in the lattice or pushed out by the newly attached crystal-building block. The fraction n_i of poisoned kinks is expressed via the impurity concentration C_i in the bulk of the mother medium, which is readily measurable experimentally. In the case of Langmuir's adsorption isotherm,

$$n_i = C_i/(A + C_i)\,, \tag{4.4}$$

where $A \propto \exp(-\varepsilon_{is}/kT)$, and ε_{is} is the adsorption energy of the impurity. Substituting (4.4) into (4.3), we obtain

$$\frac{1}{V_0 - V} = \frac{1}{V_0 - V_\infty} + \frac{A}{(V_0 - V_\infty)C_i}\,. \tag{4.5}$$

This equation implies a linear dependence of $(V_0 - V)^{-1}$ on $1/C_i$, which indeed is often observed in experiment (Fig. 4.3). Furthermore, according to (4.4), ln A must depend linearly on $1/T$, and the slope of the line gives the adsorption energy of the impurity under conditions of crystal growth. The proportionality of ln A and $1/T$ does indeed manifest itself, although only a narrow temperature range has been investigated so far (Fig. 4.3). The corresponding adsorption energies are given for some crystals and impurities in Table 4.1. Judging by the low adsorption energies (up to 10 kcal/mol), we are dealing here with physical adsorption, as would be expected for organic-dye impurities.

Expressions (4.3,4) in fact assume that the growth rate is proportional to the density of kinks (and hence of steps). In reality this dependence is more complicated, as was stated in the preceding section.

Apart from the above-described continuous, though quite steep, decrease in velocity with increasing impurity concentration, still another type of dependence $V(C_i)$ is observed. It is characterized by the existence of a critical impurity concentration such that at higher concentrations growth of crystalline faces was not observed, at least within the accuracy used in experiments. An example of a system exhibiting this effect is an aqueous solution of $NaCl + K_4Fe(CN)$

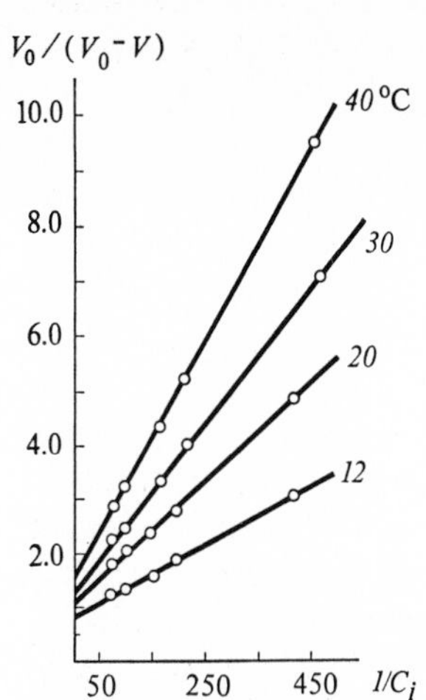

Fig.4.3. Isotherms of the dependence of the growth rate on the impurity concentration according to the data shown in Fig. 4.2 [4.19]

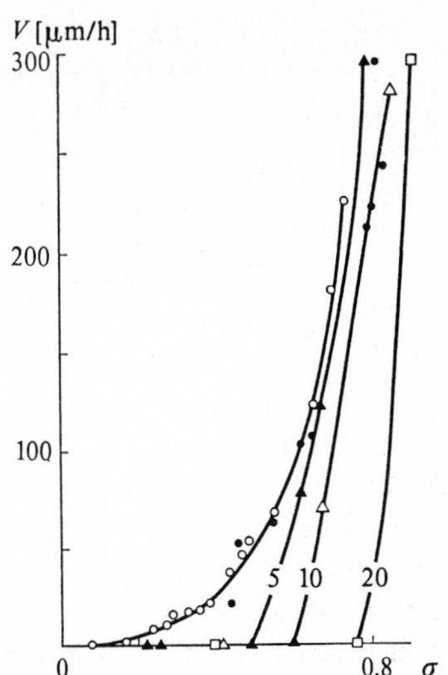

Fig.4.4. Dependence of the growth rate of crystals of $Na_5P_3O_{10}\cdot 6H_2O$ on the supersaturation in a pure solution (*left-hand curve*) and in the presence of sodium dodecylbenzylsulfonate in amounts of 5, 10, and 20 mg/l (shown by the *figures* on the curves) [4.3]

Table 4.1. Adsorption energies of impurities affecting growth rates [4.19]

Crystal	Impurity	Face	Adsorption energy [kcal/mol]
$MgSO_4\cdot 7H_2O$	$Na_2B_4O_7\cdot 10H_2O$	111	4.18
KBr	Phenol	100	5.10
$KClO_4$	PONSO3R	131	3.68
NaCl	Cd^{2+}	100	7.85[a]
			5.10[a]
NaCl	Cd^{2+}	111	8.95[a]
			4.80[a]

[a] According to the hypothesis put forth in [4.19b], the lower value for each face may be attributed to the adsorption enthalpy for the position at the step, and the higher value to that for the kink at the step (see Table 1.1). The data were obtained from adsorption isotherms measured by the radiometric method

[4.20]. Another example can be seen in the effect of sodium dodecylbenzylsulphonate ($C_{12}H_{25}$–⟨⟩–$SO_3^-Na^+$) on the growth of crystals of $Na_5P_3O_{10}$ (Fig.4.4). From the dependences of the growth rate on the supersaturation at different impurity concentrations presented in Fig. 4.4 one can see the increase in critical supersaturations with the impurity concentration. One of the possible explanations for this effect is as follows [3.74]: The surface carries impurity particles that are strongly adsorbed and sufficiently large as compared with the spacing of the main lattice. Having reached such a particle, a step stops at the site of contact and begins to move around it. A convex portion of the step forms between neighboring particles (Fig. 4.5a), and as it requires a higher supersaturation for growth than the straight portions do, its motion slows down. If the average distance between impurity particles is less than the diameter of a two-dimensional critical nucleus, the step will stop completely.[2] The interaction of macrosteps with foreign macroparticles on the surface of SiC, which is similar to the above-discussed microprocesses, can be seen from Fig. 4.5b.

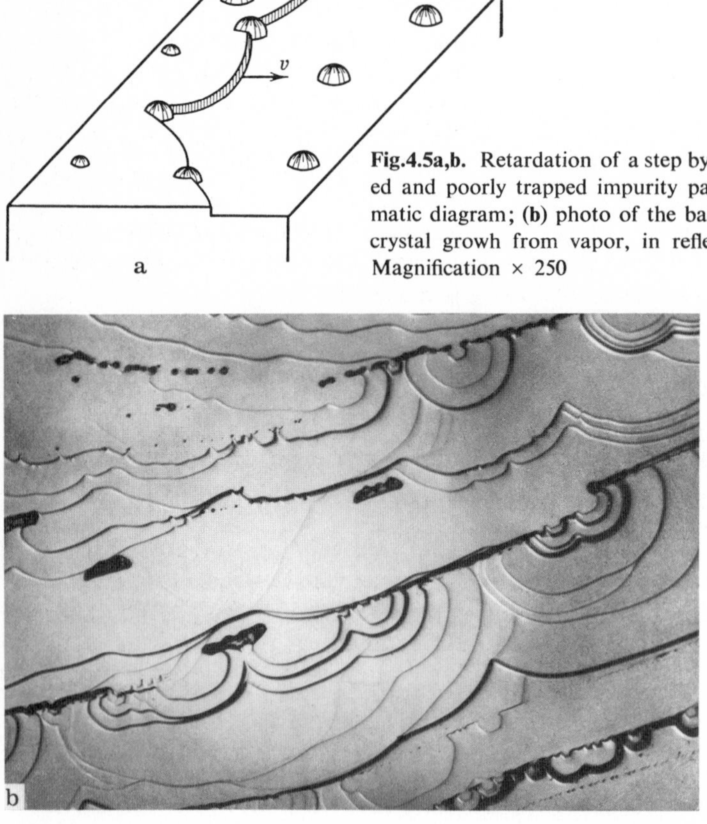

Fig.4.5a,b. Retardation of a step by strongly adsorbed and poorly trapped impurity particles. (**a**) Schematic diagram; (**b**) photo of the basal face of a SiC crystal growh from vapor, in reflected light [4.21] Magnification × 250

[2] The above-described process of step retardation is similar to the retardation of dislocations moving in a crystal with an impurity under the effect of a load (hardening by impurities). Ocassional oscillations in the growth rates [4.19c–e] of faces with unknown dislocation structure [4.19c] and of dislocation-free faces [4.19d,e] may be ascribed to this effect.

The adsorption of an impurity proceeds differently on different faces and affects the growth kinetics of a given face differently at different supersaturations. It is therefore possible to delineate, on the impurity concentration–supersaturation plane, the regions in which various forms of crystals appear. A representation of this kind is called a morphodrom [4.22]. Once adsorbed on the surface, impurities change not only the growth rates and the habit of the crystal as a whole (Sect. 5.2.1), but also the morphology of the layerwise-growing faces. For instance, the faces of a cube of alkali-halide crystals growing from aqueous solutions are covered with large irregular steps, as a result of which the crystals formed contain many inclusions (Sect. 5.2). The addition of divalent heavy ions (Pb, Cu, Zn, Cd, etc.) to the solution makes the growth surface mirror smooth [4.5, 23, 24]. Such a surface most probably grows in elementary or near-elementary layers; the crystals thus do not trap any inclusions (Sect. 5.2) and hence grow absolutely transparent to the eye. This does not, of course, exclude the presence of colloid inclusions, and even less, of an atomic impurity in them.

If the adsorption energy is high enough, the amount of impurity on the surface may not be small, even if its concentration in the bulk is low. Under these conditions, epitaxial crystallization of the impurity itself may begin on the surface, and the growth rate of the principal crystal may drop to zero.

Since the structure of the surface and the growth mechanism depend on the composition of the solvent, it is sometimes said that certain solvents can be regarded as impurities present in large amounts.

A huge body of factual material on the effect of impurities and the medium upon growth forms has been compiled in *Buckley's* book [4.25].

4.2 Trapping of Impurities: Classification and Thermodynamics

4.2.1 Classification

Homogeneous and Heterogeneous Trapping. An impurity present in the mother medium is usually trapped by the growing crystal both in the form of individual atoms, ions, molecules, or atomic-size complexes—pairs, triads (homogeneous trapping)—and in the form of colloid ($\sim 10^{-5}$–10^{-6} cm) or macroscopic inclusions (heterogeneous trapping).

Equilibrium and Nonequilibrium Trapping. Homogeneous trapping results in the formation of solid solutions (mixed crystals), whose concentration is either equal to or differs from thermodynamic equilibrium concentration at the growth temperature (Sect. 1.1), and also depends on the process kinetics. Therefore it is customary to speak of thermodynamically equilibrium or nonequilibrium trapping. Heterogeneous trapping is always nonequilibrium.

The concentration and state of a homogeneously and heterogeneously trapped impurity are seldom constant in the crystal bulk. Three main types of nonuniformity — sectorial, zonal, and structural — can be distinguished in the distribution of impurities. A nonuniform impurity distribution is always thermodynamically nonequilibrium.

Sectorial Nonuniformity. The amounts of impurities differ in different growth pyramids, i.e., in those sectors of the crystal whose material was formed as a result of the growth of different faces. A crystal nonuniformity of this kind is called a sectorial structure (Fig. 4.6). Naturally, the sectors differ not only in concentration and in the form of ingress of the impurity, but also in the concentrations of other defects. Generally speaking, the material making up the

Fig.4.6a,b. Sectorial and zonal distribution (**a**) of mother liquor microinclusions in a crystal of potash alum (the single inclusions are not visible; magnification × 100); (**b**) of an impurity in growth pyramids of the faces of a rhombohedra in a quartz crystal

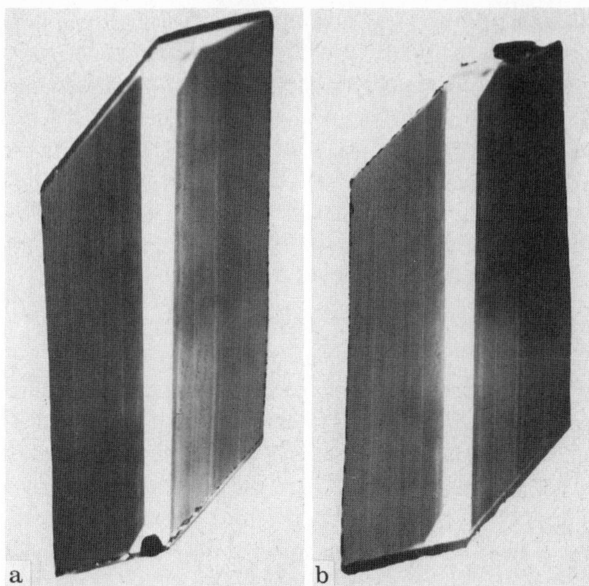

Fig.4.7a,b. Anomalous pleochroism in growth sectors of a minor rhombohedron of a synthetic quartz crystal. The *dark areas* on the photos are the growth sectors of the face of the minor rhombohedron r, and the *light band* is the seed. The differing darknesses of the *left* and *right sectors* in **(a)** and **(b)** correspond to the differing tilts of the sample around the normal to the r face [4.26]

growth pyramid "remembers" not only the crystallographic orientation of the face, but also the direction of its growth. This finer effect manifests itself, for instance, in the smoky-quartz anomalous pleochroism discovered by *Tsinober* [4.26] on the growth pyramids of small rhombohedron r $(01\bar{1}1)$ faces. Figure 4.7 shows photos of a quartz plate cut out parallel to the optic axis and perpendicular to the rhombohedra r faces. If such a plate is slightly turned about the normal to the r face, one of the growth pyramids will become brownish, and the other greenish (Fig. 4.7a). If it is turned in the opposite direction, the colors of the pyramid switch around (Fig. 4.7b). In Figs. 4.7a,b the darker field corresponds to the violet hue, and the lighter field to the greenish one.[3]

Zonal Nonuniformity. An impurity can be trapped not only differently by different faces, but also unequally at different moments in the growth of one

[3] The cause of this phenomenon is as follows: Each unit cell of the quartz structure contains three silicon–oxygen tetrahedra joined at the vertices and having three different orientations relative to the r $(01\bar{1}1)$ face. The replacement $Si^{4+} \rightarrow Al^{3+} + Na^{+}$, which leads to the appearance of color centers, occurs during the growth process with unequal probability in tetrahedra of these three orientations, which creates the "memory" of the growth direction. Indeed, according to EPR data, the population density of different tetrahedra differs in opposite growth pyramids of anomalously pleochroic quartz, while in crystals of normal quartz it is the same.

and the same face, primarily as a result of the inconstancy of the growth rate. Therefore the impurity inside each growth pyramid is distributed in layers parallel to the face or in general to the growth front. This is called zonal, or striation nonuniformity (Fig. 4.6).

Structural Nonuniformity. This is the enrichment or depletion, by an impurity, of the boundaries of twins and grains and the vicinities of dislocation, inclusions, and other structural defects of a crystal. The impurity is also concentrated along the boundaries of so-called pencil or lineage structures (Sect. 5.2)

The causes of sectorial and zonal structures are discussed in Sect. 4.3.

4.2.2 Thermodynamics

Let us now dwell in more detail on the thermodynamics of homogeneous impurity trapping. Homogeneous trapping is based on the ability of impurity particles (atoms, molecules, or ions) to enter the crystal lattice. The determining crystallochemical parameters for such an ingress are the size of the particles, their valence, and, in general, the type of the bond which arises between the impurity particles and the lattice. The types of solid solutions thus formed were discussed in [1.2a]; [Ref. 4.27, Sect. 5.6].

The thermodynamic equilibrium between crystal and medium with regard to the distribution of the impurity between them is determined by the equality of the chemical potentials of the impurity in the crystal (S) and in the medium (M). Let us write down the chemical potentials of the solute impurity in the crystal, (S), and in the medium (liquid, gas), (M), in the form:

$$\mu_{iS} = \mu_{iS}^0 + kT \ln \gamma_{iS} X_{iS} \,, \tag{4.6}$$

$$\mu_{iM} = \mu_{iM}^0 + kT \ln \gamma_{iM} X_{iM} \,. \tag{4.7}$$

Here, X_{iS} and X_{iM} are the molar fractions of the impurity in the solid and liquid (gaseous) phases, and μ_{iS}^0 and μ_{iM}^0 are the chemical potentials of the impurity in its standard state. By standard state can be meant, as we shall see below (Sect. 4.2.3), the liquid or gaseous state at a temperature T and a pressure P of crystallization of the main component. The values of μ_{iS}^0 and μ_{iM}^0 depend on P and T, but not on X_{iS} and X_{iM}. The conventional activity coefficients $\gamma_{iS,M}$, which are concentration dependent, are determined in such a way that the chemical potential of the real solutions could be represented in the form of (4.6, 7). The condition of equilibrium with respect to the impurity $\mu_{iS} = \mu_{iM}$ thus yields the equilibrium coefficient of its distribution between the phases:

$$K_0 = \frac{X_{iS}}{X_{iM}} = \frac{\gamma_{iM}}{\gamma_{iS}} \exp \frac{\mu_{iM}^0 - \mu_{iS}^0}{kT} \,. \tag{4.8}$$

In practice, the distribution coefficient defined as the ratio of the impurity concentration in the crystal and in the medium expressed in weight fractions, in grams per unit of volume, etc. is also often used. In growth from solution, when the solvent does not enter the crystal, it is convenient to determine the distribution coefficient with respect to the crystallizing substance:

$$\mathscr{D} = X_{iS}/\bar{X}_{iM} ,$$

where $\bar{X}_{iM}$ is the ratio of the number of impurity molecules contained in the solution to the total number of dissolved molecules, i.e., of the macrocomponent (the principal crystallizing substance) and the impurity. Hereafter K and $\mathscr{D}$ denote the nonequilibrium distribution coefficients due both to thermodynamic and to kinetic factors, and $\mathscr{D}_0$ and K_0 denote the equilibrium values. Because of the low impurity concentration, $\bar{X}_{iM} = X_{iM}/X_M$, where X_M is the molar fraction of the macrocomponent in solution. Hence

$$\mathscr{D} = KX_M . \tag{4.9}$$

At low impurity concentrations the ratio $X_{iS}/\bar{X}_{iM}$ is equal to the ratio of the concentrations expressed in weight fractions. The highest-quality crystals usually form at low supersaturations (just a few percent). Therefore X_{iM} and $\bar{X}_{iM}$ are close to the corresponding concentrations for saturated solutions, specifically, $X_{iM} \simeq X^0_{iM}$.

If X_{iS}, $X_{iM} \ll 1$, the solutions are regular, and the activity coefficients γ_{iS} and γ_{iM} are independent of the impurity concentration, as are the distribution coefficients. According to Fig. 1.2, this corresponds to approximating the liquidus and solidus curves by straight lines in the vicinity of the melting point of the macrocomponent. The range of constancy of K_0 or $\mathscr{D}_0$ differs in different systems and equals $\sim 10^{-3} - 10$ at. % of the impurity. Deviations from the law $K =$ const are caused by interaction between impurity particles in the mother medium and the crystal.

Determining the numerical values of equilibrium distribution coefficients at different concentrations means determining the thermodynamics phase diagram of the system. The experimental and theoretical analysis of diagrams (especially for metal alloys) is treated in numerous publications, but thus far there is no complete and yet simple theory for systems of different types. We therefore restrict ourselves to an elementary description of the principal approaches to the problem, which are of assistance in understanding the foundations of the phenomenon and arriving at qualitative estimates.

4.2.3 Equilibrium Impurity Distribution in a Crystal–Melt System

Let us first consider growth from a melt with an impurity, i.e., a two-component system, and assume that the impurity forms a substitutional solution in the crystal and the melt. We shall use the chemical potentials of the impurity in the crystal and the medium in the form of (1.1), where ε_j, s_j, and Ω_j are understood to be the energy, entropy, and specific volume of an impurity particle in the crystal and the medium. The energy of an impurity atom in a solid and liquid solution is composed of its internal energy $\varepsilon_{S,L}^{int}$, the configuration energy of the bonds with the neighbors, $\varepsilon_{S,L}^{conf}$, and the elastic energy ε^{el}: which arises in a crystal as a result of the difference in the sizes of the impurity particle and the particle proper. The entropy is also divided into the internal entropy $s_{S,L}^{int}$, which is related to the electron states of the impurity atom and to their vibrations, and the configurational entropy $s_{S,L}^{conf}$, which depends on the number of possible positions of the impurity atom as a whole in the lattice. We shall now calculate ε_S^{conf}, ε^{el}, and s_S^{conf}. Suppose each atom proper or impurity atom in the crystal is surrounded by Z_1 nearest neighbors. The energy of a single bond between atoms of the crystal macrocomponent (S), between atoms of the impurity (i), and between atoms of both will be denoted by ε_{SS}, ε_{ii}, and ε_{Si}, respectively. We further assume that the atoms of the impurity and the crystal occupy lattice points randomly. This means that the probability of finding an atom (i), for instance, is equal to $N_{iS}/(N_S + N_{iS})$, where N_S and N_{iS} are the total numbers of atoms proper and impurity atoms in the crystal. Then, as is easy to calculate, the total potential energy of all the bonds in the crystal is

$$E_{fS}^{conf} = \frac{Z_1}{2}\left(2\varepsilon_{Si}\frac{N_S N_{iS}}{N_S + N_{iS}} + \varepsilon_{SS}\frac{N_S^2}{N_S + N_{iS}} + \varepsilon_{ii}\frac{N_{iS}^2}{N_S + N_{iS}}\right).$$

The change in the potential energy of the system upon the addition of one impurity atom to it is

$$\varepsilon_{iS}^{conf} = \left(\frac{\partial E_S^{conf}}{\partial N_{iS}}\right)_{N_S=\text{const}} = \frac{Z_1\varepsilon_{ii}}{2} + (1 - X_{iS})^2 Z_1 \Delta\varepsilon_S\,, \tag{4.10}$$

where $X_{iS} = N_{iS}/(N_S + N_{iS})$, and $\Delta\varepsilon_S$ is the so-called mixing energy for the crystal per bond:

$$\Delta\varepsilon_S = \varepsilon_{Si} - (\varepsilon_{SS} + \varepsilon_{ii})/2\,. \tag{4.11}$$

It is the change in potential energy needed to replace half of two bonds — ε_{SS} and ε_{ii} — by one bond ε_{Si}. Accordingly, $Z_1\Delta\varepsilon_S$ is the energy required for the simultaneous replacement (exchange) of one impurity atom i in a crystal of the impurity substance by an atom S, and of one atom S in the crystal of the principal substance by an atom i (provided that the numbers of nearest neighbors in the two one-component crystals are the same).

The energy ε_{Si} required to calculate mixing energy $\Delta\varepsilon_S$ is expressed very approximately via the observed parameters of pure components, for example by *Allen*'s empirical rule [4.28]:

$$\frac{1}{\varepsilon_{Si}} = \frac{1}{2}\left(\frac{1}{\varepsilon_{SS}} + \frac{1}{\varepsilon_{ii}}\right). \tag{4.12}$$

The energy ε_{SS} appearing here can, according to (1.22), be expressed via the sublimation heat ΔH of the main component. As regards the impurity, ε_{ii} must be expressed in terms of the sublimation heat ΔH_i of a polymorphous modification of the impurity crystal which has the same structure as the matrix. The heats of polymorphous transformations usually do not exceed several percent of the sublimation heat, whereas the sublimation heats of different substances, even with a similar type of binding, often differ by many tens of percent. In these cases, ε_{ii} can also be estimated by means of (1.22), where ΔH must be equated to ΔH_i. Expressing the binding energy thus in terms of the sublimation heat, and taking into consideration the fact that the values of the binding energies of the mutually attracted atoms are negative, we obtain, by substituting (4.12) into (4.11):

$$Z_1 \Delta\varepsilon_S = \frac{(\Delta H - \Delta H_i)^2}{\Delta H + \Delta H_i} . \tag{4.13}$$

The evaporation heat ΔH of crystalline germanium at its melting point is equal to 90.2 kcal/mol, and that of lead to 46.3 kcal/mol. For this pair, $Z_1\Delta\varepsilon_S =$ 14 kcal/mol, according to (4.13). This value is appreciably lower for more similar substances such as tin ($\Delta H_i = 72$ kcal/mol) in germanium where $Z_1\Delta\varepsilon_S = 2.1$ kcal/mol. It should be emphasized that when ΔH and ΔH_i are similar, the reliability of estimation by (4.13) drops sharply, and it can only be concluded that $Z_1\Delta\varepsilon_S < \Delta H_S, \Delta H_i$. Moreover, approximation (4.12) leads each time to positive numerical values of $Z_1\Delta\varepsilon_S$, which is by no means always true for affine substances (for instance in the formation of chemical compounds).

Let us now turn to the energy of the elastic deformation of the matrix, ε^{el}, which is associated with the incorporation of an impurity atom into the crystal. The simplest (but also least precise) method of estimating this is carried out within the framework of the isotropic theory of elasticity. Assuming the impurity atom (ion) to be a spherical inclusion of radius r_i replacing the material proper which had occupied a sphere of radius r_S, we have, at small concentrations of the impurity atoms, when the interaction of their elastic fields is low,

$$\varepsilon_{el} = 8\pi G r_S (r_i - r_S)^2 , \tag{4.14}$$

where G is the shear modulus.

A more accurate estimation of the elastic energy involved in the incorporation of an impurity particle into the lattice employs a discrete description of the deformation of this particle's bonds with the nearest neighbors, and also a continual description (within the frame of elasticity theory) of the deformation in the remaining crystal [4.29]. The energy distribution among the discretely and continuously described bonds is such that the total elastic energy is minimal. Therefore the described approach yields a lower value of ε^{el} than expression (4.14). For instance, according to (4.14), the incorporation of a lead atom

(r_i = 12.68 Å) into the germanium lattice (r_S = 2.44 Å, $G = 6.7 \times 10^{11}$ erg/cm³) requires that work against elastic forces be performed which is equal to 41×10^4 $(r_i - r_S^2)$ erg/atom $= 2.36 \times 10^{-12}$ erg/atom $\simeq$ 34 kcal/mol, For the incorporation of lead into silicon (r_S = 3.35 Å, $G = 8.0 \times 10^{11}$ erg/cm³), the work equals 5.1×10^{-12} erg/atom = 73 kcal/mol. Calculation by equations that take the discreteness of the first bonds into account gives 10 kcal/mol and 24 kcal/mol, respectively, i.e., about three times less.

The role of the elastic energy in the trapping of an impurity by ionic crystals is accounted for in a similar way [4.30].

If r_i is much less than r_S, the impurity atom may at each instant form bonds with only some of the nearest neighbors, and expressions (4.10, 14) are altogether inapplicable.

The number of ways in which $N_S + N_{iS}$ lattice sites are filled with N_S and N_{iS} atoms of types S and i is equal to $(N_S + N_{iS})!/N_S!N_{iS}!$. Hence the system entropy associated with different modes of distribution of S and i atoms, i.e., the configurational entropy, is equal to

$$S^{\text{conf}} = -kN_S \ln \frac{N_S}{N_S + N_{iS}} - kN_{iS} \ln \frac{N_{iS}}{N_S + N_{iS}} \,.$$

This expression naturally holds true only as long as there are no correlations between the positions of the different atoms, i.e., as long as they are distributed absolutely randomly among the lattice points. In this approximation the change in entropy upon the addition of one impurity atom is

$$S_{iS}^{\text{conf}} = \left(\frac{\partial S^{\text{conf}}}{\partial N_{iS}}\right)_{N_S=\text{const}} = -k \ln X_{iS} \,. \tag{4.15}$$

Substituting (4.10, 15) into (1.1) applied to the impurity atoms in the crystal, we get

$$\mu_{iS} = \mu_{iS}^0 + kT \ln X_{iS} + (1 - X_{iS})^2 Z_1 \Delta\varepsilon_S + \varepsilon^{\text{el}} \,, \tag{4.16}$$

where the quantity

$$\mu_{iS}^0 = \varepsilon_{iS}^{\text{int}} + Z_1 \varepsilon_{ii}/2 - Ts_{iS}^{\text{int}} \tag{4.16a}$$

is the chemical potential of an impurity which is in the standard state at a temperature T. Assuming that the energy states with respect to the internal degrees of freedom of the impurity atoms in the solid solution do not differ markedly from such states in a pure crystal of the impurity substance, and identifying $Z_1\,\varepsilon_{ii}/2$ with the lattice energy of such a crystal, we can conclude the μ_{iS}^0 is close to the chemical potential of the impurity crystal under the same conditions ($\mu_{iS}^0 \approx \mu_i$). In actuality, of course, the distribution of the electron density of valence electrons is different in the crystal formed by the impurity substance than it is for the valence electrons of the impurity in the solid solution, but the bonds i – i are saturated in both cases. Furthermore, the numbers Z_1 of the nearest neighbors are not the same in the crystal of the macrocomponent as they are in the impurity. However, since the heats of polymorphous transformations are not high, as noted above (of the order of 1 kcal/mol), it is suitable, in rough estimates, to identify $Z_1\,\varepsilon_{ii}/2$ with the binding energy in the impurity crystal.

Formally, identification of the standard state of the impurity with the crystalline state corresponds in (4.16) to the equality $X_{iS} = 1$, $\Delta\varepsilon_S = \varepsilon_{SS}/2$, and $\varepsilon^{el} = 0$, as should be the case in a homogeneous crystal.

An expression similar to (4.16) at $\varepsilon^{el} = 0$ is also valid for the chemical potential of an impurity in the melt. Comparing (4.16) and its analog for the medium (melt) with (4.6, 7), we concretize the form of activity coefficients $\gamma_{i,S,M}$ for the model at hand:

$$kT \ln \gamma_{iS} = (1 - X_{iS})^2 Z_1 \Delta\varepsilon_S + \varepsilon^{el}\,, \tag{4.17a}$$

$$kT \ln \gamma_{iM} = (1 - X_{iM})^2 Z_{1M} \Delta\varepsilon_M\,, \tag{4.17b}$$

where Z_{1M} is the number of nearest neighbors of the impurity atom in the medium, and $\Delta\varepsilon_M$ is the mixing energy for the medium in terms of one bond.

The difference appearing in (4.8) is

$$\mu^0_{iM} - \mu^0_{iS} = \Delta H_i - T\Delta s_i\,, \tag{4.18}$$

where ΔH_i and Δs_i can be assumed (with due regard for the estimate of $\mu^0_i \simeq \mu_i$) to be equal to the heat and entropy of fusion of the impurity crystal, taken at the crystallization temperature of the main component, T_0. For instance, the heat and entropy of the fusion of lead are 1.22 kcal/mol and 2.03 kcal/mol deg. respectively ($T_0 = 600$ K). Therefore for the trapping of lead by germanium ($T_0 = 1210$ K), $\mu_{iM} - \mu_{iS} \simeq 1.2$ kcal/mol.

Thus at low impurity concentrations we obtain from (4.8) the following expression for the distribution coefficient:

$$K_0 = \exp\left(-\frac{Z_1\Delta\varepsilon_S + \varepsilon^{el} - (\Delta H_i - T\Delta s_i) - Z_{1M}\Delta\varepsilon_M}{kT}\right). \tag{4.19}$$

The experimentally determined total mixing heat values $Z_{1M}\Delta\varepsilon_M$ in liquid solutions (Z_{1M} is the number of nearest neighbors of the atom in the medium), which are known for certain substances, usually equal several kcal/mol, and may be either positive or negative. For liquid solutions of Sn, Pb, Sb, Bi, and In in Ge the heats of mixing are 0.72, 4.0, 0.87, 3.6, and 9.7 kcal/mol, respectively, while for Sn, Sb, Ga, Pb, and In in Si they are 5.2, 4.7, 1.6, and 5.8 kcal/mol; for As, Al, and Ga in Ge they are –1.5, –2.5, and –0.56, and for As and Al in Si, –3.5, and –2.3 kcal/mol [4.29].

Summarizing all the above estimates for lead in germanium, we can say that the most substantial contributions to the heat of formation of a solid solution are made by the binding energy (4.13) and the elastic energy; these equal 14 kcal/mol and 10 kcal/mol, respectively. The total heat of the transition of lead from the melt to the Ge crystal, which determines the exponent for the distribution coefficient (4.19), equals $14 + 10 + 1.2 - 4 = 21.2$ kcal/mol, i.e., $K_0 \simeq 1.5 \times 10^{-4}$. The experimental value of the distribution coefficient of Pb in Ge is $K_{\rm exp} = 4 \times 10^{-4}$. For Sn in Ge a similar calculation [4.29] gives $K_0 = 2 \times 10^{-2}$ at the experimental value of $K_{\rm exp} = 2 \times 10^{-2}$; for Sn in Si, $K_0 = 0.8 \times 10^{-2}$, $K_{\rm exp} = 2 \times 10^{-2}$; for Ga in Ge, $K_0 = 2 \times 10^{-2}$, $K_{\rm exp} = 10^{-1}$; for Ga in Si, $K_0 = 10^{-2}$, $K_{\rm exp} = 10^{-2}$, In other words, the calculated values differ from the experimental values by no more than one order of magnitude. The same result is obtained from calculations of K_0 for an impurity in alkali-halide crystals [4.30]. Therefore this approach [4.29–34] can be used for tentative estimates of the equilibrium distribution coefficient. We shall see below that the calculations diverge from experiment not only because of the approximate nature of the theory, but also because of the substantital nonequilbrium of trapping in many cases.

As is well known, an impurity reduces or increases the melting temperature of the macrocomponent. This shift ΔT_0 is related to the impurity concentrations in the crystal and melt by an expression following from the equality of chemical potentials for the macrocomponent. It has the following form for low concentrations:

$$\mu_{\rm S} + kT\ln(1 - X_{\rm iS}) = \mu_{\rm M} + kT\ln(1 - X_{\rm iM}) \,. \tag{4.20}$$

Hence, taking into account definition (4.8) and the expression for the difference between the chemical potentials of an impurity-free crystal and a melt, at a temperature $T_0 - \Delta T_0$, $\mu_{\rm M} - \mu_{\rm S} \simeq \Delta H \Delta T_0/kT_0^2$, we set

$$K_0 = 1 - \frac{\Delta H_{\rm S} \Delta T_0}{kT_0^2 X_{\rm iM}} \,. \tag{4.21}$$

The shift in the melting point at low $X_{\rm iM}$ is proportional to the impurity concentration, so that the ratio $\Delta T_0/X_{\rm iM}$ is concentration independent. If this ratio is known, the equilibrium coefficient of distribution can be found from the above expression.

4.2.4 Equilibrium Impurity Distribution in a Crystal–Solution System

In such a system it is customary to use the coefficient $\mathscr{D}_0$, which is related to K_0 by the expression $\mathscr{D}_0 = K_0 X_{\rm M}$ see (4.9). The chemical potential $\mu_{\rm iM}^0$ of the impurity in the standard state in solution, which appears in K_0 and hence in $\mathscr{D}_0$, can be found from the condition for the equilibrium of an impurity crystal with a macrocomponent-free, saturated solution of it in the same solvent [4.35]:

$$\mu_i = \mu_{iM}^0 + kT \ln \gamma_{iM}^0 X_{iM}^0 , \tag{4.22}$$

where μ_i is the chemical potential of the impurity crystal, and concentration X_{iM}^0 and activity coefficient γ_{iM}^0 refer to the saturated impurity solution. Expressing μ_{iM} in terms of μ_{iM}^0 with the aid of this equality and using (4.17a) for the activity of the impurity in the crystal, we obtain from (4.9, 8) at $X_{iS} < 1$:

$$\mathscr{D}_0 = (\gamma_{iM} X_M / \gamma_{iM}^0 X_{iM}^0) \exp [(\mu_i - \mu_{iS}^0 - Z_1 \Delta\varepsilon_S - \varepsilon^{el})/kT] . \tag{4.23}$$

The molar fractions X_M and X_{iM}^0 appearing in (4.23) can easily be expressed via the concentrations in moles or grams per liter of solution, in weight percent, etc., and for electrolyte solutions via the activities or the solubility products of the impurity and the macrocomponent (the principal crystallizing substance) in the same solvent (for instance water).

As noted above, the chemical potential of the impurity crystal, μ_i, does not differ markedly from that of an impurity in the standard state in solid solution. This is particularly true for isomorphous impurities. In determining μ_{iS}^0 from the water vapors over solutions of KCl–$PbCl_2$–H_20, *Ratner* and *Makarov* found, at 25°C for $PbCl_2$, $\exp(\mu_i - \mu_{iS}^0)/kT = 2.3$, i.e., $\mu_i - \mu_{iS}^0 = 490$ cal/mol. On the other hand, changes in binding energy $Z_1\Delta\varepsilon_S$, and the deformation energy of the matrix, ε^{el}, may as was mentioned above, be quite considerable and affect the distribution coefficient more strongly.

The role of the solvent in the distribution of the impurity is reflected by the ratio X_M/X_{iM}^0 in the preexponential factor of (4.23). The macrocomponent concentration X_M is, at the usual low supersaturations, close to its solubility X_M^0. Hence, according to (4.23), the lower the impurity solubility in a given liquid solvent as compared with the macrocomponent, the more easily it enters the crystal. This—Ruff's—rule can also be formulated in a more precise manner: if the solubility of the crystallizing substance in a given liquid solvent is lower than that of the impurity, i.e., $X_M^0/X_{iM}^0 < 1$, then $\mathscr{D}_0$ is also less than unity, and conversely, an impurity "not favored" by the solvent tends to transfer to the crystal in a greater amount. Ruff's rule is fulfilled in experiment better, the more isomorphous the impurity is and the smaller $Z_1\Delta\varepsilon_S$ and ε^{el} are. These last values must be small, in particular in the case of truly isomorphous impurities forming continuous series of solid solutions with the macrocomponent. Some examples of the correlation between the ratio of the solubility of the impurity (C_{i0}) to that of the macrocomponent (C_0) and the experimentally found distribution coefficient $\mathscr{D}_{exp}$ are listed in Table 4.2 (the concentrations of saturated solutions used in it are expressed in g/l, i.e., C_0/C_{i0} slightly differs from X_M^0/X_{iS}^0).

The consideration presented above for low impurity concentrations may be generalized as follows. Let us consider an equilibrium between a crystal AB and a ternary solution ABC, where C is the solvent [4.35, 36b, c]. At equi-

Table 4.2. Distribution coefficients of cations between isomorphic salts [4.36a]

Macrocomponent	Impurity	Temp. [°C]	C_0/C_{i0}	$\mathscr{D}_{exp}$
$Ni(NH_4)_2 \cdot 6H_2O$	$Fe(NH_4)_2(SO_4)_2 \cdot 6H_2O$	20	0.3	0.13
$Cu(NH_4)_2 \cdot H_2O$	$Zn(NH_4)_2(SO_4)_2 \cdot 6H_2O$	20	1.5	2.4
$MgSO_4 \cdot 7H_2O$	$NiSO_4 \cdot 7H_2O$	20	0.9	0.65
$Pb(NO_3)_2$	$Ba(NO_3)^2$	25	4.0	2.47
$PbSO_4$	$BaSO_4$	25	16.6	11.1
$PbSO_4$	$SrSO_4$	25	0.4	0.17
RbCl	KCl	20	1.9	3.2
RbCl	CsCl	25	0.7	0.05
NaCl	AgCl	20	1.7×10^4	19.1

librium the chemical potentials of A and B must be equal one to another in the crystal and the solution:

$$\begin{aligned} \mu_{AL}^0 + kT \ln a_{AL} &= \mu_{AS}^0 + kT \ln a_{AS} \,, \\ \mu_{BL}^0 + kT \ln a_{BL} &= \mu_{BS}^0 + kT \ln a_{BS} \,. \end{aligned} \tag{4.24}$$

Here a stands for activity, the subscripts A and B denote the type of substance in the liquid solution (L) and in the solid crystal (S), and the superscript 0 stands for the chemical potential of the standard state. The unknown standard potentials μ_{AL}^0 and μ_{BL}^0 may be excluded by making use of the following conditions of equilibrium between the pure crystals A and B and their saturated binary solutions in the same solvent:

$$\begin{aligned} \mu_{AL0} &= \mu_{AL}^0 + kT \ln a_{AL0} = \mu_{AS0} \\ \mu_{BL0} &= \mu_{BL}^0 + kT \ln a_{BL0} = \mu_{BS0} \,, \end{aligned} \tag{4.25}$$

where the subscript 0 stands for the quantities describing this equilibrium. By excluding μ_{AL}^0 and μ_{BL}^0 one obtains the following Ratner relation between activities:

$$\frac{a_{BS}/a_{AS}}{a_{BL}/a_{AL}} = \frac{a_{AL0}}{a_{BL0}} \exp\left(\frac{(\mu_{BS0} - \mu_{BS}^0) - (\mu_{AS0} - \mu_{AS}^0)}{kT} \right). \tag{4.26}$$

Let us now define the distribution coefficient $D_{B/A}$ of B in A, in terms of molar fractions X_A and X_B in the solid, and molalities[4] m_A and m_B in the solution (these are the conventional units in the literature on solution growth):

$$D_{B/A} = \frac{X_B m_A}{X_A m_B} \,. \tag{4.27}$$

[4] Molality is defined as the number of moles of solute in 1000 g of solvent.

Replacing in (4.26) the activities $a_{\alpha S} = \gamma_{\alpha S} X_\alpha$, $a_{\alpha L} = \gamma_{\alpha L} m_\alpha$, α = A, B, and making use of (4.27) one gets

$$D_{B/A} = \frac{\gamma_{AS}\ \gamma_{BL}\ \gamma_{AL0}}{\gamma_{BS}\ \gamma_{AL}\ \gamma_{BL0}} \cdot \frac{m_{A0}}{m_{B0}} \exp\left(\frac{(\mu_{BS0} - \mu^0_{BS}) - (\mu_{AS0} - \mu^0_{AS})}{kT}\right). \tag{4.28}$$

For ideal liquid and solid solutions, all $\gamma = 1$. For nonideal solutions, mutual compensation of γ in the preexponential factor is also possible, and gives the same result. In ideal solutions the exponent itself equals unity and we return to Ruff's rule:

$$D_{B/A} = m_{A0}/m_{B0}\,. \tag{4.29}$$

In nonideal solutions both the exponent and activity coefficients may be found either experimentally or by means of empirical relations. Some of these relations are discussed by *Urusov* [4.36b], who used them to calculate distribution coefficients for numerous mixed alkali halide crystals (NaCl: K^+ or Br^-; KCl: Na^+, Rb^+, Cs^+, Br^-, I^-; KBr: Rb^+, Cl^-, I^-; KI: Na^+, Rb^+, Cs^+, Br^-, Cl^-; RbCl: K^+; RbI: K^+ [the colon stands between the host crystal and impurity ions]), for several sulfates of divalent salts with divalent impurity cations (Sr^{2+}, Pb^{2+}, Ba^{2+}, Zn^{2+}, Ca^{2+}, Cd^{2+}) and anions (CrO_4^{2-}, Br^-), and also for some low-soluble impurities in alkali halides (NaCl and KCl with Tl and Pb; $BaCl_2$, LiCl, and NaBr with Pb). These calculations have been made for crystallization from aqueous solutions, and the closer the solution is to being ideal, the better the agreement with experiment. The theory of isomorphous mixing in nonmetallic solids is reviewed in [4.36d].

The general analysis based on Ratner's relation may be specified for dissociative ionic compounds with a common cation or anion. Let compound A dissociate, producing ν_+ cations A^+ and ν_- anions A^-, thus taking the form $A = A^+_{\nu+} A^-_{\nu-}$. Analogously, let $B = B^+_{\nu_+} B^-_{\nu_-}$ with the same ν_+ and ν_-. In this case

$$a_{AL} = \gamma^{\nu_+}_{A^+L}\, m^{\nu_+}_{A^+}\, \gamma^{\nu_-}_{A^-L}\, m^{\nu_-}_{A^-} = \gamma^{\nu}_{AL}\, m^{\nu_+}_{A^+}\, m^{\nu_-}_{A^-} = \nu^{\nu_+}_+\, \nu^{\nu_-}_-\, \gamma^{\nu}_{AL}\, m^{\nu}_{A}\,, \tag{4.30}$$

where $\nu = \nu_+ + \nu_-$, $m_{A^+} = \nu_+ m_A$, $m_{A^-} = \nu_- m_A$ and $\gamma_{AL} = (\gamma^{\nu_+}_{A^+L}\gamma^{\nu_-}_{A^-L})^{1/\nu}$ is the mean activity coefficient. The same relations may be written for the B compound. Let us suppose that the anions in A and B are the same, i.e., $A^- \equiv B^-$ (common anions). Then in the mixed (ternary) solution ABC one has $m_{A^-} = m_{B^-}$ and $\gamma_{A^-L} = \gamma_{B^-L}$, because the chemical potentials μ_{AL} and μ_{BL} (i.e., the work needed to add an A or B molecule to the solution), and thus a_{AL} and a_{BL}, depend on the overall concentration of A^- in the solution. Similar relations are valid for activities in the AB crystal. For the saturated binary solution

$$a_{AL0} = \nu^{\nu_+}_+\, \nu^{\nu_-}_-\, \gamma^{\nu}_{AL0}\, m^{\nu}_{A0}\,, \quad a_{BL0} = \nu^{\nu_+}_+ \nu^{\nu_-}_- \gamma^{\nu}_{BL0} m^{\nu}_{B0} \tag{4.31}$$

Substituting into (4.26) the first and second equalities of (4.30) for the solid and liquid, respectively, and $D_{B/A} = X_{B^+}m_{A^+}/X_{A^+}m_{B^+}$, one gets

$$D_{B/A}^{\nu_+} = \left(\frac{\gamma_{BL}\,\gamma_{AL0}}{\gamma_{AL}\,\gamma_{BL0}}\right)^{\nu}\left(\frac{\gamma_{A^+S}}{\gamma_{B^+S}}\right)^{\nu_+}\left(\frac{m_{A0}}{m_{B0}}\right)^{\nu}\exp\left(\frac{(\mu_{BS0}-\mu_{BS}^{0})-(\mu_{AS0}-\mu_{AS}^{0})}{kT}\right).$$

For isomorphous ions, mutual compensation of activity coefficients may be expected in both the liquid and the solid. The compensation in the solid should be best for hydrates where cations are surrounded by water molecules which form "buffering" and "screening" shells, thus leveling the difference between the isomorphous ions. "Buffering" and "screening" were introduced to explain *Gorshtein's* experiments [4.36a], in which the distribution coefficients $D_{B/A}$ for different hydrated salts (Table 4.2) were found to be constant at all m_A/m_B ratios in aqueous solutions. Since noncommon as well as common ions are heavily hydrated, "the electrolyte may conditionally be considered to be a binary one, i.e., $\nu = 2$." This reasoning [4.36c] reduces the above expression for $D_{B/A}$ to the following:

$$D_{B/A} = (m_{A0}/m_{B0})^2 \,. \tag{4.33}$$

The values of $D_{B/A}$ calculated according to this formula are compared with experimental data in Table 4.3, which was compiled by *Balarev* [4.36c] from numerous sources. With some exceptions, the agreement is quite good. Further analysis shows that the experimental values of $D_{B/A}$ are typically higher (lower) than the calculated values if the ionic radius of the admixture is larger (smaller) than that of the host ion.

All the assumptions made in order to obtain the final simple expressions for $D_{B/A}$ require further consideration. The most dangerous assumption appears to be that $\nu = 2$, because the real amount of ions is larger for many of the salts presented in Table 4.3. Thus it would be better to write

$$D_{B/A} = (m_{A0}/m_{B0})^{\nu/\nu_+} \,. \tag{4.34}$$

For all but the last of the sulfates presented Table 4.3, $\nu_+ = 2$, $\nu_- = 2$, and $\nu = 4$. Consequently $\nu/\nu_+ = 2$, and the calculated values of $D_{B/A}$ in the table remain valid. For other salts, other powers of m_{A0}/m_{B0} should be used which will diminish the agreement between the calculated data and the measured data. In these cases, the degree of dissociation of the salts needs to be analyzed, along with the actual values of the activities and the exponent in general expressions for $D_{B/A}$.

It should also be emphasized that the experimental values of $D_{B/A}$ are usually measured for crystals obtained in the course of growth, and not at equilibrium. Thus the distribution of impurities must be influenced by growth kinetics. The most striking manifestation of this influence is that different faces have different distribution coefficients (see Sect. 4.3).

Table 4.3. Distribution coefficients between isomorphic salt hydrates at 25°C [4.36c]

Admixed ion / Host ion	m_{A0}/m_{B0}	$D_{B/A}=(m_{A0}m_{B0})^2$	$D_{B/A\,exp}$
	$MeSO_4 \cdot 7H_2O$		
Ni/Zn	3.590/2.660	1.82	1.87
Zn/Ni	2.660/3.590	0.55	0.53
Ni/Mg	3.100/2.660	1.36	1.43
Mg/Ni	2.660/3.100	0.74	0.70
Mg/Zn	3.590/3.100	1.34	1.37
Zn/Mg	3.100/3.590	0.75	0.74
Co/Fe	1.964/2.425	0.66	0.68
	$NH_4Me(SO_4)_2 \cdot 12H_2O$		
Cr/Al	0.278/0.604	0.21	0.25
Al/Cr	0.604/0.278	4.72	4.17
Fe/Al	0.278/1.705	0.026	0.021
Al/Fe	1.705/0.278	37.6	66.14
Cr/Fe	1.705/0.604	7.97	6.67
Fe/Cr	0.604/1.705	0.13	0.15
	$MeAl(SO_4)_2 \cdot 12H_2O$		
NH_4/K	0.273/0.278	0.96	0.93
K/NH_4	0.278/0.273	1.04	1.08
Tl/K	0.273/0.177	2.38	2.40
K/Tl	0.177/0.273	0.42	0.42
Tl/NH_4	0.278/0.177	2.47	2.63
NH_4/Tl	0.177/0.278	0.40	0.38
	$MeCr(SO_4)_2 \cdot 12H_2O$		
K/NH_4	0.604/0.817	0.55	0.63
	$(NH_4)_2Me(SO_4)_2 \cdot 6H_2O$		
Ni/Mg	0.595/0.226	6.93	7.7
Mg/Ni	0.226/0.595	0.14	0.13
Ni/Cu	0.663/0.226	8.65	4.65
Cu/Ni	0.226/0.663	0.12	0.21
Ni/Co	0.539/0.226	5.95	3.5
Co/Ni	0.226/0.539	0.17	0.29
Zn/Fe	0.930/0.425	4.73	4.17
Fe/Zn	0.425/0.930	0.21	0.24
Ni/Zn	0.425/0.226	3.56	3.50
Zn/Ni	0.226/0.425	0.28	0.29
Co/Fe	0.930/0.539	2.97	3.00
Fe/Co	0.539/0.930	0.34	0.33
Zn/Cu	0.663/0.425	2.44	2.50
Cu/Zn	0.425/0.663	0.41	0.40
Mg/Fe	0.930/0.595	2.44	1.62
Fe/Mg	0.595/0.930	0.41	0.62
Cu/Fe	0.930/0.663	1.98	1.56
Fe/Cu	0.663/0.930	0.50	0.64
Zn/Co	0.539/0.425	1.60	1.29
Co/Zn	0.425/0.539	0.62	0.75
Co/Cu	0.663/0.539	1.52	1.53
Cu/Co	0.539/0.663	0.66	0.65

Table 4.3. (Continued)

Admixed ion / Host ion	m_{A0}/m_{B0}	$D_{B/A}=(m_{A0}m_{B0})^2$	$D_{B/A\,exp}$
	$Me(HCOO)_2 \cdot 4H_2O$		
Mg/Ni	0.146/1.188	0.02	0.04
Mg/Co	0.168/1.188	0.02	0.02
Mg/Zn	0.366/1.188	0.09	0.08
Mn/Ni	0.146/0.462	0.10	0.27
Mn/Co	0.168/0.462	0.13	0.25
Mg/Mn	0.462/1.188	0.15	0.12
Cd/Mn	0.462/0.687	0.45	0.86
Mn/Zn	0.366/0.462	0.62	0.67
Zn/Mn	0.462/0.366	1.60	1.50
Mn/Cd	0.687/0.462	2.21	1.19
Mn/Mg	1.188/0.462	6.60	8.33
Co/Mn	0.162/0.168	7.57	4.00
Ni/Mn	0.462/0.146	10.0	3.70
Zn/Mg	1.188/0.366	10.5	12.5
Co/Mg	1.188/0.168	50	50
Ni/Mg	1.188/0.146	66	25
	$Me(CH_3COO)_2 \cdot 4H_2O$		
Mg/Co	1.424/4.609	0.09	0.05
Co/Mg	4.609/1.424	10.5	20
Mg/Ni	0.747/4.609	0.026	0.02
Ni/Mg	4.609/0.747	38.1	50
	$Me(NO_3)_2 \cdot 6H_2O$		
Zn/Mn	7.393/6.204	1.42	1.41
Mn/Zn	6.204/7.393	0.70	0.71

The experimental data presented in Table 4.3 clearly demonstrate the validity of general rules for isomorphous salts [4.36a] which follow from general expressions for $D_{B/A}$:

$$D_{B/A} = D_{A/B}^{-1}, \quad D_{B/A} = D_{B/E} \cdot D_{E/A}, \tag{4.35}$$

where E stands for a third compound isomorphous to both A and B.

4.2.5 Equilibrium in the Surface Layer

We have considered the distribution coefficient between the mother medium and the crystal bulk, i.e., a three-dimensional solid solution. But the concentrations of an impurity differ in different atomic positions at the crystal–medium interface, even at thermodynamic equilibrium. The impurity atoms (or molecules or ions) have different concentrations in the surface layer, in a step, and at kinks (positions *5*, *4*, and *3* in Fig. 1.7), where they form two-, one-, and zero-dimensional solid solutions. Their heats of formation differ, firstly, due to the difference in the numbers Z_1 of the nearest "crystalline" neighbors. Secondly, the incorporation of an impurity particle into the surface layer or into a step requires less work for lattice deformation than incorporation into the crystal bulk. Even if we assume that the lattice deformation due to an impurity particle in the surface layer or in the step is the same as that caused by a particle in the bulk, the elastic energy obtained by integration over the crystal volume is, accordingly, $\simeq 7\pi\, Gr_S(r_S - r_f)^2$ and $\simeq 5\pi Gr_S(r_S - r_f)^2$ instead of $8\pi Gr_S(r_f - r_S)^2$ for the three-dimensional solution, see (4.14). The indicated differences can lead for instance, to a situation where for a bismuth impurity in germanium, for which the heat of formation of the three-dimensional solution is 25 kcal/mol according to the theoretical estimate presented in Sect. 4.2.3, the two-dimensional solution requires 23 kcal/mol to form and the one-dimensional solution 19 kcal/mol. Consequently, the concentrations of a one- or two-dimensional solution at $T \simeq 1200$ K exceed that of a three-dimensional solution by 2 and 10 times, respectively [1.18].

4.2.6 Mutual Effects of Impurity Particles

If the impurity concentration in a system is high enough, the impurity particles begin to interact, and this results in a dependence of the distribution coefficient on the concentration. The statistical interaction in the absence of a correlation — accidental formation of pairs, triads, etc., from impurity particles in a crystal and a liquid — is reflected by the terms $(1 - X_{iS})^2\, Z_1 \Delta\varepsilon_S$ and $(1 - X_{iM})\, Z_{iM}\Delta\varepsilon_M$ in the expressions for the chemical potentials. This interaction becomes appreciable when $X_{i,S,M} \gtrsim 10^{-1}$ and, of course, depends on the mixing energies. The probability that groups of impurity atoms (ions, molecules) will be formed increases strongly if their formation from isolated dissolved atoms is energetically advantageous. Furthermore, the incorporation of an impurity particle into a lattice already deformed by another, identical impurity particle requires more work than incorporation into an undeformed lattice. This elastic repulsion lowers the solubility in the crystal and reduced the distribution coefficient, starting with certain concentrations.

In semiconductors, impurity particles become ionized and interact with each other also through the reservoir of intrinsic and impurity electrons and holes, with the result that the distribution coefficient decreases with concentration at high impurity concentrations [4.37]. Let us consider this effect in more detail. We denote an impurity atom in a mother medium as I_M, an impurity atom in the crystal as I_S, a negative ion in the crystal as I_S^-, an electron as e^-, and a hole as e^+. The concentrations of particles will, by convention, be denoted by brackets: $[e^+]$, $[e^-]$, and $[I_S^-]$. Then, if an impurity atom is ionized with the formation of ion I_S^- and hole e^+, the impurity exchange between the phases is written as a chain of reactions $I_M \rightleftarrows I_S \rightleftarrows I_S^- + e^+$. The mass action law, as applied to the extreme states of this chain, has the form:

$$X_{iS}[e^+]/X_{iM} = \mathscr{K}_0, \tag{4.36}$$

where $\mathscr{K}_0$ is the equilibrium constant of the reaction. The above equality assumes that the impurity atoms in the crystal are fully ionized, i.e., that $X_{iS} = [I_S^-]$. If the crystal has no other sources of electric charges apart from the ionized impurity atoms, the fact of ionization affects the value of the distribution coefficient only through the expressions for intereaction energy ε_i, elastic energy ε^{el}, and the other parameters appearing in (4.19). In actuality, however, semiconductors have electrons and holes arising from the ionization of proper atoms of the material and causing its intrinsic conductivity. Here, the neutrality of each macrovolume of the crystal is ensured by the concentrations not only of the impurity (extrinsic) holes and ions, but also of the intrinsic holes and electrons. Hence, the neutrality condition has the form $[e^+] = [e^-] + [I_S^-]$, i.e., under conditions of full ionization of the impurity, $[e^-] = [e^+] - [X_{iS}]$.

According to the mass action law for the ionization of proper atoms of a semiconductor, the product of the concentrations of electrons and holes in a semiconductor with an impurity must be the same as in a purely intrinsic semiconductor at a given temperature. If the carrier concentrations in the intrinsic semiconductor are denoted $[e^+]_{in} = [e^-]_{in}$, then in a crystal with an impurity we must have

$$[e^-][e^+] = ([e^+] - X_{iS})[e^+] = [e^+]_{in}^2. \tag{4.37}$$

Thus (4.36, 37) determine the concentrations of holes, $[e^+]$, and of impurity atoms, X_{iS}, in a semiconductor with an impurity if the impurity concentration in the medium X_{iM} is given. Solving these equations, we have

$$X_{iS} = \mathscr{K}_0 X_{iM}/\sqrt{[e^+]_{in}^2 + \mathscr{K}_0 X_{iM}}.$$

If the concentration of the impurity in the crystal is lower than that of the intrinsic carriers, $\mathscr{K}_0 X_{iM} < [e^+]_{in}$, the latter do not affect the trapping of the impurity, so that according to (4.36) the distribution coefficient is $K = K_{iS}/X_{iM}/ = \mathscr{K}_0[e^+]_{in} = K_0$, i.e., the equilibrium coefficient, which does not take the effect of the electron-hold subsystem of the crystal into account. In the general case

$$K = K_0/\sqrt{1 + K_0 X_{iM}/[e^+]_{in}}. \tag{4.38}$$

Thus when the concentration of a fully ionized impurity in a crystal $\sim K_0 X_{iM}$ becomes comparable to the concentration $[e^+]_{in}$ of the intrinsic carriers at the crystallization temperature, and further introduction of the impurity is perceptibly impeded by the fact that the number of ionized proper atoms must fall appreciably. As a result, the electron-hole subsystem of the crystal prevents a further increase in impurity concentration, and the distribution coefficient diminishes [4.38].

An expression of the type of (4.38) correctly describes, for instance, the concentration dependence of the coefficient of trapping of Zn by crystals of GaAs during growth from the gas phase. But the value $[e^+]_{in} = 4 \times 10^{18}\ \mathrm{cm}^{-3}$, necessary for agreement with experiment, exceeds the carrier density obtained from the Hall effect ($7 \times 10^{17}\ \mathrm{cm}^{-3}$). The cause of the divergence may lie in the nonequilibrium of the trapping processes (Sect. 4.3), the defect structure of the crystal, its nonstoichiometry, or the inadequacy of the carrier density measured on the basis of the Hall effect and the total amount of trapped impurity.

The correctness of the described physical picture of the phenomenon is also confirmed by the experimentally observed mutual influence of two different impurities if their bulk concentrations C_{iS} in the crystal reach values of $\gtrsim 10^{17}$–$10^{18}\ \mathrm{cm}^{-3}$. Thus, simultaneous introduction of two donor or two acceptor impurities into the crystal reduces the distribution coefficient of each of them in accordance with the above reasoning and a ratio of the type of (4.38). Simultaneous introduction of one donor and one acceptor impurity, on the other hand, increases the coefficient of distribution of each of them.

There is also reason to believe that the interaction of impurities via the electron subsystem in defective crystals depends on the concentration of charged defects rather than on that of intrinsic carriers $[e^+]_{in}$) as in the ideal crystal [4.39].

A mutual effect is also observed when a crystal simultaneously traps several impurities, complexes of which have a lower heat of formation of a solid solution than the sum of the heats corresponding to the trapping of each impurity individually. This is manifested most prominently when a dielectric crystal traps two ions such that the sum of their valences is equal to the valence of the replaced ion. For instance, in quartz the Si^{4+} ion is replaced by complexes $Fe^{3+} + H^+$, $Fe^{2+} + 2H^+$, $Al^{3+} + H^+$, $Al^{3+} + Li^+$, etc., In other words, the presence in the mother liquor of a trivalent or divalent cation trapped by the crystal increases the amount of hydrogen or lithium entering the crystal. The equilibrium coefficient of pairwise replacement can be found to be the same as the equilibrium constant for the corresponding chemical reaction, e.g.,

$$(M^{2+})_M + 2(H^+)_M \rightleftarrows (M^{2+}\,2H^+)_S \,. \qquad (4.39)$$

where the subscripts M and S indicate, as before, that the ion belongs to the medium and the crystal [4.40].

4.3 Trapping of Impurities: Kinetics

4.3.1 Surface Processes

In this section we will discuss the mechanism and kinetics of the trapping of impurities by the surface of a layerwise-growing crystal, and also the subsequent relaxation processes. If the growth rate is high enough, the crystal traps the impurity in an amount differing from the equilibrium amount, and there is no time for subsequent relaxation to take place. The result is nonequilibrium trapping and the appearance of metastable structures.

a) Statistical Selection [4.41]
The existence of surface solid solutions must affect the amount of impurity trapped during layer growth. A new layer laid by a step buries the impurity atoms of the surface layer, transferring them from a position "in the surface" to a position "in the bulk" (Fig. 1.7). But even equilibrium concentrations of an impurity in the surface and in the bulk are different (Sect. 4.2.5), and therefore the amount of impurity in just "buried" regions of the crystal bulk will not be equilibrium, generally speaking. A similar situation may occur on the step: building up new atomic rows during the motion of kinks and "burying" the impurity atoms in the step rise lead to a nonequilibrium concentration of the impurity in the surface layer. Finally, the impurity concentration in the step rise rows which are laid in the course of kink motion may also differ from equilibrium. Let us consider this last process of impurity trapping for the simplest case of an isolated kink (Fig. 4.8). The kink moves along the step as a result of individual acts of attachment and detachment of particles. We assume that these events are accidental and the impurity concentration is low. The frequencies with which particles arrive at the kink and are detached from it depend on the state of the medium and the impurity concentration. Due to the consecutive acts of attachment and detachment of particles, the kink executes random walks (i.e., it diffuses) along the step. In the course of growth, a systematic forward motion is superimposed on this diffusion, and during dissolution (evaporation, melting), a reverse motion is superimposed. Suppose now that at a certain moment an impurity particles finds its way into the kink (Fig. 4.8a, shaded cube). Only two outcomes are conceivable: either the impurity leaves the kink and the kink returns to its initial state, or a new particle is added (Fig. 4.8b). Since there are many more host particles in the medium than there are impurity particles, this new particle will almost certainly be a host particle. Then there are two further possibilities: either the kink loses the new particle and returns to the configuration of Fig. 4.8a, or it adds on one more particle, also of the main crystallizing substance, etc. The impurity particle may leave the crystal during the entirety of this time, but the most favorable times for this are periods when the impurity particle is "open" (Fig. 4.8a). The probability that the impurity particle will be preserved in the kink during all the subsequent

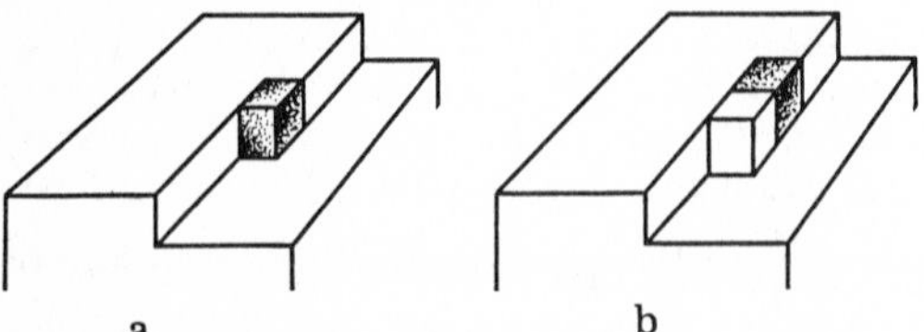

Fig.4.8a,b. Consecutive acts of entrapment of an impurity particle (*shaded cube*) during the building of a row with host particles (*open cubes*)

events is equal to zero at equilibrium and increases with deviation from equilibrium.[5] The higher the energy of adsorption of the impurity in the kink, the closer the probability of particle preservation is to unity. Therefore at a fixed frequency of addition of a strongly adsorbing impurity to the kink, its concentration in the atomic row laid by the kink will drop with increasing supersaturation: the higher the supersaturation, the larger the number of host particles deposited in the time between the consecutive acts of addition of impurity particles. For weakly adsorbed particles it is the probability of preservation that is most important, and the distribution coefficient increases with the supersaturation [4.41]. Statistical selection at rough interface and resulting phase diagrams were analyzed, mainly numerically, in [4.42b,c].

A similar process of statistical "natural selection" of the crystal composition also takes place during the motion of a step over the surface of the layer deposited by the preceding step. The impurity atoms incorporated into the surface layer are stoppers for the step in question. As a result, the step is retarded in front of such a stopper and fluctuates over it, opening and closing it, so that the impurity particle gets a better chance to leave the crystal [4.42a]. Here, too, the statistical kinetics of fluctuations determines the concentration of the trapped impurity, in this case in the second surface layer of the crystal. If the growth proceeds extremely slowly, a greater number of trials and errors for the final filling of each lattice site will ensure the equilibrium value of the impurity concentration in the crystal. If, however, the number of trials is insufficient, the distribution coefficient will differ from its equilibrium value, and will depend on the supersaturation or supercooling in the system. Specifically, a theoretical analysis of statistical selection by kinks on the step yields [4.41]:

$$K_{st} = \frac{K_{st0}}{1 + \Delta\mu/\Delta\mu^*}, \tag{4.24}$$

where K_{st0} is the equilibrium distribution coefficient at the step corresponding to the equilibrium concentration in the surface layer, $\Delta\mu$ is the degeee of deviation from equilibrium and $\Delta\mu^*$ is some constant depending on the elementary frequencies of attachment and detachment of host particles and impurity particles. The value of $\Delta\mu^*$ is positive for an impurity adsorbing more strongly in kinks than particles of the crystallizing macrocomponent, and negative in the opposite case. Accordingly, in the first case the distribution coefficient reduces with increasing supersaturation, and in the second it increases. A dependence

[5] This does not, of course, imply that the distribution coefficient at equilibrium is equal to zero. Indeed, the distribution coefficient is the ratio of the number of impurity particles entering the crystal and remaining in it to the number of identical particles of the substance proper. At equilibrium, the probability of remaining in the crystal is zero for particles of both kinds, i.e., the time-average values of the fluxes of these particles into the crystal are equal to zero. The flux ratio is thus an indeterminate form whose evaluation gives the equilibrium distribution coefficient.

similar in nature is obtained for selection of the impurity by the step in the surface layer.

Finally, in considering the kinetics of selection on steps one should bear in mind that impurity particles find their way into the step not only directly from the solution bulk, but also from the adsorption impurity layer on the surface. At sufficiently low lifetimes of impurity particles on the surface (compared with h/R, where R is the normal growth rate of the face, and h is the step height), it is this adsorption layer that will be the mother medium from which the selection is made. In equilibrium trapping, however, the presence of such a layer is immaterial, because the concentration in the crystal depends on the chemical potential of the impurity, which is the same for all its states in the system.

Thus, statistical selection leads to results which can be summarized as follows: At every small deviations from equilibrium the statistical slection on the step ensures equilibrium surface concentration in each new surface layer. The deposition of each subsequent layer proceeds so slowly that the transformation of the preceding surface layer into a bulk layer is accompanied by a change in the amount of impurity in it to the value corresponding to bulk equilibrium between crystal and medium. The distribution coefficient will ultimately be equal to the thermodynamic equilibrium value for the bulk (Sect. 4.2). This situation is evidently realizable in practice only during growth from the melt.

At higher supercoolings or supersaturations the statistical selection on the step can ensure equilibrium concentration of the impurity in each new surface layer, but burying of the surface layers during the motion of the subsequent steps proceeds so rapidly that almost all the impurity atoms in the boundary layer of the crystal are preserved, and are thus found to be in the crystal bulk. In this case the distribution coefficient of a given face will be equal to the equilibrium distribution coefficient K_{st0} between the melt (solution) and the surface atomic net of the given crystal face. The value of K_{st0} corresponds to the concentration of a two-dimensional surface solid solution, which may, in accordance with Sect. 4.2.5, differ by several times from the equilibrium concentration of a bulk three-dimensional solution. The values of K_{st0} are different for faces of different crystallographic indices, and therefore the amounts of impurity trapped by these faces will be different — the so-called sectorial structure of the crystal will arise.

At still greater deviations from equilibrium, the concentration of impurity trapped by each newly deposited layer may due to the deterioration of statistical selection, correspond to the equilibrium concentration in the one-dimensional solution in the step rise (Sect. 4.2.5), or it may not even correspond to any equilibrium concentration, being determined by purely kinetic phenomena. As a result, the distribution coefficient for a given face will depend not only on the index of this face, but also on the crystallographic orientation of the steps on it. Since one face almost always has steps of different orientations (for instance on vicinal growth hillocks), the growth pyramid of even one face will be spatially inhomogeneous as regards the impurity concentration. This situation is possible in growth from solution. We observed this effect recently by in situ x-ray topography and also in slices cut parallel to the dipyramidal face of ADP.

Finally, in the case of maximum growth rates the composition of the crystal will repeat that of the mother medium (Sect. 4.3.1.d).

Each of the described characteristic cases is realized in its own range of growth rate V. In the intervals between neighboring ranges of V the distribution coefficient changes from one characteristic value to another. For growth from solution such transitions have been thoroughly investigated by *Melikhov* [4.43, 44].

b) Diffusional Relaxation

As has already been mentioned, the ingress of an impurity particle into the surface or even into deeper layers does not necessarily mean that it will stay in the crystal. Indeed, let us consider a crystal layer just deposited during the motion of the step. The impurity concentration in this surface layer is generally different from the equilibrium value for such a layer. Furthermore, the deposited layer has "buried" the layer deposited by the preceding step, transfering it into the crystal bulk. The concentration in it is, in turn, different from the equilibrium value for the crystal bulk. Both these deviations from equilibrium will cause diffusion flows of the impurity in the crystal and across its boundary at a characteristic rate D_{iS}/h, where D_{iS} is the impurity diffusivity either in the crystal or across the phase boundary (i.e., from the surface layer of the crystal to the mother medium), depending on the process under consideration, i.e., on the value of h. Each impurity atom in the surface layer is in an "open" state until the approach of the next step, i.e., within the time $\lambda/v = h/R$, where h is the layer thickness (rise height), v and R are the tangential and normal growth rates of the face, and λ is the spacing between steps. The probability that the trapped particle will be removed from the surface layer before the arrival of the next step is $\exp(-h/R\tau_i^{\ddagger})$, where $\tau_i \simeq h/^2D_{iS}$ is the lifetime of the impurity particle in the surface layer.[6] Therefore

$$K = K_0 + (K_{st} - K_0)\exp(-D_{iS}/Rh)\ ; \tag{4.41}$$

i.e., at low growth rates, when

$$R < D_{iS}/h\ , \tag{4.42}$$

the bulk distribution coefficient will have its thermodynamic equilibrium value. In the opposite case anything that has been trapped by the step will remain in the crystal bulk. Furthermore, the impurity buried in the surface layer while the step moves over it also has an opportunity, although not a very great one, for diffusion exchange with the medium, now through the deposited layer.

---

[6] This estimate neglects the selection made by the step fluctuating at the impurity particle to be buried (Sect. 4.3.1a).

This relaxation is described by an expression of the same type as (4.41), but with different h and D_{iS}.

Statistical selection and diffusional relaxation in normal growth have not been investigated, but inequality (4.42) is suitable for estimation in this cases, too.

From (4.41) it follows that the amount of impurity remaining in the crystal depends on the height of the step, i.e., it is not indifferent to whether the given growth rate R is ensured by slowly moving but high steps or by rapidly propagating thin layers. Physically, this is related to the dependence of characteristic diffusion rate D_{iS}/h on step height h.

In growth from the melt at high temperatures, when the diffusion in the solid is quite fast, inequality (4.42) may hold for the growth rates used in laboratory practice. For instance, for the diffusion of impurities in crystalline germanium near the melting point, $D_{iS} \simeq 10^{-9}$–10^{-12} cm^2/s, and at $h \simeq 10^{-7}$ cm we have $R < 10^{-5}$–10^{-2} cm/s. In growth from solution at room or moderate temperatures ($\lesssim 500$°C—hydrothermal synthesis), on the other hand, the diffusion coefficients are much lower. Thus in alkali halide crystals under these conditions, $D_{iS} \simeq 10^{-15}$–10^{-17} cm^2/s, i.e., the trapping will be equilibrium only at $R \ll 10^{-8}$–10^{-10} cm/s ($\ll 10^{-2}$–10^{-4} mm/d). Such low velocities are realized only under special conditions of very low supersaturation, for instance in experiments on recrystallization ("ripening") of fine-crystal powders, where a very low supersaturation (10^{-4}–10^{-5}) is achieved due to the difference in the solubilities of small crystals of different sizes (Sect. 1.4) or to periodic alternation of growth and dissolution of crystals.

c) Sectorial Structure

As demonstrated above, the concentrations of an impurity trapped by different faces, that is, the concentrations in different growth pyramids, correspond to the equilibrium concentrations in the surface layers of these faces (in growth from the melt this is the most probable case) or in the rises of the corresponding steps; or, at the highest growth rates, they are determined by the statistics of selection by kinks. The more efficiently the growth conditions ensure the fulfillment of inequality (4.42), the less pronounced is the sectorial structure of the crystal with respect to the homogeneously trapped impurity.

A crystal surface layer having a thickness equal to one lattice spacing very seldom consists of a single plane net such as that in a crystal with a simple cubic lattice. Even in the diamond lattice, (111) layers of thickness equal to one lattice spacing are formed by two plane nets. Each atom in the external net of such a layer has three neighbors in the crystal and one neighbor in the liquid (if the melt has the same short-range order as the crystal), whereas each atom of the second, internal net is linked by four "crystalline" bonds. This non-equivalence causes, firstly, the very propagation of such a layer as a whole rather than of each net separately: the upper net would have a higher tangential velocity than the lower one, and would catch up with it. Secondly, a difference between the impurity concentrations in neighboring nets becomes possible.

The nonequivalence of positions in the structure of the surface layer of the minor quartz rhombohedron discussed at the beginning of Sect. 4.2.1 is still more pronounced (Fig. 4.7). The nonequivalence of atomic positions within a single layer causes inequality both in the probabilities of entrapment of the impurity in the course of primary selection during the deposition of this layer and in the probabilities of subsequent diffusion exchange with the medium. Therefore the crystal bulk can "memorize" not only the type of the face that formed it, but also the growth direction of this face, or even the direction of the step motion on the face.

A distinctive example of sectorial structure in impurity distribution is presented by faces on a rounded growth front which are formed when crystals with melting entropies of $\sim$ 2–4 eu are grown from the melt (so-called "faceting effect", Sect. 5.2). Under steady-state growth conditions the velocity of such a face is equal to that of the rounded front. Therefore the comparatively sparse steps on the face move at enormous velocities, of the order of tens of cm/s (Sects. 3.1, 5.2), and the opportunities for "selection" of the impurity are very limited. At the same time, on rough areas of the surface the number of trials before the final filling of a lattice site may reach values of $\gtrsim 10^6$ (the ratio of the number of particles arriving at the crystal surface per unit time to that of the particles permanently added to it in the course of growth). Therefore selection ensures an impurity concentration close to the equilibrium value for the crystal bulk. Since even the equilibrium concentration of the impurity (for which $K < 1$) in the surface layer on the face exceeds that in the bulk, the amount of impurity trapped by the face is greater than that trapped by the rounded growth front (Fig. 4.9a). The ratios of the corresponding coefficients of distribution for In and Sb in Si are 1.3 and 1.4, for P and As in Ge 2.5 and 1.8, and for Sn and Te in InSb 3.7 and 8, respectively [4.46]. All these examples are for growth from the melt. In addition, sectorial structure is practically unavoidable for impurity distribution in crystals grown from solution at moderate temperatures, and must always be taken into account in experimental investigation of the distribution coefficient. Therefore careful researchers analyze individual growth pyramids or (if possible) take advantage of conditions under which a crystal was formed by faces of the same index only. Although this is convenient (and in quantitative analysis of an impurity in small crystals it is practically the only method used), it is not always possible and is inconsistent with the change in growth conditions which alters the crystal habit.

d) Vicinal Sectoriality

The amount of impurity trapped by an echelon of steps forming a vicinal face (vicinal) should, generally speaking, depend on the azimuthal orientation, the velocity and the density of steps. Indeed, the steps of various orientations possess various atomic structures and thus should have various non-equilibrium distribution coefficients. These coefficients depend evidently on step velocities. Finally the time alotted for impurity adsorption and diffusion from and to the

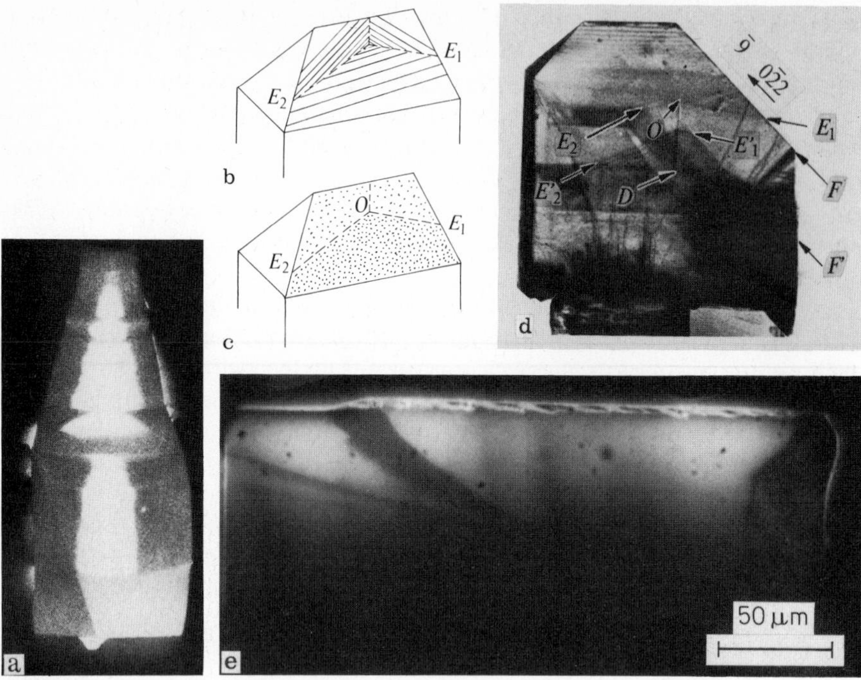

Fig.4.9a–e. Sectoriality in the trapping of impurties. (**a**) Radioautograph of the thin section of an InSb crystal grown by the Czochralski technique along the normal to an octahedron face (in the photo—from top to bottom) in the presence of radioactive Te [4.45]. The Te effective distribution coefficient is $K_{\text{eff}} = 0.6$–0.7 for the rounded part of the growth front, and $K_{\text{eff}} \simeq 4$ for the part formed by the octahedron face. Accordingly, the central region of the crystal formed by the faceted part of the growing interface contains more impurity and looks brighter than the periphery formed by the rounded interface; (**b–e**) Vicinal sectoriality: schematic drawing of a vicinal hillock (**b**) on the dipyramidal face of an ADP crystal where the *lines* symbolize the vicinal forming steps. The corresponding difference in the impurity content and lattice spacing is shown by *dots* in **c**. Photo **d** presents an x-ray topograph of an ADP crystal grown from aqueous solution: E_1, E_1' and E_2' are contrasts due to planes matching neighbouring vicinal sectors of a hillock around dislocation D. The line F′is parallel to the dipyramidal face F and presents a trace of its position on an earlier stage of growth. The interference contours parallel to E_1', E_2' within the *black region* below E_1', E_2' and F′ apper due to a sandwich-like structure of the vicinal hillock (see text) (**e**) photoluminescene image of the cross section of a GaP layer doped with Zn-O during liquid phase epitaxy. The *upper bright line* is the profile of the growth surface. The photoluminescence comes from Zn-O additive: there is a deficit of Zn-O in the dark trace associated with the growth macrostep on the surface (Courtesy of T. Tanbo, K. Pak, T. Nishinaga)

surface layer of each interstep singular terrace is inversely proportional to the step density and step velocity.

The vicinals on a growing singular face appear most often as slopes of growth hillocks around dislocations. The single crystalline body of each pyramidal hillock possessing a triangular cross-section is built up of three sectors the materials of which are deposited by three different vicinal slopes (Fig 4.9b). Thus the lattice constants and other properties in these sectors should be different (Fig. 4.9c). This phenomenon which we call vicinal sectoriality was experimentally found by in-situ x-ray topography of ADP crystals growing from aqueous solution. Matching the sectors of various lattice constants causes a contrast along the edges separating neighbouring vicinals of a hillock. One of the edges, E_1, depicted in Fig. 4.9d by an arrow, is terminating in the point O where the hillock-forming dislocation D crosses the dipyramidal face of the ADP crystal, on which the hillock is developed. The contrast from another edge, E_2, is very poor in Fig. 4.9c. The same edges in the earlier stages of growth give the contrast along the lines E_1' and E_2'. The line F′, parallel to the dipyramidal face F is a trace of the face in an earlier period when the face F did not grow. The face was immobile at relative supersaturation $\sigma = 0.006$. When rising the supersaturation up to the value of $\sigma = 0.072$, the face F starts to grow again and leaves the trace F′. This increase in supersaturation has led to the deposition of new macroscopic layers above all vicinals of the vicinal hillock around the dislocation D. These new layers possess lattice constants different from the constants in the underlying sectors which have been deposited at $\sigma = 0.006$. The resulting sandwich gives the interference contrast in the region below the lines F′, E_1', E_2'.

This interference contrast at higher supersaturation ($\sigma = 0.072$ vs. $\sigma = 0.006$) continues the independently observed tendency that the contrast increases along the edges of type of E_1 and E_2 with an increase in supersaturation. In other words, the higher the supersaturation the lager is the difference in lattice parameters in adjacent vicinals. On the other hand, the higher the supersaturations the shorter is the time $\lambda/v = h/R$ during which each of the interstep terraces is exposed to the solution (Sect. 4.3.1b). Thus the amount of impurities adsorbed on each terrace should be the lower the higher the supersaturation is. Consequently the adsorption of an impurity seems not to be an important stage in its trapping. We may thus expect that the impurity is captured by steps directly from solution, but the amount trapped changes during exposure time λ/v towards the equilibrium value in the surface layer of a singular face. The latter value is evidently equal for all vicinals on the same singular face.

If a bunch of steps appears on the vicinal due to an instability of the latter, the supersaturation, the step velocities and the exposition times within the bunch are different from these values on the smooth part of the vicinal. For these reasons, a bunch of elementary steps should engulf the impurity in a concentration different from that captured by the surface growing via rarer deposition of elementary layers. An illustration of this possibility is presented in Fig. 4.9e.

e) Rapid Diffusionless Crystallization

If the growth rate is so high that statistical selection is inefficient, we have so-called diffusionless (selectionless) crystallization, in which $K = 1$ and a metastable homogeneous solid solution is formed [4.47, 41]. Since particles are bound on the surface by cooperative interaction, transition from the regime with $K > 1$ or $K < 1$ to that with $K = 1$ occurs abruptly; at a certain value of the growth rate it is as if a "phase transition" from the formation of one modification ($K \gtrless 1$) to the formation of another ($K = 1$) took place [4.41]. Here, the critical parameter is not the temperature, as in ordinary phase transitions, but the growth rate (more precisely, the degree of deviation from equilibrium: supersaturation or supercooling). Therefore the described transitions were termed kinetic phase transitions [4.41]. Kinetic phase transitions must take place not only at an infinitely large, but also at a finite velocity of diffusion mixing in the mother medium [4.48].

An abrupt change in the distribution coefficient is shown in Fig. 4.10. Alloys of Al with Mg and Cu with Sn and Sb were crystallized by splat cooling. The cooling rates varied from 10 to 10^7 deg/s, resulting in a dendrite structure of the solidified alloys. The concentration of Mg, Mn, and Sb in the cores of the dendrites (not the average concentration in the solid!) was determined from the lattice spacings, which were measured by means of Debye diffraction patterns. In all cases the concentrations changed discontinuously at cooling rates above 10^6 deg/s. At these rates the distribution coefficients changed to $K = 1$ from $K \simeq 0.5$ (Mg in Al), $K \simeq 0.3$ (Sn in Cu), and $K \simeq 0.15$ (Sb in Cu).

Kinetic phase transitions must lead not only to the formation of crystals anomalously enriched with impurity, but also to the appearance of new polymorphous or even amorphous [4.49–52] modifications of a pure crystallizing substance which are metastable at those temperatures and pressures at which the growth takes place; they must lead, furthermore, to the appearance of modifications differing from stable ones both in composition and structure. In such cases the impossibility of selection pertains not only to the composition, but also to the rearrangement of the structure of the atomic complexes present in the mother phase. The larger these complexes, the lower is their mobility,

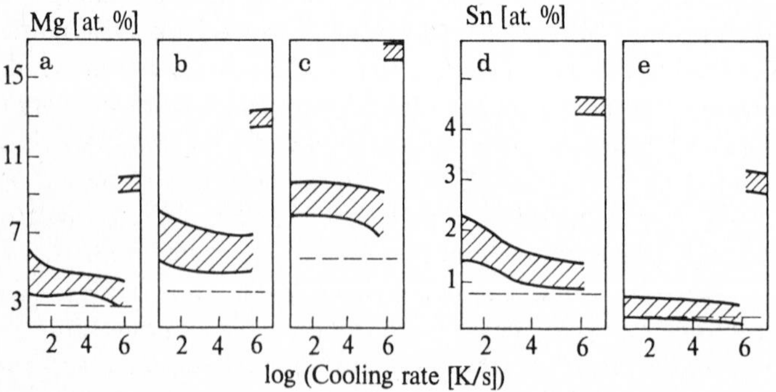

Fig.4.10a–e. Composition of the central parts of dendrite crystals (*ordinate*) grown at different cooling rates (*abscissa*). (**a**) Al + 9.6 at. % Mg; (**b**) Al + 13.0 at. % Mg; (**c**)Al + 16.4 at.% Mg; (**d**) Cu + 4.5 at.% Sn; (**e**) Cu + 2.9 at.% Sb. A discontinuous increase in the concentration of the solid solutions is seen at cooling rates $\simeq 10^6$K/s. [4.47] Note that the distribution coefficient K = 1 at higher rates

and the easier it is for them to freeze, forming an amorphous solid. That is why the amorphous metals usually have a rather complex composition and include such polyvalent elements as boron and phosphorus.

For diffusionless growth of simple substances, for instance metals, from the melt, growth rates of $\gtrsim$ 10 m/s are required. (The growth rate when each atom, molecule, or complex freezes into a solid just as the result of a first attempt, i.e., when there is no statistical selection, equals, in order of magnitude, $a/\tau_0 \simeq 10^4$ cm/s, if the lattice spacing is $a \simeq 3 \times 10^{-8}$ cm and the thermal vibration period is $\tau_0 \simeq 3 \times 10^{-12}$ s). In complex solutions with large, sluggish, and long-lived complexes and/or considerable activation energy for surface growth processes, and thus a low frequency of transitions across the phase boundary (Sect. 3.1), these velocities are correspondingly much lower.

4.3.2 Pulse Annealing

Pulse annealing was invented in 1974 in Kazan and Novosibirsk (see [4.53–56] and references therein), where initially laser pulses were used. Later, laser and electron-beam annealing attracted the attention of many scientists [4.57–59]. This interest was caused by the unique features of pulse annealing: a record-high solidification rate (up to meters per second) and distribution coefficient of impurities (up to 500 times the equilibrium value), high structural perfection of annealed layers on semiconductor surfaces, improved mechanical properties of annealed metal surfaces [4.60], and resistance to corrosion.

In pulse laser annealing, a sample surface $\simeq 10^{-4}$ to 1 cm in diameter is illuminated during a time $\tau \simeq 10^{-11}$–10^{-7} s or longer. The typical energy density per pulse is several J/cm^2. In electron-beam annealing, analogous parameters are used. In it, the sample must be in a vacuum, but the advantages of manipulation with an electron beam (e.g., scanning), can be utilized; intensive non-coherent light sources could be used as well.

The radiation is absorbed in the surface layer of the sample, providing the desired heating. In this way, the surface layer may be heated well above the melting temperatures of metals and semiconductors. After the end of a pulse, the heated or melted layer is cooled, the heat fluxes being directed at the bulk of the sample (cooling by thermal diffusivity) and at the external media (radiation cooling, which is usually less important). If the surface layer was melted, it crystallizes during the cooling period. If the material which remains unmelted is a single crystal, epitaxial overgrowth occurs.

The electronic mechanism of pulse laser annealing of semiconductors is based on the assumption that a very dense electron-hole plasma appears in the irradiated material. As a result, the lattice bonding becomes weaker, making possible an intense diffusion of interstitials and thus an improvement in an initially defective structure [4.61]. The lattice may also be softened via a decrease in the melting temperature and the potential barrier for crystallization events at the interface. Up until now, however, no reliable evidence of the electronic mechanism has been reported [4.62, 63].

Laser pulses also cause stress waves with pressures as high as 10 GPa on metal and alloy films covered with other films that are transparent for laser radiation. Such pressure waves may be an important factor causing changes in the properties of the target material [4.64].

Let us briefly consider the main, thermal mechanism of laser annealing [4.65]. The depth of the layer where the laser radiation is absorbed is of the order of α_λ^{-1} (α_λ is the absorption coefficient). For metals, $\alpha_\lambda \simeq 10^6$ cm^{-1}, and for semiconductors, $\alpha_\lambda \sim 10^4$–10^5 cm^{-1}, depending on the radiation wavelength λ and the electrical conductivity of the material, i.e., on the type of substance and its doping level and temperature. The quantity α_λ^{-1} may also determine the thickness of the surface layer where the temperature rises substantially. However, the characteristic absorption length α_λ^{-1} is often less than the depth $2\sqrt{a_s\tau}$ up to which the heat penetrates by heat diffusion. Here a_s is the sample thermal diffusivity, and τ is the duration of the radiation pulse. At $\tau \simeq 10^{-9}$ s, $a_s \simeq 0.7$ cm^2/s we have $2\sqrt{a_s\tau} \simeq 5.10^{-5}$ cm. The molten film may be thinner or thicker than that, depending on the maximum temperature reached at the heated surface. Numerical calculations [4.66] of molten film thickness and the liquid state duration in give, for Si, values of about 3000 Å and 100 ns for a 20-ns ruby laser pulse bringing the energy of 2 J/cm^2. The molten film duration increases with the duration and energy of the laser pulse. Nevertheless, typical cooling rates may be roughly estimated with the use of the simple thermal conductivity model [4.65]. A heated layer of thickness $2\sqrt{a_s\tau}$ should be cooled during a time of the order of $(2\sqrt{a_s\tau})^2/4\,a_s = \tau$, i.e., the pulse duration. Thus, if the layer is heated up to $T_{max} \simeq 2 \times 10^3$ K by a pulse with $\tau \simeq 10^{-8}$ s, one would expect a cooling rate $T_{max}/\tau \simeq 2 \times 10^{11}$ K/s. For $\tau \simeq 2 \times 10^{-11}$ s, the cooling rate is $\simeq 10^{14}$ K/s. These rates are much higher than the maximum rates of $\simeq 10^9$ K/s achieved by conventional techniques of rapid quenching, e.g., by splashing droplets on a cold substrate ("splat cooling").

The duration of the molten film and the regrowth velocity of silicon have been measured experimentally using the reflection of an additional weak laser beam from the annealing surface [4.68, 69]. One of the techniques is based on the difference in reflectivities between the solid and liquid surfaces [4.68]. As soon as the main heating beam reaches the silicon surface, the reflectivity sharply increases from its value for the solid state to the value typical for the liquid state. Then the reflectivity remains unaltered during the period τ_L when the surface is in the liquid state, and falls at the end of this period during the time τ_f. The intensity of the reflected probe light is presented schematically as an insert in Fig. 4.12. Liquid phase durations for different pulse energy densities, materials, and wavelengths are presented in Fig. 4.11. The initial laser pusle duration was $\tau = 30$ ns for $\lambda = 530$ nm and $\tau = 40$ ns for $\lambda = 1060$ nm (Nd: glass Q switched laser). The points in Fig. 4.11 present experimental data, and the curves give the calculated durations of the liquid film for the liquid reflectivity values $R_l = 0.57$ and 0.72. The insert in Fig. 4.11 gives the vicinity of the origin in enlarged form and demonstrates the existence of an energy threshold for

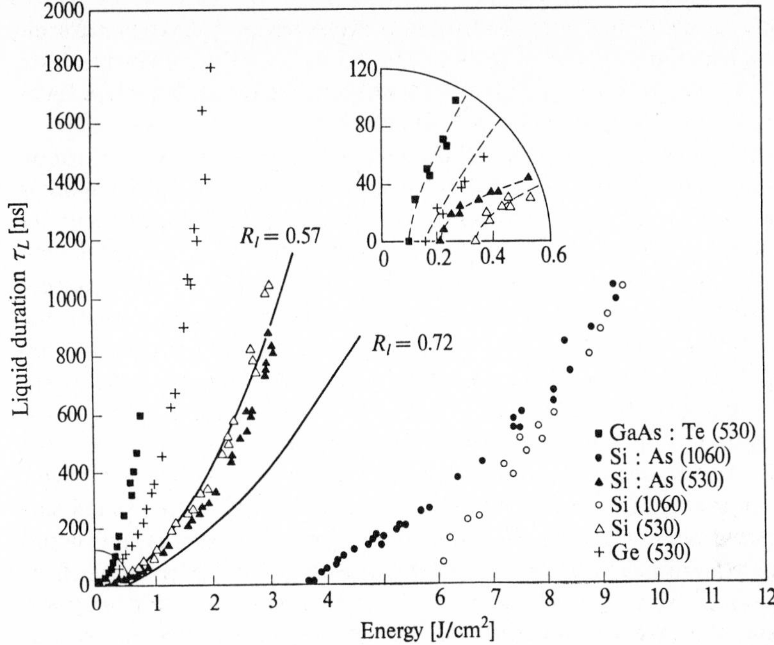

Fig.4.11. Duration of the liquid film on the surfaces of different materials after a laser pulse of the wavelength [nm] indicated in parentheses in the *lower right corner* — as a function of the density of the pulse energy [4.68]

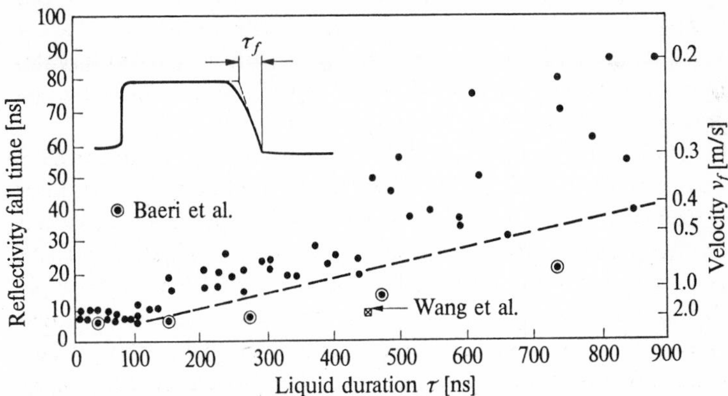

Fig.4.12. Reflectivity fall time τ_f and regrowth velocity V_f as functions of melt duration. Silicon (100) surface. ●—experimental data, ⊙—calculations [4.66]. The *dashed line* corresponds to the maximal growth rates [4.68]

melting which varies from 0.1 to 0.3 J/cm² for different materials and wavelengths. The time τ_f during which the reflectivity returns to the value typical of the solid surface is the time when the moving solid–liquid interface covers the last interval, $\simeq \alpha_{\lambda}^{-1}$ in length, to the free sample surface. Thus, by measuring τ_f one could estimate the growth rate.

The reflectivity fall times τ_f and the growth rates V_f through the last 170 Å of the liquid silicon layer are given in Fig. 4.12 [4.68]. It can be seen that $V_f \simeq$ 2 m/s and does not depend on the liquid duration τ at $\tau \lesssim 100$ ns. At longer duration ($\tau \simeq 900$ ns) the growth rate decreases down to 0.2–0.4 m/s. In these experiments, the second harmonics ($\lambda = 530$ nm) of a Q switched Nd: glass laser were used, with a pulse duration of 30 ns.

Another technique for measuring the growth rate is based on interference of the additional laser beam reflected from the external sample surface and from the moving solid–liquid interface [4.69]. This interference results in intensity oscillations in the reflected probe beam when the molten film decreases in thickness during freezing.

According to [4.70] the growth rate increases from 1.5 to 6 m/s when the sample temperature decreases from 300 K to 77 K.

Impurities are usually implanted in the form of ions accelerated to energies of tens to hundreds of keV into the surface layer subjected to pulse annealing. The implantation causes matrix disordering, leading to the creation of an amorphous layer. Subsequent laser annealing may produce a perfectly crystalline and even dislocation-free surface layer epitaxially oriented on the undamaged single-crystalline substrate. On the other hand, annealing a single-

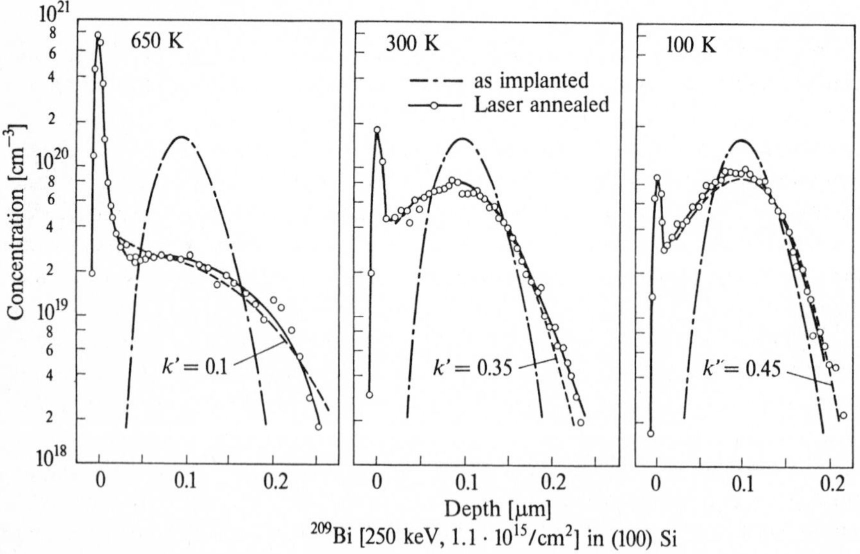

Fig.4.13. Distribution of ^{209}Bi in the (100) Si surface layer. *Dot-dashed line*—as implanted in the form of ions accelerated to 250 keV to the dose $1.1 \cdot 10^{15}$ cm^{-3}; *dots*—experimental measurements after annealing; *solid curves*—calculations according to diffusion equations with effective distribution coefficients indicated for each sample temperature [4.70]

crystalline Si surface by very short ($\tau \simeq 3 \times 10^{-11}$ s) laser pulses results in an amorphous silicon layer [4.67].

The impurities implanted in the surface layer obey a Gaussian distribution as regards the distance from the surface, and reach maximal concentrations near the middle of the layer, as shown by the dot-dashed curves in Fig. 4.13. The maximum becomes wider after annealing and may even be transformed into a plateau. The growth front, by pushing the impurities to the free surface, also causes an additional concentration maximum at the surface (Fig. 4.13). The impurity distribution after annealing may be quite accurately approximated (dashed curve in Fig. 4.13) by a numerical solution of the diffusion equation for the liquid layer ahead of the moving phase boundary, with the effective non-equalibrium distribution coefficient K being introduced at the phase boundary. Different values of K corresponding to different substrate temperatures (650,300, and 100 K) and thus to different regrowth rates are indicated in Fig. 4.13.

The value of K strongly exceeds the equilibrium coefficient K_0. For example, solidification of Si at the rate $V \simeq 4.5$ m/s results in complete trapping of As: $K = 1$, whereas $K_0 = 0.3$. Analogous experiments with Bi, Ga, and In give $K = 0.4$, 0.2, and 0.15 as supposed to $K_0 = 7 \times 10^{-4}$, 8×10^{-3}, 4×10^{-4}, respectively. For Sb, at a growth rate $V = 2.7$ m/s, $K = 0.7$, while $K_0 = 0.023$. Thus the K/K_0 ratios are 3.3(As), 25(Ga), 30(Sb), 375(In), and 571(Bi) [4.70]. The ion channeling technique for 2.5-MeV H^+ ions exhibits sharp minima in the $\langle 110 \rangle$ and $\langle 111 \rangle$ directions, and allows one to conclude that 89% to 99% of the atoms of the impurities discussed above are trapped in the substitutional position [4.71].

All the impurities discussed above have retrograde solubility in silicon, and their absolute concentrations in the solid after laser annealing exceed the corresponding equilibrium solubility limits Thus the high values of K/K_0 cannot be explained by the continuation of the solidus and liquidus lines to the low-temperature region [4.72, 73]. These high values may be interpreted as a result of nonequilibrium burying of the impurity atoms in the surface, the steps, and the kinks (discussed in the previous section). This mechanism is connected with the impurity enrichment of the surface layer and thus may result in $K < 1$ even for $K_0 < 1$.

However, for the experimental conditions discussed, the K value still does not exceed unity and may be explained simply by assuming that the impurity atoms have no time to escape from the growth front and thus should be buried independently of the enrichment of the surface layer [4.74].

The maximum impurity concentrations which can be trapped at the maximum growth rate in a given system with retrograde solubility have been estimated from a curve presenting the relationship between impurity concentration and temperature at which Gibbs free energies of liquid and solid solutions are equal [4.73]. This curve in the phase diagram begins at the melting point of the pure matrix and descends between the solidus and liquidus curves. Its intersection with the abscissa $T = 0$ gives the absolute solubility limit we are looking for. For the solubility limit of As, Ga, In, and Bi in Si the calculations predict values of (5, 6, 2, and 1) $\times$ 10^{21} cm^{-3}, respectively. The maximum concentrations achieved in experiments [4.70] are given by close values of (6, 0.88, 0.28, and 1.1) $\times$ 10^{21} cm^{-3}, respectively.

4.3.3 Diffusion in the Mother Medium

When studying impurity trapping experimentally, one actually measures the effective distribution coefficient K_{eff}. It is defined as the ratio of the impurity concentration in the crystal to its average concentration in the bulk of the stirred mother medium rather than to the concentration immediately shead of the growth front. The latter concentration differs from the average concentration, because the interface serves either as a source ($K < 1$) or as a sink ($K > 1$) of the impurity for the mother medium. The strength of this source is equal to $C_i(0)(1 - K)V$, where $C_i(0)$ is the impurity concentration at the growing interface. The condition of impurity balance on the growing surface has the form:

$$D_{iM} \frac{\partial C_i}{\partial z} = - C_i(0)(1 - K)V \qquad \text{for} \qquad z = 0\,, \tag{4.43}$$

where D_{iM} is the diffusivity of the impurity in the medium, and z is the coordinate reckoned from the growth front along the normal to it in the direction of the mother medium.

The removal of the impurity from the growth front (at $K < 1$) or its transportation in the opposite direction (at $K > 1$) is effected by diffusion. If the mother medium (liquid, gas) is being mixed by forced stirring or by natural convection (Sect. 5.1), then using the simplest model of transport phenomena, one says that purely diffusional transport dominates only within a boundary layer of thickness δ, while in the entire remaining volume the average impurity concentration $C_{i\infty}$ is constant: [7]

$$C_i|_{z \geq \delta} = C_{i\infty}\,. \tag{4.44}$$

In a stagnant medium the diffusion, uncomplicated by the motion of the medium, proceeds throughout the bulk ($\delta \to \infty$). The same occurs during the growth of a crystal in the solid phase, for instance during the decomposition of supersaturated solid solutions.

Within the boundary layer the impurity concentration satisfies the diffusion equation. In a reference system moving together with the flat growing interface at a constant velocity V, this equation can be written as follows [see 5.1.2, (5.9)]:

$$D_{iM} \frac{\partial^2 C_i}{\partial z^2} + V \frac{\partial C_i}{\partial z} = 0\,. \tag{4.45}$$

[7] Solute distribution in laminar flow in the vicinity of an absorbing flat interface without assumption of the stagnant layer δ was found in [3.12].

The solution of this steady-state equation with the boundary conditions (4.43, 44) is

$$C_i(z) = C_{i\infty} \frac{K + (1 - K)\exp(-Vz/D_{iM})}{K + (1-K)\exp(-V\delta/D_{iM}}\ . \tag{4.46}$$

The amount of impurity trapped by the crystal is equal to $KC_i(0)$. Therefore the effective distribution coefficient is

$$K_{eff} = \frac{KC_i(0)}{K_{i\infty}} = \frac{K}{K + (1 - K)\exp(-V\delta/D_{iM})}\ . \tag{4.47}$$

If the mother medium—melt, solution, or gas—is stagnant, then $\delta \gg D_{iM}/V$, and according to (4.46), the concentration in the mother medium at the growing interface is $C_i(0) = C_{i\infty}/K$, i.e., at $K < 1$ the impurity concentration at the growing interface exceeds the initial one. When $K \ll 1$, this increase in impurity concentration at the growth front can be so great that it may cause crystallization of the impurity itself, of some of its compound, or of eutectics on the interface of the macrocomponent.

The effective distribution coefficient in a stagnant medium ($\delta \to \infty$) is equal to unity for the one-dimensional problem considered above. Vigorous stirring of the mother medium, on the other hand, reduces the thickness of the boundary layer, and at $\delta \gg D_{iM}/V$ one has $C_i(0) = C_\infty$, so that $K_{eff} = K$, i.e., the trapping efficiency depends exclusively on surface growth processess. In the general case of an arbitrary δ, the dependence of the actually observed distribution coefficient on the growth rate (on supersaturation or supercooling) depends both on transport processes, in accordance with (4.47), and on the processes of incorporation of particles into the lattice, in conformity with (4.40, 41). Therefore growth rate fluctuations in time cause variations in the impurity content of the crystal, i.e., its zone structure or striations [4.75–77]. Zones of increased (or decreased) impurity content in the crystal delineate the contours which the crystal had at the moment of increase or decrease in its growth rate, and thus supply important information on the growth processes. Zonal nonuniformity in impurity distribution gives rise to dislocations, internal stresses, and other defects in crystals, and also modulates the refractive index, thus deteriorating the working characteristics of, for instance, semiconductor and electrooptical crystals. To eliminate the striation one must suppress the fluctuations in the growth rate. The most effective and widespread cause of such fluctuations is natural or forced convection and/or crystal rotation in the furnace thermal field, which is usually slightly asymmetric. The amplitude of the fluctuations may exceed the average growth rate, because periods of growth alternate with periods of melting [4.81–83]. "Freezing" the convection in metals by switching on a magnetic field [4.78] abruptly reduces the striation. The absence or at any rate the weakening of convection intensity under conditions of microgravitation in space has reduced by several times the striation amplitude in germanium grown in Skylab [4.79].

4.3.4 Observed Distribution Coefficients

It follows from the foregoing that the actually observed coefficients of impurity distribution (even of homogeneous trapping) are the result of a number of processes. Therefore in order to control trapping one must attempt to determine the limiting stage, and first of all to isolate the effect of the bulk transport and that of the surface processes on the distribution coefficient. Furthermore, it is important to establish whether trapping has equilibrium character in which case sectorial structure is nonexistent or only weakly manifested. Sharply defined striation corresponds to fluctuations of the local values of K_{eff} by several times. It has also been shown [4.78] that fluctuations in the growth rate bring the average value of K_{eff} closer to unity (provided $K \neq 1$).

Sectorial and zonal nonuniformities sometimes change K_{eff} by a factor of 10 or more. Nevertheless, K_{eff} correlates with the thermodynamic equilibrium value K_0. Moreover, the equilibrium values of the distribution coefficient in the surface layer and the step, and also the elementary frequencies for the attachment and detachment of particles proper and impurity particles, depend on the same basic parameters which determine K_0, namely the binding energies, lattice deformation energies, etc. Therefore the experimentally observed coefficients of homogeneous distribution are essentially dependent on the above-mentioned thermodynamic parameters both in the equilibrium and nonequilibrium cases. Figure 4.14 illustrates the effect of the difference between the ionic radii of $Ca^{2+}(r_S)$ and the impurity cation (r_i) on the distribution coefficient of the latter in $CaMoO_4$. The parabolic nature of the maximum at $r_S = r_i$, which follows from (4.14), is clearly visible here. If, however, the difference between the radii is considerable, a linear dependence of log K on r_i^3 is observed (Fig. 4.15). This seems strange at first glance, since according to (4.19), the equilibrium distribution coefficient in the bulk, and also in the surface layer, the step, etc., depends not only on the size of the ions, but also on the binding energy. In actuality, however, the binding energy itself correlates with the ionic radii. The role of

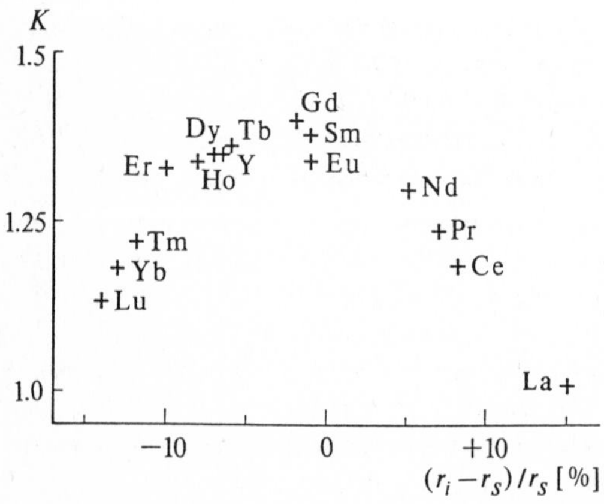

Fig.4.14. Dependence of the distribution coefficients K of impurities (ions of rare-earth elements replacing Ca^{2+} in crystals of $CaMoO_4$) on the relative difference between the radii of Ca^{2+} and the rare-earth ions [4.80]

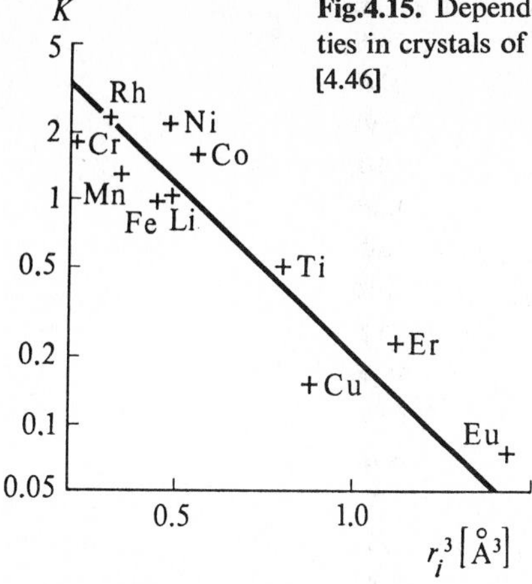

Fig.4.15. Dependence of the distribution coefficients K of impurities in crystals of $ZnWO_4$ on the volume of the impurity ions (r_i^3) [4.46]

the elastic energy and the binding energy of the impurity in the crystal is manifested in the correlation of the distribution coefficients of one and the same impurity in crystals of one and the same substance grown from solution and from the melt (Table 4.4). The summary table of distribution coefficients (Table 4.5) gives an idea of the experimental distribution coefficients in different systems.

Table 4.4. Comparison of the effective distribution coefficients in solution and melt growth [4.46]

Macro-component	Impurity	Distribution coefficient	
		Growth from aqueous solutions	Growth from the melt
NaCl	Li	0.007 ±0.004	0.21; 0.20±0.05; 0.19
	K	0.005 ±0.001	0.20; 0.008±0.003
	Br	0.047 ±0.005	0.6
	I	4×10^{-4}	0.06
KCl	Na	6×10^{-4}	0.03; 0.11±0.02; 0.31
	Rb	0.113 ±0.005	0.68; 0.6±0.1; 0.70
	Cs	0.0040±0.0006	0.16; 0.21
	Br	0.189 ±0.003	0.75; 0.71
	In	0.001	0.14
KBr	Rb	0.334 ±0.004	0.78; 0.4±0.1; 0.75
	Cl	0.453 ±0.005	0.86; 0.85
	I	0.039 ±0.004	0.5; 0.52
KI	Rb	0.82 ±0.03	0.76±0.02
	Cs	0.03 ±0.01	0.31±0.01
	Cl	0.015 ±0.02	0.39±0.02
	Br	0.42 ±0.02	0.79±0.02
	NO_3	0.071 ±0.004	0.43±0.01

Table 4.5. Effective distribution coefficients k for melt growth systems [4.46]

Macro-component	Distribution coefficient			
AgBr	Cd 20	Cu 0.7	K 0.04	Na 0.01
AgCl	Br 0.8	Ca 2.3	Cd 2.7	Cu^{+} 0.2
	Ni 1.4	Pb 0.4	Rb 0.01	S 0.04
Al	Ag 0.64	Be 0.1	Co 0.14	Cr 0.8
	Ge 0.1	La <0.1	Mg >0.5	Mn ~ 1
	Sm 0.67	Ti 1.4	V 3.7	W 0.32
Al_2O_3	$k \sim 1$ for Ag, Cr, Cu, Mg, Si; $k \ll 1$ for B, Ga, In, Mn, Na, Pb;			
AlSb	Ag 0.1	B 0.01	Co 0.003	Cu 5×10^{-4}
	Ni 0.01	Se 0.3	Si 0.04	Sn 1
As	Sb 0.8	Se 0.55		
Bi	Ag 0.2	Cd 0.5	Cu 0.6	Pb 0.4
	Zn 0.5			
CaF_2	Ce 0.88	Co 0.56	Eu 0.81	Gd 1.07
$CaWO_4$	Dy 0.44	Ho 0.3	Nd 0.24	Pr 0.36
CdTe	Ag 2×10^{-6}	Al 8×10^{-4}	As 1×10^{-3}	Au 7×10^{-5}
	Fe 5×10^{-3}	Ge 6×10^{-4}	Hg 0.05	In 6×10^{-3}
	Sb 3×10^{-3}	Se 2×10^{-1}	Sn 1×10^{-3}	TeI 2×10^{-3}
Fe	Al 0.92	B 0.14	C 0.16	Co 0.9
	O 0.022	P 0.17	S 0.04	Ta 0.43
GaAs			Al 3	C 0.8
			Ge 0.01	Mg 0.1
			Se 0.1	Si 0.1
$GaCl_3$	Fe 0.14	Mn 0.02	Na 0.1	Zn <0.01
GaP	Zn 10			
GaSb	Cd 0.02	Ge 0.2	Se 0.4	Sn 0.03
Ge	Al 0.07	As 0.02	B 17	Bi 4.5×10^{-4}
	Se 1×10^{-6}	Si 5.5	Te 1×10^{-6}	Zn 4.10^{-4}
GeI_4	Fe 0.14	Mn 0.02	Na 0.1	Zn <0.06
H_2O	D_2O 1.021	HF 10^{-4}	NH_3 0.17	MH_4F 0.02
InAs	Ag 4.7×10^{-5}	Cd 0.26	Cu <0.05	Fe 0.5
	Sn 0.06	Te 0.44	ZnO 0.77	
InP	Ge 2.4×10^{-2}	Si 7.5×10^{-4}	Sn 2.2×10^{-2}	
InSb	Ag 4.9×10^{-5}	As 5.4	Cd 0.26	Cu 6.6×10^{-4}
	S 0.1	Se 0.35	Sn 0.06	Te 0.5
KBr	see Table 4.4	Ca 0.35	Cs 0.23	Mn 0.01
KCl	see Table 4.4	Ca 0.1	Cu 0.44	Cr 0.18
KI	see Table 4.4	Ba 0.3	Ca 0.38	K 0.39
KNO_3	Ca 0.04	PO_4 0.03	SO_4 0.1	Na 0.28
LiI	B 0.63	Ca 0.78	Cu 0.77	Fe 0.56
Mg	Al 0.33	Cu 0.05	Fe 0.17	Mn 6.1
	Zn 0.26			
NaCl	see Table 4.4		Ba 0.03	Ca ~ 0.7
$NaNO_2$	Al 0.56	Ca 0.6	Cu 0.6	Fe ~ 1
$NaNO_3$	Cu 0.02	Sr 0.05		
Nb	Ta 1.4			
NH_4NO_3	Ag 0.3	Ba 0.54	Ca <1	Cs 0.56
	Sr 0.34			
Pb	Ag 0.04	Au 0.015	Sn 0.05	
Sb	Al <1	Ag <1	As 0.32	Bi 0.5
	S <1			
Sn	Ag 0.03	Bi 0.5	Cu <1	Fe <1
Si	Al 2×10^{-3}	As 0.3	B 0.8	Bi 7×10^{-4}
	P 0.35	Sb 0.023	Te 4×10^{-6}	Zn 1×10^{-5}
$SiCl_4$	Al 0.07	B 0.16	Cu 0.64	Fe 0.15
$Y_3Ga_5O_{12}$	$k \sim 0.1$ for Ag, Al, Mg, Sn;			
U	Al ~ 1	Co 0.1	Cr 0.2	Fe 0.1
	Pd 0.2	Ru ~ 1		
ZnS	Ag 2.0×10^{-2}	Cu7.5×10^{-2}	K 0.4	
$ZnWO_4$	see Fig. 4.11			

Rb 0.1	S 0.24	Zn 0.15	
Cu^{2+} 0.4	Fe$\sim$0.7	K 0.04	Na 1.5
Zn 0.14			
Cu 0.06	Dy 0.02	Fe 0.21	Ga$<$0.1
Mo 2.3	Nb 1.6	Sc 0.17	Si 0.14
Zn 0.4	Zr 2.5		
$k\sim0.1$ for Fe; $k>1$ for Ni, W			
Fe 0.02	Ge 1.2	Mg 0.1	Mn 0.01
Ta 8×10^{-6}	Ti 0.01	V 0.01	
Sb 1.45	Sn 0.13	Te 0.31	Tl 0.7
Nd 1.06	Sm 0.98	Tb 1.1	Yb 0.88
Tm 0.32			
Bi 9×10^{-5}	Ca 8×10^{-4}	Cr 0.2	Cu 8×10^{-7}
Mg 2×10^{-4}	Mn 0.7	Ni 9×10^{-4}	Pb 1×10^{-3}
Tl 2×10^{-4}			
Cr 0.95	Cu 0.56	Mn 0.84	Ni 0.8
Ca 2×10^{-3}	Co $<2\times10^{-2}$	Cu $<2\times10^{-3}$	Fe 3×10^{-3}
Mn $<4\times10^{-5}$	Ni $<2\times10^{-2}$	P 2	S 0.3
Sn 3×10^{-3}	Te 0.05	Zn 1.9	
Te 0.4	Zn 0.3		
Cd 1×10^{-5}	Ga 0.09	P 0.08	S 1×10^{-5}
Mg 0.7	Mn 0.05	S ~1	Si 0.4
Fe 0.04	Ga 2.4	Ni 6×10^{-5}	P 0.16
Zn 2.3			
Na 0.35			
Na 0.32	PO_4 0.43	SO_4 0.34	Sr 0.36
Li 0.03	Sr 0.5		
Sr 0.2	Y 0.3		
Mg 0.23	Na ~1	Si 0.48	
Ni 0.015	Pb 0.35	Si 0.04	Sn 0.32
Cd 0.16	Rb 0.04	Sr 0.2	
Mg 0.28	Pb <0.7	Si 0.06	Ti$<$0.7
Cu 1	K 0.45	La 0.12	Rb 0.64
Ca <1	Cu 0.06	Fe 0.2	Pb 0.09
In 0.5	Pb <1	Sb 1.5	Zn 0.05
Cd 1×10^{-4}	Cu 4×10^{-4}	Ga 8×10^{-3}	Ge 0.33
Mg 0.16	Mn 0.09	Ti 0.91	
$k\sim1$ for Cu, Er, Si, Yb; $k<1$ for Fe, Mn, Ni, Pb			
Mn 0.25	Mo 1.7	Nb 0.6	Ni 0.15

5. Mass and Heat Transport. Growth Shapes and Their Stability

Mass and heat transport (Sect. 5.1) inevitably occurs in the bulk of both the growing phase and the mother phase. It often plays a substantial role in growth kinetics, and influences the quantity and character of defects arising in crystals during their formation, impurity distribution and phase composition.

The effect of interrelated surface growth processes and of the distribution of temperature and/or concentration around the growing crystal determines the shapes it takes on and the stability of these shapes with respect to specific perturbations. The kinetics and crystallographic principles involved in the origin of growth shapes are discussed in Sect. 5.2; Sect. 5.3 is devoted to shape stability.

5.1 Mass and Heat Transfer in Crystallization

Crystal growth does not include only surface processes; a necessary stage is also the transport of the crystallizing components to the phase boundary and the removal of the heat of crystallization. In a vacuum, this transport is effected by molecular beams. During growth from condensed phases—solutions and melts—and also from sufficiently dense gases, stirring or natural convection influences the transport. However, a liquid or a dense gas remains relatively stagnant within the boundary layer adjacent to each solid surface. Within this layer the transport can be assumed to occur by ordinary diffusion or thermal conduction. Purely diffusional transport throughout the entire bulk of the mother medium is realized, for instance, during the growth of crystals in supersaturated solid solutions, in gels, or under zero gravity or artificially created conditions, for example when the mother liquor and the growing crystal are sealed between two parallel pieces of glass spaced so closely that convection in the liquid is excluded. The theory of diffusion transport in crystallization is discussed in [1.2a, 5.1, 5.2].

During last decade a substantial interest has been maintained in the transport and growth phenomena associated with convection of the liquid phase, as can be seen from the sections on fluid dynamic and microgravity in the Proceedings of the 5th and 6th International Conferences on Crystal Growth [5.3, 4] and other recent publications such as [1.2a, 5.2, 5–7]. Here, the related problems are touched upon only very briefly (Sects. 4.3.3, 5.1.2, 10.2.5).

5.1.1 Stagnant Solution. Kinetic and Diffusion Regimes

Purely diffusional heat and mass transport is encountered only under the above-mentioned special conditions. However, an analysis of such transport elucidates the role of transport processes in crystallization and is therefore given below.

Heat and mass transport in stagnant media is described by identical heat and mass diffusion equations, and thus the results obtained in one case are often extended to the other. The main difference is that diffusion problems of crystallization are generally restricted to consideration of the mother medium, while thermal problems are formulated for both medium and crystal.

To show the fundamental aspects of the effect which transport phenomena have on crystal growth from solution, we consider the simplest problem: quasi-steady-state diffusion around a crystal in the isotropic approximation (Fig. 5.1) We assume the crystal to be a sphere of radius R, on whose surface the growth is determined by isotropic kinetic coefficient β. The concentration of solution around the crystal, $C(r, t)$, which depends on the distance r to its center and on time t, satisfies the diffusion equation

$$\partial C/\partial t = D\Delta C \, . \tag{5.1}$$

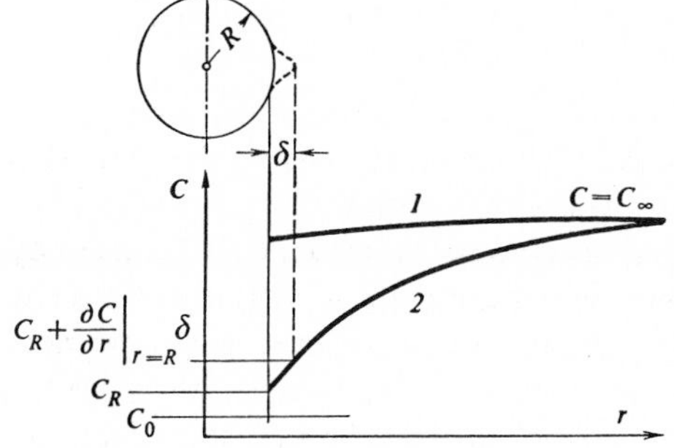

Fig.5.1. Distribution of the concentration of the mother liquor around a growing spherical crystal; $\beta l/D \ll 1$ is the kinetic regime (*1*), and $\beta l/D \gg 1$ the diffusion regime (*2*)

Let us consider an unstirred solution in which the concentration far from the crystal is C_∞:

$$C(r, t) = C_\infty \quad \text{at} \quad r \to \infty \tag{5.2}$$

Quantity C_∞ may be either lower or higher than the equilibrium value C_0 of the saturated solution.

The condition at the phase boundary is defined by the law of conservation of the crystallizing substance at the moving growth front:

$$D\partial C/\partial n = (\rho_S - C)V(\boldsymbol{n}, \Delta\mu) \qquad \text{for} \qquad r = R \, , \tag{5.3}$$

where ρ_S is the crystal density expressed in the same units as C; the linear growth rate V of a given area of the surface [cm/s] depends on its crystallographic orientation $\boldsymbol{n}$ and the deviation $\Delta\mu$ from equilibrium over this area; and $\partial/\partial n$ is a derivative with respect to the normal to the growth front. In this approximation $\partial/\partial n = \partial/\partial r$. For a complete foımulation of the diffusion problem we also require the initial conditions at time $t = 0$.

Let us analyze the characteristic qualitative features of the solution to this problem in the quasi-steady-state approximation. The characteristic time within which the change in the distribution of concentration around the crystal must occur is R/V, so that the left-hand side of (5.1) is approximately equal to $V(C_{\text{surf}} - C_\infty)/R$, and the right-hand side, also within an order of magnitude, equals $D(C_{\text{surf}} - C_\infty)/R^2$, where $C_{\text{surf}} = C(R) \equiv C_R$ is the concentration at the crystal surface. Hence if the growth rate

$$V \ll D/R\,, \tag{5.4}$$

the left-hand side of (5.1) is much smaller than the right-hand side, and (5.1) transforms to the Laplace equation

$$\Delta C = 0\,. \tag{5.5}$$

Physically, condition (5.4) means that the diffusion field manages to "tune in" to the change in crystal size. The smaller the crystal, the smaller is the region ($\simeq R$) in which the tuning must take place, the faster is the tuning, and the smaller is the restriction imposed on the growth rate. At $R \simeq 1$ cm and $D \simeq 10^{-5}$ cm^2/s, condition (5.4) requires $V \ll 10^{-5}$ cm/s.

Let us now investigate the change in the size of the crystal with time in the steady-state approximation (5.5). According to (5.3, 3.5), we have at the growth front (if concentration C and density ρ are expressed in terms of the number of particles per cm^3):

$$D\partial C/\partial n = \beta(1 - \Omega C_{\text{surf}})(C_{\text{surf}} - C_0)\,. \tag{5.6}$$

The distribution of concentration around the crystal satisfying (5.5, 6, 2) at $\Omega C_{\text{surf}} \ll 1$ has the form:

$$C = C_\infty - (C_\infty - C_0)\frac{\beta R/D}{1 + \beta R/D}\frac{R}{r}\,, \tag{5.7}$$

and the growth rate is

$$V = \frac{\Omega\beta(C_\infty - C_0)}{1 + \beta R/D}\,. \tag{5.8}$$

Taking into account factor $(1 - \Omega C_{\text{surf}})$ in (5.6), subject to condition $\Omega(C_{\text{surf}} - C_0) \ll 1$, leads to the same euqations (5.7, 8) after replacement $\beta \to \beta(1 - \Omega C_0)$.

From the above equations it follows that there exist two crystal size ranges characterized by different growth laws. At $\beta R/D \ll 1$ we have $C_{\text{surf}} \simeq C_\infty$, so that the growth rate is determined almost completely by the surface kinetics processes and is independent of the crystal size, which increases linearly with time: $R \simeq \beta(C_\infty - C_0)\Omega t$. This is the so-called kinetic regime of growth. If $\beta R/D \gg 1$, then $C \simeq C_\infty - (C_\infty - C_0)R/r$, the rate decreases with time as $V \simeq D\Omega(C_\infty - C_0)/R$, and the crystal size grows as $R \simeq \sqrt{2\Omega(C_\infty - C_0)t}$; i.e., the growth is completely limited by bulk diffusion and is independent of the kinetics of surface processes. This is the diffusional growth regime. The distribution of concentration near a growing spherical crystal in the kinetic and diffusion regimes is schematized in Fig. 5.1.

The differentiation between the kinetic and diffusion regimes with the aid of parameter $\beta R/D$ is also valid qualitatively for more complex growth forms, in particular for faceted crystals. The kinetic and diffusion growth regimes can also be differentiated during the formation of oxide films on a metal or, in general, for any reactive diffusion process. The oxidation processes themselves occur at the metal–oxide or oxide–gas interface. In either case one of the reagents (oxygen or metal) is transported to the growth front by diffusion of the substance through the film already formed. It is this transport that defines the growth rate of the film in the diffusion regime.

The dimensionless parameter determining the distinction between the kinetic and diffusion regimes in the case of cylindrical symmetry of the diffusion field is $(\beta R/D)\ln(L/R)$, where L is the characteristic size of the vessel or the diffusion area. Since the logarithmic dependence of this parameter on R is weak as compared with the linear one, we obtain practically the same dependences $R(t)$ for a cylinder as for a sphere.

5.1.2 Stirred Solution. Summation of Resistances

If a mother liquor is stirred, the average concentration remains virtually constant throughout its bulk. However, the thin boundary layer of the liquid adjacent to a face of the growing crystal is an exception. Within this layer the mass transport remains purely diffusional.

Let us now consider the concentration field over a growing crystal face whose size greatly exceeds boundary layer thickness δ. Here, the solution concentration depends only on one coordiante z perpendicular to the growing face. Suppose this face grows at a constant rate V, and a steady-state concentration distribution has been established in the diffusion layer, i.e., concentration C at a given distance z from the growth front (in other words, in a moving reference system associated with the front) is time-independent. This means that in the laboratory frame of reference z' the concentration can be written as $C(z', t) = C(z' - Vt) = C(z)$, $z = z' - Vt$, so that diffusion equation (5.1) reduces to

$$D\frac{d^2C}{dz^2} + V\frac{dC}{dz} = 0\,. \tag{5.9}$$

Solving this equation with boundary conditions (5.6) at $z = 0$ and $C = C_\infty$ at $z = \delta$ and assuming $\Omega(C_{surf} - C_0) \ll 1$, we have, for low supersaturations, $(C_\infty - C_0)/(\rho_S - C_0) \ll 1$:

$$C = C_\infty - (C_\infty - C_0)\frac{\exp(-Vz/D) - \exp(-V\delta/D)}{1 - \exp(-V\delta/D) + V/\beta}\,. \tag{5.10}$$

Substituting (5.10) into (5.6) and restricting ourselves to low concentrations and solubilities, when $V\delta/D \ll 1$ and $(C_\infty - C_0)(\rho_S - C_0) \ll 1$, we get

$$V = \frac{\beta(C_\infty - C_0)}{(1 + \beta\delta/D)(\rho_S - C_0)}\,. \tag{5.11}$$

For dilute solutions, when $C_0 \ll \rho_S = \Omega^{-1}$, (5.11) transforms into (5.8).

Equation (5.11) can be rewritten in the form:

$$(\rho_S - C_0)V = \frac{1}{(1/\beta) + (\delta/D)}(C_\infty - C_0) = \beta_{eff}(C_\infty - C_0)\,,$$

i.e., the "crystallization current" $(\rho_S - C_0)V$ is proportional to "crystallization potential difference" $(C_\infty - C_0)$. The effective kinetic coefficient (system "conductivity") β_{eff} reflects both the diffusional resistance δ/D to the transport of the substance and the resistance β^{-1} to crystallization at the phase boundary. The total resistance β_{eff}^{-1} is (as in electrical engineering) the sum of the resistances in the crystallization circuit:

$$1/\beta_{eff} = 1/\beta + \delta/D\,. \tag{5.12}$$

If the substance is transported through a diffusion layer of thickness $\sim R$, which is the case in an unstirred solution, see (5.8), the effective resistance to crystallization is

$$1/\beta_{eff} = 1/\beta + R/D\,. \tag{5.12a}$$

According to (3.17), boundary layer thickness δ decreases with increasing velocity u of the liquid near the surface proportionally to $1/\sqrt{u}$. Therefore growth rate V must increase with the flow rate: $V \propto \sqrt{u}$. This law is indeed obeyed, for instance, by the growth rates of crystals of copper sulfate ($CuSO_4 \cdot 5H_2O$) in an aqueous solution (Fig. 5.2). The growth rate increases with u by the parabolic law if V is limited by the transport. At high velocities of solution stirring when $\delta \ll D/\beta$, the growth rate is determined by the surface kinetics, and V ceases to depend on u.

The perfection of a crystal usually decreases with increasing supersaturation on the growing surface. Therefore the quality of a crystal may serve as a rough measure of supersaturation. Using this relationship, it is possible to demonstrate

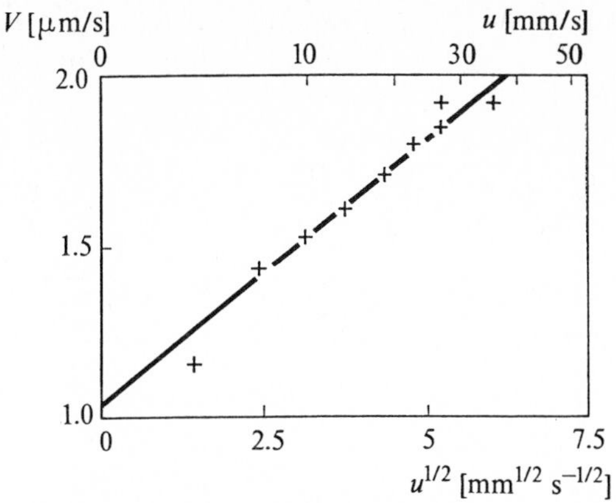

Fig.5.2. Dependence of the growth rate of $CuSo_4 \cdot 5H_2O$ crystals on the velocity of motion of the solution relative to the crystal [4.46]

qualitatively the effect of stirring the solution on supersaturation in the immediate vicinity of the growing surface with the aid of the two experiments described in the following. In one of them, crystals of $NaClO_3$ grew freely on the bottom of a vessel filled with a supersaturated unstirred solution (the supersaturation was achieved by supercooling the saturated solution). In the other, the growth took place in a setup where an intensive stream of supersaturated solution flowed past the crystal. In the first experiment the solution supersaturation at which the formation of inclusions and skeletal growth began was ~5 K, and in the second, macroscopic inclusions appeared on supersaturation as low as ~0.3 K.

Another example can be given of the effect of the motion of the solution on the crystal's perfection. During convection or stirring the liquid is not quite homogeneous: it has regions with slightly differing concentrations and/or temperatures. These inhomogeneities are eliminated by ordinary diffusion within a time proportional to the square of the size of the region divided by the diffusion coefficient, and fail to disappear completely within the time in which the solution moves from the cooling (saturation) zone to the crystal surface. Therefore the supersaturation (supercooling) on the surface fluctuates in time. In some systems, for instance in a horizontal boat for growing from the melt (Sect. 10.2), at a sufficiently low temperature gradient, the changes in supercooling at the front are of a periodic nature. The amplitude of the abovementioned oscillations and fluctuations may approach one degree (Fig. 10.46). The growth rate fluctuates accordingly. If, in addition to the principal component, the solution contains an impurity whose trapping is sensitive to the growth rate, these fluctuations in rate result in a zone structure (Sect. 4.2.3).

5.1.3 Kinetic and Diffusion Regimes in the Melt

Kinetic regimes of growth from a pure, one-component melt are regimes in which the growth rate is limited by surface phenomena, whereas in diffusion regimes it is limited by heat removal (heat diffusion) processes in the bulk of the contacting phases.

In growth from a pure melt the relative role of surface processes and heat removal can be established by analyzing the condition of thermal balance at the growth front:

$$\kappa_L \frac{\partial T_L}{\partial n} - \kappa_S \frac{\partial T_S}{\partial n} = \Delta H \rho_S V , \tag{5.13}$$

where T_L and T_S are the temperatures in the liquid and solid phases, respectively; κ_L and κ_S [cal/cm s K] are their thermal conductivities; ΔH[cal/g] is the latent heat of crystallization per unit mass; ρ_S[g/cm^3] is the crystal density; and the growth rate is reckoned in a reference system associated with the crystal. If we compute the growth rate with respect to the liquid, then ρ_S in (5.13) is replaced by melt density ρ_L. Condition (5.13) can be simplified by assuming that the densities and heat capacities of melt and crystal are equal. Then, at the phase boundary,

$$a_L \frac{\partial T_L}{\partial n} - a_S \frac{\partial T_S}{\partial n} = T_Q V . \tag{5.14}$$

Here, a_L and a_S are the heat diffusivities of the liquid and solid phases [cm^2/s], and quantity $T_Q = \Delta H/\tilde{c}$ is the ratio of the latent heat of fusion to heat capacity $\tilde{c}$, T_Q is the temperature increase which must occur after instantaneous adiabatic crystallization of an arbitrary melt element. Typically, T_Q varies between tens and hundreds of degrees. In actual fact, an increase in temperature by T_Q is only possible when melt supercooling $T_0 - T_\infty > T_Q$, where T_∞ is the temperature far away from the crystal. If this inequality is fulfilled, the temperature of each newly crystallized element of the bulk may increase to a value $T_\infty + T_Q$, which is still below the melting point. Here, crystallization would continue even without heat removal from the growth front, and the supercooling at the front would be equal to $T_0 - T_\infty - T_Q$. Heat removal from the surface of a spherical crystal slightly increases this value, which continues to remain finite, as is also indicated by rigorous mathematical analysis [5.8]. Accordingly, the growth rate is defined by this supercooling and the kinetics of the surface processes, and ceases to depend on time for instants which are not very close to the beginning of growth. If, however, the supercooling is small ($T_0 - T_\infty < T_Q$), the latent heat release at the growth front without heat removal by thermal conductivity would raise the temperature at the front to values exceeding the melting point, which is incompatible with crystallization. Therefore at $T_0 - T_\infty < T_Q$ the supercooling on the crystal surface drops nearly to zero almost immediately after growth begins, and subsequently goes on dropping proportionally to $t^{-1/2}$, i.e., according to the law determined by the heat removal. The growth rate decreases accordingly;

$V \propto t^{-1/2}$. Let us consider this process in more detail for an isolated crystal placed in a supercooled melt and having a characteristic size R (for instance a spherical crystal of radius R). Here, the temperature inside the crystal can be assumed to be constant (the heat is not removed through the crystal) and equal to the temperature of its surface, T_{surf}. Thus the problem of the temperature distribution in the melt in this approximation is similar to the diffusion problem discussed above. Since the thermal field around the crystal extends to a distance $\sim R$, boundary condition (5.14) can be rewritten approximately as

$$a_L \frac{\partial T_L}{\partial n} \simeq a_L \frac{T_{surf} - T_\infty}{R} \simeq T_Q \beta^T (T_0 - T_{surf}) .$$

By determining T_{surf} from this equation it is easy to find the supercooling at the growth front,

$$T_0 - T_{surf} \simeq \frac{T_0 - T_\infty}{1 - (\beta^T T_Q R / a_L)} , \tag{5.15}$$

and the velocity of its motion,

$$V = \beta^T (T_0 - T_{surf}) \simeq \frac{\beta^T (T_0 - T_\infty)}{1 + \beta^T T_Q R / a_L} . \tag{5.16}$$

Equation (5.16) is similar to (5.8), and is obtained from it by replacement

$$\Omega (C_\infty - C_0) \to (T_0 - T_\infty)/T_Q , \qquad \beta \to \beta^T T_Q , \qquad D \to a_L . \tag{5.17}$$

Parameter $\beta^T T_Q R/a_L$ determines the kinetic ($\beta^T T_Q R/a_L \ll 1$) and diffusion ($\beta^T T_Q R/a_L \gg 1$) regimes, as in growth from solution. For a small crystal of potassium, $\beta^T \simeq 40$ cm/s K, $T_Q = 75$ K, $a_L \simeq 1.4$ cm²/s, and hence at $R \gtrsim 10^{-3}$ cm the growth rates are already defined by the heat removal, and surface processes are substantial only for crystals with a size of about a micrometer or less.

The growth rate and the temperature distribution can easily be estimated in the same way when the heat of crystallization is removed through the crystal, for instance during the solidification of a flat layer on a cooled substrate. This model is suitable for describing the growth of an epitaxial thin film on a substrate and a polycrystalline layer of metal on the wall of a mold if the layer thickness is small as compared to the size of the mold. Here the role of crystal size R is played by the layer thickness, and a in dimensionless parameter $\beta^T T_Q R/a$ must be represented by the thermal diffusivity of the crystal, a_S, or some combination of a_S and a_L. Since, however, a_L and a_S are of the same order of magnitude, their difference is immaterial for qualitative estimates.

As follows from (5.8, 16), the growth rate decreases with crystal size. Integration of the equation $dR/dt = V$ of the motion of the phase boundary clearly shows that at small R, i.e., in the kinetic regime, $R \simeq \beta^T(T_0 - T_\infty)t$, and in the diffusion regime, $R \simeq \sqrt{2(T_0 - T_\infty)a_L t/T_Q}$. Thus in the former case crystal size R increases linearly with time and is determined by the surface processes, and in the latter $R \propto \sqrt{t}$ and depends only on the coefficients of diffusion and temperature conductivity.

The release of the latent heat increases the surface temperature in growth from solution also, and thereby reduces the growth rate. But since the mass diffusivity is usually about 3 orders of magnitude less than the thermal diffusivity, the indicated effect is important only for systems in which the compositions of the crystal and liquid phases are similar with respect to both components. Such systems include alloys. An analogous situation is given with Rochelle salt, which grows from a mixture containing up to 90 wt. % salt and 10 wt. % water. Because the heat removal from the vertices of a Rochelle salt crystals more intensive than that from its faces, a temperature difference of ~3–5 K, depending on the growth rate, arises along the face [5.9].

5.1.4 Diffusion Field of a Polyhedron

Suppose a crystal has the shape of a polyhedron and grows in a stagnant solution, remaining similar to itself in shape. Such growth is only possible when the density of the diffusion flux of the crystallizing substance to all the points of each of the crystal's faces is constant. In other words, along the ith face,

$$\frac{\partial C}{\partial n} = q_i = \text{const}\,.$$

The values of the fluxes towards the different faces, i.e., values q_i, are generally different. By making use of this boundary condition, we can solve the diffusion equation and find the concentration distribution around the polyhedron [5.10]. Thus we can express through q_i the supersaturation over each point of the growing face, including the point (points) generating layers on this face [5.11, 12]. If the growth rate is low, so that condition (5.4) is fulfilled, the diffusion equation can be replaced by Laplace's equation, and the problem is greatly simplified. Precisely this case is discussed below. Knowing the supersaturation and the strength of the layer generator, we can find the local density of the steps and the slope of the surface p, i.e., kinetic coefficient $\beta(p)$ in (5.6). Substituting now q_i for $\partial/C\partial n$ into the left-hand side of (5.6) and supersaturation $C_{\text{surf}} - C_0$, expressed via q_i, into its right-hand side, we obtain the condition for determining q_i, and together with it the complete solution of the problem on the concentration field around the polyhedron. In the kinetic regime, and also in a diffusion regime that is not very pronounced, it is almost spherically symmetric. For the two-dimensional case we naturally have cylindrical symmetry. Such a diffusion field can be seen in Fig. 5.3. It shows the interference pattern near a crystal of $NaClO_3$ growing from an aqueous solution sealed between parallel glass plates, the solution layer thickness being $\simeq 10^{-2}$ cm. The crystal size (edge length) is $2l \simeq 3 \times 10^{-2}$ cm. Each of the interfering beams has an optical path proportional to the refractive index, which in turn is proportional to the solution concentration at the point under consideration. Therefore the interference bands arising in the solution-filled gap are lines of equal concentrations. In Fig. 5.3 they are similar to concentric circles, although the crystal has a square cross section. The solution concentration decreases

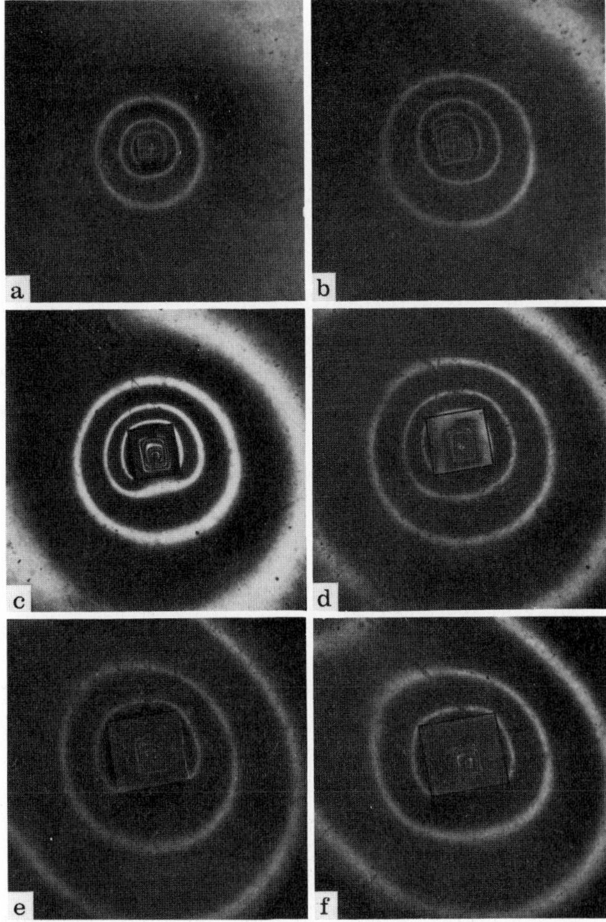

Fig.5.3a–f. Diffusion field around a crystal of $NaClO_3$ in consecutive stages of growth [5.13]. Magnification × 40

towards the crystal. The intersection of the interference bands with the crystal surface indicates that the concentration on the growing surface is not constant: it is higher near the crystal vertices and lower at the face centers. This important circumstance agrees with the solution of the diffusion problem for the polyhedron. It is one of the main causes of layer generation in the vicinity of vertices, and results in the formation of skeletal growth forms (Sect. 5.2).

The increase in crystal size with time in experiments of the type shown in Fig. 5.3 is demonstrated in Fig. 5.4, and confirms the $l \propto \sqrt{t}$ dependence obtained above. Another interferometrial technique to determine diffusion field around ADP crystal was cesed in [5.13b].

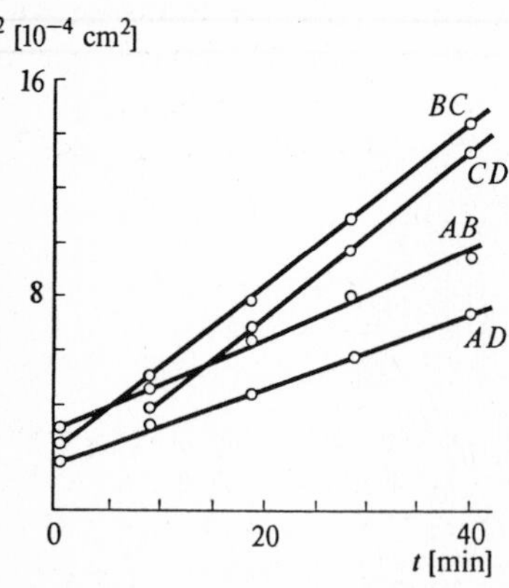

Fig.5.4. Square of the distance l from the crystal center to the face as a function of time t elapsed from the initial moment of crystal growth. The four *straight lines* correspond to the four different faces of the crystal (see Fig. 5.3) [5.13]

If kinetic coefficient β increases abruptly, as is the case when the supersaturation is increased (Sect. 5.3), the concentration on the surface approaches its equilibrium value, in accordance with (5.6), and the diffusion field changes drastically: the equal-concentration lines near the crystalline polyhedron repeat its shape almost precisely. Away from the crystal, the concentration distribution naturally remains spherically (or cylindrically) symmetric.

The nonequivalent conditions of mass and heat transport near the crystal vertices, edges, and face centers are common to all growing polyhedra. During growth from solutions the vertex environment is fed from a large solid angle, and therefore the supersaturation near it on the crystal surface is higher than over the central parts of the faces. Similarly, during the growth of an isolated crystal in the melt the supercooling near its vertices is higher than in the other portions of the growth front.

In some cases, however, we have the opposite situation, i.e., the supersaturation near the vertices is lower. For instance, crystals of calcite ($CaCO_3$) grow from a $H_2O - CO_2 - CaCO_3$ solution, where the calcite solubility reduces with increasing CO_2 concentration. During the growth of the calcite crystal, carbon dioxide diffuses from the growth front, and the CO_2 concentration near the vertices is lowest. Hence the solubility of calcite near the vertices may prove to be higher, and the supersaturation $C - C_0$ lower, than at the other points of the face. This results in an accelerated growth of the central parts of the faces, which thus become convex. The same picture can be observed during growth in the presence of any impurity that is pushed out by the crystal and reduces its solubility.

The nonuniformity of the nutrition of the growing surface or of heat removal from it exists almost always in a moving mother medium as well. Thus if a crystal is located in the flow of a supersaturated solution, the supersaturation on the "head" faces is highest, and they grow more rapidly than the "tail" faces. This nonequivalence naturally remains important as long as the diffusion through the boundary layer affects the growth rate, i.e., in the diffusion regime. Therefore the indicated nutrition nonuniformity is manifested particularly prominently at low solution velocities, for instance when the crystal is in contact with concentration flows (natural convection) created by itself. The nonuniformity of boundary layer thickness δ along the surface also causes nonuniformity of supersaturation along each face. As a result the supersaturation is again usually higher at vertices and edges than at the face centers.

5.2 Growth Shapes

The shape which a crystal acquires in the course of its formation — the growth shape — is highly sensitive to growth conditions, and reflects the growth mechanism, as does the surface morphology [5.14a]. Therefore growth shapes make it possible to judge the conditions of formation and to correct the growth parameters when crystals are grown artificially.

5.2.1 Kinematics

Polyhedral shapes arise when each face, in growing, remains parallel to itself, i.e., when the growth rate depends exclusively on the surface orientation and is equal (on the average) for all its points: $V = V(\boldsymbol{n})$. It can be demonstrated [5.14b] that under the described conditions the steady-state growth shape is determined by *Wulff's* rule with the aid of function $V(\boldsymbol{n})$ in precisely the same way as the equilibrium shape is determined with the aid of specific surface energy $\alpha(\boldsymbol{n})$. Analytically, the steady-state growth shape is described by the envelope of the family of planes given by the equation

$$\boldsymbol{n}\boldsymbol{r} = V(\boldsymbol{n})t \,, \qquad (5.18)$$

which is obtained from (1.41) by replacement $\alpha(\boldsymbol{n}) \to V(\boldsymbol{n})$, $2\Omega/\Delta\mu \to t$, where t is the time from the nucleation of the crystal. Experssion (5.18) assumes that in the course of growth the shape of the crystal is always similar to the original one. If this condition is fulfilled, the shape will remain unaltered even if the growth rate is time dependent — all we have to do in the latter case is to replace the right-hand side of (5.18) by $\int V dt$. The equation of the envelope of family (5.18) can be written, by analogy with (1.40), as:

$$(1/R_1)(V + \partial^2 V/\partial\varphi_1^2) + (1/R_2)(V + \partial^2 V/\partial\varphi_2^2) = \text{const}\,, \tag{5.19}$$

where the constant in the right-hand side is independent of the coordinates, but is an arbitrary function of time. For orientations corresponding to singular surfaces, function $V(\boldsymbol{n})$ or which is the same $V(\varphi_1, \varphi_2)$, has sharp minima, where the first derivatives $\partial V/\partial\varphi_1$ and $\partial V/\partial\varphi_2$ are discontinuous, and the second derivatives appearing in (5.19) are infinite. Consequently, on the areas of the crystal surface with singular orientations, surface curvature radii R_1 and R_2 must also be infinite. This means that singular surfaces must be represented in the crystal habit by macroscopically flat faces. The existence of vicinal hillocks on them corresponds to the flattening of the singular minimum within small angles $p < p_0$ (dashed line in Fig. 3.7a). The transformation of an atomically smooth surface into a rough one means a rounding out of the sharp mimimum of dependence $V(\boldsymbol{n})$ [or $R(\boldsymbol{n})$ — dashed line in Fig. 3.7a]. The rounded minimum of $V(\boldsymbol{n})$ corresponds to finite second derivatives in (5.19), and hence to finite curvature radii. This means that the atomically rough, normally growing surfaces are represented by rounded sections on the stationary growth shape. We are thus enabled to make judgments about the atomic mechanism of growth from the macroscopic shape of the crystal.

The size of a given face or the length of a rounded section with a given set of orientations on the growth form is determined in accordance with the analog (5.18) of the *Curie-Wulff* rule. From the geometric procedure of constructing Wulff's envelope (Fig. 1.11) it follows that the higher the growth rate of a face (or of a rounded area), the smaller its size on the convex steady-state growth shape must be. Indeed, let us consider three neighboring faces belonging to the same zone, the continuations of both extreme faces intersecting behind the middle one (i.e., outside the crystal). Suppose now that the growth rate of the middle face alone has increased, for instance because of an increase in supersaturation near it, the addition of a selectively acting impurity, or the appearance of a defect—an active layer generator—on this face. It is obvious that the size of the accelerated face will then be reduced, and that it will ultimately disappear (so-called wedging out of the face). Analogously, faces with maximal growth rates should form the concave growth shape (i.e., with the cavity decreasing in size).

5.2.2 Determination of Crystal Habit by the PBC Method

As noted in Sect. 1.3, during growth from the gas phase and solution most crystals have atomically smooth surfaces and hence faceted growth shapes. When describing such forms it is first necessary to establish the crystallographic indices of the faces and their relative sizes, i.e., the crystal habit. On the basis of modern concepts one can indicate qualitatively, with sufficient accuracy, the sequence of sizes of the most probable faces during growth from the gas phase and, with less certainty, during growth from solutions and melts.

In accordance with the Curie-Wulff rule in the form of (5.18), the faces with the lowest growth rates will be the most developed on the crystal surface. These lowest rates should be expected for faces parallel to the maximum number of chains of the strongest bonds, i.e., for atomically smooth F faces (Sect. 1.2.2). Indeed, the growth of such faces requires either that the potential barrier for the formation of two-dimensional nuclei be overcome, or that dislocations or other defects for step generation be present. Both of the indicated generation processes are slower, the higher the specific linear energy of the steps is. This energy is higher, the stronger are the PBCs (Sect. 1.2.2) crossing the rises of the variously oriented steps on the given face, i.e., parallel to this face. Furthermore, during growth due to the arrival of particles from the adsorption layer to the steps the growth rate is higher, the greater the density of this layer is. The layer density reduces exponentially with increasing energy of the bond between the adsorbed building-block particle and the surface; i.e., it must again be least for the faces crossed by the least number of the most energetic PBCs.

The above guidelines qualitatively justify the main precept of the PBC theory developed by *Hartman* and *Perdok* (Sect. 1.2), which states that the growth rate of a face is lower, the fewer chains of strong bonds cross this face. For the surface energy this dependence stems from the very definition of energy, whereas for growth rates it can be considered only approximate. Specifically, the growth rates of the various faces depend differently on the supercooling and on the temperature at a given supercooling, which no longer fits into the framework of the PBC theory. Nevertheless, in many cases this theory describes the most general features of the crystal habit rather well.

The above-mentioned correlation of the surface energy and the growth rate results from the crystal structure, which determines both quantities for any orientation. The angular dependence of the growth rate has minima for the same faces as does the angular dependence of the surface energy. The deepest minima of $V(\boldsymbol{n})$ correspond to the deepest minima of $\alpha(\boldsymbol{n})$. It follows that the relative sizes of the faces on the growth form and on the equilibrium form must be similar under given conditions. This by no means implies that the growth form is determined directly by the surface energy via the *Gibbs-Thompson-Herring* relation (1.40). This would be the case if the driving force $\Delta\mu = \Omega(\alpha + \partial^2\alpha/\partial\varphi_1^2)/R_1 + \Omega(\alpha + \partial^2\alpha/\partial\varphi_2^2)/R_2$, which is due to the surface energy, were comparable to the driving force of growth, which determines the growth rate (Sect. 1.4). At actual values of supersaturations or supercoolings this is possible only for crystals less than 1 μm in size (Sect. 1.4.4). Nevertheless, the similarity between the equilibrium form and the growth form enables one to use, in determining the main faces of the growth form, methods used to find the equilibrium form, in particular the mean detachment works method discussed in Sect. 1.4.3.

In accordance with the above-formulated postulate of PBC theory, the "specific weight" of a face in the habit decreases in the order F, S, K (flat, stepped, kinked). For instance, fibrillary silicates made up of atomic chains, inside which the bonds are much stronger than between the chains, grow in the shape of fine fibers, i.e., they have an acicular habit (the lateral faces are much larger than the end faces). Crystals of SbSI also have the strongest bonds only along the *c* axis and also grow in the shape of needles both from the gas phase and from the melt. The acicular habit of these crystals is not related to whisker shapes.[1]

Similarly, crystals with a clearly defined layer structure — graphite, mica, a number of crystals of organic substances — have at least two PBC systems in the plane of these layers and grow in the shape of platelets.

Finally, crystals which have at least three systems of PBCs not lying in the same plane are isometric, as a rule. Let us find, for instance, using the PBC method, the habit of fcc crystals with nonpolar bonds, taking into account only interaction between the first nearest neighbors. In this approximation a close-packed fcc lattice has only one type of PBC — along all the possible $\langle 110\rangle$ directions. It is easy to see that there exist only two types of faces parallel to two (or more) distinct systems of $\langle 110\rangle$ PBCs; these are $\{111\}$ faces, containing three such systems, and $\{100\}$ faces, containing two systems. It is they that should be expected in the habit of a growing crystal, $\{111\}$ faces being preferred.

Taking into account interactions between neighbors of the next order ($\langle 100\rangle$ chains in this case) means including new PBCs, which are, generally speaking, nonparallel to the preceding ones. Accordingly, there appear faces containing at least two systems of PBCs of the first or second nearest neighbors, and the expected shape of the crystal becomes more complicated. In the case considered these are $\{110\}$ faces. The size and the very appearance or nonappearance of new faces on the equilibrium form depend on the ratio of the energies of PBCs of different orders. The sequence of faces thus found is given in decreasing order of their importance in Table 5.1, which was obtained by *Stransky* and *Kaishev's* method of mean detachment works (Sect. 1.4.3).

When determining the indices of F faces for ionic crystals use is also made of the intial postualate on the proportionality between the growth rate of the face and the energy of addition of a new plane atomic net to the crystal, i.e., actually between growth rate and surface energy. This energy is calculated for ionic crystals by summing up all the *Coulomb* bonds by the *Madelung* method. More rigorous calculations also take Born repulsion between neighboring ions into account. The faces compared must have no dipole moments perpendicular to

[1] In whiskers the radial sizes are typically 1–10 μm, and in axial crystals, they are 10 to 100 times larger. This shape is acquired under special conditions of supersaturation, temperature, composition of the mother medium, and presence of impurities, by many crystals whose structure presupposes isometric growth forms, i.e., forms having similar sizes along the principal crystallographic axes. Whiskers are known for a large number of metals, alkali halide and other salts, oxides, and other substances. For VLS growth of whiskers see Chap. 8.

Table 5.1. Indices of faces expected on equilibrium and growth shapes[a] [1.9]

Lattice type	Faces with allowance for:		
	1st nearest neighbors	2nd nearest neighbors	3rd nearest neighbors
Simple cubic	100	110, 111	211
Body-centered cubic	110	100	211, 111
Face-centered cubic	111, 100	110	311, 210, 531
Diamond	111	100	110, 311
Hexagonal close-packed	0001, $10\bar{1}\bar{1}$, $10\bar{1}0$	$11\bar{2}0$, $10\bar{1}2$	

[a] The face indices taking only the first nearest neighbors into account are given in the first column; the faces with inclusion of the first and second nearest neighbors are listed both in the first and second columns, etc.

the face plane, since otherwise the face would have had a somewhat higher electrostatic energy. An example is the (111) NaCl face, which consists of ions of the same sign; it must become microrough, being divided into microareas made up of (100) faces. At the boundary with the electrolyte solution such charged faces may be stabilized by the adsorption of impurity ions or polar molecules of the solvent, and may appear on the growth shape.

Being adsorbed differently on different faces, the impurities change their growth rates and hence the crystal habit, also in different ways (Sect. 4.1).

Faces with orientations lying between those of the neighboring F faces are S and K faces; S faces usually belong to the same zones as F faces, while K faces occupy general positions on the stereographic projection. Thus S faces are stepped, and K faces are made up of kinks (Sect. 1.2). It must be remembered that the stepped structure of the S faces, for instance of the (110) face of a simple cubic lattice, results, in the first-nearest-neighbor approximation, from the crystal structure rather than from the generation of steps by nuclei or dislocations. The same is true of kink-covered, structurally rough K faces. This roughness of structural origin exists even at $T = 0$. In the first-nearest-neighbor approximation, the problem of formation of steps and kinks is nonexistent for S and K faces. In actual fact, however, these faces can also grow by deposition of layers parallel to them; but the linear energies α_l of the corresponding steps are due exclusively to higher-order PBCs and are therefore lower than for the F faces. This does not imply negligible smallness of α_l under actual conditions: cube faces (100) in the diamond lattice are, in the first-nearest-neighbor approximation, K faces, but artificial diamonds, for instance, have well-developed (100) faces growing by layers. Spiral layers have been observed as well (Fig. 3.5a).

5.2.3 The Bravais-Donnay-Harker Rule

The PBC method is actually a generalization of the earlier and simpler empirical rules of *Bravais* and *Donnay-Harker* for determining the crystal habit. According to Bravais the minimum growth rates are characteristic of faces with the $\{hkl\}$ orientation, which are parallel to those atomic nets of the lattice between which the distance d_{hkl} is the largest (i.e., between which the strength of bonds is the least, in qualitative agreement with Hartman's postulate). As is well known, $d_{hkl} \propto (h^2 + k^2 + l^2)^{-1/2}$, i.e., the largest d_{hkl} correspond to planes with the simplest indices, which in actuality often form the crystal habit. But faces with the simplest indices are usually those formed by the closest-packed atomic nets, i.e., faces with the highest reticular density. Indeed, the reticular density in an (hkl) net is equal to S_{hkl}^{-2}, where S_{hkl} is the area of a unit loop in this net;

$$S_{hkl}^2 = h^2 S_{100} + k^2 S_{010} + l^2 S_{001} + 2(hk S_{100} S_{010} \cos\lambda + kl S_{010} S_{001} \cos\mu + lh S_{001} S_{100} \cos\nu) ;$$

S_{100}, S_{010}, and S_{001} are the areas of the unit loops in the nets denoted by the indices; and λ, μ, ν are angles between (100) and (010), (010) and (001), and (001) and (100), respectively. Consequently, the larger h, k, and l are for a given net, the lower is the atomic density in it. Thus the principle of the maximum of d_{hkl} is equivalent to that of the maximum of the reticular density, which means that the crystals must be formed by the closest-packed faces.

The equation for d_{hkl} defines the distance between crystallographically equivalent planes (hkl). If, however, the crystal has a screw-symmetry axis of the n-th order perpendicular to a given plane, the actual distance between the neighboring plane nets (no longer equivalent, but turned by $2\pi/n$ relative to one another) will be d_{hkl}/n. Accordingly, the area of faces with such an orientation on the crystal growth shape will be smaller. A similar conclusion for several faces can be drawn from the existence of glide-reflection planes parallel to these faces. The center of symmetry in the unit cell also halves some interplanar distances.

In general, in order to find the growth shape one should use the distance between the nearest planes in which the atoms reside, instead of d_{hkl}. In determining these distances it is necessary to take into account the space-symmetry elements inherent in crystal structure, as was discussed above. This is the essence of Donnay and Harker's correction to Bravais's rule, which eliminated the contradiction between the latter and the actually observed habit of some crystals. For instance, according to Bravais's rule, the quartz basal face (0001) must be the closest packed and hence slowest growing and most developed form of the habit. It should have been followed by $(10\bar{1}0)$, $(10\bar{1}1)$, $(11\bar{2}0)$, $(10\bar{1}2)$, and $(11\bar{2}1)$ in order of decreasing size. In reality, however, the basal face grows rapidly, wedges out, and is nonexistent on the final shape. The contradiction disappears if we recall that a third-order screw axis passes along the c axis normal to (0001), i.e., d_{0001} must be divided by 3. With due allowance for the symmetry operations in accordance with *Donnay* and *Harker* we obtain the sequence of the faces in order of decreasing size: $(10\bar{1}0)$, $(10\bar{1}1)$,

$(11\bar{2}0)$, $(10\bar{1}2)$, $(11\bar{2}1)$, $(20\bar{2}1)$, and $(11\bar{2}2)$. This series, with the exception of $(11\bar{2}0)$, corresponds to the habit observed.

Thus, in contrast to the PBC method, those of Bravais and Donnay-Harker operate only with geometric parameters and therefore permit an unambiguous conclusion on the shape of the crystal, provided its space group is known. But since the values of the binding energies are neglected, the answer obtained is not always accurate.

5.2.4 Effect of Growth Conditions

The structure of the surface and the mechanism of growth essentially depend on the supersaturation, temperature, and composition of the environment (Sects. 1.3; 3.1, 2), and therefore the habit of the crystal depends not only on its structure, but also on the above-mentioned factors. For instance, at supersaturations of $\lesssim 1\%$ the crystals of potash alum have only the faces of an octahedron, and at supersaturations of $\gtrsim 1\%$, those of an octahedron, a cube, and a rhombododecahedron. Crystals of NaCl have a cubic habit at low supersaturations and an octahedral habit at high ones. At increased supersaturations, crystals of CsCl acquire (100) faces in addition to (110) faces, etc. In general, the habit of crystals growing with singular faces simplifies with decreasing supersaturation or supercooling. The reason is that singular faces have clearly defined nonlinear dependences of the growth rate on the supersaturation, and different kinetic coefficients, even on linear sections of their $V(\sigma)$ dependences (Sect. 3.3). The rates of layer generation by homogeneous and heterogeneous formation of two-dimensional nuclei and of that due to the effect of dislocations depend on the linear energies of the steps and the spectrum of defects on the faces; they differ substantially for the various faces. Therefore at small deviations from equilibrium and, accordingly, at low growth rates, the face velocities differ widely on nonlinear sections of $V(\sigma)$. At large deviations from equilibrium, when the acts of incorporating particles at kinks and steps, as well as transport over the surface and in the bulk (i.e., phenomena less anisotropic than layer generation) increasingly become the limiting stages of the process, the differences in growth rates even out. The ratios of the velocities of the (100) and (111) faces of gallium crystals are an example (Fig. 3.21).

The effect of the mother medium on the crystal habit is manifested even on such obviously anisotropic crystals as mica: Figure 5.5a shows a mica crystal grown from the melt, and Fig. 5.5b one grown from the flux. The melt is much more viscous than the flux. Therefore the convection in it is retarded, and heat removal from the growing crystals is worse in the melt than in solution, while the supersaturation at the growth front is lower. Accordingly, the platelet shape of crystals grown in the melt is more pronounced.

At the same time, under equal transport conditions, the trend towards the formation of a more isometric shape is also in evidence during the transition between growth from solution and growth from the melt effected by an increase

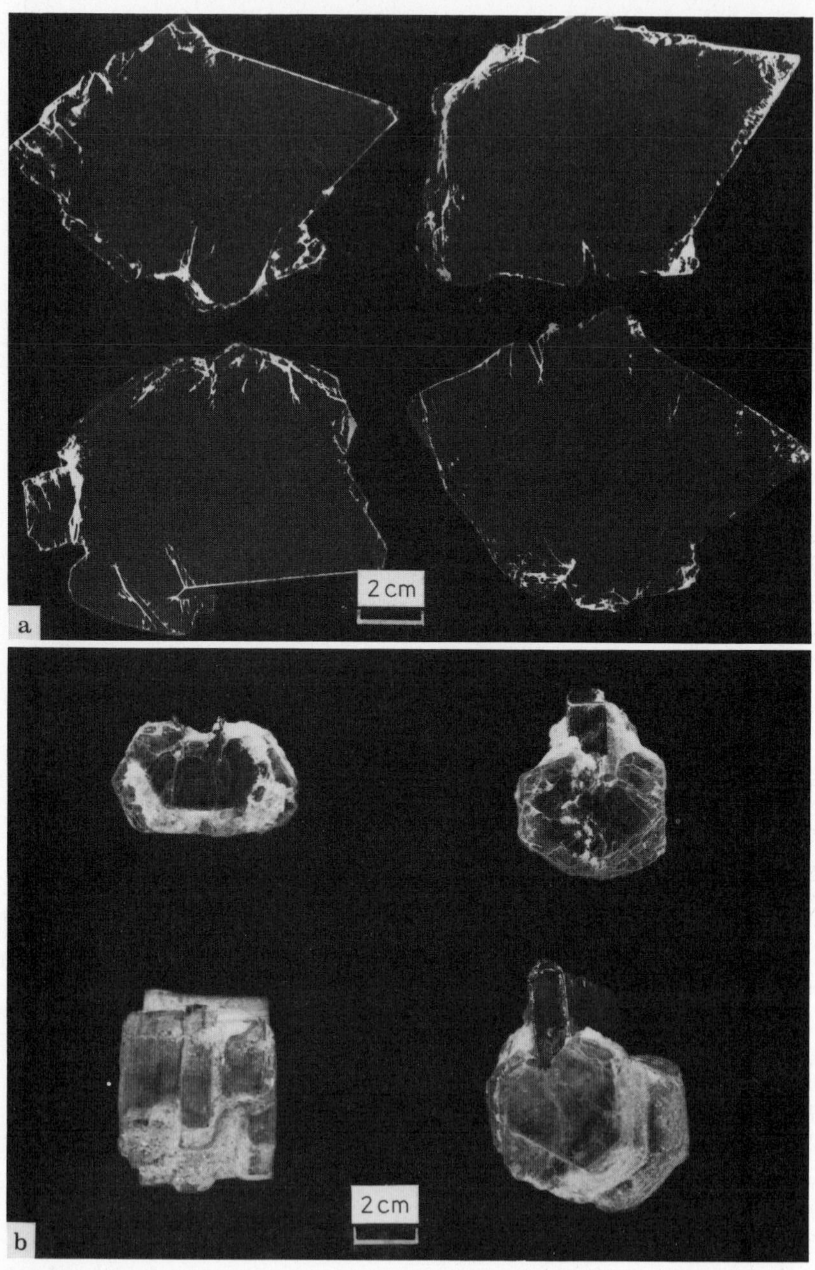

Fig.5.5. Crystals of mica (fluor-phlogopite) grown (**a**) from the melt; (**b**) from a high-temperature solution. Courtesy of I.N. Anikin

in the concentration of the saturated solution and (in accordance with the shape of the liquidus line) by an increase in the temperature. The change in the growth mechanism and hence in the habit of Al crystals with a change in solution composition and crystallization temperature in the Al–Sn system [5.15] is shown schematically in Fig. 5.6. It follows that the crystals grow layerwise at low concentrations (3%) of Al in Sn and practically normally at high ones (20%). Accordingly, in the former case the microcrystals are shaped mainly from cubic faces; in the latter the crystals are rounded, i.e., they have all faces in their habit; and in the intermediate case they show a rather intricate habit. In the limiting case when any of the surfaces are atomically rough, the growth shape depends exclusively on heat removal or diffusion. The mass and heat diffusion in the mother liquor are isotropic, and the anisotropy of the thermal conductivity of most crystals is small (in cubic syngony crystals it is zero). Therefore the stationary growth shape of a crystal growing from a point seed and having an atomically rough surface is near spherical.

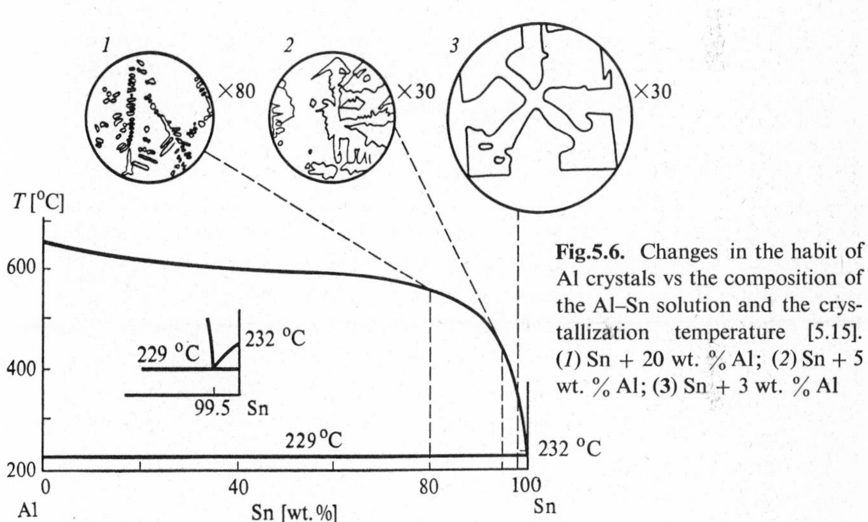

Fig.5.6. Changes in the habit of Al crystals vs the composition of the Al–Sn solution and the crystallization temperature [5.15]. (*1*) Sn + 20 wt. % Al; (*2*) Sn + 5 wt. % Al; (3) Sn + 3 wt. % Al

A growing crystal sets up a certain temperature distribution around it, which, in turn, defines the supercooling and the heat flux at the growth front at each of its points and hence the growth rate and shape of the crystal. Then a stationary, time-invariant growth form of a crystal is possible only when the growth front is a plane, an ellipsoid, or a paraboloid. This theoretical result [5.16b], which is given here without proof, holds true for infinitely rapid surface kinetics, and neglects the shifts in equilibrium temperature due to the curvature of the growth front (Sect. 1.4). The effect of these factors was established by *Temkin* [5.17] for the growth of a paraboloid-shaped crystal from the melt.

This theory gives a relationship for the product of the paraboloid growth rate along its axis and the paraboloid curvature at the apex. The second relationship between these two quantities was chosen to be an unproved condition of the maximal growth rate. Recently, *Langer* and *Müller-Krumbhaar* [5.18] demonstrated that the desired second equation follows from the condition of paraboloid stability (see Sect. 5.3) with respect to the dendritic side-branch formation. The theory fits quantitatively with experiment [5.19].

5.2.5 Faceting Effect

In addition to purely faceted and purely rounded shapes, one and the same crystal may have a combination of both. This is the case, for instance, in Czochralski pulling of crystals from the melt that have fusion entropies $\Delta H/kT \simeq 2$–4, when only the closest-packed faces remain atomically smooth. Silicon is a typical example. In Czochralski pulling (Fig. 5.7), the heat flux is directed from the melt through the crystal to the holder and also to the environment (gas) through the lateral surfaces of the crystal. The isotherms in the crystal may be either convex or concave, depending on the ratio between the axial and radial thermal fluxes. On atomically rough areas of the surface, in accordance with (5.15) at $\beta^T T_Q \delta / a_{S,L} \gg 1$, one has $T_{surf} = T_0$; this means that the growth front coincides with isotherm $T = T_0$, which is called the crystallization isotherm (here, δ is the thickness of the boundary layer in the melt). Figures 5.7a, b demonstrate cases of a convex and a concave isotherm and, hence, a concave and a convex growing interface. The plane face, which truncates the rounded convex growth front and forms a circle of diameter d, intersects isotherms $T < T_0$, which are situated inside the crystal. Therefore at the center of the circle the supercooling is $\simeq h \partial T_S / \partial n$ [5.11] where $\partial T_S / \partial n$ is the temperature gradient at the growth front, and h is the height of the segment cut off by the face from the spherical growth front. A more rigorous analysis which takes into account the heat transfer both in the solid and liquid phase shows that

$$\Delta T_{max} = \mathcal{G} h = \mathcal{G} d^2/8R\,, \tag{5.20}$$

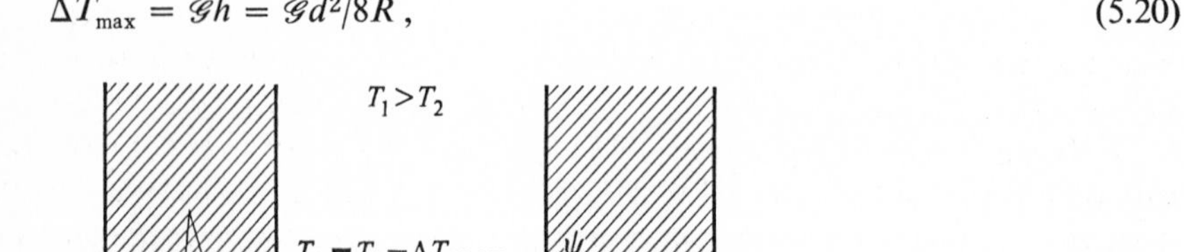

Fig.5.7a,b. Formation of a singular flat face at a rounded growth front. **(a)** Convex front, **(b)** concave front

where

$$\mathscr{G} = \frac{\kappa_S(\partial T_S/\partial n) + \kappa_L(\partial T_L/\partial n)}{\kappa_S + \kappa_L} \tag{5.20a}$$

is the generalized (weighted) temperature gradient at the growing interface, and R is the curvature radius of the rounded growing interface. From (5.20) it follows that the size of the face is

$$d = 2\sqrt{\frac{2R\Delta T_{max}}{\mathscr{G}}}, \tag{5.21}$$

i.e., d diminishes $\propto \mathscr{G}^{-1/2}$ with increasing temperature gradient $\mathscr{G}$ at the growth front and increases with the curvature radius R of the rounded front. Supercooling ΔT_{max} is determined by the condition that the face must have the same velocity along the crystal axis as the rounded front, i.e., the crystal pulling velocity. Knowing the dependence $V(\Delta T)$ it is possible to find ΔT_{max} from the given pulling velocity V with the aid of (5.21) and, hence, the size $d(V)$ for different growth mechanisms: to ensure one and the same growth rate the face of a dislocation-free crystal requires higher supercooling than a face containing dislocation step sources. Therefore, in conformity with (5.21), the facets on dislocation-free crystals have larger diameters. This effect is observed on silicon, germanium and garnet (Fig. 10.49) and has been used to determine the specific linear step energy and step kinetic coefficient for Si: $\alpha_l = 2 \cdot 10^{-6}$ erg/cm, $\beta^T = 50$ cm/s K.

By similar reasoning it is easy to obtain the width of the face strip arising by *Voronkov* [5.16c] on the periphery of a concave growth front (Fig. 5.7b):

$$d = \frac{\Delta T_{max}}{\mathscr{G}\cos\phi} = \frac{\Delta T_{max}}{\mathscr{G}_r}, \tag{5.21a}$$

where angle ϕ (Fig. 5.7b) is formed by isotherm $T = T_0$ and the side of a crystal cylinder, and $\mathscr{G}_r$ is the generalized radial temperature gradient $\mathscr{G}\cos\phi$.

A face on a rounded growing interface captures an impurity in a concentration different from that trapped by the rounded portions of the interface (Sect. 4.3): the grown crystal turns out to be inhomogeneous, since it contains the impurity-enriched core — the so-called faceting effect arises (Fig. 4.9). The most radical way to exclude the faceting effect is to obtain a flat crystallization isotherm or, on the other hand, to use a concave isotherm and neglect the periphery of the crystal. The value of d can also be reduced by increasing the temperature gradient at the growth front (Sect. 10.3.2).

If a crystal is pulled from the melt (Fig. 5.7) a contour appears where the Solid, liquid and gas phase meet. This contour is termed the three-phase line.

Mutual orientation of the three interfaces at this line is governed by the balance between the crystal-gas, crqstal-liguid and liquid-gas free interfacial energies. This balance (depending on the temperature at the line) results in a definite mutual orientation of these surfaces at the three-phase line. Changing the height of liquid meniscus, we change the orientation of liquid-gas interface and thus the orientation of the crystal-gas interface. The later causes a change in the diameter of pulled crystal. Thus the shape of the lateral crystal-gas interface varies with the crystallo-graphic orientation position and temperature of the growing crystal-melt interface. In particular, these variations are due to the appearence or disappearence of a facet on the crystal-melt interface, appearence or total disappearence of dislocation outcrops on this facet. These phenomena are considered in detail in [5.16d-f] and in the papers on shaped growth [5.16g-i].

5.3 Stability of Growth Shapes

The above-discussed crystals with convex rounded or faceted forms remain similar to themselves in the course of growth only under certain conditions. In an unstirred mother medium such growth is possible as long as the crystal size and the deviation from equilibrium do not exceed certain values [5.11, 20]. Otherwise the crystals acquire a so-called skeletal or dendritic shape. The former is shown in Fig. 1.14a, and the latter (for the same ammonium chloride crystal) in Fig. 5.8. The dendrite develops from each skeleton stem as a result of the appearance of branches of the second order, third order, etc.

5.3.1 Sphere

Skeletons and dendrites result from the instability of the initial convex form of a crystal with respect to accidental perturbations. Let us consider the stability of a sphere growing from solution. The distribution of solution concentration around a spherical crystal is schematized in Fig. 5.1. Suppose a bump of height $\delta \ll R$ arises accidentally on the crystal surface (Fig. 5.1), so that the curvature of the surface at the top of the bump has increased by $\sim M\delta/R^2$, where the number M is greater, the sharper the bump is. The top of the bump finds itself in a more supersaturated solution with a concentration $C_{\text{surf}} + (\partial C/\partial r)\delta$, where $C_{\text{surf}} = C_R$ and $\partial C/\partial r$ is obtained from (5.7) for $r = R$. The increase in concentration at the top of the bump stimulates its further increase to macroscopic dimensions and transforms the sphere into a skeletal crystal, i.e., it leads to instability.

On the other hand, the equilibrium concentration C_0 at the top of the bump with a larger curvature is also higher than that over the rest of the crystal surface, in accordance with Laplace's equation (1.34):

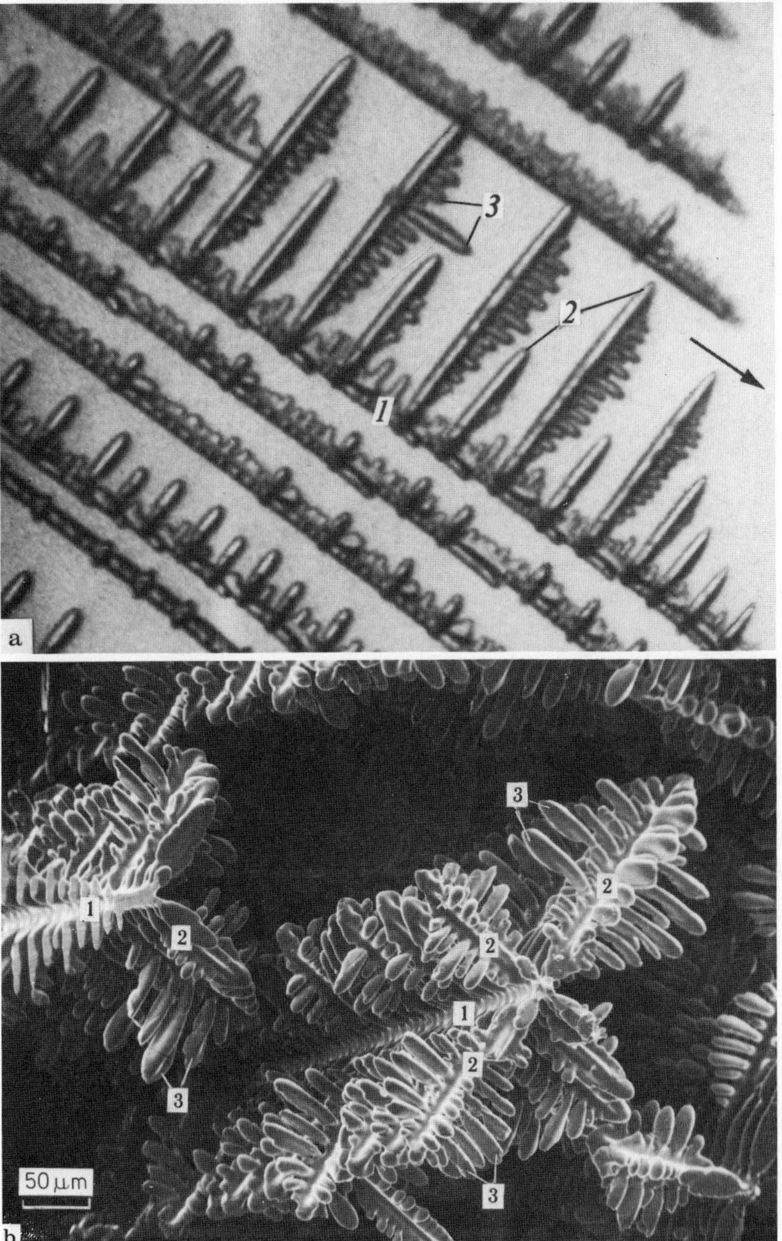

Fig.5.8a,b. Dendritic crystals. (**a**) NH_4Cl in aqueous solution. Optical microscope, magnification × 200. Courtesy of M.O. Kliya. (**b**) Co crystals in the alloy Co–TR–Cu. Micrograph taken in a scanning electron microscope from a sample obtained by grinding and subsequent etching out of the matrix [5.21]. The *numbers* (*1–3*) denote the dendrite branches of the corresponding order. The tops of the stems (the first-order branches) in (**b**) were cut during grinding; in (**a**) the direction of their growth is indicated by the *arrow*

$$C_0 \simeq \bar{C}_0\left(1 + \frac{2\Omega\alpha}{kTR} + \frac{2\Omega\alpha}{kTR^2}\, M\delta\right),$$

where $\bar{C}_0$ is the equilibrium concentration of solution over an infinite flat surface. Thus, by increasing the value of C_0, the surface energy reduces the supersaturation at the top of the bump, and thus impedes its further elongation. This means that the surface energy helps to conserve the shape of the surface.

The stability criterion must reflect the conditions under which the "kinematic force," which elongates the bump, is less than the thermodynamic Laplace force acting in the opposite direction. In other words, the velocity of the bump top relative to the unperturbed front, ΔV, must be negative:

$$\Delta V = \beta\Omega\left[(\partial C/\partial r)\delta - \bar{C}_0\frac{2\Omega\alpha}{kTR^2}\, M\delta\right] < 0\,. \tag{5.22}$$

In deriving (5.22) we took advantage of relation (3.5), in which C_0 denoted the equilibrium concentration of solution over the curved surface related to the curvature of this surface by Laplace's equation. C_0 has the same meaning in (5.6, 7). Substituting $\partial C/\partial r$ from (5.7) into (5.22) and solving the resultant equation for R, we have the stability criterion:

$$R < R_{\rm cr} \simeq \tfrac{1}{2}MR_{\rm r}(1 + \sqrt{1 + 4D/M\beta R_{\rm c}})\,, \tag{5.23}$$

where R_c is the radius of the critical nucleus [(2.2), Sect. 2.1]. A rigorous analysis performed by *Mullins* and *Sekerka* [5.20, 22, 23] used perturbation in the form of different harmonics of the spherical functions superimposed on the initial spherical shape. This analysis was made for infinitely fast interface kinetics ($D/\beta R < 1$). It leads to the conclusion that the critical radius $R_{\rm cr}$, above which the growing crystalline sphere becomes unstable, is given by [5.22]:

$$R_{\rm cr} = R_{\rm c}[1 + \tfrac{1}{2}(m + 1)(m + 2)]\,, \tag{5.23a}$$

i.e., in order of magnitude, (5.23) correctly reflects the principal dependences if we put $2M = 2 + (m + 1)(m + 2)$, where m is the number of the perturbation harmonic. The value $m = 2$ corresponds to the transformation of a sphere to an ellipsoid; $m = 3$ corresponds to the symmetry of a tetrahedron; $m = 4$ corresponds of a cube, etc. The higher the perturbation symmetry, i.e., the larger m is, the more stable is the sphere with respect to such a perturbation. The physical cause for the appearance and development of perturbations of a particular symmetry is the anisotropy of crystal growth kinetics: even for rough surfaces, the symmetry of the angular dependence of the kinetic coefficient is evidently slightly different from the spherical one.

Similar reasoning can easily be presented for a crystal growing in a supercooled melt. The corresponding criterion is obtained from (5.23) by replacing (5.17).

According to (5.23), at high rates of the surface processes ($\beta R_{cr}/D \gg 1$ or $\beta^T T_Q R_{cr}/a_L \gg 1$), a sphere is stable only if its radius does not exceed MR_c, which at $m = 2$, equals 7 R_c. Even at low supercoolings, which exclude spontaneous nucleation in the melt, but ensure an appreciable growth rate, R_c is of the order of $10^{-6} - 10^{-4}$ cm, i.e., $R_{cr} \simeq 10^{-5} - 10^{-3}$ cm. For instance, for iron ($\Omega = 1.2 \times 10^{-23}$ cm^3, $\alpha = 204$ erg/cm^2, $\Delta s/k = 1.97$) at a supercooling $\Delta T = 10$ K a spherical crystal is stable only up to a size of $\simeq 2.5 \times 10^{-5}$ cm. With increasing supercooling the critical stability radius falls off as $1/\Delta T$.

The lower the rate of the surface processes, the larger is the critical stability radius; at $\beta R_c/D \ll 1$ and $\beta^T T_Q R_c/a_L \ll 1$ the critical size reduces with increasing supersaturation as $1/\sqrt{\Delta T}$, because $R_{cr} \simeq 0.5\sqrt{MDR_c/\beta}$, or $R_{cr} \simeq 0.5 \sqrt{Ma_L R_c/\beta^T T_Q}$. For growth from solution at $\alpha \simeq 50$ erg/cm^2, $\Omega \simeq 3 \times 10^{-2}$ cm^3, $D \simeq 10^{-5}$ cm^2/s, and $\beta \simeq 10^{-4}$ cm/s, for supersaturation $\Delta C/C \simeq 10^{-2}$ we have $R_c = 7 \times 10^{-6}$ cm and $R_{cr} \simeq 1.6 \times 10^{-3}$ cm, whereas at $\beta R_c/D \gg 1$ the critical size is $7R_c \simeq 4 \times 10^{-5}$ cm.

The initial stage in the loss of stability and transformation into a skeleton of a cyclohexanol crystal is illustrated in Fig. 5.9.

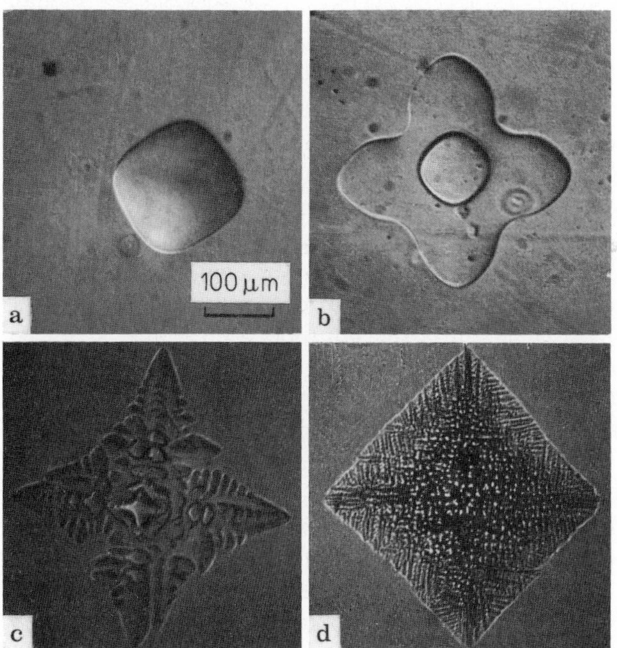

Fig.5.9. Consecutive stages in the loss of stability and development of a multibranch dendrite of a cyclohexanol crystal [5.24]

Each stem of a developed skeleton grows independently of the others and has a near-paraboloid shape; its growth rate is defined by the curvature at the top of the paraboloid. At a certain distance from the top, the curvature radius of the paraboloid surface exceeds the critical R_{cr}, and this surface also loses its stability: side branches of a dendrite appear (Figs. 5.8, 9). For a detailed anlaysis see [5.18, 3.33b].

5.3.2 Polyhedron

Experiment shows that the perfect faceted shapes described in Sect. 5.2 arise when the supersaturation or supercooling at the growth front is rather low. With increasing deviation from equilibrium a crystal changes its habit, turning into a skeleton or a dendrite.

In the first stage of transformation the crystal, remaining a polyhedron, begins to trap macroscopic inclusions of the mother medium and therefore becomes turbid. Inclusions usually begin to appear beneath the central parts of the faces, and the transverse dimension of the area containing the inclusions is larger, the higher the supersaturation is. The formation of inclusions is preceded by the appearance of macroscopic steps which have a height of $\sim 10^{-4} - 10^{-2}$ cm and are visible to the naked eye. A further increase in supersaturation results in skeletal forms. At still higher supersaturations, dendrites ultimately arise.

Let us consider the possible causes for the described evolution of the faceted growth shapes [5.11, 12]. The most important external factors determining the development of the different forms are the absolute value of supersaturation over the growing surface and its nonuniformity. The supersaturation distribution over a face can be found by solving the equation of diffusion or thermal conductivity with boundary conditions giving the constancy of the mass or heat flow to each point of the face. For the sake of definiteness, we shall now consider growth from solution. The simplest approximation, but one which is at the same time fairly good, for small values of the kinetic coefficient, is the spherically symmetric approximation of the diffusion field around a polyhedron. Indeed, a near-cylindrical concentration field is set up near a crystal growing from solution sealed between two parallel glass platelets (Fig. 5.3). In such a diffusion field each face is intersected by several lines of equal concentrations (or isotherms, in melt growth), and the lines corresponding to large supersaturations or supercoolings pass near the vertices. Since according to (5.7), $\partial C/\partial r \propto \beta/D$ on the crystal surface, the difference between the supersaturations at the center of the face and at the vertex is $\propto \beta l/D$. Therefore if the crystal size l is so small that $\beta l/D \ll 1$ (purely kinetic regime) the supersaturation on the surface is practically constant and equal to that in the bulk of the mother liquor. The same situation must take place in an ideally stirred solution for crystals of any size. Here, the growth shape is determined by the kinetic analog of Wulff's rule (5.18) (Fig. 5.10a).

As the crystal increases in size, the supersaturation over the central parts of the faces becomes lower than over the vertices and edges (Fig. 5.10d). Nevertheless, experiment shows that the surface remains macroscopically flat. Conse-

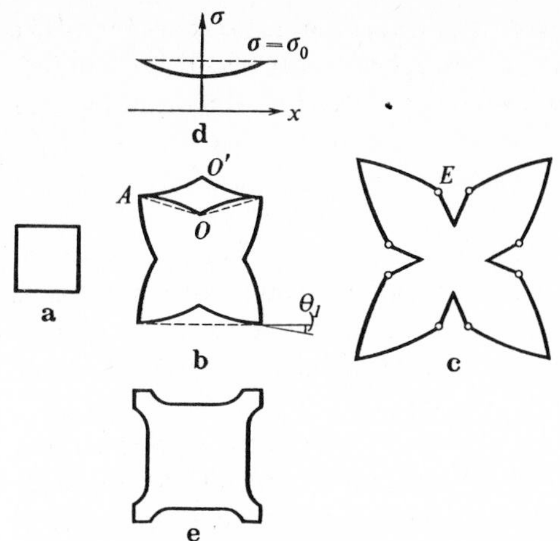

Fig.5.10. Consecutive stages (**a–c**) in the development of a skeletal crystal from a crystal polyhedron due to the inconstancy of supersaturation along the faces (**d**), σ_c is the critical supersaturation for the formation of two-dimensional nuclei on the face; (**e**) profile of the surface during the formation of kinematic waves of step density near the vertices and edges of a polyhedron. In (**c**), E represents the edge of the shock waves of steps emanated by the upper left apex

quently, there must be a mechanism compensating for the inconstancy of the supersaturation. Compensation can be achieved, for example, by a nonuniform distribution of a growth-retarding impurity, also with a minimum at the center of the face. A distribution of this type may arise if the impurity is not trapped by the crystal bulk in the course of growth and does not leave its surface. Another factor promoting compensation is a temperature distribution along the face with a minimum near its center. Such a distribution results if the crystal contains a "cooler." But the most general and most probable factor maintaining the polyhedral shape of a growing crystal is the anisotropy of $V(\boldsymbol{n})$.

In contrast to Sect. 5.3.1, we neglect here the effect of the surface energy, which may be done for the rather large size ($\gtrsim 10^{-2}$ cm) of stable polyhedra. The measure of anisotropy of the growth rate is parameter $\Theta = \partial \ln b/\partial p$, see (3.21). In this case the inconstancy of the supersaturation may be offset by a slight curvature of the face as shown in Fig. 5.10b. Indeed, if a face takes a shape whose profile is of type AO′B or AOB (Fig. 5.10b), the step density and hence the kinetic coefficient for its central, most curved areas is higher than for a face periphery (see Fig. 3.7a). With a strong dependence $V(\boldsymbol{n})$, i.e., at $\Theta \gg 1$, a deviation of about one degree from the orientation of the singular face is sufficient to keep a velocity R constant for any point on the curved face, see (3.6, 19, 32):

$$R(p, C - C_0) = b(p(x))\,[C(x) - C_0] = \text{const}\,, \tag{5.24}$$

where kinetic coefficient $b(p) = \beta(p)\sqrt{1 + p^2}$. The inconstancy of supersaturation $C(x) - C_0$ along the face (Fig. 5.10d) is offset by the inconstancy of the local orientation [step density $\propto p(x)$], which causes the above-mentioned inconstancy of kinetic coefficient $b(p(x))$. The constant equal to $b(p_1)(C_1 - C_0)$ is determined physically by the activity of the center generating the growth layers on the face.

If the supersaturation is below the critical value σ_0 necessary for the formation of two-dimensional nuclei on a given face, including sites near the vertices, the layers will be produced by screw dislocations originating, as a rule, from a seed, and therefore emerging at the surface in the central parts of the faces. Here, p_1 is the slope at the top of the vicinal hillock on the most active dislocation.

If the supersaturation exceeds the critical value in the vicinity of vertices, is the latter that will be the sources of growth steps. The layers propagating from the vertices will cause the face to divide into large vicinal hillocks with centers at the vertices. In this case p_1 is the local slope of the surface near the vertex ($p_1 = \tan \theta_1$; see Fig. 5.10b). At near-critical supersaturations the leading layer sources may also be dislocations arising in the course of growth and emerging at the surface in the vicinity of the vertices, where the supersaturation is the highest. Assuming that the relative changes in supersaturation along the face, and also the changes in local slopes, are small, and expanding (5.24) with respect to these changes, it is easy to obtain the difference in slopes at the center of face p_2 and near vertex p_1;

$$p_0 - p_1 = \frac{(\partial R/\partial \sigma)(C_1 - C_0)}{(\partial R/\partial p)} = \frac{C_1 - C_2}{(C_1 - C_0)\Theta} . \tag{5.25}$$

If the relative change in supersaturation is $(C_1 - C_2)/(C_1 - C_0) = (\sigma_1 - \sigma_2)/\sigma_1 \simeq 0.2$ and $\Theta \simeq 10$, then $p_2 - p_1 \simeq 2 \times 10^{-2}$; i.e., to compensate for supersaturation inconstancy it is sufficient to change the local slopes by $\sim 1°$, as mentioned above.

The larger the crystal as compared with D/β, the greater must be the curvatures of its faces to offset the inconstancy of the supersaturation.

Buckling at the center of the face as a result of high supersaturations leads to further deterioration in the nutrition of this area. This, in turn, lowers the growth rate in the face center, and causes it to lag behind the vertices. A fact of much greater importance is that the curvature leads to the appearance on the surface of areas with considerably higher values of the kinetic coefficient and therefore with substantially higher values of parameter $\beta l/D$ and supersaturation nonuniformity, and also to an abrupt drop in the anisotropy of kinetic coefficient $\Theta = (1/b)(\partial b/\partial p)$. Experimental evidence of an abrupt increase in the kinetic coefficient at the beginning of skeletal growth has been presented by *Goldsztaub* et al. [5.25]. Indeed, as can be seen from Fig. 3.7a, derivative $\partial b/\partial p$ falls off rapidly as we recede from the orientation of the singular face and approach the maximum of $b(p)$, where $\partial b/\partial p = 0$. As a result of these changes the derivation from the singular interface increases still further; this, in turn, leads to a further fall of $\partial b/\partial p$, and so on. Thus the attainment of certain critical values of local slopes leads to an avalanchelike loss of stability, or more precisely, it becomes impossible for the growing polyhedron to remain similar to itself in the course of growth. Depending on the type of $b(p)$ anisotropy the

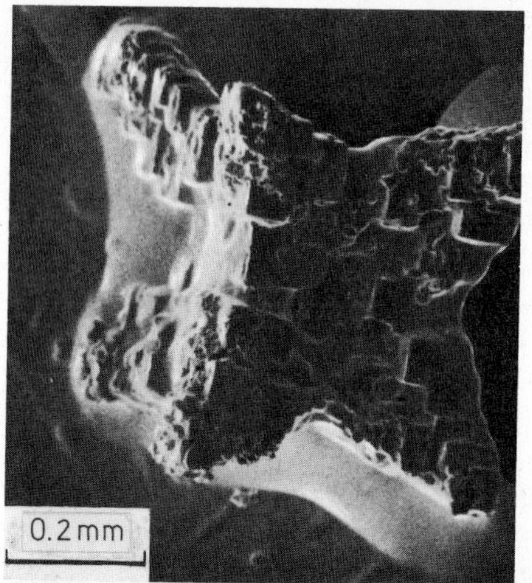

Fig.5.11. Skeletal crystal of NaCl grown from an aqueous solution in the presence of potassium ferrocyanine. The photo was taken in a scanning electron microscope [5.26]

Fig.5.12. Skeletal crystal of $Pb_3NiN_2O_9$ in polarized light. Courtesy of M.O. Kliya. Magnification × 70

▼

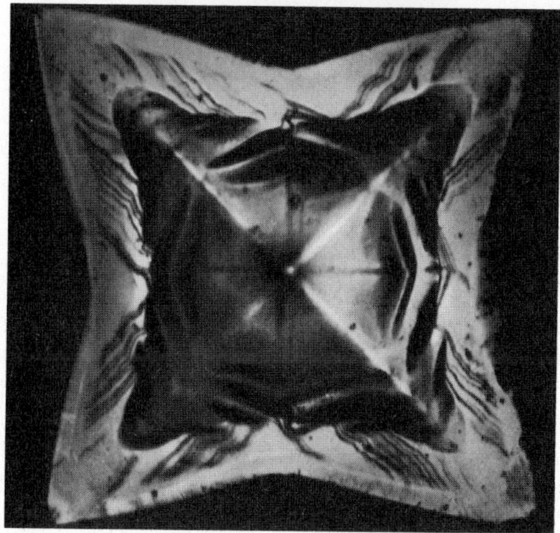

progressive deepening of the pit at the center of the face may either occur by smooth curving of the face or be accompanied by the appearance of a new face surrounded by the edge E shown in Fig. 5.10c.

The skeletal growth shapes of faceted crystals of NaCl and $Pb_3NiNb_2O_9$ presented in the photos in Figs. 5.11, 12 correspond to the model discussed (Fig. 5.10b, c).

The kinetic coefficient increases abruptly with increasing local slope of the surface, i.e., with increasing step density at each point, as long as the step density is low (Sect. 3.3), i.e., with nonlinear dependence of the growth rate on supersaturation. At larger slopes the overlapping of the diffusion fields of the steps is sufficiently strong, and the kinetic coefficient of the face is practically independent of its orientation. This range corresponds to the linear dependence $V(\sigma)$.

According to morphological and kinetic data [5.25, 27], the range of the strong dependence $\beta(p)$ extends to the values of local slopes $p \sim 10^{-2} - 10^{-1}$. Therefore the above-mentioned values can be used as the critical value of the maximum slope of the surface at the face center, p_{cr}, at which skeletal growth begins.

Calculation shows [5.11, 12] that a slope p_{cr} is achieved at the center of the face when the crystal size is

$$l_{cr} = N(\Theta)(p_{cr} - p_1)D/b(p_1), \qquad \Theta = \frac{1}{b}\frac{\partial b}{\partial p}\bigg|_{p=p_1}. \tag{5.26}$$

This is the maximum crystal size, above which skeletal growth begins. For melt growth, the analogous criterion has the form

$$l_{cr} = N(\Theta)(p_{cr} - p_1)a_L/b^T(p_1)T_Q. \tag{5.27}$$

Here, function $N(\Theta) \simeq 2.5\,\Theta$ at $\Theta \gg 1$, and $N(\Theta) \simeq 1$ at $\Theta \lesssim 1$.

The crucial importance of bulk diffusion for the formation of the skeletal shape has been clearly demonstrated in experiments with zinc [5.28] and ice [5.29] crystals grown from their vapors. Indeed, crystals of every size remained polyhedral when growing in a vacuum and became skeleton-shaped when the growth chamber was filled with an inert gas (argon for zinc and air for ice). It was also shown that the critical size at which the skeletal shape arose was inversely proportional to the pressure of the inert gas, i.e., to the diffusivity of the crystallizing substance, in accordance with (5.26).

Let us now discuss in more detail the dependence of the critical size on the deviation $\Delta\mu$ from equilibrium, i.e., on the supersaturation or supercooling. If the supersaturation or supercooling throughout the interface is lower than the critical one necessary for face growth by nucleation, i.e., $\Delta\mu < \Delta\mu_c$, then, as noted above, the sources of steps must be dislocations. Dislocations in the crystal body often begin to form on a seed at the crystal surface, and emerge at the surface in the central parts of the faces. Therefore in the range $\Delta\mu < \Delta\mu_c$ one should expect face profiles similar to AO′B in Fig. 5.10b. As the crystal increases in size during growth in the diffusion field, the absolute value of supersaturation on the surface diminishes, which reduces the rate of step generation by the dislocation as well as the growth rate. If the defect in the center of the face (at point O′ in Fig. 5.10b) remains the only source of layers in the course of face growth, the face must remain macroscopically flat, although its growth gradually slows down. In other words, in growing, the crystal will always remain similar to itself even in the diffusion regime. And conversely, if the supersaturation is sufficient for the formation of two-dimensional nuclei, or if the leading layer sources are defects located near the crystal vertices, the crystal can retain the macroscopically flat faces only as long as it remains within the critical size l_{cr}, which is determined by (5.26) or (5.27).

The dependence of l_{cr} on deviation $\Delta\mu$ from equilibrium is also included in (5.26, 27) via the anisotropy parameter $\Theta = \Theta(p_1)$. Indeed, according to Fig. 3.7a, the value of $\Theta(p)$ reduces with increasing deviation p of the face from the nearest singular orientation $p = 0$. In the case at hand, the "working point" $p = p_1$ on the plot $b(p)$, and hence $\Theta(p_1)$ as well, is defined by the step generator strength, which in turn depends on $\Delta\mu$. For dislocation sources at small $\Delta\mu$ and p_1, we have $p_1 \propto \Delta\mu$, i.e., $\Theta(p_1) \propto p_1^{-1} \propto \Delta\mu^{-1}$ and, accordingly, $l_{cr} \infty \Delta\mu^{-1}$. If the nucleation mechanism of layer generation operates, intensive nucleation begins at $\Delta\mu \gtrsim \Delta\mu_c$. At $\Delta\mu \ll \Delta\mu_c$ the probability of nucleation is practically zero and increases rapidly with increasing $\Delta\mu$ in the range $\Delta\mu \simeq \Delta\mu_c$. If $\Delta\mu$ is so large that the surface becomes kinetically rough, then Θ is small, and the critical size is no longer determined by the anisotropy of the surface kinetics, but by the surface energy (Sect. 5.3.1).

Steps propagating from the vertices may merge together, forming higher, sometimes macroscopic steps, which usually trap the mother medium in the form of macroscopic inclusions (Sect. 6.1.2 and Figs. 6.2–4). It is evidently for this reason that the appearance of skeletal forms is often preceded by the formation (mentioned above) of inclusions beneath the central parts of faces.

The formation of macroscopic steps and shock waves sometimes occurs in the immediate vicinity of vertices and edges of a growing crystal, and then profiles of the type presented in Fig. 5.10e arise. Such kinematic (shock) waves of step density (Sect. 3.4.2) were observed during the growth of zinc crystals from vapors in an argon atmosphere [5.28].

The formation of macrosteps in the immediate vicinity of a vertex was also observed by *Papapetrou* [5.30] on KCl crystals. It began with the rapid growth, from a cube vertex, of a needlelike crystal branch oriented along the cube diagonal. Then the needle, which had achieved a length of several tenths of a millimeter, ceased to grow and began to thicken, giving rise to macrosteps on the faces or to a small cubic crystal. When this crystal attained some critical size, the process was repeated from the beginning, leading to a large-step skeleton of the type shown in Fig. 5.11. The length of the needle, and hence the height of the steps, reduces with decreasing supersaturation.

In the presence of impurities of divalent ions (Pb, Cd, Zn, etc.) (Sect. 4.1), no more "shooting off" of macroscopic needles is observed (at least in an optical microscope), and the crystals acquire smooth shapes similar to those shown in Fig. 5.12.

The above-discussed mechanism of changes in growth forms also extends to growth in a stirred solution: the diffusion layer is thinnest near the vertices and edges, which also results in the maximum supersaturation there, causing layer generation at these sites. But otherwise the nonuniformity of supersaturation over different parts of the face is considerably less for a stirred solution than for pure diffusion. Therefore with vigorous stirring, even large crystals develop practically no skeletal forms: the process is terminated in the stage at which turbid polyhedral crystals with vicinals and macrosteps are formed.

5.3.3 Plane

Let us now consider the stability of a flat growth front. An infinite flat growth front (Fig. 5.13a) is unstable if relief irregularities arise in it which exist permanently during growth (Fig. 5.13b). The front as a whole, i.e., the envelope of the relief that has formed, then often remains flat. A flat growth front is the surface of a spherical crystal of infinite radius. Therefore according to (5.23), a flat front with isotropic surface kinetics in a supercooled melt or a supersaturated solution is always unstable and turns into an agglomerate of parallel paraboloids or dendrites. We now consider the conditions for the development of instability in more detail.

Instability stems from improved nutrition of the top of an accidental bump (or, during growth from the melt, from improved heat removal from its top), i.e., from an increase in supersaturation or supercooling in the mother medium with the distance from the main growth front (as in Fig. 5.1). The distribution will be inverted if we remove the heat of crystallization through the crystal. The typical temperature distribution in a crystal (T_S) and a melt ($T_L^{(1)}$ or $T_L^{(2)}$) in this case is shown in Fig. 5.14a. It is achieved in growing by the Czochralski process, and also in all methods where the container holding the crystal and the melt moves (horizontally or vertically) through the temperature-gradient zone. If the temperature distribution is such that the generalized temperature gradient (5.20) is positive ($\mathscr{G} > 0$), then, as can be shown, the supercooling at the top of an accidental bump will be lower than at the unperturbed front, and the bump will disappear, i.e., the flat growth front will be stable. Condition $\mathscr{G} > 0$ generally means that the melt is superheated (Fig. 5.14), although the requirement $\mathscr{G} > 0$ is usually compatible with condition $\partial T_L/\partial z < 0$.

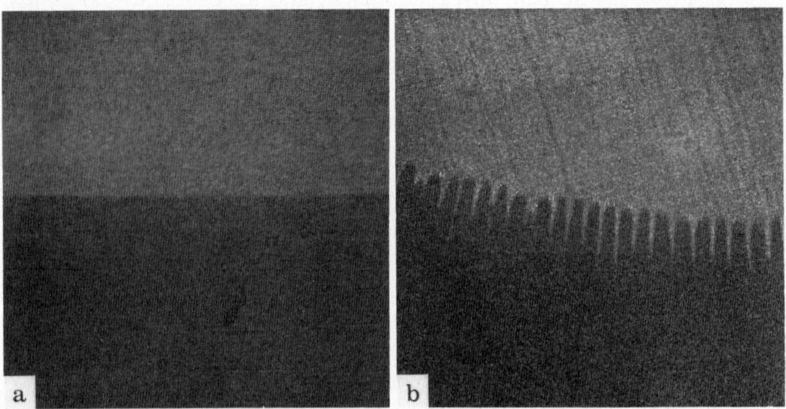

Fig.5.13a,b. Morphology of the solidification front of a thin tin layer enclosed between two parallel glass platelets. Polarized light, magnification × 100. **(a)** Flat solidification front. The conditions are close to equilibrium ones; the melt (in the *upper part* of the photo) is superheated; **(b)** cellular front (on the average flat) at $\Delta T < 0.03°$. The melt (in the *upper part* of the photo) is supercooled [5.31]

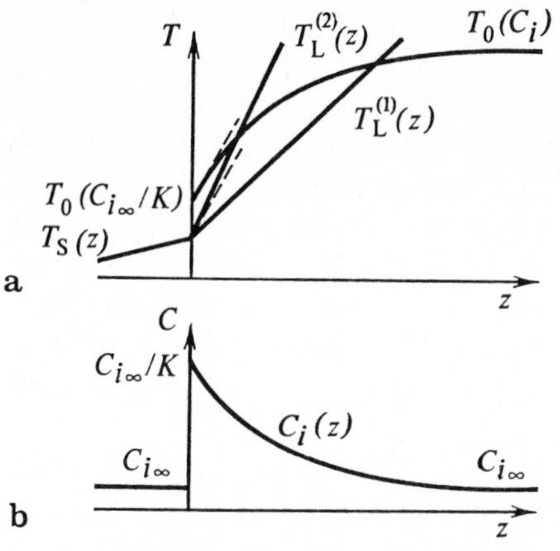

Fig.5.14a,b. Distribution of the temperature (**a**) and impurity concentration (**b**) ahead of the growth front. $T_L^1(z)$ and $T_L^2(z)$ are the actual temperatures in the melt at different gradients; $T_0(C_i)$ is the equilibirum temperature corresponding to the impurity distribution (**b**). The *dotted lines* show the gradient of equilibrium temperature $T_0(C_i)$ in the liquid at the growing interface $z = 0$

If the growing plane is unstable, it will split into cells, thus acquiring a cobblelike shape. Let us estimate the cell size and demonstrate in an elementary way the stability condition. Let the normal growth rate of the flat interface under consideration be $V = \beta^T(\bar{T}_0 - \bar{T})$, where $\bar{T}_0$ and $\bar{T}$ are the melting temperature and actual temperature of the flat interface, and β^T is the kinetic coefficient. Suppose that the originally flat interface $z = 0$ undergoes a perturbation $\delta z = z_0 \sin(2\pi/\lambda)x$, where z_0 is a small amplitude and λ is the wavelength of the surface goffer, the x axis being within the original plane. The melting temperature at the ruffled interface is, according to the Laplace equation,

$$T_0 = \bar{T}_0 + \delta\bar{T}_0 = \bar{T}_0 + (\Omega\alpha/\Delta s)\partial^2\partial z/\delta x^2\,, \tag{5.28}$$

where

$$\partial^2\delta z/\partial x^2 = -z_0(2\pi/\lambda)^2 \sin(2\pi/\lambda)x$$

is the interface curvature. If $\delta z(x)$ is convex, its melting temperature is lower than $\bar{T}_0$. The actual temperature at the perturbed interface is $\bar{T} + \delta\bar{T} = \bar{T} + \mathscr{G}\delta z$.

Thus the variation in growth rate due to perturbation is $\delta V = \beta^T(\delta\bar{T}_0 - \delta\bar{T}) = \beta^T[-(\Omega\alpha/\Delta s)(2\pi/\alpha)^2 - \mathscr{G}]\delta z$. The growing interface is stable if $\delta V/\delta z < 0$. This condition is fulfilled when

$$\lambda < \lambda_{cr} = 2\pi\sqrt{\Omega\alpha/(-\mathscr{G})\Delta s}\,. \tag{5.29}$$

The critical wavelength λ_{cr} gives the characteristic size of the cells into which the unstable interface splits. For iron, $\Omega = 1.2 \times 10^{-23}$ cm^3, $\alpha = 204$ erg/cm^2, $\Delta s = 1.97$ K, and if $\mathscr{G} = 50$ K/cm, one has $\lambda_{cr} = 2.6 \times 10^{-3}$ cm.

An analogous approach is also applicable to solution growth at isotropic interface kinetics [5.32]. A typical average size of the "cobble" structure on the hydrothermally grown pinacoid quartz face found in this way is in agreement with experiment [5.33].

A considerable effect on stability is exerted by impurities, which shift the melting point. The impurity concentration is distributed ahead of the front in accordance with curve $C_i(z)$ in Fig. 5.14b, e.g. (4.46). Therefore the distribution of the equilibrium temperature (liquidus temperature) corresponds to curve $T_0(C_i)$ in Fig. 5.14a. If the variation in the true melt temperature near the crystal surface corresponds to straight line $T_L^{(2)}$, the supercooling diminishes when the distance from the front increases. On the other hand, if the temperature gradient at the growing interface is lower and the melt temperature is represented by straight line $T_L^{(1)}$, the supercooling increases with the distance from the front. In the latter case the growth front is unstable. The supercooling caused by the shift in equilibrium temperature due to the presence of an impurity ahead of the growth front is termed constitutional. If the rate of the surface processes is infinite ($T_{surf} = T_0$), constitutional supercooling exists when, roughly speaking, at the growth front (for $z = 0$)

$$\frac{\partial T_L}{\partial z} < \frac{\partial T_0}{\partial C_i}\frac{\partial C_i}{\partial z} = -\frac{\partial T_0}{\partial C_i}\frac{C_i(1-K)V}{D}, \tag{5.30}$$

where $\partial T_0/\partial C_i$ is taken along the liquidus line in the phase diagram at the solution concentration existing in the immediate surface vicinity, and K is the impurity distribution coefficient (a nonequilibrium one, generally speaking). A detailed analysis of thermal fields not only in the liquid, but also in the solid phase, leads in (5.30) to the replacement $\partial T_L/\partial z \to \mathcal{G}$ and to the following criterion of stability of a flat interface growing in a melt with impurities:

$$\frac{\mathcal{G}}{V} > -\frac{mC_{i\infty}(1-K)}{DK}, \tag{5.31}$$

where $m = \partial T_0/\partial C_i$ and $C_{i\infty}$ is the impurity concentration in the liquid bulk. Criterion (5.31) neglects the stabilizing effect of the surface energy. According to (5.31), at given values of the temperature gradient at the growth front and of the impurity concentration, the growth rate should not exceed a certain value. Conversely, a stable growth front is possible at a given growth rate only when the temperature gradient at the front is high enough. The higher the impurity concentration, the higher must be the temperature gradient and the lower the growth rate (Fig. 5.15, dashed line). If the stabilizing effect of the surface energy at the top of the bump is taken into account [5.34], the stability range slightly expands (solid line in Fig. 5.15).

It is worth mentioning here that in Czochralski-Kyropoulos, Stockbarger-Bridgman, and Stepanov growth techniques, zone melting, etc. (Sect. 10.2), the temperature gradient at the growth front and the growth rate are assigned independently of each other by the furnace heaters and by the rate at which

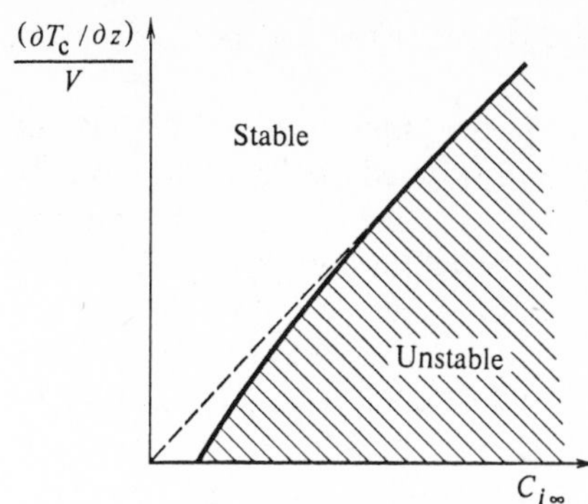

Fig.5.15. Ranges of the values of the temperature gradient in the melt near the growth front, $\partial T_M/\partial z$, growth rate V, and impurity concentration $C_{i\infty}$, at which the flat growth front is stable or unstable. Both the *dashed line* and the *solid line* are drawn according to the stability criterion; the former neglects the surface energy, and the latter takes it into account

the crystal is pulled (*Czochralski, Kyropoulos, Stepanov*) or of the ampule motion (*Stockbarger-Bridgman*, zone melting). In these cases the growth front automatically finds, in the thermal field of the furnace, the position corresponding to the assigned growth rate. Conversely, during the growth of a crystal simply immersed into a melt, or in general in systems where the thermal field is set up by itself as the crystal grows, the values of $\mathscr{G}$ and V are in one-to-one correspondence.

As the growth conditions approach the line separating the stability and instability ranges in Fig. 5.15, the surface undergoes changes which are clear from Figs. 5.16, 17. At the beginning, when the temperature gradient is sufficiently high, the initial growth front, which is smooth with the exception of the three grain boundaries visible in the frame of Fig. 5.17a, shows individual depressions—"pocks"—from which filaments with a defective structure extend into the crystal body (Figs. 5.16a, 17b). Then, with a decrease in the temperature gradient or an increase in the growth rate, grooves (crosslinks) leaving behind defective "ribbons" appear between the "pocks" (Figs. 5.16a, 17c). This is followed by the formation of a system of long parallel grooves—a line structure (Figs. 5.16b, 17d). Its width λ_{cr} was discussed above. In the next stage in the development of instability, cross links appear between the grooves (Fig. 5.17e), a hexagonal network of grooves ultimately forms on the surface (so-called

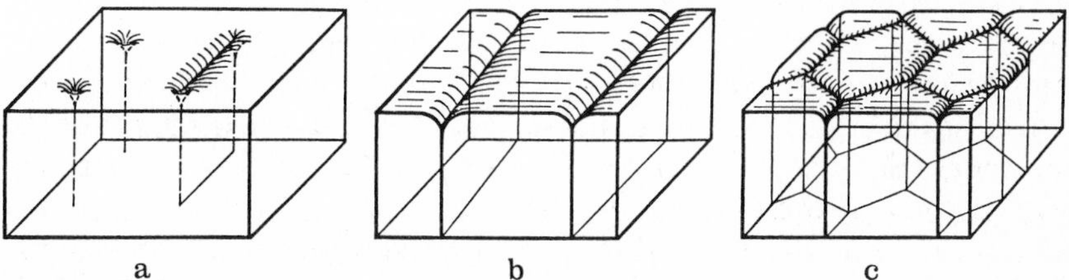

Fig.5.16a–c. The sequence of structures accompanying the gradual loss of stability by a flat growth front. **(a)** Individual pits and isolated grooves on the surface, and filaments and bands in the bulk; **(b)** line (platy) structure; **(c)** cellular ("pencil") structure

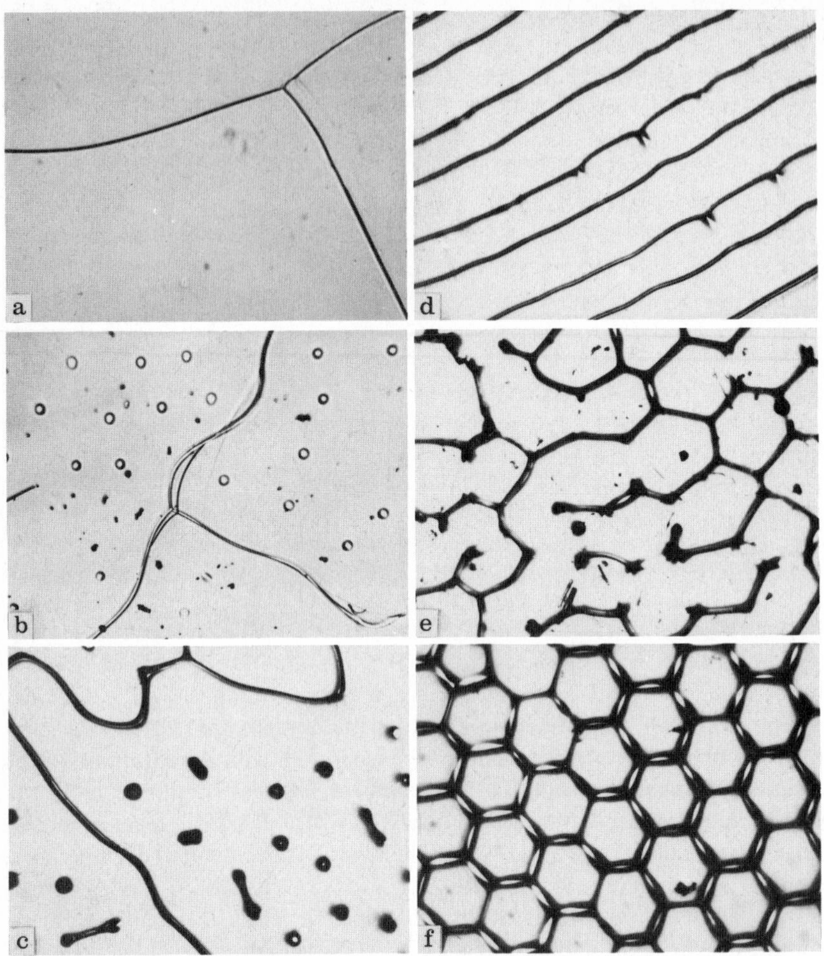

Fig.5.17a–f. Growth front of aluminium crystals (purity 99.997%) photographed after rapid removal (decantation) of the melt at conditions characterizing consecutive stages in the loss of stability. The temperature gradient in the melt [K/cm] and the growth rate [cm/s] take on the following values, respectively: (**a**) 33.5 K/cm, 1.7×10^{-3} cm/s; (**b**) 34.6, 6.2×10^{-3}; (**c**) 31.2, 6.1×10^{-3}; (**d**) 10.0, 5.0×10^{-3}; (**e**) 30.0, 15.5×10^{-3}; (**f**) 19.5, 23.7×10^{-3} [5.35]

cellular structure), and the corresponding "pencil structure" in the bulk appears (Figs. 5.16c, 17f).

Let us now consider the effect of anisotropy on stability. If the flat growth front is a singular face or, more precisely, if its average orientation corresponds to a singular face, the stability of the front increases considerably. The growth rate anisotropy causes faster growth of the perturbation hillock slope and thus raises the temperature everywhere on the hillock, including the point generating the growth layers. This rise in temperature, in turn, reduces the intensity of step generation, and therefore the hillock, which has become shaper accidentally, must flatten out. The effect of the increasing kinetic coefficient is naturally superimposed on the factors that were mentioned at the beginning of this section and that determine the stability or instability of the flat front at isotropic kinetics. The anisotropy of the kinetic coefficient exerts a stabilizing effect on the flat growth front, as well as on the polyhedron. The stability criterion of the flat front, which is similar to (5.31), has the following form, with due allowance for the anisotropy [3.21, 5.36]:

$$\frac{\mathcal{G}}{V} > \frac{(1-K)mC_{\mathrm{i}\infty}}{DK}(1-\Theta) - \frac{T_Q\Theta}{a_\mathrm{L} + a_\mathrm{S}}\,. \tag{5.32}$$

Consequently, at $\Theta > 1$ the singular front may be stable at $\mathcal{G} < 0$ as well. Anisotropy parameter Θ is taken at a vicinal-hillock slope corresponding to supercooling at the front. The higher the supercooling, the lower is Θ and the closer (5.32) is to that for the isotropic case (5.31).

6. Creation of Defects

Imperfection in a crystal affects many of its properties; such properties are therefore referred to as structure sensitive. For instance, the electrical resistance of metals at low temperatures can change by several orders, depending on the quantity of grains, dislocations and impurities present in the crystal. Impurities often determine the electrical properties of semiconductors. Dislocations, vacancies, impurity atoms, phase segregations (precipitates) and other elements of the real structure of a crystal determine all phenomena associated with its plasticity and strength. Under normal practical conditions these same defects define the shape of the magnetization and repolarization curves. The observed optical absorption often has its origin in impurities. Defects in the quartz lattice (dislocations, impurities in the form of OH^- ions, etc.) change the Q factor of the radio-frequency stabilization elements by several times and reduce optical transmission in the infrared region. Finally, crystals with numerous macroscopic (10^{-4} cm or more) or even colloidal (10^{-6}–10^{-5} cm) inclusions and cracks may lose their practical value altogether.

The interest in defects that arise during growth is intensified by the irreproducibility of a faulty structure and hence the spread of the characteristics of the structure-sensitive properties from one specimen to another and within one and the same crystal. Furthermore, by no means all such defects can be eliminated by subsequent annealing or other treatment. The zonal and sectorial structures of most crystals are among these stable defects. Therefore highly perfect crystals must be "born perfect."

The most common growth defects are the following:

1) Heterogeneous inclusions, i.e., inclusions of another phase of macroscopic size. These range from colloidal ($\sim 10^{-6}$ cm) to several millimeters or more across.
2) Point defects, primarily impurity atoms and ions, and vacancies, interstitials, and aggregates of these defects ($< 10^{-6}$ cm).
3) Twins, stacking faults, dislocations, grain boundaries.
4) Internal stresses.
5) Nonuniformities in the distribution of the defects listed above: zonal, sectorial, "pencil" structures, etc.

The mechanism of impurity trapping, and the formation of zonal, sectorial, cellular, and other structures are discussed in Sects. 4.2, 5.3. Here we shall dwell briefly on the formation of inclusions (Sect. 6.1), dislocations, and internal stresses (Sect. 6.2).

6.1 Inclusions

6.1.1 Inclusions of the Mother Liquor

Trapping of the mother medium results from any loss of stability by the crystal growth form. In dendritic growth, parts of the mother medium remain between the dendrite branches, the stems of a cellular structure or the platelets of a line structure. These parts communicate with the bulk of the mother medium only through a network of narrow ($\sim 10^{-1}$–10^{-5} cm) channels between the dendrite branches. Therefore the bulk substance and the impurity, as well as the solute in solution growth, cease almost immediately to be transported to and from the inclusion when a channel or a slit appears at the growth front. As a result, in solution growth a near-equilibrium concentration of the solute and a surface steady-state concentration of the impurities that are pushed ($K < 1$) or trapped ($K > 1$) by the crystal is achieved in the inclusion. Similarly, during melt growth, the material of the liquid phase in the channel is enriched in impurities at $K < 1$ and depleted of them at $K > 1$. After the liquid in the channel solidifies, an alloy is formed which has a different composition and a different melting point than that of the basic melt. Therefore the melt always freezes at a lower temperature in inclusions and between dendrite branches. Conversely, when a specimen with inclusions is heated, these will liquefy before the crystal as a whole begins to melt.

Inclusions are also trapped when the growth front as a whole is stable. If this front is a singular face growing layerwise, inclusions arise when growth takes place by deposition of macroscopic rather than elementary steps. During growth from solution, the nutrition of the step rise is better in the vicinity of the outer edge than it is in the vicinity of the inner edge forming the reentrant angle (Fig. 6.1a). Therefore for reasons similar to those discussed in Sect. 5.3 in connection with Fig. 5.10, the rise of a macrostep cannot remain flat while propagating at high supersaturations, and a layer overhanging the growing face will appear (Fig. 6.1b). This overhanging will give rise to a flat inclusion parallel to the growing face. A similar situation takes place in rapid growth from a melt enriched in impurities. The inclusions may be filled with the mother liquor or a melt of a different composition than the basic crystal, or may contain gas bubbles, if they are present ahead of the growth front (see below). The thickness of flat inclusions is comparable to the height of the macrosteps,

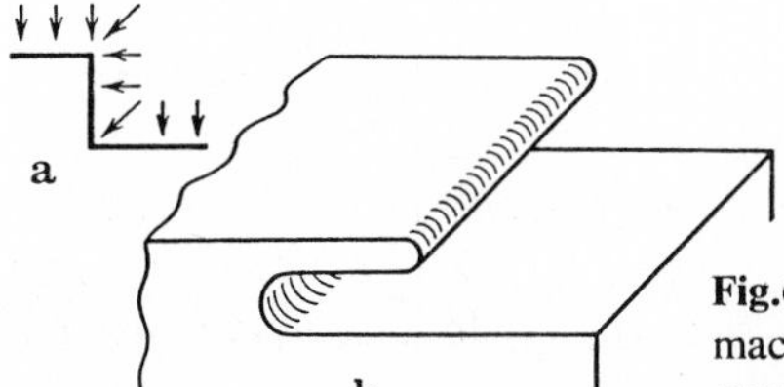

Fig.6.1a,b. Nonuniformity in the nutrition of a macrostep rise (**a**) and the resulting formation of an overhanging layer and a flat inclusion under it (**b**)

Fig.6.2. Flat inclusions under the face of an octahedron of a synthetic diamond. The *bright bands* result from the interference of light in inclusions. Plan view, reflected light. Courtesy of M.O. Kliya Magnification × 150

i.e., $\sim 10^{-4}$–10^{-2} cm or slightly greater. If the inclusion is gaseous or the inclusion material is etched out in subsequent treatment of the specimen, light interference leading to patterns of the type shown in Fig. 6.2 arises between the surfaces bounding the flat inclusion.

Macrosteps often have rounded shapes in the plane of the face along which they propagate. This means that the growth rate of the macrostep rise is isotropic with respect to azimuthal rotation of the rise about the normal to the face, and the rise can be regarded as an atomically rough, nonsingular interface. On the other hand, when the step is low, its thermal field or field of diffusion is similar to that of a thin filament. Bends in such a filament do not materially affect the intensity of the mass or heat flux towards or away from it. Indeed, deterioration of the nutrition (heat removal) from the side where the step is

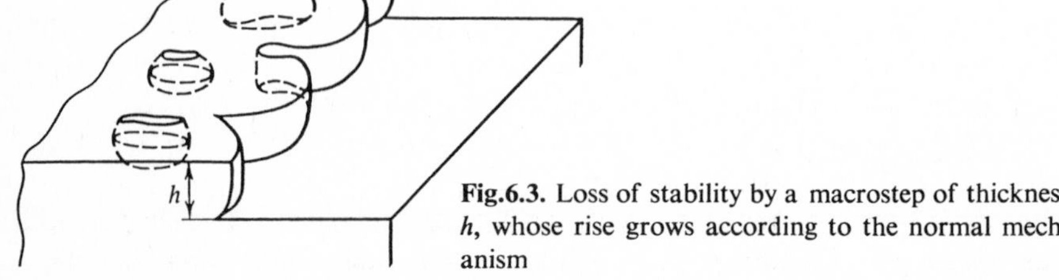

Fig.6.3. Loss of stability by a macrostep of thickness h, whose rise grows according to the normal mechanism

Fig.6.4. Loss of stability by macrosteps on the surface of a sucrose crystal growing from an aqueous solution. The steps are in motion from left to right. Inclusion channels can be seen behind the middle step [6.1]. Magnification × 3

concave is offset by improved nutrition from the side where it is convex. Suppose now that the step height h (Fig. 6.3) becomes appreciable as compared with parameter D/β (during growth from solution) or $a/\beta^T T_Q$ (during growth from the melt). The diffusion field near such a step ceases to be cylindrical and approaches the field of the flat front as the step height increases. If the step rise continues to grow normally, it will lose its stability like any flat front, as shown in Fig. 6.3. The formation of inclusions with such morphology behind macrosteps was observed by *Sheftal'* on a sucrose crystal (Fig. 6.4). In his experiments the appearance of macrosteps, and then inclusions, ended in the cracking of the crystal. This sequence of events was an indication that macrosteps herald the imminent deterioration of the growing crystal.

At sufficiently high supersaturations (supercoolings) steps are emitted by vertices or edges of the crystal (Sect. 5.3). Steps of different heights have different velocities (Sect. 3.3), and thus in moving towards the center of a face, they catch up with each other and merge together. Therefore the average step height increases with the distance from the vertex (edge) of the crystal, i.e., as the steps approach the face center. When the average height of the steps reaches the value corresponding to the loss of stability by the step rise via the mechanisms shown schematically in Figs. 6.1, 3, numerous inclusions begin to form. Steps reach this critical height at a certain critical distance (which depends on the conditions) from the vertices and edges. Therefore as the crystal increases, inclusions are formed first in the central parts of the face. In the course of growth, the area they occupy expands, and turbid growth pyramids filled with inclusions arise in the body of the crystal. The conditions under which macrosteps form and under which they lose their stability are naturally different on faces of different indices, and the crystals also have a sectorial structure with respect to the distribution of inclusions. A temporary increase in supersaturation and growth rate leads to zonal distribution of inclusions.

6.1.2 Inclusions of Foreign Particles

The gases surrounding a mother liquor usually dissolve in it in greater amounts, the higher their pressure is. On the other hand, many gases are poorly soluble in crystals and therefore are repulsed by the growth front. The gas supersaturation arising in the boundary layer adjacent to the growing interface is eliminated by the formation of gas bubbles in this layer. Under favorable conditions the bubbles coagulate, float up to the surface of the liquid, and do not affect crystal growth. But quite often (for instance if the bubbles are small, the growing face is oriented downwards, and the stirring is not vigorous enough) no such cleaning of the front occurs, and the bubbles are trapped by the crystal. Colloid particles formed within the mother liquor and solid foreign particles are also accumulated ahead of the growth front. For instance, melts of semiconductors and metals enclosed in graphite crucibles may contain pieces of graphite tens or hundreds of micrometers in size.

Let us consider the interaction of a foreign particle with a growth front, beginning with growth from the melt [6.2]. What kinds of forces exist between the front and the particle? Archimedean and capillary[1] forces always act. However, the former does not depend on the gap width h between the particle and the interface, and the latter depends on it only if $h \lesssim R$. After all, if the particle is small (see below) and $h \ll R$, both forces are in many cases much weaker than the hydrodynamic force discussed later is this section. Therefore Archimedean and capillary forces are not taken into account here.

As long as the particle is located far from the growing interface, this interface does not interact with the particle. As the front approaches the particle to a distance of $\simeq 10^{-5}$–10^{-7} cm, appreciable molecular forces begin to operate between them (footnote, p. 8). If the material of the particle wets the crystal surface well enough, mutual attraction between particle and front becomes possible, and the particle will be trapped by the crystal. In the opposite case

---

[1] Effective capillary force acting on a gaseous or liquid particle of radius R is of the order of $-\nabla[4\pi R^2 \alpha^*(T, C)] = -4\pi R^2[(\partial\alpha^*/\partial T)\nabla T + (\partial\alpha^*(\partial C)\nabla C]$, where ∇T and ∇C are the temperature and concentration gradients. Thus, the particle should move to the region having a higher temperature and a higher concentration of a surface-active impurity, C, where the particle interface energy $\alpha^*(T, C)$ is lower [6.3].

the particle begins to be repulsed by the front and move together with it relative to the melt. If a viscous melt flows past a spherical particle of radius R at growth rate V, it presses the particle against the flat front with a force similar to Stokes' force: $6\pi\eta VR^2/h$, where η is the melt viscosity, and h is the minimum width of the gap between sphere and front. The chemical potential of the melt in a thin film between the particle and the growing interface is lower than its bulk value due to molecular forces (disjoining pressure). As a result, the driving force for crystallization under the particle is lower, the front buckles under the particle, and therefore the pressing ("sucking") force grows still stronger. The repulsive force due to the disjoining pressure also increases. The shape of the buckling front is determined by surface energy α at the crystal–melt interface for small particles with $R \lesssim 100$ μm or by the temperature gradient at the front for large particles with a radius $R \gtrsim 500$ μm. As the growth rate increases, the pressing force increases more rapidly than the repulsive force. As a result, at growth rates exceeding a certain critical value V_{cr} the particle will be trapped, and at $V < V_{cr}$ it will be repulsed by the front and move together with it [6.5]. For small particles at a disjoining pressure depending on the thickness as Bh^{-3}, calculation yields [6.2, 4]

$$V_{cr} = \frac{0,1B}{\eta R}\left(\frac{\alpha}{BR}\right)^{1/3}, \tag{6.1}$$

where B is a constant characterizing the disjoining forces. For Van der Waals forces, $B \simeq 10^{-14}$ erg. Figure 6.5a summarizes experimental data along with theoretical dependences $V_{cr}(R)$ found for various types of interaction contributing the disjoining pressure [6.2].

For large particles, the critical velocity depends on the thermal processes, i.e., on the temperature gradient at the growth front and the particle-to-melt thermal conductivity ratio. In other words, if the thermal conductivity of the particle exceeds that of the melt and the melt is superheated, the temperature at the front beneath the particle increases, and a depression of the front appears. Conversely, if the thermal conductivity of the particle is lower than that of the melt, the front beneath the particle forms a bump. In the former case, the crystal may grow over the particle together with the liquid interlayer adjacent to it. In the latter case, either repulsion of the particle or the "piercing" of this interlayer by the bump may take place, after which the particle "sticks" to the front and is trapped.

An impurity contained in the melt strongly affects the trapping of particles. The reason is that because of the difficulties involved in diffusion removal of the impurity from the narrow clearance between the growth front and the repulsed particle, many more molecules of an impurity with $K < 1$ accumulate there than over the free front. At $K > 1$, on the other hand, there is less impurity in the gap. In both cases the impurity reduces the melting temperature beneath the particle, and the growth front buckles more prominently and enwraps the

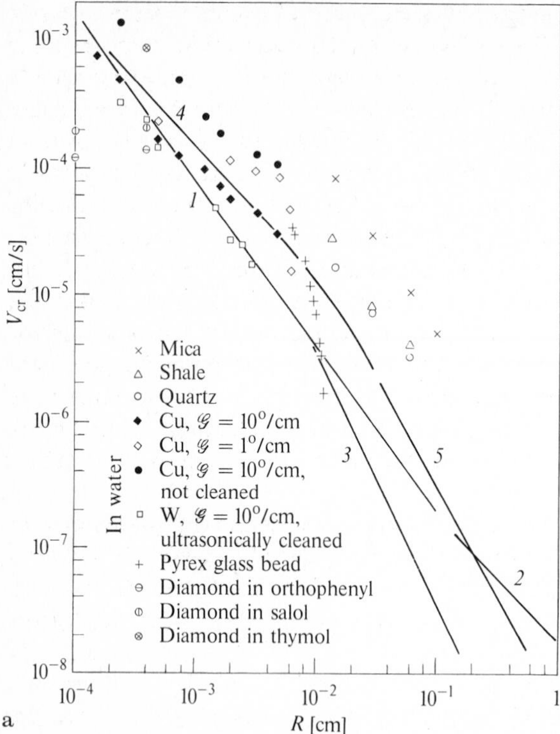

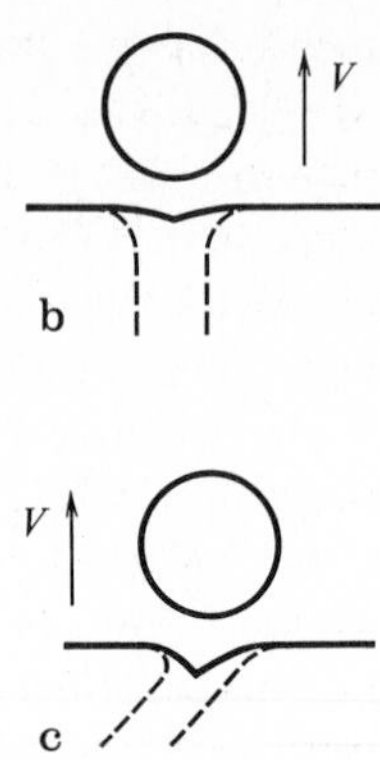

Fig.6.5. **(a)** Trapping of inclusions: The critical growth rate, V_{cr}, above which a particle of radius R is trapped by the ice, orthophenyl, salol and thymol crystals growing from the corresponding melts-experimental points. *Curves 1–5* are calculated according to various models of particle-melt film-crystal interaction. *Curve 1*-according to equation (6.1) with $B_3 = 10^{-14}$ erg, $\alpha = 20$ erg/cm^2, $\eta = 2 \cdot 10^{-2}$ g/cms, $\Delta s = 2 \cdot 2 \cdot 10^{-16}$ erg/K, $\Omega = 3 \cdot 10^{-23}$ cm^3; 2-interface under particle is stabilized by temperature gradient rather than by surface energy; 3-impurity in the interface-particle gap and surface energy are taken into account, 4, 5-the crystal-partice repulsion is determined by electrostatic Debye layers adjacent to the crystal and the particle. For further imformation see Ref. [6.2]. **(b, c)** Buckling of a singular growth front (*solid line*) and formation of a mother liquor inclusion (*dashed lines*) due to the screening of the mass flux by a foreign sphere. **(b)** The dependence $R(p)$ of the growth rate on the orientation (see Fig. 3.7a) is symmetric relative to the plane normal to the growing interface and the drawing plane; **(c)** $R(p)$ is nonsymmetric

particle more tightly there, making it easier for the particle to be trapped. As a result the critical growth rate is much lower for an impure melt than for a pure one. For instance, at an impurity concentration in the melt $C_{i\infty}$ such that $mC_{i\infty}/K \simeq 10$ K, the critical velocity drops by an order of magnitude. For a slope of the liquidus line $m = 3$ K/wt. % and $K = 0.2$, concentrations $C_{i\infty} \simeq 0.6$ wt. % are sufficient for this.

The general mechanism by which impurities are trapped in the presence of foreign particles during growth from solution has also been studied only insufficiently, but is believed to be similar to the one outlined above for the melt [6.6, 7]. During solution growth the force pressing the foreign particle to the growth front must not only have a hydrodynamic component, but must also be associated with diffusiophoresis (i.e., with the effect of entrainment of a body by the diffusion flow). The disjoining pressure in solutions, especially in those of electrolytes, is also of a more complex nature than in nonpolar liquids. A foreign particle located in solution at a certain distance from the growing face screens its surface from the diffusion flux. Therefore the supersaturation beneath the particle (bubble) becomes lower than over the entire remaining surface. As a result, a *V*-shaped depression appears beneath the particle (Fig. 6.5b, c), with the slopes of the depression having a larger coefficient of crystallization than the remaining face (Sect. 5.3.2). If the distance between the particle and the surface exceeds the radius of the particle or is comparable to it, the depletion of the solution beneath the particle is small. Such a slight decrease in supersaturation can be offset by the anisotropy of the kinetic coefficient of crystallization. With an increase in the radius of the particle and a decrease in the distance from it to the face, the depth of the cavern increases, and, finally, when the face beneath the particle loses its stability, a channel filed with mother liquor trails behind the particle (Fig. 6.5b, c).

Figure 6.6 presents a series of cineframes taken by *Kliya* which show the consecutive stages in the trapping of petroleum droplets of different diameter by a KNO_3 crystal growing from an aqueous solution. It can be seen that beneath large droplets the channel begins to form as soon as the droplet touches the front. The two small droplets at the top of the left-hand side (Figs. 6.6a–c), on the other hand, are trapped only after being repulsed by the front for a long time. The beginning of their entrapment may be caused by an accidental increase in growth rate to above-critical values. When the content of a droplet is completely transferred to the channel, the droplet disappears, and the inclusion is sealed (Figs. 6.6d–f).

In other experiments, cases were also observed in which foreign particles (e.g., petroleum or mercury droplets, lycopodium particles) only caused the formation of a channel with mother liquor in crystals of borax ($Na_2B_4O_7$) and potash alum ($KAl(SO_4)_2 \cdot 12H_2O$), but were not trapped [6.8, 9].

In chemically complex solutions from which crystals grow, liquation is possible. For instance, in the hydrothermal system SiO_2–NaOH–H_2O, a so-called "heavy phase" enriched in sodium arises. This liquid phase is released in the solution bulk and may be adsorbed on the surface as separate islands. The islands, having a different composition, may be adsorbed more firmly on the faces than the remaining solvent and be trapped by the crystal. During subsequent heating of the grown crystal the colloid inclusions cause microcracks in the crystal, i.e., turbidity in the specimen. The trapping of colloid inclusions depends on the growth rate and on the index of the face, and therefore these inclusions have a sectorial and zonal distribution (Fig. 6.7). Let us estimate the growth rate at which the crystal begins to trap such inclusions. Suppose the colloid phase precipitate on the growing surface has a size l along the normal to the surface and a lifetime τ_s in the state of adsorption. Then the time in which the colloid precipitate is "buried" by the front (cf. Sect. 4.3) moving at velocity

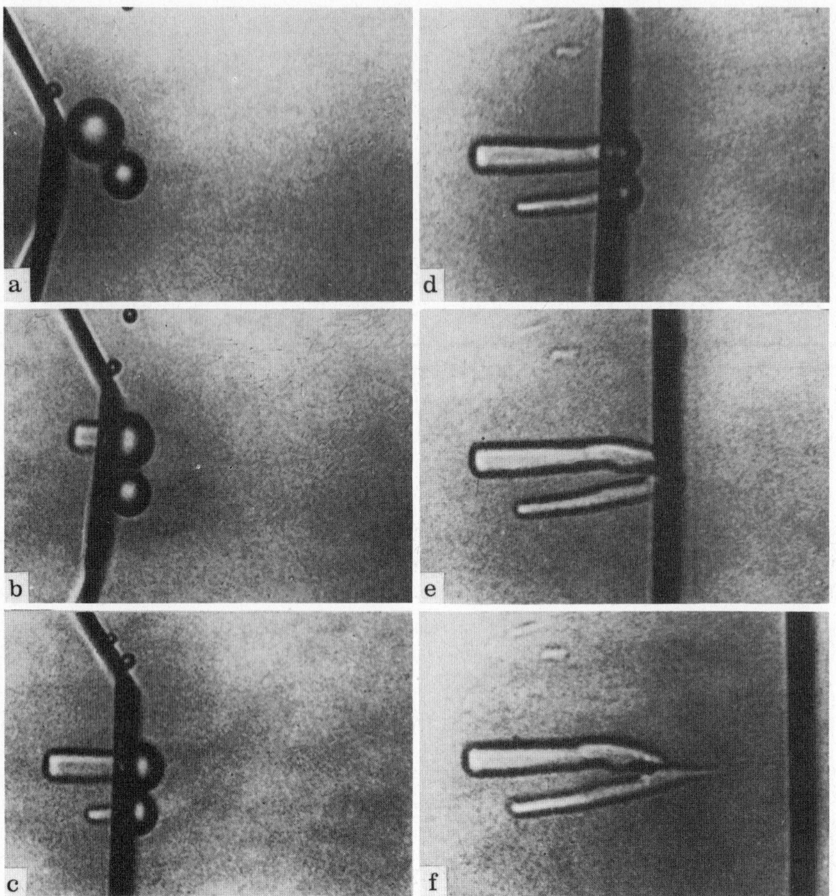

Fig.6.6a–f. Consecutive stages in the trapping of petroleum droplets by a KNO_3 crystal during its growth from an aqueous solution. The crystallization front is in motion from left to right [6.8]. Magnification × 430

V will be $\sim l/V$. Accordingly, the probability of incorporation that is, the concentration of the colloid particles in the crystal, is $\propto \exp(-l/V\tau_s) = \exp(-V_{cr}/V)$. Hence the trapping of colloid inclusions also occurs only at velocities exceeding some critical one ($V_{cr} = l/\tau_s$), as is seen from Fig. 6.8. Adsorption time τ_s and, hence critical velocity V_{cr}, must vary exponentially with the temperature, which agrees with the data in Fig. 6.9 [6.10].

Foreign particles and voids may also form in the body of the crystal after crystal growth is completed, as a result of decomposition of the supersaturated solid solution of impurity atoms, or of vacancies trapped by the crystal in the course of growth. The supersaturation is due firstly to the fact that the crystal traps a thermodynamically nonequilibrium number of impurity atoms and

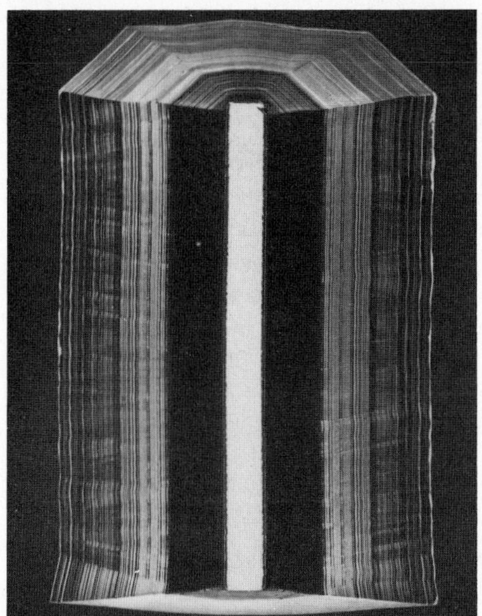

Fig.6.7. Growth zones and sectors, enriched in colloid inclusions, in a synthetic quartz crystal. After growth the specimen was subjected to annealing, causing microcracks around the inclusion which made the growth zones and sectors visible (*light areas* in the photo) [6.10]. Magnification × ca. 6

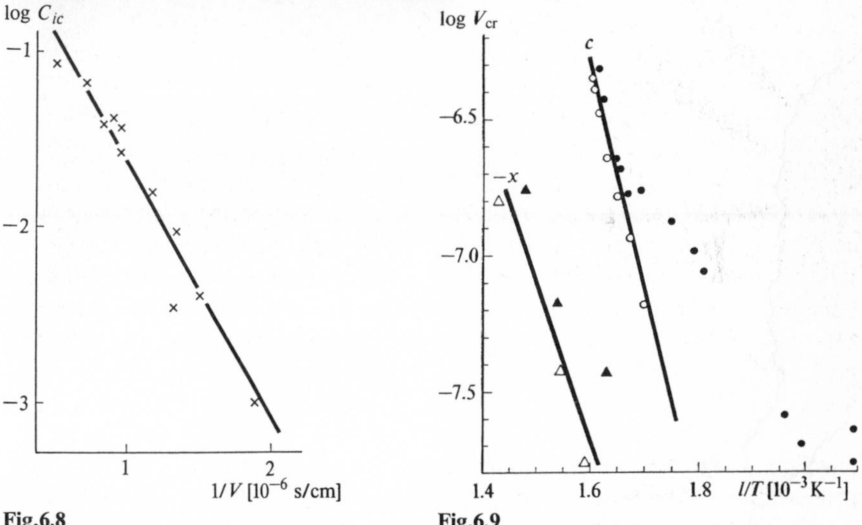

Fig.6.8 Fig.6.9

Fig.6.8. Dependence of the concentration of colloid inclusions in quartz on the growth rate. The concentration of inclusions was assumed to be proportional to the average concentration C_{ic} of the Na_2O impurity in the crystal as determined by chemical analysis. The diameters of the inclusions in the crystal varied between 20 and 400 Å; the average number density of the inclusions was $\sim 3 \times 10^{13}$ cm^{-3} [6.10]

Fig.6.9. Growth rates and temperatures at which quartz crystals entrap colloid inclusions (*shaded symbols*) and are free of them (*open symbols*). The intermediate values give the dependence $V_{cr}(\Delta T)$. The *triangles* and *circles* refer to the growth sectors of prismatic ($-x$) and basal (c) faces, respectively [6.10]

vacancies, and secondly to the cooling of the newly formed crystal layers. Furthermore, if melts crystallize near the eutectic point or below it, the second phase may not segregate directly during crystal growth. As a result, the crystal will be enriched in the component of which this second phase largely consists. The decomposition of the solid solution and the appearance of precipitates of the new phase are possible only at high temperatures, with sufficiently large mobility of the impurity atoms and vacancies; i.e., they are characteristic of crystals obtained from the melt. The particles of the new phase are released primarily on dislocations, grain boundaires, and other defects of the crystal; in other words, they decorate these defects. The decorating both of nonuniformities on the crystal surface and of defects in the bulk is a very common method of investigating the real structure of a crystal (Figs. 6.23b, c).

Post-growth processes are responsible, for example, for so-called swirl defects in melt-grown silicon [6.11, 12] (Sect. 6.2.5).

6.2 Dislocations, Internal Stresses and Grain Boundaries

6.2.1 Dislocations from a Seed

The simplest possibility for dislocations to be created in a growing crystal is the prolongation of dislocations from the seed into the body of a crystal, which continues the seed lattice. Inheriting of dislocations is observed in growth from a gas, from the melt, and from solution on a clean seed surface free of any interlayers.

A seed often has a slightly different impurity composition than the growing material, and shows internal stresses. Therefore the lattice parameters of the seed and the new crystal are somewhat different. This divergency, or so-called heterometry, is eliminated by dislocations situated along the boundary of the seed with the new material (misfit dislocations; Sect. 2.3). The ends of some dislocations leave this plane, emerge at the growing surface, and hence continue in the crystal body. During growth from the melt the growing layer is plastic and nearly stress free: heterometry is removed almost completely by dislocations. During growth from solution, when the crystal is not plastic, the growing layer and the seed are under stress, which sometimes results in their cracking [6.13].

Finally, the initial stages of growth often occur irregularly due to impurity poisoning of the seed surface, the irregularities of its relief, and also various surface defects. As a rule, numerous inclusions appear, which also give rise to dislocations or stacking faults. This mechanism is discussed in the next section.

6.2.2 Creation of Dislocations in Surface Processes

During growth from solutions in the temperature range where the plastic flow of the crystal is practically nonexistent, the appearance of dislocations is mainly associated with the kinetics and mechanism of the surface processes of growth. These processes are also a contributing factor during growth from the melt, but they are less important than the effect of thermal stresses under conditions in which the crystal is highly plastic (Sect. 6.2.4).

The formation of screw dislocations in thin platelet crystals was discussed in Sect. 3.4.2. A similar mechanism of dislocation formation operates when a platelet crystal grows around a foreign particle [3.69] (Fig. 6.10).

Figures 6.10–12 show similar mechanisms by which dislocations are formed as a result of the instability of macrosteps and subsequent formation of overhanging layers and inclusions. In all the examples shown in Figs. 6.10–12, dislocations will arise only if the layers enclosing the inclusion are bent, so that a misfit arises at their joint (the Burgers circuit is open). The causes of this bend-

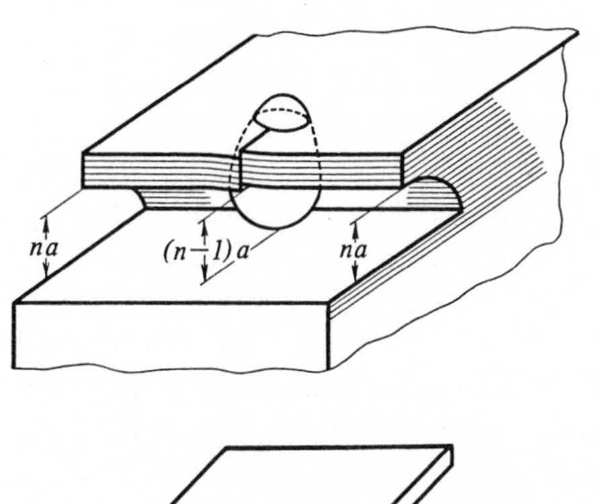

Fig.6.10. The formation of dislocations when a foreign macroparticle is trapped by a plate-like crystal or an overhanging layer. The two parts of the closed-up layer are shifted with respect to one another by the value of the interplanar distance [6.14]

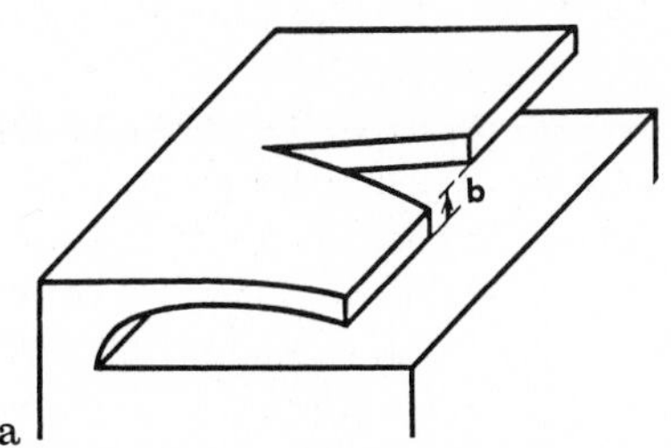

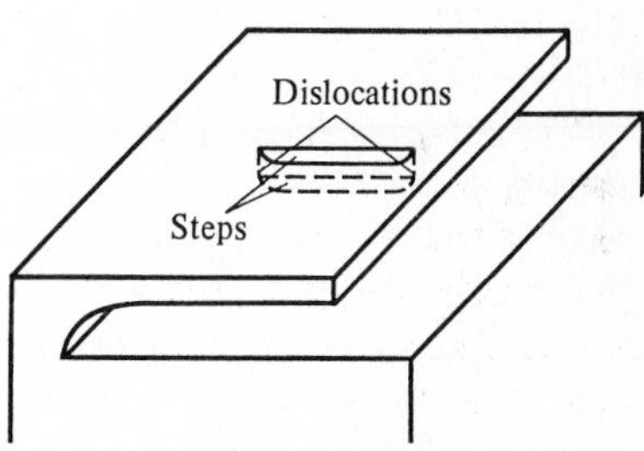

Fig.6.11. Reentrant angle on a thin overhanging layer (**a**) and its overgrowth with the formation of dislocations (**b**) according to the mechanism of Figs. 3.25, 26; ***b*** is the displacement Burgers vector

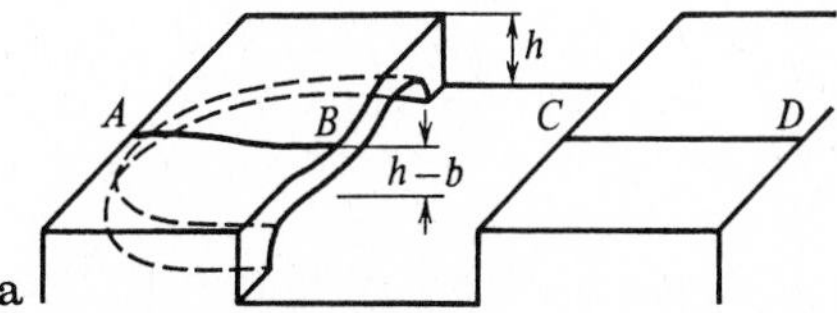

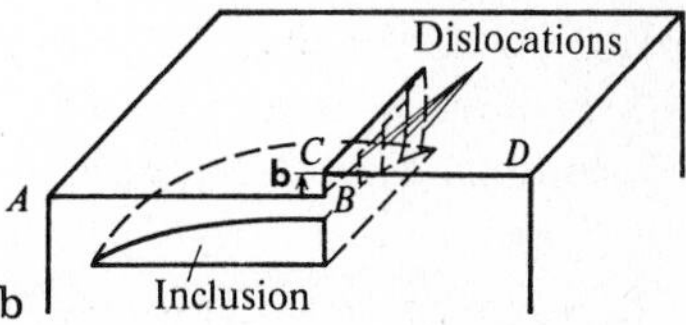

Fig.6.12a,b. Creation of dislocations accompanying the trapping of mother liquor. (**a**) Buckling of the overhanging layer; (**b**) closing up of the inclusion when macrosteps meet with forming dislocations (section on ABCD); ***h*** is the step height, and ***b*** the Burgers vector

ing have not been studied, although it has been proved experimentally that the dislocations start at the inclusions. The layers sealing the inclusion of the mother solution may become deformed as a result of a decrease in pressure inside the inclusion, caused by crystallization of the solute on the inner walls of the inclusion (Fig. 6.12a). It becomes more difficult for the solution to be sucked into the inclusion, and this sucking may not ensure that the pressure increases to the pressure value in the solution bulk. Another possible cause of bending is the variation of the temperature along the growth front.

Figure 6.13 shows the possibility that dislocations will form upon the trapping of a solid foreign inclusion or even a sufficiently viscous liquid inclusion. The curving of the atomic planes and the incongruence when they join over the inclusion are caused by adhesion (or epitaxy) of the growing material onto the surface of the inclusion. This mechanism extends both to layerwise and to normal growth. Unfortunately, there have been no direct microscopic observations of the consecutive stages in trapping corresponding to Figs. 6.10–13. The main confirmation of the efficiency of the indicated mechanisms is the morphology of the finished crystals.

Figure 6.14 shows an x-ray topographic picture of the plate (100) of a KDP crystal. It can be seen that dislocations indeed originate on inclusions,

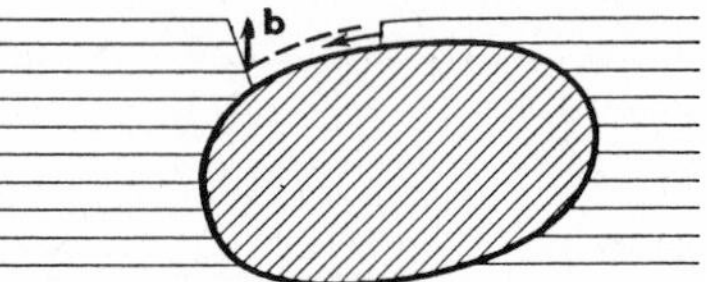

Fig.6.13. Formation of dislocations upon the trapping of a foreign particle

Fig.6.14. Dislocation bundles from inclusions located on one of the growth zones (shown by *arrows*) in a KDP crystal grown from solution. X-ray topography [6.14]. Magnification × 4

forming bundles which fan out. The inclusions, in turn, are concentrated predominantly along planes that trace the halts and acceleration of the growing face (in Fig. 6.14 these are the lines running from the bottom left to the top right). A particularly large number of dislocation bundles begins from the seed surface. In accordance with the law of conservation of the Burgers vector, the resulting vector of all the dislocations of a bundle is zero, and hence the dislocations in the same bundle inevitably have at least two different (opposite) Burgers vectors. In actual fact, the number of possible vectors and hence dislocations beginning on one inclusion is greater, and depends on the lattice structure. Let us now find, following the approach of *Indenbom* [6.15], and *Klapper* and co-workers [6.16], the orientations of the dislocations in a bundle.

6.2.3 Orientation of Dislocations

The free energy γ_d per unit dislocation length is anisotropic. It depends on Burgers vector $\boldsymbol{b}$ and on the dislocation orientation described by unit vector $\boldsymbol{\tau}$ along the dislocation. The magnitude and nature of this anisotropy is determined by the elastic moduli of the crystal in accordance with the well-known equations [6.16]. Let us consider a dislocation with a given $\boldsymbol{b}$ intersecting the growth front, whose orientation at the point of emergence is assigned by normal $\boldsymbol{n}$ (Fig. 6.15). We shall now find the dislocation orientation $\boldsymbol{\tau} = \boldsymbol{\tau}_0$ at which the increase in energy due to dislocation elongation in the course of growth would be the least, i.e., the most advantageous orientation of the dislocation with respect to the growth front. With displacement of the front by dh (Fig. 6.15) the dislocation energy increases by $\gamma_d\, dh/\cos(\boldsymbol{n\tau})$. Consequently, the sought-for orientation must ensure the minimum of function $\gamma_d(\boldsymbol{b}, \boldsymbol{\tau})/\cos(\boldsymbol{n\tau})$ or the maximum of the reciprocal value $\cos(\boldsymbol{n\tau})/\gamma_d(\boldsymbol{b}, \boldsymbol{\tau})$. To find it, we construct a polar diagram of reciprocal specific linear energy $1/\gamma_d(\boldsymbol{b}, \boldsymbol{\tau})$ as a function of orientation $\boldsymbol{\tau}$ at a fixed $\boldsymbol{b}$ (Fig. 6.16). We draw a straight line along vector $\boldsymbol{n}$ through diagram center O. Then,

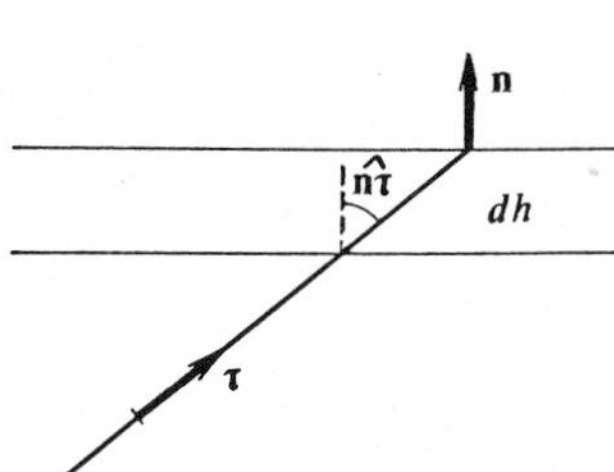

Fig.6.15. Displacement of a growing face by dh and the corresponding elongation of the dislocation along $\boldsymbol{\tau}$ by $dh/\cos(\boldsymbol{n\tau})$

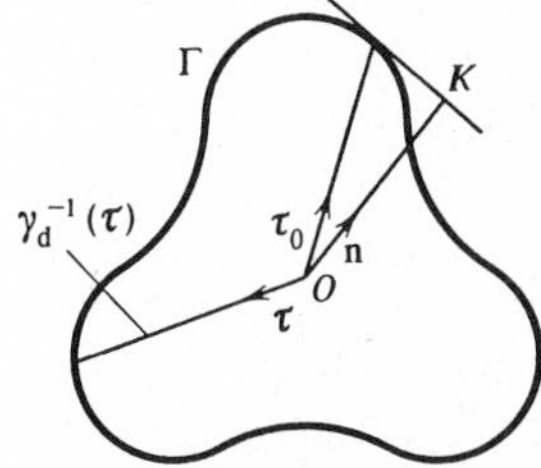

Fig.6.16. A dislocation emerging onto a face with a normal $\boldsymbol{n}$ has minimum energy if its orientation is determined by vector $\boldsymbol{i}_0$. Contour Γ is the section of the surface of the reciprocal linear dislocation energy $\gamma^{-1}(\tau)$ at a fixed Burgers vector

for arbitrary $\boldsymbol{\tau}$, segment OK $= \cos(\boldsymbol{n\tau})/\gamma_d(\boldsymbol{b}, \boldsymbol{\tau})$. This quantity reaches a maximum at $\boldsymbol{\tau} = \boldsymbol{\tau}_0$ corresponding to the point of tangency of diagram $\gamma_d^{-1}(\boldsymbol{b}, \boldsymbol{\tau})$ by a plane perpendicular to vector $\boldsymbol{n}$ from the side to which it points. Since γ_d depends not only on $\boldsymbol{\tau}$, but also on $\boldsymbol{b}$, different diagrams correspond to different $\boldsymbol{b}$, and hence to different most advantageous orientations $\boldsymbol{\tau}_0 = \boldsymbol{\tau}_0(\boldsymbol{b})$. This set of orientations $\boldsymbol{\tau}_0(\boldsymbol{b})$ actually determines the configuration of the bundle issuing from the inclusion. In the isotropic case γ_d = const and $\boldsymbol{\tau}_0 \parallel \boldsymbol{n}$, i.e., the dislocations are perpendicular to the growth front.

The described method of determining the set of orientations of dislocations with different Burgers vectors was applied to bundles of dislocations in KDP crystals (Fig. 6.14) and yielded orientations similar to the experimental ones [6.16].

6.2.4 Thermal Stresses

In a crystal growing from the melt the temperature is usually not only varying, but also has no constant gradient. The complex temperature distribution along the crystal leads to nonuniform thermal expansion of its different parts, elastic interaction between them and, as a result, to the appearance of stresses in the crystal even when its external surface is free. Let us consider the physics involved in the formation of temperature stresses and dislocations [6.17–19].

We shall first discuss the effect of such temperature stresses in a just-formed dislocation-free crystal; in other words, we shall consider essentially its quenching. If the shear stresses along the principal glide planes do not at any point exceed the critical value $\hat{\sigma}_{cr}$ for the nucleation of dislocations or for the growth of existing very small dislocation loops, the crystal will remain macroscopically dislocation-free.

In ideal crystals, $\sigma_{cr} \simeq (0.1 - 0.01)G$, where G is the shear modulus. In semiconductor crystals containing microloops, at the melting point, $\sigma_{cr} \simeq (10^{-5} - 10^{-6})$ G. The critical stress, σ_{cr}, increases as the temperature decreases: for pure GaAs, σ_{cr} = 1, 12, 35, 350 G/mm^2 at T = 1238, 1100, 800, 500°C, respectively; for pure Ge, σ_{cr} = 15 and 100 G/mm^2 at T = 936 and 620°C; for Si, $\sigma_{cr} \simeq 50 - 100$ G/mm^2 at $T = T_0$ = 1415°C (data obtained by *Mil'vidsky* and *Osvensky* [6.20]; see also [6.21]).

The stresses near foreign inclusions in the crystal may exceed the average stresses over the crystal by many tens of percent. Therefore such stress concentrators are the most dangerous sites for the nucleation of dislocations. In silicon, particles of graphite (from the crucible) trapped by the crystal, and/or phase segregations of the impurity serve as concentrators. If the critical stress is achieved anywhere, the nucleated (or previously existing) dislocation loops will expand, the dislocations will begin to multiply, and plastic deformation will set in and propagate throughout the crystal. The only sites to remain dislocation free will be those at which the stresses are not sufficient for the front of the dislocation tangle, or even the individual dislocations, to propagate, in other words, those at which the stresses are below the elasticity limit in the crystal containing the dislocations.

It should be recalled that plastic flow is associated not only with the slipping of dislocations, but also with processes of viscous flow, including the climb of dislocations and the transport of vacancies and interstitials. The gradients of the diagonal components of the stress tensor will also cause fluxes of impurity atoms whose atomic radii differ from the matrix.

Plastic deformation and the diffusion of point defects occur at much lower stresses at temperatures close to the melting point than they do at low temperatures. Therefore the magnitude and shape of the region in which the dislocations propagate and nucleate, and in which the point defects are redistributed, depend not only on the distribution of thermal stresses, but also, and to no lesser extent, on the absolute temperature.

The plastic flow redistributes the material in the crystal in such a way as to damp out the stress that caused it. It terminates when the total energy of the thermoelastic stresses and dislocations attains a minimum. A balance between the elastic energy and the dislocation energy is generally an indication that the stresses are incompletely relieved. But the remaining macroscopic stresses cannot exceed the elasticity limit for temperatures close to the melting point, i.e., they are sufficiently small. Hereafter we assume them to be equal to zero.

If a crystal with thermal stresses contains dislocations from the beginning, the nucleation stage is naturally eliminated, and the multiplication and propagation of dislocations will occur as in the preceding case.

The above-described process by which thermal stresses are relieved with the help of plastic flow of the material is achieved in the inhomogeneous temperature field in the crystal. Let us now cool the crystal down to a temperature constant throughout its bulk and sufficiently low to exclude plastic deformation. Then the regions which had been compressed before the commencement of plastic deformation and had hence lost part of their material will become expanded after a constant temperature is set up, and vice versa. Thus, the cooled crystal will acquire so-called residual internal stresses, which are close in value but opposite in sign to the initial thermal stresses that caused the plastic deformation. The sources of residual internal stresses are residual deformations, i.e., the distribution of dislocations and point defects that was established as a result of the flow at high temperatures.

The described formation of internal quench stresses occurs in any nonuniformly heated solid (crystal or glass), whether it grows from the surface or not.

The formation of dislocations in the immediate vicinity of the growing surface of the crystal is caused by the same trend towards a reduction in the elastic energy. Here, however, the material is not relieved of the stresses, but instead grows immediately in the unstressed state. In the course of growth the crystal moves relative to the temperature field which exists in the furnace and in the crystal itself, as, for instance, during growth by the Czochralski and Stockbarger processes. Gradually, each element of the surface layer becomes part of the bulk and undergoes plastic deformation, relieving those stresses which would have occurred at a given point already deep in the crystal. In the long run the given element ends up in a region of sufficiently low temperature with no plasticity, where deformation is no longer possible. Consequently the grown crystal will have, in each element of its volume, the defect structure which had formed in this element by the moment at which it departed from the plasticity zone in the course of growth. After the crystal has been cooled to a constant temperature, the defects thus "frozen in" will cause stresses equal in magnitude and opposite in sign to those which were relieved by these defects in the course of growth. If the material is plastic only within a narrow temperature range near the melting point, the plasticity boundary will pass parallel to the growth front and at a small distance away from it. As a result, the stresses (with opposite sign) which existed at the growth front in the course of crystallization will be frozen into the crystal. The stresses at the growth front depend in turn on the temperature field in the crystal bulk. This temperature field is generally very complex. For instance, during Czochralski pulling of a cylindrical crystal, both radial and axial heat flows are present in the crystal. The former is associated with the cooling of the crystal through its side and the latter also with a heat flow from the melt to the holder-cooler through the crystal. The radial heat flux corresponds to the cooling of the already formed cylinder (quenching), while the axial flow is essentially related to the growth process itself. An analysis, the results of which are reproduced below in connection with Figs. 6.19, 20, shows that the axial and radial gradients produce residual stresses differing both in magnitude and sign [6.17, 18].

Thus, determining the residual stresses in a grown crystal is equivalent to determining the thermoelastic stresses in the course of growth and cooling.

The properties of the crystal change under the effect of both the internal stresses and the residual deformations that produce them, i.e., of the distribution of vacancies, interstitial atoms, dislocations, grain boundaries, etc. In many cases the contribution from dislocations is the determining factor. Thus, before we discuss the nature of residual stresses, we shall describe dislocation structure and its relationship to macrodeformations [6.18]. It is obvious that the source of macroscopic deformations must be a distribution of dislocations such that the total Burgers vector of the dislocations intersecting the variously oriented areas at the specimen point in question is nonzero. Indeed, the deformations and stresses from the other dislocations only spread over distances of interdislocational order, and disappear after the dislocations are annihilated. The density of such "background" dislocations may be quite considerable, however. For instance, in well-annealed germanium crystals it may reach $\sim 10^4$ cm^{-2}.

Dislocations contributing to macroscopic deformations are described by dislocation density tensor $\hat{\beta}$ [Ref. 4.27, Sect. 5.5]. The component β_{ik} of this tensor is equal to the k component of the total Burgers vector of the dislocations intersecting a unit area perpendicular to axis x_i. Thus, $\hat{\beta}$ characterizes only the minimum dislocation density compatible with the given macrodeformation (macrostresses).

If the crystal structure allows for only three unit Burgers vectors $\boldsymbol{b}^{(s)}$ ($s = 1, 2, 3$) i.e., unit dislocations of only three types s, then knowing tensor $\hat{\beta}$, we can find the dislocation flow $\boldsymbol{N}^{(s)}$. Vector $\boldsymbol{N}^{(s)}$ has components $N_i^{(s)}$, which are equal to the number of type s elementary dislocations intersecting a unit area perpendicular to axis x_i. The sign of $N_i^{(s)}$ depends on that of the projection of the unit vector along dislocation $\boldsymbol{\tau}$ onto axis x_i. In determining the dislocation flow it is convenient to proceed from the representation of $\hat{\beta}$ in the form of the diad product of vectors:

$$\hat{\beta} = \boldsymbol{N}^{(s)} \cdot \boldsymbol{b}^{(s)} . \tag{6.2}$$

Let us introduce the triad of vectors $\tilde{\boldsymbol{b}}^{(s)}$($s = 1, 2, 3$) reciprocal to triad $\boldsymbol{b}^{(s)}$, in accordance with the rules for constructing the reciprocal lattice [Ref. 6.22, Chap. 3]. Then, multiplying (6.2) from the right by $\tilde{\boldsymbol{b}}^{(j)}$, we obtain

$$\boldsymbol{N}^{(j)} = \beta \cdot \tilde{\boldsymbol{b}}^{(j)},$$

i.e.,

$$N_i^{(j)} = \beta_{ik} \tilde{b}_k^{(j)} . \tag{6.3}$$

With a large number of unit Burgers vectors (e.g., in an fcc lattice, four vectors $\boldsymbol{b} = \frac{1}{2}\langle 111 \rangle$), one and the same total density $\hat{\beta}$ is ensured by different sets of nonannihilating dislocations. In such cases one must restrict oneself, in estimating the minimum dislocation density, to expression (6.3) taken only in the order of magnitude.

Complete deformations and rotations of each element of the crystal volume are described by the total-distortion tensor $\hat{u} = \{u_{ik}\}$, which is expressed in terms of displacement vector $\boldsymbol{u}$: $u_{ik} = \partial u_k / \partial x_i$, where u_i is the component along the x_i axis ($i = 1, 2, 3$). Consequently, the total-distortion tensor can be divided into symmetric and antisymmetric parts having the form

$$u_{ik} = \frac{1}{2}\left(\frac{\partial u_k}{\partial x_i} + \frac{\partial u_i}{\partial x_k}\right) + \frac{1}{2}\left(\frac{\partial u_k}{\partial x_i} - \frac{\partial u_i}{\partial x_k}\right) = \varepsilon_{ik} + \Omega_{ik} ,$$

where ε_{ik} is the symmetric total-deformation tensor, and Ω_{ik} is the antisymmetric lattice-rotation tensor. As is well known, the latter is equivalent to lattice-rotation vector ω:

$$\Omega_{ik} = e_{ikl} \omega_l , \tag{6.4}$$

where e_{ikl} is the unit antisymmetric tensor. Total deformation $\hat{\varepsilon} = \{\varepsilon_{ik}\}$ consists of elastic $\hat{\varepsilon}^{el}$, residual (plastic) $\hat{\varepsilon}^0$, and thermal $\hat{\alpha}T$ parts, where $\hat{\alpha}$ is the thermal-expansion tensor:

$$\hat{\varepsilon} = \hat{\varepsilon}^{el} + \hat{\varepsilon}^0 + \hat{\alpha}T . \tag{6.5}$$

As has already been noted, at high temperatures the elastic stresses are relieved by plastic deformations, so that in the plasticity region, $\hat{\varepsilon}^{el} = 0$, and

$$\hat{u} = \hat{\varepsilon}^0 + \hat{\Omega} + \hat{\alpha}T . \tag{6.5a}$$

After cooling to a temperature constant along the length of crystal, residual deformations $\hat{\varepsilon}^0$ (and elastic deformations caused by them) are fixed within the crystal body. In the absence of external forces, the lattice rotations reduce to thermal ones, so that the residual distortion in the cooled crystal is $\hat{u}^0 = \hat{\varepsilon}^0$, i.e.,

$$\hat{u} = \hat{u}^0 + \hat{\Omega} + \hat{\alpha}T . \tag{6.6}$$

Since $u_{ik} = \partial u_k/\partial x_i$, we have the familiar

$$\begin{gathered}(\mathrm{rot}\,\hat{u})_{ij} \equiv (\nabla \times \hat{u})_{ij} \equiv e_{ikl}\nabla_k u_{lj} = 0 , \\ \nabla \equiv \{\partial/\partial x_1, \partial/\partial x_2, \partial/\partial x_3\} \equiv \{\partial/\partial x, \partial/\partial y, \partial/\partial z\} .\end{gathered} \tag{6.7}$$

As has been shown [Ref. 4.27, Sect. 5.5, Eq. (5.72)], density tensor $\hat{\beta}$ of the dislocations, and the residual distortion $\hat{u}^0$ caused by them, are related by the expression:

$$\hat{\beta} = \mathrm{rot}\,\hat{u}^0 . \tag{6.8}$$

Substituting (6.6) into (6.7) and taking advantage of (6.8), we obtain the relationship of the density of the dislocations, $\hat{\beta}$, to the temperature field $T(r)$ that causes them and to the lattice rotations Ω induced by this field under given conditions of deformation:

$$\hat{\beta} = -\mathrm{rot}\,\hat{\alpha}T - \mathrm{rot}\,\hat{\Omega} = \hat{\alpha} \times \mathrm{grad}\,T + \hat{\kappa} - \hat{I}\,\mathrm{tr}\{\hat{\kappa}\}. \tag{6.9}$$

Here, $\hat{I}$ is a unit tensor, $\kappa_{ij} = \partial\omega_i/\partial x_j$ is the lattice-curvature tensor characterizing the change in the angle ω_i of lattice rotation in transition from point to point in the crystal. The trace of this tensor is $\mathrm{tr}\{\hat{\kappa}\} \equiv \partial\omega_i/\partial x_i$, i.e., the sum of its diagonal components.

Let us now look at some examples. The theoretically simplest case is free temperature expansion, when no elastic stresses exist, and hence

$$\hat{u}_0 = \hat{\varepsilon}^0 = \hat{\varepsilon}^{el} = 0 \tag{6.10}$$

and, in accordance with (6.8), $\hat{\beta} = 0$. Free expansion occurs if the temperature in the crystal changes linearly from point to point, i.e., grad $T = \{\partial T/\partial x, \partial T/\partial y, \partial T/\partial z\}$ is constant along the length of the crystal. This well-known result of elasticity theory follows from the equations of consistency of total deformations $\hat{\varepsilon}$:

$$\mathrm{rot}(\mathrm{rot}\ \hat{\varepsilon})^* = \nabla \times \hat{\varepsilon} \times \nabla = 0\,. \tag{6.11}$$

Indeed, in the absence of plastic deformation

$$\hat{\varepsilon} = \hat{\varepsilon}^{\mathrm{el}} + \hat{\alpha}T = \hat{S}\hat{\sigma} + \hat{\alpha}T\,, \tag{6.12}$$

where $\hat{S}$ is a fourth-rank tensor consisting of so-called compliance coefficients in Hooke's law [Ref. 6.23, Chap. 2]. Thus, stresses $\hat{\sigma}$ satisfy the equations

$$\nabla \times \hat{S}\hat{\sigma} \times \nabla = -\nabla \times \hat{\alpha}T \times \nabla\,. \tag{6.13}$$

The right-hand side of (6.13) contains only the second derivatives of the temperature with respect to the coordinates, so that with a linear temperature field it is equal to zero.[2] In this case the stresses $\hat{\sigma}$ satisfy a set of homogeneous equations and boundary conditions (the surface is free), and hence are equal to zero. Accordingly, $\hat{\varepsilon}^{\mathrm{el}} = 0$ and $\hat{\varepsilon} = \hat{\alpha}T$. Specifically, in the isotropic case, $\alpha_{ik} = \alpha\delta_{ik}$ and $\varepsilon_{ik} = 0$ for $i \neq k$, $\varepsilon_{xx} = \varepsilon_{yy} = \varepsilon_{zz} = \alpha(\partial T/\partial z)z$, if the z axis is chosen along the temperature gradient. By using equalities $\varepsilon_{ik} = 0$ as differential equations for displacement components $\boldsymbol{u}$, it is easy to show that upon heating, each plane perpendicular to the temperature gradient bends into a sphere of radius $1/\alpha(\partial T/\partial z)$, i.e., of curvature $\alpha(\partial T/\partial z)$: free temperature bending occurs (Fig. 6.17). The typical value of α is $\simeq 10^{-5}$–10^{-6} deg^{-1}, so that even at a comparatively large gradient $\partial T/\partial z \simeq 10^3$ deg/cm, the lattice curvature at the base of a

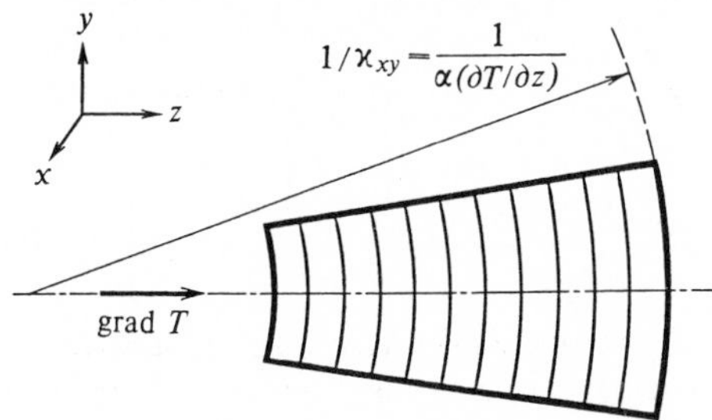

Fig.6.17. Curving of the atomic planes of a free crystal in a linear temperature field. The sample is stress-free

---

[2] For instance, with isotropic thermal expansion $\alpha_{ik} = \alpha\delta_{ik}$,

$$(\nabla \times \hat{\alpha}T \times \nabla)_{ik} = \alpha\left(\frac{\partial^2 T}{\partial x_i \partial x_k} - \delta_{ik}\frac{\partial^2 T}{\partial x_j^2}\right),$$

where $\delta_{ik} = 1$ for $i = k$, and $\delta_{ik} = 0$ for $i \neq k$.

cylindrical crystal in the axial temperature gradient is 10^{-2} – 10^{-3} cm^{-1}, i.e., much smaller than the curvature of the front during growth from the melt, e.g., by the Czochralski method.

In the general anisotropic case, the lattice curvature is found from the conditions rot $\hat{u} = 0$, $\hat{u} = \hat{\alpha}T + \hat{\Omega}$; thus,

$$\text{rot}\,\hat{\Omega} \equiv -\hat{\kappa} + \hat{I}\,\text{tr}\{\hat{\kappa}\} = -\,\text{rot}\,\hat{\alpha}T = \hat{\alpha} \times \text{grad}\,T\,. \tag{6.13a}$$

Applying operation tr to both sides of the equality relating the second and fourth expressions in (6.13a), and bearing in mind that $\text{tr}\{\hat{I}\} = 3$, we get

$$\hat{\kappa} = -(\hat{I} - \tfrac{1}{2}\hat{I}\text{tr})\,(\hat{\alpha} \times \text{grad}\,T)\,. \tag{6.14}$$

Substituting (6.14) into (6.9) yields $\hat{\beta} = 0$, as does (6.8) directly, at $\hat{u}^0 = \hat{\varepsilon}^{el} = 0$.

In practice, free-bending conditions are best realized for sufficiently thin needle-shaped crystals in which the radial temperature gradients are low and the smallness of the diameter makes it possible to set up a temperature distribution with $\partial T/\partial z = \text{const}$ at the growing end. Under these conditions dislocations neither arise nor multiply.

The ease of creating free-bending conditions for thin specimens is utilized in growing dislocation-free crystals. A thin cylindrical seed is thoroughly annealed, with care taken to ensure that the dislocations emerge at its side, and then sharpened, thus further reducing the probability that the dislocations will emerge at the tip of the seed. Then growth by the Czochralski method is begun, using the minimum possible area of the sharp point of the seed. In this way the probability that the growing crystal will inherit the dislocations is minimized. If, in addition, the thermal conditions do not differ greatly from the free-bending condition ($\partial T/\partial z = \text{const}$), no dislocations are nucleated either,

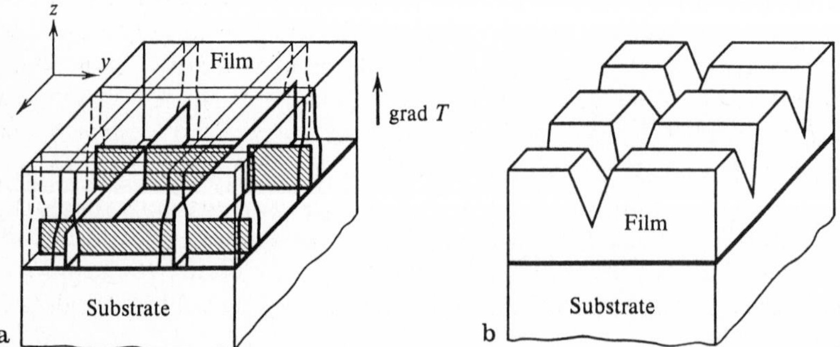

Fig.6.18a,b. Dislocations (**a**) and crevasses (**b**) in a film grown on an unbending substrate at a temperature gradient grad T (the substrate is cooler, the film warmer). In (a), extra atomic planes parallel to coordinate planes xz and yz are shown

and the crystal begins growth free of dislocations. This method made it possible to obtain dislocation-free silicon for the first time [6.24a].

Quite a different situation obtains when no crystal bending is possible, i.e., when $\hat{\kappa} = 0$. Precisely this state is achieved in a film (plate) that is infinite along the x and y axes and is glued to an unbendable substrate of the same material (Fig. 6.18). Here, elastic stresses in the film arise even with a linear temperature distribution, $\partial T/\partial z = \text{const}$. If the film grows from a free surface, we have $\partial T/\partial z > 0$, so that compression stresses would be expected in the newly formed layers. These stresses, however, are relieved by dislocations with a density determined from (6.9):

$$\hat{\beta} = \hat{\alpha} \times \operatorname{grad} T\,. \tag{6.15a}$$

To put it differently:

$$\hat{\beta} = \{\beta_{ik}\} = -\{e_{ilm}\nabla_l T\alpha_{mk}\} = -\{e_{izm}\alpha_{mk}\}(\partial T/\partial z)$$

$$= \begin{Bmatrix} \alpha_{yx} & \alpha_{yy} & \alpha_{yz} \\ -\alpha_{xx} & -\alpha_{xy} & -\alpha_{yz} \\ 0 & 0 & 0 \end{Bmatrix} \frac{\partial T}{\partial z}\,, \tag{6.15b}$$

i.e., the dislocations lie in the xy plane, as shown in Fig. 6.18a.

For crystals of cubic and tetragonal symmetly with one of the principal axes perpendicular to the substrate, $\alpha_{ik} = 0$ for $i \neq k$, and only terms α_{xx} and α_{yy} remain in matrix (6.15b). Accordingly, the minimum total Burgers vector of the dislocations intersecting the yz plane is directed along y and is equal to α_{yy} $(\partial T/\partial z)$, and the Burgers vector of the dislocations intersecting the xz plane is directed along x and is equal to $-\alpha_{xx}$ $(\partial T/\partial z)$. By constructing the Burgers circuit for dislocations and running through it in the direction of a right-handed screw, it is easy to see that the signs of the elements in matrix (6.15b) correspond to the extra half planes inserted from the side of the substrate (Fig. 6.18a). When the film has been cooled to a constant temperature it acquires a stress, which increases from zero near the substrate to $\alpha_{ik}(\partial T/\partial z)z$ $(i, k = x, y)$ at a distance z from it. If the growing crystal layer is thick enough and the temperature gradient was sufficiently large in the course of growth, the stresses in the free surface of the cooled crystal may exceed the ultimate strength: a network of cracks, shown schematically in Fig. 6.18b, appears on the surface. The described "gradient" stresses are caused by the temperature gradient along the normal to the growth front. They have no bearing on the stresses resulting from the difference in the thermal expansion coefficients of the crystal and the substrate, although these stresses clear in their origin, may greatly exceed the "gradient" stresses in thin films.

The results obtained are also applicable to an infinite crystal plate grown along two opposite planes. Here, the zero curvature of the lattice is ensured by the symmetry of the thermal field.

Lattice curvature is also completely nonexistent (i.e., $\bar{\kappa} = 0$) with a radially symmetric temperature distribution in a sphere and an infinite cylinder. Such distributions arise, for example, when these bodies are cooled from the surface. Figure 6.19a schematizes the distribution of temperature $T(r)$ across a section of an infinite cylinder. With such a distribution the hotter core of the cylinder is elastically compressed, because the cooler peripheral layer restrains thermal expansion. The peripheral layer is thus stretched by the core. The crystal is relieved of these elastic stresses by the formation of a system of dislocations in accordance with (6.8). In the quenching of an infinite cylinder discussed here, only the radial temperature gradient is nonzero, and therefore all the $\beta_{zk} = 0$ ($k = r, \varphi, z$). In other words, the total Burgers vector of the dislocations oriented along the radial directions is equal to zero. The distribution of dislocations is clear from Fig. 6.19c. Following cooling to a constant temperature, these dislocations give residual pressures $-\hat{\sigma}$, which corresponds to extension of the core and compression of the peripheral layers (Fig. 6.19d). Let us now cut the quenched cylinder normally to its z axis. The extended core tends to compress, and the compressed peripheral layer to elongate along the z axis and decrease in radial direction. As a result, a force moment arises along the contour of the base, which "draws" the base into the crystal (Fig. 6.19c), compresses it, and imparts a barrellike shape to the crystal.

In other words, compressive stresses $\sigma_{xx} \simeq \sigma_{yy} < 0$ arise at the base center. Calculations and measurements show that in general, at the base centre, $\sigma_{xx} = -k\sigma_{zz}$, where σ_{zz} is the stress on the cylinder axis before cutting, and coefficient k varies within the range 0.6 – 1.4 for different shapes of cylinder cross sections. Thus, $k \simeq 1$ can be assumed in rough estimates.

The stresses considered above are of a quenching nature, as they are caused by a decrease in temperature from center to periphery such as occurs in cooling from the surface. Inverted temperature distribution in the cooling of a cylinder or a sphere through the center yields stresses reciprocal to those indicated in Figs. 6.19c, d. During the building up of a sphere, when the heat is removed from its center (the situation is similar in the Kyropoulos method), the peripheral layers of a grown and cooled crystal are elongated and the core compressed. Hence it is possible for cracks to propagate from the surface into the crystal.

We now return to the stresses in a cylinder. This time, however, we shall consider not radial quenching, but building up of the cylinder from the base by the Czochralski or Verneuil methods. Here, the crystal has both a radial and an axial temperature gradient. The measure of their relative value is the curvature of the growth front: with noticeably convex and concave fronts the radial gradients have signs that differ from or are comparable to the axial gradients, but usually smaller; with a flat crystallization isotherm the axial gradient is the only one at the growth front (provided, of course, that the front coincides with the isotherm).

The effect of the crystal base upon the thermoelastic stresses extends along the z axis over distances of the order of crystal diameter d. Therefore two cases exist.

The first is when the temperature range of plasticity is sufficiently wide and the axial temperature gradient in the crystal is moderate, so that the crystal material is plastic up to distances $z \gtrsim d$ from the base. This is the usual situation for semiconductors. Thermoelastic and, hence, residual stresses must depend predominantly on the radial gradients, as discussed in connection with Fig. 6.19a. It should be noted, however, that with a concave growth front the temperature of the peripheral layers of a growing cylinder exceeds that of the core. If such a distribution persists in the quenching region, then, as noted above, the signs of the thermoelastic and residual stresses will be opposite to those given in Fig. 6.19a.

The second case corresponds to a narrow region of plasticity ($z \ll d$) adjacent to the growth front. Here the dislocation density and residual stresses

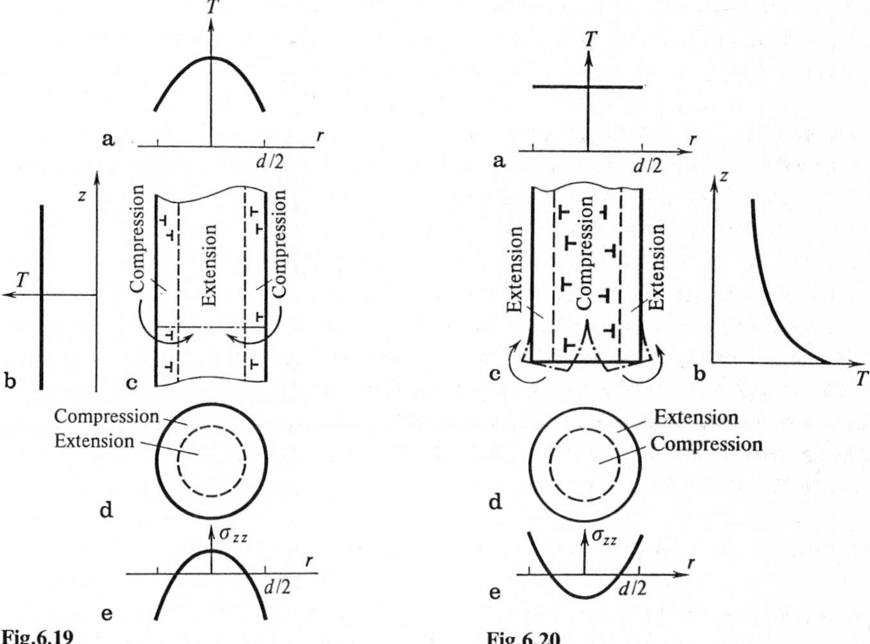

Fig.6.19 **Fig.6.20**

Fig.6.19a–e. Formations of quenching stresses in radial cooling of an infinitely long cylinder of diameter d; the radial (**a**) and axial (**b**) temperature distribution and that of residual stresses (**c, d**) in a cylindrical crystal. The *dashed line* shows the neutral (zero-stress) surface (**c, d**). The dislocations in the crystal periphery layer are shown in (**c**). The *arrows* indicate the force moments which arise when the crystal, having been cooled to a constant temperature, is cut perpendicular to the z axis along the *dot-dashed plane*. (**e**) Radial distribution of the residual axial stresses σ_{zz} after cooling of the specimen

Fig.6.20a–e. The formation of stresses during the growth of a cylinder (of diameter d) from the base; radial (**a**) and axial (**b**) distribution of the temperature, and the distribution of residual stresses (**c, d**) in a cylindrical crystal. The *dashed line* shows the neutral surface (**c, d**). Dislocations in the central part of the crystal are shown in (**c**). The *arrows* indicate the force moments arising at the end face of a crystal cooled to a constant temperature; the *dot-dashed* line indicates possible cracking of the specimen under the effect of these moments. (**e**) Radial distribution of the residual axial stresses σ_{zz} after cooling of the specimen

depend, in accordance with (6.9), on the temperature gradient and lattice curvature at the base of the crystal, rather than in its bulk as in quenching. If the growth front curvature is not large, the decisive role is played by axial temperature distribution $T(z)$, and the $T(r)$ dependence can be neglected; i.e., $T(r)$ can be assumed to be constant (Fig. 6.20a).

Under conditions ensuring that axial temperature gradient $\partial T/\partial z$ remains constant at distances z, of the order of a diameter or more from the base, the base region undergoes free temperature bending virtually independently of the temperature distribution in the rest of the crystal. Accordingly, the crystal will be free of stresses and grow dislocation-free, provided that the seed was free of dislocations and that large stresses were not allowed during the cooling of the grown crystal [6.24a]. In practice, such conditions are easier to attain, the smaller the diameter of the crystal is.

When the axial gradient is essentially inconstant even at distance $z < d$ from the base, stresses would be expected to arise in the near-front zone in the absence of plasticity. But the growing layer and the material in the plastic zone adjacent to it have been relieved of stresses at the expense of the formation of the corresponding dislocations, which actually determine the residual stresses in the grown and cooled crystal. The nature of these stresses is qualitatively clear when the temperature falls off abruptly near the base in a layer which is narrow as compared with the diameter, but wide as compared with the plastic zone, and changes only slightly at large z (Fig. 6.20b). With this kind of temperature distribution we are again dealing with the problem of growth on an unbendable substrate, the role of the substrate now being played by the crystal region with a slightly changing temperature. Consequently, tensile stresses ($\sigma_{xx} > 0$, $\sigma_{yy} > 0$) rather than compressive ones as in radial quenching (Figs. 6.20c, d) will act at the base center of the grown crystal. The tensile stresses along the base correspond to the moments shown in Fig. 6.20c, which can split the crystal along the z axis (this is similar to the splitting of a film; see Fig. 6.18b). Corundum crystals grown by the Verneuil method sometimes split this way. Additional heaters, which reduce the value of $\partial^2 T/\partial z^2$ (and not simply the temperature gradient $\partial T/\partial z$, as is sometimes thought), increase the absolute value of the temperature in the furnace, and expand the plastic zone, serve to reduce the residual stresses and prevent cracking. Such furnaces are used not only for growing crystals, but also for annealing them and thus relieving some of the stresses which have nevertheless arisen.

The characteristic parameters that define a particulur temperature distribution in a crystal with a given system geometry are, at low and medium temperatures ($T \lesssim 500°$C), the thermal conductivity in the crystal body and the coefficient of heat transfer across its boundary to the medium filling the growth chamber (a gas, a high-temperature solution, etc), At high temperatures ($T \gtrsim 1000°$C), the parameters are the corresponding values of raditive transport.

To conclude the discussion of thermal stresses and associated dislocations, we present some patterns of the distribution of temperature and stresses at the

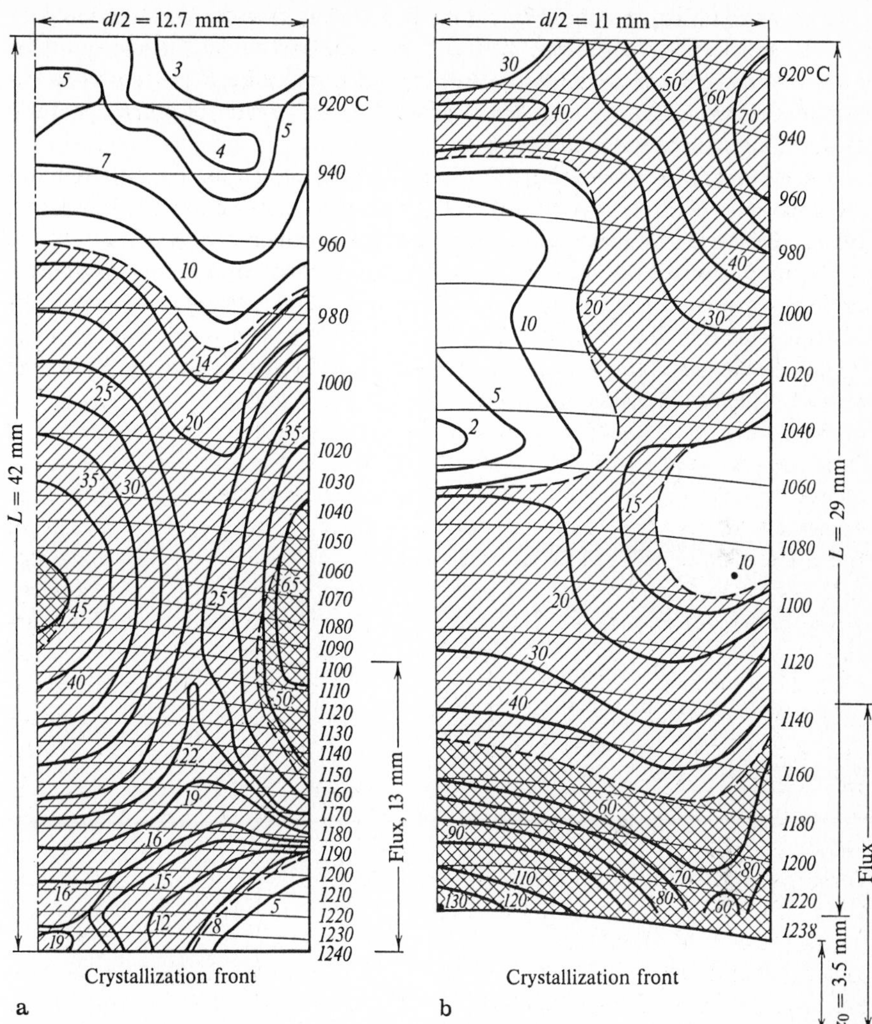

Fig.6.21a,b. The distribution of temperatures and stresses calculated for a cylindrical crystal of GaAs with a diameter d and length L under conditions corresponding to growth by the Czochralski method (Sect. 10.2) under a layer of flux—B_2O_3 melt. The figure shows cross sections of two crystals cut by a plane parallel to the cylinder axis. The crystal axis is to the left and the free surface to the right. The *numerals* at the right are values of the temperature on the *thin* isotherms. The *heavy lines* represent the lines of equal shear stresses σ_{rz}, and the *numerals* on them the stress values in G/mm². The *hatching* indicates the regions where the stresses exceed the critical values for the formation of dislocations at the appropriate temperatures. A comparison of (**a**) and (**b**) reveals that the stresses and hence the density of dislocations decrease when the temperature gradients are reduced. Courtesy of M.G. Mil'vidsky and V.B. Osvensky

base of a cylindrical crystal (Fig. 6.21a, b). The relevant thermal conductivity and thermoelasticity equations were solved numerically under the assumption that the growth front coincides with the melting point isotherm and that the heat removal through the side of the cylinder occurs according to Kirchhoff's law, although there is no radiative transport inside the cylinder. The regions where the shear stresses exceed the critical stresses have the densest hatching. A comparison of these results with the conditions for the formation of dislocation-free crystals of germanium and silicon yields satisfactory results.

Thermal stress and the corresponding formation of dislocations during liquid-encapsulated Czochralski growth of GaAs single crystals was calculated and discussed in [6.24b].

Numerical calculations are extremely laborious, and therefore the following relation is helpful in estimating the stresses in the base region:

$$\sigma \simeq \alpha E L^2 \partial^2 T/\partial z^2 \, , \tag{6.16}$$

where α is the thermal expansion coefficient, and E is the Young modulus. The characteristic length L, according to the estimates of *Indenbom* et al. [6.19] in the case of $\partial^2 T/\partial z^2 \approx$ const, is $\approx 0.2 - 0.5d$. At $\alpha \simeq 10^{-6}\ \mathrm{K}^{-1}$, $E = 10^{12}$ dyn/cm^2, $\partial^2 T/\partial z^2 \simeq 10\ \mathrm{K/cm}^2$, for a crystal of diameter $d = 5$ cm we have $\sigma \simeq (1 - 2.5) \times 10^7$ dyn/cm$^2 \simeq 10 - 25$ kp/cm^2. This is much less, for instance, than the critical stress for the formation of dislocations in silicon.

At low thermal stresses and high growth rates the dislocations present in a seed may have such low glide velocities that the growing interface will "escape" from the dislocation network, which remains frozen in the seed. In this case the crystal will grow dislocation-free; some dislocations should spread into the crystal during sufficiently long exposure at temperatures close to the melting point. Evidence of such "escape" [6.24d] was found in space-grown Ge [6.24e].

Thermal stress may be estimated analytically in an infinitely long isotropic cylindrical crystal pulled with constant velocity V if the temperature in the crystal is azimuthally isotropic, is distributed parabolically with respect to the radial coordinate z, and follows a known law $T = T(z)$ along the cylinder axis. In this case the diagonal components of the stress tensor in the cylindrical coordinates (r, φ, z) are

$$\sigma_{rr} = \frac{\alpha E}{16}(R^2 - r^2)\left[\frac{VR}{(1-\nu)a_\mathrm{s}}\frac{\partial T}{\partial z} - \frac{1}{1+\nu}\frac{\partial^2 T}{\partial z^2}\right], \tag{6.17}$$

$$\sigma_{\varphi\varphi} = \frac{\alpha E}{16}(R^2 - 3r^2)\left[\frac{VR}{(1-\nu)a_\mathrm{s}}\frac{\partial T}{\partial z} - \frac{1}{1+\nu}\frac{\partial^2 T}{\partial z^2}\right], \tag{6.18}$$

$$\sigma_{zz} = \frac{\alpha E}{8}(R^2 - 2r^2)\left[\frac{VR}{(1-\nu)a_\mathrm{s}}\frac{\partial T}{\partial z} + \frac{1}{1+\nu}\frac{\partial^2 T}{\partial z^2}\right], \tag{6.19}$$

where ν is the Poisson ratio and a_s is the thermal diffusivity in the crystal. $R = d/2$ is the radius of the cylinder. The maximal shear stress equals $(\sigma_{zz} - \sigma_{\varphi\varphi})/2$ [6.24c].

6.2.5 Dislocations Related to Vacancies and Impurities

The density of crystals usually exceeds that of the corresponding melts by a few per cent. Therefore during growth crystals may trap vacancies in numbers exceeding the thermodynamic equilibrium numbers at the melting point. Cooling a crystal increases the supersaturation of the vacancy solid solution, or, if vacancies are trapped in the amount corresponding to equilibrium at the melting temperature, produces supersaturation. The decomposition of the vacancy solid solution results in the formation of micropores, particularly disc-shaped ones. The collapse of each such disc leads to the formation of a dislocation loop with a Burgers vector normal to the disc plane. A supersaturated solid solution of interstitials appears after cooling in melt-grown silicon. Clusters of these interstitials are revealed by electron microscopy. They are viewed as loops of different shapes with sizes in the range of $5 \times (10^{-5}–10^{-3})$cm [6.12a] and might be interpreted as impurity-stabilized dislocation loops. Some of the observed loops exhibit stacking fault contrast. The larger loops are called A-type defects and the smaller ones B-type defects. Macroscopically, the clusters manifest themselves and were discovered as so-called swirl defects in dislocation-free crystals [6.11, 12]. These defects can be seen as a spiral (or concentric) nebula after etching on cuts perpendicular to the growth direction; hence their name. Electron microscopy shows that this nebula is an agglomerate of interstitial or stacking fault dislocation loops of the order of $\lesssim 10^{-4}–10^{-5}$ cm in size. Macroscopically, the region containing these loops has the outline of a helicoid screwed around the growth axis and often occupying the entire central region of the crystal. This geometry of precipitations corresponds to fluctuations in the growth rate. These include periodic remelting due to the rotation of the pulled crystal in a slightly asymmetric temperature field when the meet temperature and composition at the growing interface fluctuate. These fluctuations are caused by melt convection and external heating instabilities [6.25–27]. In other words, swirls are closely related to growth processes. The swirls are observed if the growth rate is within the range of ~ 2 to ~ 5 mm/min. Additional annealing of crystals grown at rates above the upper limit also causes swirls to appear. The presence of impurities, especially carbon, is also important in the creation and detection of swirl defects. Thus loop creation may be treated as a precipitation process in a supersaturated solid solution of silicon interstitials [6.12a] and vacancies. Annihilating, they produce clusters of various types [6.24f].

If a mechanical stress is applied to the crystal, the dislocation loops give rise relatively easily to dislocations of macroscopic length. Thus the plastic deformation in crystals containing swirls begins at a much lower critical stress than that in genuinely dislocation-free crystals, where nucleation of dislocation loops is necessary for plastic deformation.

Dislocations are also formed as a result of nonequilibrium impurity trapping. An impurity trapped homogeneously by a crystal changes its lattice parameters. Thus we have "impurity expansion" (or contraction), which can be likened to temperature expansion. This field of macroscopic stresses in a crystal with an impurity can be found with the aid of thermoelasticity equations. In place of temperature deformations $\hat{\alpha}T$, concentration deformations $\hat{\alpha}_C C$ are introduced, where C is the impurity concentration, and the second-rank tensor $\hat{\alpha}_C$ characterizes the concentration expansion (or contraction) of the lattice. Accordingly, the linear distribution of the impurity concentration throughout a crystal with a free surface will lead to a free concentration bend with a curvature determined by (6.14) after replacement $T \to C$, $\hat{\alpha} \to \hat{\alpha}_C$. Under conditions which exclude the bending of the lattice, the dislocation density related to the gradient of the impurity concentration in the crystal will be

$$\hat{\beta} = \hat{\alpha}_C \times \operatorname{grad} C \,. \tag{6.20}$$

The formation of dislocations in connection with zonal and sectorial structure (Sects. 4.2, 3) is very typical. The concentration stresses caused by an impurity-enriched zone are relieved by dislocation networks at the boundaries of and/or inside this zone (Figs. 6.22a, b). Photos of networks on impurity zones are presented in Fig. 6.23. Dislocation networks are very stable formations, as no unit of the network can move independently of the others. This stability is enhanced by the presence of an impurity, which creates a "potential valley" for the networks. Conversely, impurities cannot leave the zone, since dislocations would then cause stresses, which in turn would draw the impurities back. As a result, it is often impossible to eliminate growth zonality.

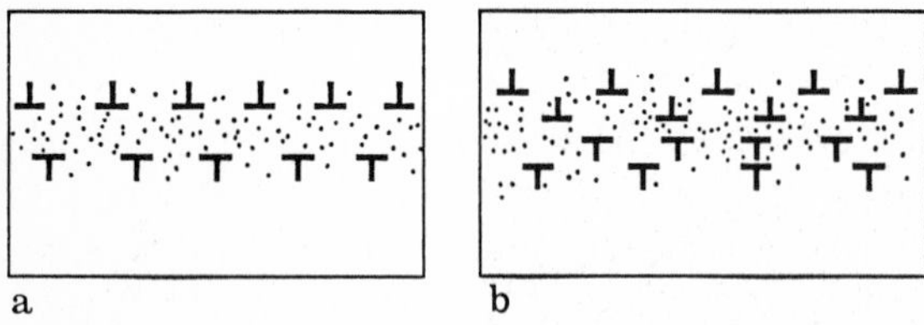

Fig.6.22a,b. Formation of dislocations around a zone enriched in an impurity having an atomic size exceeding that of the matrix atoms. **(a)** Dislocations on the periphery of the zone; **(b)** dislocations distributed on the periphery of the zone and inside it

The concentrations of impurity atoms and other point defects differ in different growth pyramids, and hence so do the lattice spacings. For this reason, contact between such pyramids leads to the formation of internal stresses and misfit dislocations at the boundaries of the growth pyramid. Stresses are observed in polarized light owing to the birefringence caused by them. Misfit disloca-

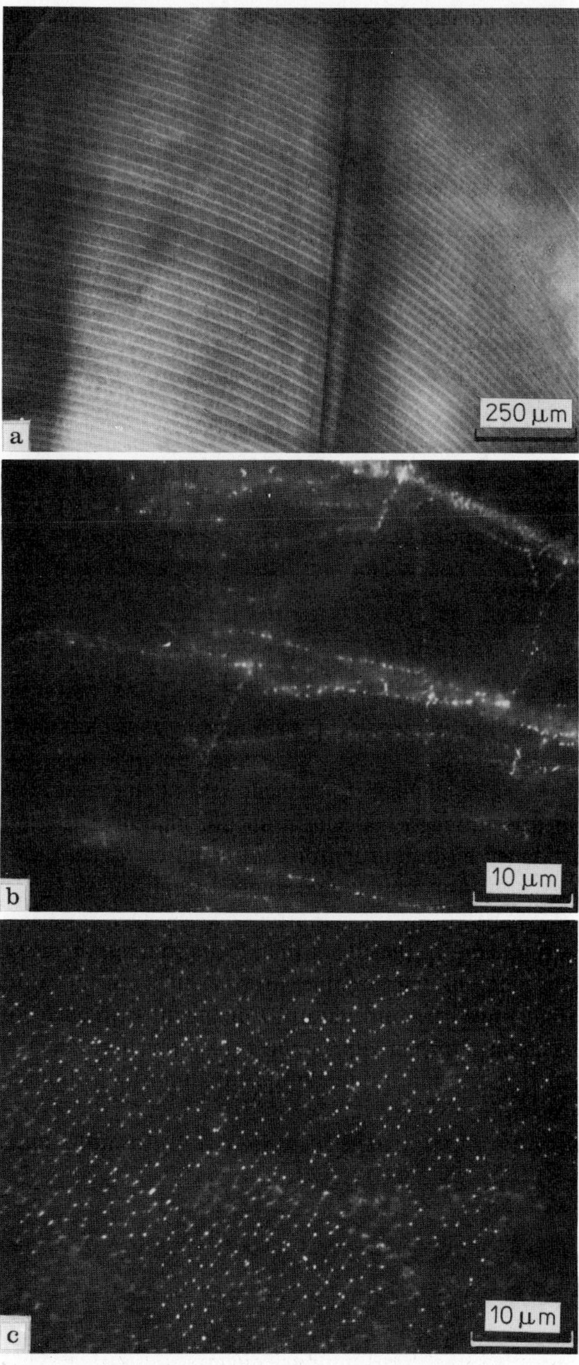

Fig.6.23a–c. Dislocation networks associated with the impurity zones in ruby [6.28]. (**a**) Side view of the impurity zones (*light bands*) with low magnification; (**b**) three impurity zones positioned almost horizontally, side view. Decorated dislocation lines are seen as sequences of light dots; (**c**) dislocation net lying within a single impurity zone, plan view

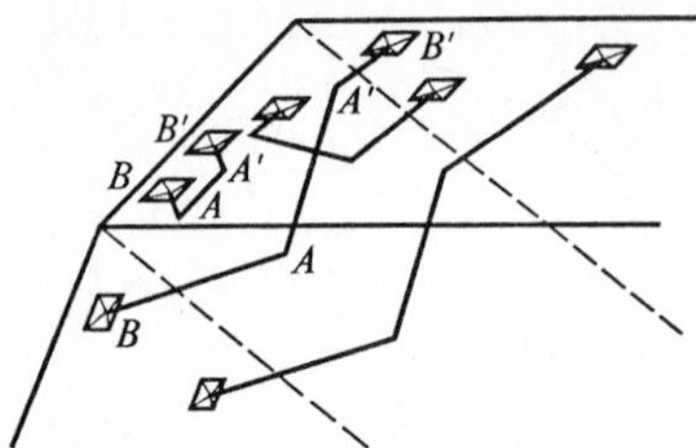

Fig.6.24. Dislocations nucleated at the boundary of two growth pyramids (denoted by the *dashed line*). Dislocations are shown leaving the boundary between the growth pyramids at points A and A′ and emerging at the horizontal face. The emergences of the dislocations onto the surface are indicated by schematically represented *etch pits*

tions at the boundaries of growth pyramids (sectors) may, like epitaxial misfit dislocations (Sect. 2.3.3), lie completely within the boundary plane or leave the plane and emerge at the growing faces. The dislocations which leave the boundary plane are depicted schematically in Fig. 6.24. Segments AA′ and similar ones lie in the boundary plane, while segments AB and A′B′ emerge at free surfaces at points B and B′, where they can be revealed, for instance, by etching. Etch pits are also shown schematically in Fig. 6.24.

6.2.6 Grain Boundaries

Individual dislocations attract each other due to the interaction of their elastic fields, and join together to form grain boundaries. This attraction becomes effective when the distances between the dislocations are small enough, i.e., when the dislocation density is sufficiently high. In ruby, the joining of dislocations into grain boundaries begins with densities of $\sim 10^4$ cm^{-2}. It has been repeatedly noted that grain boundaries begin on inclusions, this happens for instance on macroparticles ($\sim 10^{-4}$–10^{-2} cm) of an incompletely fused charge trapped by a crystal grown by the Verneuil technique.

Investigations into the morphology of ruby crystals show that the orientations of the grain boundaries are close to the basal planes and the planes of a rhombohedron, but they do not strictly adhere to these planes, deviating from them by up to about ten or twenty degrees. In such crystals, the grain boundaries usually intersect the growth front at right angles, so that a "fan" of grain boundaries corresponds to a convex growth front.

The mutual orientation of the grain boundary and the growth front is determined by the same consideration as the orientation of individual dislocations. Let φ and ϑ be two angles which determine the mutual misorientation of two lattices contacting at the grain boundary, and let $\boldsymbol{\nu}$ be the normal to the grain boundary. Then the specific free energy of the grain boundary, γ, is completely assigned by φ, ϑ, and $\boldsymbol{\nu}$, i.e., $\gamma = \gamma(\varphi, \vartheta, \boldsymbol{\nu})$. The orientation of the growth front intersected by the grain boundary will be assigned, as before, by a normal $\boldsymbol{n}$, so that $(\boldsymbol{n}\boldsymbol{\nu})$ is the angle between front and boundary (Fig. 6.25). If the front has shifted by dh, the increase in boundary energy per unit length of the line of its intersection with the front is $\gamma dh/\sin(\boldsymbol{n}\boldsymbol{\nu})$ (Fig. 6.25).

Thus the sought-for, most advantageous boundary orientation with respect to various $\boldsymbol{\nu}$ is defined by the condition

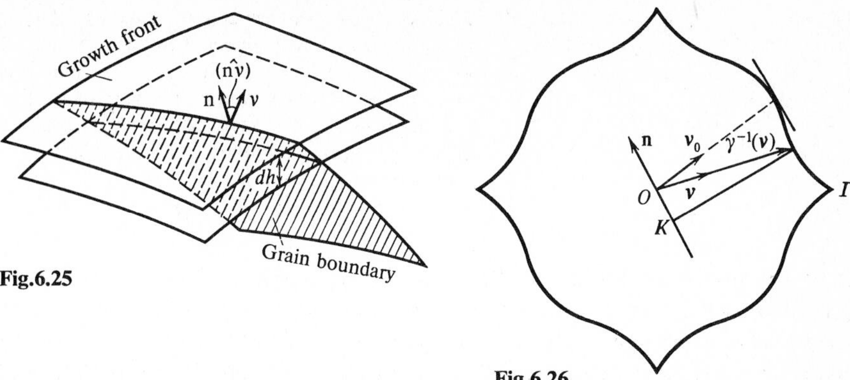

Fig.6.25

Fig.6.26

Fig.6.25. A grain boundary and two consecutive positions of the growth front spaced at dh; $\boldsymbol{\nu}$ is the normal to the grain boundary and $\boldsymbol{n}$ the normal to the growth front

Fig.6.26. A grain boundary emerging onto a face with normal $\boldsymbol{n}$ has minimum energy if its orientation is determined by vector $\boldsymbol{\nu}_0$. Contour Γ is a section of the reciprocal grain boundary energy surface $\gamma^{-1}(\boldsymbol{\nu})$, where the energy corresponds to a fixed misorientation angle between the two lattices contacting at this boundary

$$\frac{\sin(\boldsymbol{n\nu})}{\gamma(\varphi, \theta, \boldsymbol{\nu})} = \max . \tag{6.17}$$

Figure 6.26 presents a schematic polar diagram of reciprocal energy $\gamma^{-1}(\varphi, \theta, \boldsymbol{\nu}) = \Gamma$ as a function of $\boldsymbol{\nu}$ at constant φ and θ. The left-hand side of (6.17) is equal to the length of segment ΓK, and reaches a maximum at orientation $\boldsymbol{\nu} = \boldsymbol{\nu}_0$. This orientation corresponds to the point of tangency of the γ^{-1} diagram with a plane parallel to $\boldsymbol{n}$ and farthest from point 0 (Fig. 6.26). If γ = const, i.e., $\gamma(\boldsymbol{\nu})$ is a sphere, then $\boldsymbol{\nu} \perp \boldsymbol{n}$, which means that the grain boundary is normal to the growth front.

The lattice misorientations at the neighboring grain boundaries are as a rule not accidental, but reflect systematic rotations of the lattice in the crystal as a whole. For instance, ruby boules grown by the Verneuil technique often consist of two "half cylinders" screwed in opposite senses about the boule axis.

Systematic lattice rotations are also observed in connection with the replacement of some originally assigned growth front orientations by other orientations during growth. This spontaneous rotation of the lattice about the axis parallel to the growth front may be attributed to the anisotropy of elastic moduli and to the minimization of thermoelastic stresses in an inhomogeneous temperature field at the growing crystal base. The same circumstance evidently explains rotation of the crystal lattice that brings the lattice symmetry and the temperature field of the furnace into correspondence even if the crystallographic orientation of the growth front remains unaltered (lattice rotation about the normal to the growth front). The correspondence between the symmetry of the crystal and that of the thermal field is essentially achieved in the first of the above-mentioned cases also. These phenomena have not yet been sufficiently investigated.

7. Mass Crystallization

Mass crystallization is the nucleation and growth of a large number of usually small crystals ($\sim 10^{-3} - 10^{-1}$ cm) in one and the same area of space. Examples of it are the formation of metal ingots and kidney stones, the solidification of concrete, and the production of granulated fertilizers, medicines, sugar, and salt. It has been used and studied for hundreds of years. In mass crystallization, special attention is usually given to the purity, size, and shape of the crystals, and the aim is to achieve a maximum strength of intercrystallite bonds (as in concrete and metals) or to obtain a fine-grained, loose, noncaking product (sugar, salt, fertilizers). The fine defects of individual crystals are usually given much less attention than in the growing of single crystals. Total rates of output and other economic considerations are extremely important.

The wide use of mass crystallization is reflected in the term "industrial crystallization," a synonym that is commonly encountered in the literature but that usually does not include metallurgy, which is also based on mass formation of crystals. Mass crystallization has been treated in many books [7.1–5], and for this reason we shall only give a general idea of selected problems in this extensive field.

In melt solidification, the kinetics of mass crystallization is described by the fraction of the substance crystallized in a given time. In crystallization from a solution, the main kinetic characteristics are the mass of the deposited crystals or their fraction relative to the solute, or else simply the supersaturation of the solution. Particularly in the late stages of growth, the deposited and growing crystals interact with each other both through their diffusion and thermal fields and mechanically, by direct collisions. As a result, a theoretical description of mass crystallization kinetics is very complicated, and experimental results strongly depend not only on fundamental parameters (supercooling, the type of substance and solvent), but also on technological parameters (crystallizer geometry and design, the way in which the supersaturation is introduced, the type of stirring, etc.).

7.1 Solidification Kinetics and Grain Size

Mass crystallization of the melt has been considered, under the most general assumptions, by *Kolmogorov* [7.6a] and independently by *Johnson* and *Mehl*

[7.6b], and *Avrami* [7.6c], who used different approaches. A critical review of the present state of the problem is given by *Belen'ky* [7.7a].

The process can be described schematically as follows: We have a sufficiently large (as compared with the size of the crystallites to be produced) homogeneous volume of a melt, in which nuclei are formed at random times and points at the rate of $J(t)$ [$\mathrm{cm}^{-3}\,\mathrm{s}^{-1}$]. These nuclei then grow at a normal rate $V(t)$, where t is the time. It is assumed that either all the crystals are spherical, or have the same shape and are oriented parallel to each other. The temperature at all points of the crystallizing volume is assumed to be constant, but may change with time according to an arbitrary law reflected in the given dependences $J(t)$ and $V(t)$. The requirement of temperature constancy in space is equivalent to averaging it over the space, i.e., to neglecting the detailed temperature distribution around each crystallite. Our goal is to determine the volume fraction $p(t)$ crystallized by moment t after the beginning of crystallization.

In the initial stage the crystals are not large, do not contact each other, and thus have the same shape. Therefore, for spherical crystals

$$p(t) = \frac{4\pi}{3}\int_0^t \left[\int_{t'}^{t} V(t'')dt''\right]^3 J(t')dt' , \tag{7.1}$$

where $\frac{4\pi}{3}[\int_{t'}^{t} V(t'')dt'']^3$ is the volume that a crystal nucleated at moment t' has attained by moment t.

In general, one must take contact among the crystals into account, as well as the corresponding fantastic shapes of the contact surfaces. To do this, we introduce $q(t) = 1 - p(t)$ as the probability that a point of the melt taken randomly at moment t does not belong to the crystallized part of the volume. Consequently, the quantity $-dq(\tau) = -[q(\tau + d\tau) - q(\tau)]$ is the probability of crystallization at the point concerned within the period from τ to $\tau + d\tau$. In order for this last event to take place, it is necessary that within the indicated time period $d\tau$ the point in question be reached by the crystallization front propagating from some center which arose previously, between moments t' and $t' + dt'$. The crystals appearing at moment t' will reach the point under consideration within the period between τ and $\tau + d\tau$ only if they appeared at distances of $\int_{t'}^{\tau} V(t'')dt''$ to $\int_{t'}^{\tau+d\tau} V(t'')dt''$, respectively, from the point. Thus nuclei leading to solidification at this point must be disposed in a spherical layer of volume

$$4\pi\left[\int_{t'}^{\tau} V(t'')dt''\right]^2 V(\tau)d\tau . \tag{7.2}$$

The probability of nucleation in this layer within the period between t' and $t' + dt'$ is equal to

$$4\pi\left[\int_{t'}^{\tau} V(t'')dt''\right]^2 V(\tau)J(t')dt'd\tau .$$

Hence the probability of appearance of a nucleus which leads to crystallization at a randomly selected point during the period of time between $t' = 0$ and $t' = \tau$ is

$$4\pi \int_0^\tau \left[\int_{t'}^\tau V(t'')dt''\right]^2 V(\tau)J(t')dt'd\tau .$$

The front of growth from one such nucleus will cause crystallization at the chosen point at moment τ, $\tau + d\tau$ only if the liquid at this point has not solidified previously. The probability that the liquid state will be conserved at the selected random point by moment τ is $q(\tau)$. Therefore

$$dq(\tau) = -q(\tau)\cdot 4\pi \int_0^\tau \left[\int_{t'}^\tau V(t'')dt''\right]^2 V(\tau)J(t')dt'd\tau , \tag{7.2}$$

whence, integrating again and changing the sequence of integration, we obtain the solution to the problem:

$$\begin{aligned} p(t) &= 1 - q(t) = 1 - \exp\left\{-4\pi \int_0^t d\tau \int_0^\tau dt' \left[\int_{t'}^\tau V(t'')dt''\right]^2 V(\tau)J(t')\right\} \\ &= 1 - \exp\left\{-\frac{4\pi}{3}\int_0^t J(t')\left[\int_{t'}^t V(t'')dt''\right]^3 dt'\right\}. \end{aligned} \tag{7.3}$$

If the temperature changes only slightly in the course of crystallization, the growth and nucleation rates are time independent, and (7.3) yields

$$p(t) = 1 - \exp\left(-\frac{\pi J V^3 t^4}{3}\right). \tag{7.4}$$

By similar reasoning it is easy to obtain the kinetics of the overgrowing of the surface and of the straight line, in other words, the solutions for two-dimensional and one-dimensional space. The corresponding expressions are obtained from (7.3) by replacing the quantity $4\pi[\int_{t'}^\tau V(t'')dt'']^2$ with $2\pi \int_{t'}^\tau v(t'')dt''$ for two dimensions (v is the tangential front velocity along the surface to be covered by crystals) and with unity for one dimension. At J = const and v = const we have, in place of (7.4), in the two-dimensional case

$$p(t) = 1 - \exp\left(-\frac{\pi J v^2 t^3}{3}\right), \tag{7.5}$$

and in the one-dimensional case

$$p(t) = 1 - \exp\left(-\frac{J v t^2}{2}\right). \tag{7.6}$$

Different models for the two-dimensional case have been considered to describe the initial stages of single-crystal growth on substrates [7.7b]. In some models, crystallites of different shapes (discs, reversed cups, flat epitaxially oriented polygons) are supposed to exhibit no liquidlike coalescence after touching each other during growth. In this case the deposition kinetics is described by equations analogous to (7.5) with numerical factors in the exponent which differ slightly from $\frac{\pi}{3}$ and depend on the shape of the crystallites. In the other models, the two crystallites, upon touching each other, merge together and form one crystallite having the same shape and a volume equal to the total volume of the merging crystallites. In this case the general behavior of the $p(t)$ curve is still described by (7.5). However, this behavior is violated by the local maximum and minimum which appear on the $p(t)$ curve in the interval of time when the majority of growing crystallites reach each other and coalesce.

The solution given above for the problem of the filling of a plane can be generalized by taking into account the possibility that the next layer may nucleate on the surface of the preceding one. At high supersaturations this results in a kinetically rough surface (Sect. 3.3.1). Probabilistic characteristics of the surface structure have not been studied in detail, but it is obvious that the normal growth rate of such a surface is inversely proportional to the average time in which one atomic layer is filled. According to (7.5) this time is $\sim(Jv^2)^{-1/3}$, i.e., $V = h(Jv^2)^{1/3}$ to within a factor of the order of unity, which means that we come back to (3.23). Similar consideration of the one-dimensional case allows us to determine the elementary step velocity due to one-dimensional nucleation at the step rise, or the velocity of a dislocation due to double nucleation kinks on this dislocation [7.8]. It turns out that the step velocity is $v = (e/2) Jv_{\text{kink}}$, where v_{kink} is the kink velocity along the step. The numerical factor was obtained not analytically, but from computerized solutions of equations corresponding to a multilevel model for a step. However, a nonrigorous heuristic analysis, leads to a value of $e/2$, which is very close to the one calculated by computer ($e = 2, 718\ldots$).

But let us return now to three-dimensional mass crystallization. According to (7.4), kinetic solidification curve $p(t)$ is S-shaped, the characteristic time being $\propto(JV^3)^{-1/4}$. The experimental data of many authors fit straight lines in coordinates ln ln $q(t) - t$. The slope of these lines, however, does not always correspond to the fourth power of the time, giving lower values (from 2 to 4). One of the reasons for this is preferential nucleation on foreign particles in the melt. If the activity of these particles is high, i.e., if nuclei arise on them immediately after supercooling is applied at moment $t = 0$, then $J(t) = n_0\delta(t)$, where n_0 is the number of crystallization centers per unit volume of the melt. In this case, according to (7.3) we have, at $V = \text{const}$,

$$p(t) = 1 - \exp\left(-\frac{4\pi n_0 V^3 t^3}{3}\right). \tag{7.7}$$

In other words, the exponent of t is less by unity than that in (7.4). A further substantial effect is exerted on the process kinetics, in that the growing crystals act on each other by means of the regions of increased temperature and impurity concentration which surround them and overlap (at $K < 1$). The above calculation neglects this mutual effect of small crystals.

The fraction of the volume that is uncrystallized by moment t is $q(t)$, and hence the number of nuclei which appeared within the period t, $t + dt$ is $q(t)\,J(t)dt$. Therefore the total number density n of the grains making up the polycrystalline specimen is $\int_0^\infty q(t)J(t)dt$. At J = const and V = const,

$$n = \frac{1}{4}\,\Gamma\left(\frac{1}{4}\right)\left(\frac{3}{\pi}\right)^{1/4}\left(\frac{J}{V}\right)^{3/4}. \tag{7.8}$$

The value of the numerical coefficient in (7.8) is ~ 1.3. This pertains only to spherical crystals, but the values are close to unity for other sufficiently isometric forms as well. The dependence of n on J and V also follows directly from considerations of dimensionality and holds true for any crystal. Thus the average grain size in the specimen is $\simeq (V/J)^{1/4}$. When crystals are formed only around existing crystallization centers, the average grain size is $n_0^{-1/3}$.

During growth from stirred solutions whose concertration is not too high (as compared with 100%, i.e., with the melt), the crystals usually occupy a considerably smaller fraction of the volume than the solvent, and the growth of each crystal can be regarded as geometrically independent of that of the others. But the supersaturation at which each crystal grows depends on the entire crystallite community (especially with stirring). As in melt solidification, the total mass of the crystals deposited from solution changes with time according to an S-shaped curve. However, the deceleration of this process is connected with the removal of supersaturation, rather than with the filling of the entire space by crystals. The analytical expression of this curve and its quantitative parameters naturally differ from those of the melt.

During growth from an unstirred solution the crystals, having reached a size of ~10–100 μm, descend to the bottom of the crystallizer, from a bed of deposit there, and come into intimate contact with one another. They subsequently grow at a much slower rate, and may form a solid mass.

7.2 Geometric Selection and Ingot Formation

The solidification of a metal poured into a mold begins on its cooled walls, where the first crystals appear and begin to grow. The continuous front which is subsequently formed by neighboring dendritic tips moves into the melt, and the number of crystals emerging at the front gradually decreases, with the front area of each of them increasing accordingly. In this competitive struggle for a place at the front, only those crystals survive whose fastest growth directions are close to the normal to the substrate or the front [7.9]. Accordingly, at distances from the wall which appreciably exceed the average distance between the nuclei, all the crystals in the ingot have an elongated shape and are nearly parallel to each other; a so-called columnar texture arises.

Let us take a closer look at the competition which results in the described crystal selection (termed geometric selection). Suppose we have a substrate (mold wall) on which crystals of various random orientations have arisen (Fig. 7.1), and these crystals have the shape of prisms with rapidly growing pyramidal tops, so that the vertices of the pyramids recede from the crystallization center (point of nucleation) at the highest velocity, and the prism faces at the lowest velocity. The reasoning given below also applies to dendritic crystals. Let us consider two neighboring, arbitrarily oriented crystals which formed simultaneously at points O_1 and O_2 on the substrate (Fig. 7.1). The growth rate of crystal O_1 along the normal to the substrate is higher than that of crystal O_2, so that the former will be the first to reach any preassigned height over the substrate. Therefore by the time vertex O_2 reaches the height corresponding to the intersection of the crystals, the space ahead of vertex O_2 will

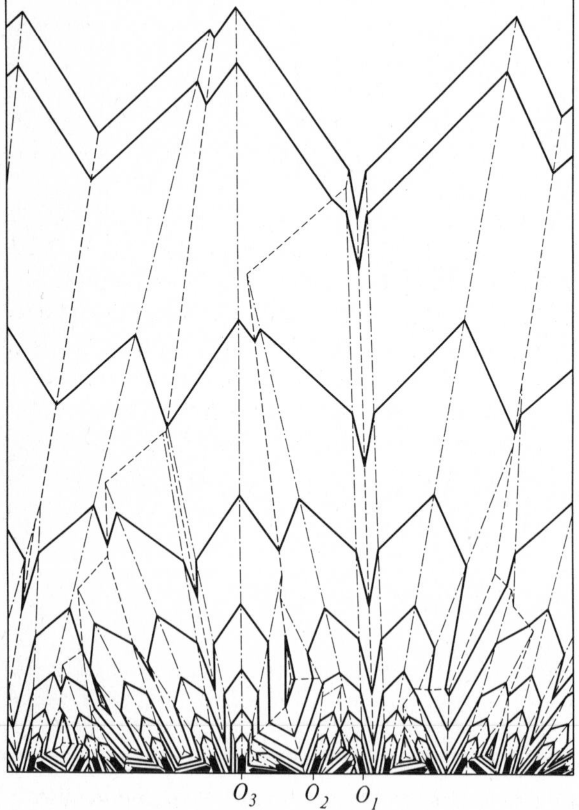

Fig.7.1. Consecutive stages in the geometric selection of crystals [7.9]

be occupied by the body of crystal O_1. As a result, the fastest growth direction of crystal O_2 will be blocked, and the entire crystal O_2 will ultimately be buried under crystals O_1 and O_3, whose maximum growth rate directions are closer to the normal to the substrate than that of O_2. In further growth crystal O_1 will be stopped by crystal O_3, in which the axis of the prism forms a smaller angle with the normal, etc. Thus the farther the growth front is from the substrate, the smaller the number of its constituent crystals is, and the closer is the maximum growth rate direction to the normal to the substrate. Thus geometric selection indeed leads to a columnar texture, i.e., to so-called orthotropism in the ingot crust. Quantitative experiments on geometric selection were conducted by *Lemmlein* on thymol crystals [7.9]. The thymol melt was placed between two glasses and supercooled. Then a piece of thymol was passed along one of the edges of a flat preparation and crystals were thus nucleated. Randomly oriented crystals grew into the melt layer, and no new nuclei arose in the interior of the liquid. When the growth concluded, the crystals intersecting lines parallel to the nucleation edge and spaced at different distances h from it were counted. It was found that in the flat preparation the linear population density of surviving crystals was $n(h)$ [1/cm] $\propto 1/h$. The probabilistic theory developed by *Kolmogorov* [7.10] leads to the same result for the two-dimensional case and to the law $n(h) \propto 1/\sqrt{h}$ for the three-dimensional case. In both laws the proportionality factor depends on the crystal shape.

As the ingot crust grows thicker, its thermal resistance increases, and the cooling of the central liquid part of the metal filling the mold slows down. Accordingly, the velocity of the growth front (polycrystalline in this case) decreases with time t as $t^{-1/2}$ (Sect. 5.1).

Metallurgists usually attempt to prepare fine-grained ingots, and for this purpose they treat the crystallizing melt, say with ultrasonic waves, to obtain a large number of nuclei in the ingot bulk. Irradiation with powerful ultrasonic waves produces cavitation bubbles, which collapse on the surface of the crystals, breaking them down and providing new crystallization centers. Intensive convection flows promote the cooling of the melt and increase the number of crystallization centers. These flows, whose velocity may reach tens of centimeters per second, break off the dendritic branches of the nucleated crystals and carry the debris (crystallization centers) through the entire volume, thus speeding up solidification. The number of nuclei can also be increased by introducing additives.

7.3 Heat and Mass Transfer

Heat transfer in a solidifying ingot was discussed above; it is effected by thermal conduction through the mold wall and the crust of the crustals adjoining it.

In the central, liquid part of the material, convective heat transfer takes place. This and related questions are elucidated in the abundant literature on metallurgy [7.11].

We shall now dwell briefly on the specific features of mass transfer during mass crystallization from solution. These features are associated with the complex motion of the solution with the crystals suspended in it. Each crystal in such a suspension follows a complex trajectory along with the flow. Owing to inertia, however, it also moves relative to the flow at places where the latter accelerates and decelerates, because the density of the crystal differs from that of the solution. For the same reason, the crystal velocity u with respect to the solution will be higher for heavier, large crystals than for lighter, small ones. The nature of the liquid's motion is also of great importance: the lower the turbulence scale, the higher is velocity u (sometimes called the slipping velocity). If the velocity u of the relative motion is known, the mass transfer rate depends on the effective thickness δ of the boundary layer near each crystal. For isometric particles, δ is estimated with the aid of (3.17), where x must be replaced by the characteristic size of the crystal, l (for a spherical crystal, by its radius). The resulting equation is sometimes used in the dimensionless form by introducing the coefficient of effective diffusion transport, $\beta_{tr} = D/\delta$:

$$\frac{l}{\delta} = \frac{\beta_{tr} l}{D} = \text{const}\left(\frac{\nu}{D}\right)^{1/3}\left(\frac{ul}{\nu}\right)^{1/2}. \tag{7.9}$$

Quantity $\beta_{tr} l/D$ is called the Sherwood number (Sh), ν/D the Schmidt number (Sc), and ul/ν the Reynolds number (Re). With the aid of these dimensionless relations, the characteristic rate of diffusion mass transport in a moving solution is also written [Ref. 7.1, p. 204] in the form of the Frössling relation generalizing (7.9):

$$\text{Sh} = 2 + \text{const Re}^{1/2}\text{Sc}^{1/3}, \tag{7.10}$$

where the addend 2 indicates the possibility of mass transfer in a stagnant solution (Re = 0), and the constant in (7.10) depends on the nature of the motion, the shape of the particles, and the characteristics of the particle–medium interface. Transforming (3.17) to the form of (7.9) gives const ~0.2. An analysis of the flow past a sphere leads to an expression of the type of (7.10), which is different from that for the flow past a plate. Accordingly, the constant in the Frössling equation (7.10), written for a sphere, takes the value const = 0.72, which agrees with experiment in the range of Reynolds numbers $20 < \text{Re} < 200$. The typical value is $\text{Sc} \simeq 10^3$, so that with intensive stirring of the solution, when $\text{Re} \simeq 10^3$, we can restrict ourselves to the second term in (7.10), thus returning to (7.9).

To be able to use the Frössling relation (7.10) during random mixing of the solution it is necessary to know, in addition to the constant in (7.10), the relationship between relative velocity u and the observed characteristics of the suspension being mixed. This relationship is discussed in [7.12–14]. One approach, based on the theory of isotropic turbulence, gives an idea of the nature of the above-mentioned relationship:

$$u = 0.055\,(l/\nu)^{1/3}(LP/M)^{4/9}(\nu_0/\nu)^{4/9}(\Delta\rho/\rho_c)^{2/3}\,, \qquad (7.11)$$

where slipping velocity u is expressed in m/s, crystal size l in m, kinematic viscosity ν and its value ν_0 at 25°C in m²/s, stirrer diameter L in m, its power consumption P in W, the solution mass in the crystallizer M in kg, solution density ρ_L and the difference between the crystal and solution densities $\Delta\rho$ in kg/m³. The method for checking relations of the type of (7.11) developed in works by *Hughmark* [7.12] and *Nienow* et al. [7.13] is as follows: First, special experiments are run to measure the rate of growth or dissolution of the crystals. The crystals are either fixed in a holder inside a moving solution, or are kept in a suspended state by an upwardly moving flow of the mother liquor (fluidized bed crystallizer). The velocity of the solution flow with respect to the crystal is readily measurable in experiments of both types, and the same growth rates are obtained for the same solution flow velocities u. Then crystals of the same substance are mass grown in the stirred suspension, and their growth rate is determined. The corresponding "slipping" velocity of the solution u relative to the crystal is calculated by checking expressions of the type of (7.11). Thus, the dependence of the growth rate on the flow velocity is obtained for experiments with measured and calculated growth rates. A comparison yields satisfactory agreement of (7.11) with experiment for systems consisting of boric acid and water, benzoic acid and water, and zinc and dilute hydrochloric acid. But for the growth of ammonium alum, (7.11) gives results which are correct only in order of magnitude. Here, the points having measured growth rates of alums for their ordinates and the slipping velocities calculated by (7.11) for their abscissas lie above the curves obtained in experiments permitting direct measurement of the flow velocity. The growth rate differs by ~1.5–2 times, and the slipping velocity by up to 3 times. The improvements in (7.11) introduced in [7.13] reduce the divergencies to ~30% and ~80%, respectively.

In conclusion we give, for the sake of orientation, the orders of magnitude of the slipping velocities calculated for ammonium alum. For particles with an average size of 470 μm at a stirrer rotation velocity of 3.2 rps (the diameter of the stirrer is 7 cm, the solution mass 2.3 kg) the slipping velocity is 3 cm/s according to (7.11), and 4.7 cm/s according to the improved equations of *Nienow* and co-workers. The growth rate is 0.12 kg/m²s at an absolute supersaturation of 0.01 kg of crystal hydrate per kg H_2O. At a rotation speed of 6.7 rps the calculated slipping velocities increase to ~8 cm/s and 7 cm/s, respecitively. The growth rate is 0.14 kg/m²s. For particles having a diameter of 1740 μm, at a rotation speed of 4 rps the results are 6.5 cm/s and 22 cm/s, respectively, and at 6.7 rps they are 12.7 cm/s and 28 cm/s. The growth rate for the last two rotation speeds at the same supersaturation is 0.18 kg/m²s.

7.4 Ripening (Coalescence)

Ripening (or coalescence[1]) is a gradual increase in the average size of the crystals in a saturated liquid or solid solution, or in equilibrium with the vapor or melt. It is a recrystallization process in which substance is transported from the smaller to the larger crystals. One of the first researchers of this phenomenon was *Ostwald*, who experimented with particles of HgO in aqueous solutions. Ripening reduces the total surface energy of the system, and this reduction is actually the thermodynamic driving force of the process (at least in ensembles of particles $\lesssim 10^{-2}$ cm in size). The coalescence theory was originated by *Ostwald* and *Smolukhovsky* and developed further by *Todes* [7.15].

The most general and consistent approach to the problem was developed by *Lifshits* and *Slyozov* [7.16], who obtained a rigorous solution for coalescence in dilute solid solutions. Let us consider an ensemble of particles of different sizes (small crystals in a solution or vapor, droplets in a cloud, pores in a solid, etc.). For the sake of definiteness we assume that the particles are spherical, that the space between them is filled by a solid solution, and that the average distances between the particles are much greater than the sizes of the particles themselves. If the growth or dissolution of the particles proceeds in the diffusion regime, the concentration of solution over each particle of radius R is defined by the Gibbs-Thomson formula

$$C_R = \bar{C}_0 \exp\left(\frac{2\Omega\alpha}{kTR}\right), \quad \text{or} \quad C_R \simeq \bar{C}_0 + \frac{2\Omega\bar{C}_0\alpha}{kTR}, \tag{7.12}$$

where $\bar{C}_0$ is the concentration of the saturated solution over the flat surface.

Each particle is in the field of concentration C that is created by all the other particles and that depends on the coordinate of the observation point. Because, however, the population density of the particles is assumed to be low, the value of C can be replaced by the average value $\langle C\rangle$, which is constant throughout the solution bulk with the exception of the immediate vicinity of each crystal, $\simeq R$, and depends only on the time: $\langle C\rangle = \langle C(t)\rangle$. Thus, in accordance with (5.8), where we assume $\Omega\bar{C}_0 \ll 1$ and $\beta R/D \to \infty$, we have

$$\frac{dR}{dt} = \frac{D\Omega}{R}\left[\langle C(t)\rangle - \bar{C}_0 - \frac{2\Omega C_0\alpha}{kTR}\right]. \tag{7.13}$$

---

[1] By coalescence we mean the joining of crystals whereby large crystals grow as a result of the dissolution of small ones. The terms "coalescence" and "ripening" are equivalent here.

The total statistical characteristic of the ensemble is distribution function $f(R, t)$, such that $f(R, t)dR$ is the number of crystals with radii in the range of R, $R + dR$ enclosed in a unit volume of solution. Accordingly, $\int_0^\infty f(R, t)dR$ is the total crystal population density at time t. The total number of atoms (molecules) which make up the crystals and which are in both the solid solution and the crystals, is (per unit volume of solution)

$$\langle C(t)\rangle + \frac{4\pi}{3\Omega}\int_0^\infty R^3 f(R, t)dR = \text{const} ; \tag{7.14}$$

i.e., the number is time independent throughout the ripening. In the ripening process small particles decrease in size, and large ones increase, so that the distribution function of particles with respect to their size is deformed with time and displaces to the right along the size axis, towards the larger sizes. This displacement, i.e., the shift of particles from one size to another, is described by the particle number balance condition:

$$\frac{\partial f}{\partial t} + \frac{\partial}{\partial R}(fV) = 0 , \quad V = \frac{dR}{dt} . \tag{7.15}$$

By adding the initial condition $f(R, 0) = f_0(R)$ to (7.13–15), we get a complete set of equations for $f(R, t)$ and $\langle C(t)\rangle$. The average crystal size is expressed via the distribution function

$$\langle R(t)\rangle = \int_0^\infty Rf(R, t)dR \Big/ \int_0^\infty f(R, t)dR . \tag{7.16}$$

The solution of (7.13–15) is rather complicated; therefore we do not give it here, but restrict ourselves to estimating dependence $\langle R(t)\rangle$, which we present below, focusing our attention on the process mechanism and using, broadly, the approach of *Lifshits* and *Slyozov* [7.16]. Corresponding to the supersaturation $\langle C(t)\rangle - \bar{C}_0$, which exists in the system at each moment, is critical size $R_c = 2\bar{C}_0\Omega\alpha/(\langle C(t)\rangle - \bar{C}_0)$. Particles with a below-critical radius $(R < R_c)$ will dissolve, and in accordance with (7.13) the rate of dissolution will be higher, the smaller the particle radii. Particles with an above-critical radius, on the other hand, will grow. Particles of critical size have zero growth rates at each moment, but the critical radius itself increases with time, since the average size of the particles in the system increases and the supersaturation drops off accordingly. It is intuitively clear, and can, indeed, be rigorously proved, that as the process continues, distribution function $f(R, t)$ ceases to depend on the initial distribution of $f_0(R)$; it "forgets" about it and is completely determined by the ripening process itself.

As was mentioned above, this process consists in the diminution, going as far as complete disappearance, of small particles, and in the advancement of large ones into the large-size region, with the boundary at $R = R_c$. Therefore the steady-state distribution function is dependent upon time only via critical particle size: $f(R, t) = f(R/R_c)$. Consequently, the average value of the particle radius raised to an arbitrary power k, i.e., $\langle R^k \rangle$, is expressed through the critical size and the kth moment M_k of the distribution function:

$$\langle R^k \rangle = R_c^k \left[\int_0^\infty x^k f(x) dx / \int_0^\infty f(x) dx \right] = R_c^k M_k / M_0 \; ;$$

$$M_k = \int_0^\infty x^k f(x) dx \, , \quad x = R/R_c \, . \tag{7.17}$$

Hence the average particle size ($k = 1$) in the system is proportional to the critical size. The distribution function reduces to zero at $R = 0$ and as $R \to \infty$, because small particles dissolve rapidly and leave the system, while infinitely large ones cannot appear within any finite time. Therefore, using (7.15), we obtain from (7.16)

$$\frac{d\langle R \rangle}{dt} = \frac{d}{dt} \left[\int_0^\infty R f(R, t) dR / \int_0^\infty f(R, t) dR \right] = \left\langle \frac{dR}{dt} \right\rangle . \tag{7.18}$$

Expressing supersaturation $\langle C \rangle - \bar{C}_0$ in (7.13) in terms of the critical nucleus radius R_c, we get

$$\frac{dR}{dt} = \frac{2D\Omega^2 \bar{C}_0 \alpha}{kTR} \left(\frac{1}{R_c} - \frac{1}{R} \right) . \tag{7.19}$$

We now average the expression obtained, multiplying both sides by $f(R/R_e)/\int_0^\infty f(R/R_e) dR$ and integrating with respect to R from zero to infinity. Taking (7.18, 17) into account, we have as a result

$$\frac{dR_c}{dt} = \frac{2D\Omega^2 \bar{C}_0 \alpha}{kTR_c^2} \, \frac{M_{-1} - M_{-2}}{M_1} \, . \tag{7.20}$$

If M_{-1} were less than M_{-2}, then according to (7.20), the critical (and hence the mean) size would reduce progressively with time, i.e., all the particles would dissolve with time. This means that the solution would change from being supersaturated ($R_c > 0$) to being undersaturated, which contradicts the law of conservation of matter (7.14) and is therefore impossible. Consequently, $M_{-1} \geq M_{-2}$. Equality is achieved only for a strictly monodisperse system, when all the particles have the same size $R = R_c$, i.e., $f(x) = \delta(x - 1)$, and do not grow larger. But in any real system ideal monodispersity is impossible, so that $M_{-1} > M_{-2}$.

Integrating (7.20) yields

$$R_c^3 = \frac{6(M_{-1} - M_{-2})}{M_1} \frac{D\Omega^2 \bar{C}_0 \alpha t}{kT} . \tag{7.21}$$

When a sufficiently long period of time has elapsed since the beginning of coalescence, particle size distribution function $f(R/R_c)$ ceases to depend on the initial distribution, as noted previously. A rigorous analysis shows that this asymptotic function is such that the critical radius is equal to the average radius, i.e., $R_c = \langle R \rangle$, and the numerical factor in (7.21) is equal to 8/9. Thus, in place of (7.21) we have

$$\langle R \rangle^3 = \frac{8 D\Omega^2 \bar{C}_0 \alpha t}{9kT} . \tag{7.22}$$

The above theory correctly describes the coalescence of pores in crystals of NaCl. Figure 7.2 shows the dependence of $\langle R \rangle^3$ on time t at a temperature of 500°C. The linearity of this dependence corresponds to (7.22), and the slope of the line in Fig. 7.2 gives the value of the coefficient $8D\Omega^2\bar{C}_0\alpha/9\ kT \simeq 7 \times 10^{-17}$ cm³/s. Putting $\alpha \simeq 500$ erg/cm², $\Omega = 1.15 \times 10^{-23}$ cm³ (volume of the vacancy in NaCl), $kT = 10^{-13}$ erg, we find that the reduced diffusion coefficient is $D\Omega\bar{C}_0 \simeq 10^{-9}$ cm²/s, which corresponds in order of magnitude to the diffusion along the grain boundaries: the coalescing pores were disposed along the grain boundaries in the experiment described.

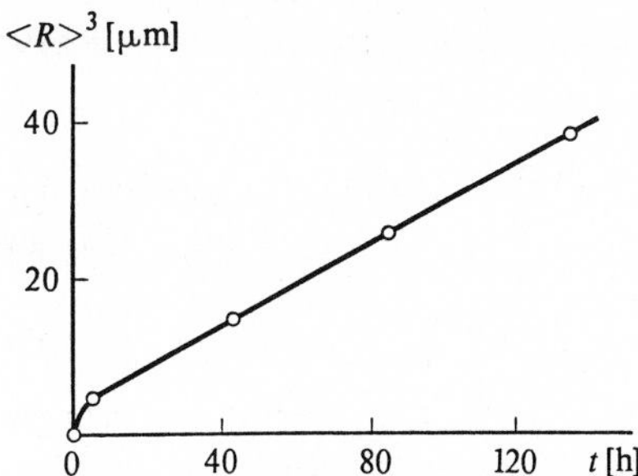

Fig.7.2. Dependence of the average size (cubed) of pores in a NaCl crystal on the time of annealing at 500°C [7.17]

The above reasoning is applicable to ensembles in which the particles are widely spaced. If the distances between particles are comparable with their sizes, or if the particles are in contact, it is less justified to introduce average solution concentration $\langle C \rangle$. Nevertheless, (7.22) can be used for rough estimates of the order of magnitude. The inversion of (7.22) yields the time in which the particles ripen to average size $\langle R \rangle$:

$$t = \frac{9kT\langle R \rangle^3}{8 D\Omega^2 \bar{C}_0 \alpha} . \tag{7.23}$$

The actual ripening time is shorter. The factor $1/R$ in the right-hand side of (7.19), which reflects the characteristic distance to which the diffusion field around the particle extends, must now depend on the size distribution of the particles, their packing, the thickness of the liquid film separating the crystals at sites of contact, etc. This factor must have an effectively larger value which in turn will lead to an increase in the diffusion flux between the particles and speed up the ripening as compared to that described by (7.22). There is no analytical analysis of the above-mentioned factors available as yet, but it is believed that the period of ripening in a dense ensemble of particles will be as many times less than (7.23) as the average interparticle distance is less than $\langle R \rangle$. Indeed, the diffusion flux between the particles will in this case be determined by interparticle gap width Δ rather than by particle size R. Consequently, repleacement $R \to \Delta$ has to be made in the first factor of the right-hand side of (7.19). Thereafter, the reasoning used to obtain (7.23) from (7.19) can be repeated. Assuming the numerical factor to be of the order of unity, as in (7.23), one gets an expression differing from (7.23) only by replacement $\langle R \rangle^3 \to \langle R \rangle^2 \Delta$. In other words, the ripening time in dense ensembles is Δ/R times shorter than in ensembles of sparse particles.

To obtain a numerical estimate of the ripening time in liquid solutions we use the following values in (7.23): $D \simeq 10^{-5}$ cm²/s, $\Omega \simeq 3 \times 10^{-23}$ cm³, $\Omega \bar{C}_0 \simeq 2 \times 10^{-1}$, $\alpha \simeq 50$ erg/cm², $kT \simeq 4 \times 10^{-14}$ erg ($T \simeq 300$ K). Then $t \simeq 1.5 \times 10^{13} \langle R \rangle^3$ s if $\langle R \rangle$ is expressed in centimeters. Thus, about 6 months are required for an ensemble of sparse particles suspended in a liquid to ripen to a size of $\sim 10^{-2}$ cm. For an ensemble where the interparticle spacing is $\sim 10^{-3}$ cm, the expected ripening time should be about one order of magnitude less ($\sim$20 days). The ripening time decreases very rapidly with the final size needed: ripening up to $\langle R \rangle = 10^{-3}$ cm should take only $\sim$4 h in a sparse ensemble of particles.

Let us now look into the problem of ripening in a liquid solution with stirring. We neglect the fact that stirring causes the small crystals to collide with one another and with the propeller and walls. (Such collisions result, in turn, in the rounding of the crystals, particularly because of strain hardening and subsequent etching of the damaged sites [7.17b]). It is realistic enough to neglect collisions if the crystals are small and occupy a volume appreciably smaller than that of the solvent.

In estimating the transformation time with stirring, we assume that each crystal is fed through a boundary layer having thickness δ, which is defined by (3.17) or (7.9, 10). Then, in a diffusion (rather than kinetic) regime of growth and dissolution, (7.19) must be rewritten

$$\frac{dR}{dt} = \frac{2D\Omega^2\bar{C}_0\alpha}{kT\delta}\left(\frac{1}{R_c} - \frac{1}{R}\right). \tag{7.24}$$

Repeating the reasoning which led from (7.19) to (7.21) and using (3.17) with $x = 2R$, we find that

$$R_c^{5/2} = \frac{5}{4,5\sqrt{2M_1}}[M_{-1/2} - M_{-3/2}]\frac{D\Omega^2\bar{C}_0\alpha t}{kT(D/\nu)^{1/3}(\nu/u)^{1/2}} . \quad (7.25)$$

Since (7.24) is different from (7.19), the limiting asymptotic function of particle size distribution in a strired solution must be different in type from that in purely diffusional transfer. Therefore equality $R_c = \langle R \rangle$ may not hold, but according to (7.17), one can write $R_c^{5/2} = M_0 \langle R^{5/2} \rangle / M_{5/2}$, and from (7.25) follows the time dependence of $\langle R^{5/2} \rangle$. We assume approximately that here, too, $R_c = \langle R \rangle$, and that as before, the numerical coefficient is near unity in order of magnitude. Then the time of ripening to average size $\langle R \rangle$, which considerably exceeds the initial size, can be estimated in order of magnitude from the relation

$$t \simeq \frac{kT\nu^{1/6}\langle R \rangle^{5/2}}{D^{2/3}\Omega^2 C_0 \alpha u^{1/2}} .$$

For $\nu = 10^{-2}$ cm^2/s, $u = 10$ cm/s (vigorous stirring) and for the above-indicated values of the other parameters, we have $t \simeq 4.2 \times 10^{10} \langle R \rangle^{5/2}$ s (if R is expressed in cm). Thus ripening to $\langle R \rangle \simeq 10^{-2}$ cm requires $\sim$ 5 days, i.e., much less time than in a stagnant solution.

Isothermal ripening to sizes of the order of millimeters takes quite a long time: to obtain crystals with $\langle R \rangle \simeq 0.5$ cm in a stirred solution one would need over 230 years! Indeed, under isothermal conditions virtually no ripening takes place in suspensions consisting of particles over 10^{-2} cm in size.

An altogether different picture is observed in systems which are isothermal only on the average, with the temperature undergoing accidental or artificially induced oscillations. For instance, it takes ammonium alum crystals about 10 h to ripen to $\sim$ 1 cm in a boiling saturated solution. Indeed, measurements in such experiments show that the temperatures of the convecting solution near the heater differ from those near the free surface by approximately 10 K, and thus temperature oscillations must occur.

Figure 7.3 shows a curve for the increase in the average weight of small crystals of Rochelle salt with time in a slowly stirred solution whose temperature was periodically raised and lowered artificially with an amplitude of 0.6 – 0.7 K and a period of $\sim$ 1.3 h. In these experiments by *Gordeyeva* and *Shubnikov* [7.18], the solution was placed in a hermetically sealed crystallizer, so that

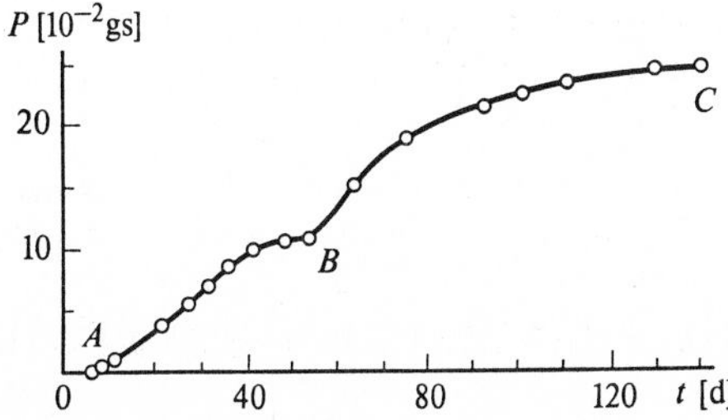

Fig.7.3. Increase in the average weight P of small crystals of Rochelle salt with time t under conditions of periodically varying temperature. At the moment corresponding to point B, a new portion of small crystals was added

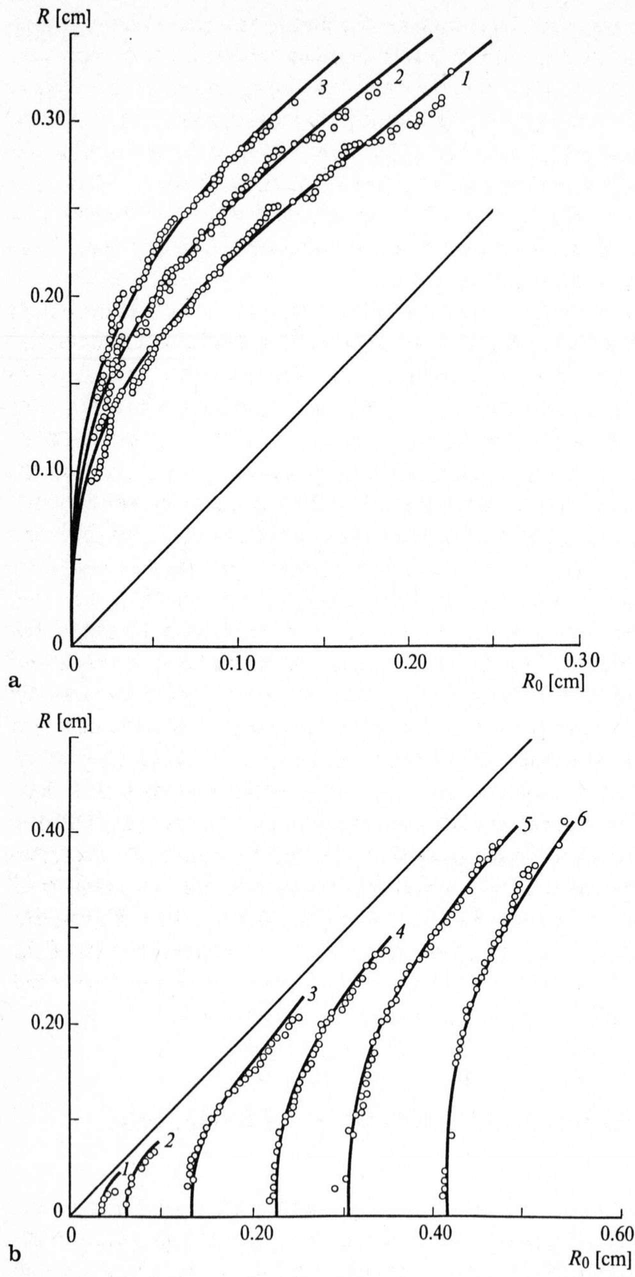

Fig.7.4a,b. Dependence of the effective radius R of small crystals of ammonium alum on their initial size R_0 (**a**) after growth for 26, 34.5, and 43.4 h (*curves 1, 2,* and *3*, respectively) and (**b**) after dissolution for 1.2, 1, 5, 10, 18, and 30 min (*curves 1–6*) [7.20]

the solvent could not evaporate. The absence of evaporation was also confirmed by the fact that the total weight of the crystals remained unaltered after recrystallization. The initial material was a fine powder obtained by crushing the crystals in a mortar. After 6 days the crystals had reached an average size of ~ 1 mm, and after 50 days they grew to ~ 3 mm. Their number decreased accordingly. Fifty-five days after the experiment was started (point B in Fig. 7.3), a new portion of suspension with a fine powder was added to the crystallizer. As a result, the sizes of the existing crystals increased further at the expense of the added fine particles which disappeared.

In experiments conducted by *Bazhal* [7.19, 20], each of several hundred small ammonium alum crystals sized 0.2 – 2 mm was inserted into a separate cell in the form of a through hole in an organic glass disc. Kapron sieves of high penetrability to solution were drawn over the holes on both sides. The disc was placed in a saturated, stirred solution. A thermostat raised or reduced the temperature linearly, thus making the solution undersaturated or supersaturated. Each crystal was weighed before and after the dissolution or growth experiments. The results were used to find the effective radius R, i.e., that of a spherical crystal equal in weight to the actual crystal. The dependences of radius R after growth or dissolution experiments on initial radius R_0 are presented in Fig. 7.4. The bisector of the coordinate angle corresponds to the absence of growth or dissolution: $R = R_0$. Figure 7.4a, b gives the curves for different growth and dissolution times, respectively, both for different ranges of initial sizes. If the growth and dissolution rates were size independent, we should have obtained straight lines parallel to the bisector of the coordinate angle and passing below it (dissolution) or above it (growth), rather than the curves depicted in Figs. 7.4a, b. Deviations from linearity show that the smaller the crystal, the faster it dissolves, but the slower it grows. The effect disappears with vigorous agitation, i.e., in the kinetic regime of growth or dissolution. The mechanism of ripening with temperature fluctuations has not yet been established. The experiments also showed that the rate of ripening (increase in the average size of the crystals) grows linearly with the frequency of temperature fluctuations, and increases with their amplitude.

7.5 Principles of Industrial Crystallization of Nonmetals

Industrial crystallization of nonmetals — salts, oxides, or organic substances — is aimed primarily at obtaining a product in the form of a crystal powder having the desired purity, and grains of a specific size and shape. The grain may be either monocrystalline (common salt, sugar) or polycrystalline (ammonium nitrate used as fertilizer). In growing crystals for this purpose the same types of raw materials are used with various solvents. A further goal in growing such crystals is to obtain new products by chemical reaction.

Another common application of mass crystallization is in the purification of substances, i.e., the separation of components. The simplest system of removing impurities is as follows: The nutrient is dissolved in a pure solvent, such that the coefficient of impurity distribution between solution and crystal is less than unity ($D < 1$). The solvent may be selected by using Ruff's rule (4.29). Crystals grown from a prepared solution have a lower impurity content than the initial reagent. The product of this first purification stage is then dissolved again in a pure solvent, crystallized once more, etc. The efficiency of purification is closer to the maximum possible at a given D, the smaller the number of inclusions and aggregates in the resulting product. Therefore the efficiency increases with reduced supersaturation and also with the intensity of stirring. The powder obtained is usually rinsed.

Since the mother liquors spent in the different stages are not reused in this method, the product yield is not high. The efficiency is increased by using solutions spent in the later stages as solvents for the ealier ones, in which the material to be purified contains a larger amount of impurities.

Crystallization installations for purifying the product and obtaining powders of the desired dispersity from the initial reagent of the same composition are based on various supersaturating methods. The most widespread methods are evaporation and cooling, which form the basis for the design of the corresponding crystallizers. In order to attain, within a minimum amount of time, high supersaturations leading to the formation of a fine-grained product ($\sim$0.1 mm), so-called vacuum crystallizers are used. The supersaturation is achieved by adiabatic evaporation of part of the solvent into a vacuum (25–40 torr). The whole mass of the solution comes to a boil (the boiling point of water at the indicated pressure is 24 – 34°C), so that the loss of solvent and cooling occur macroscopically throughout the bulk, and not only at the surface.

The solution can also be supersaturated by adding substances to it which reduce the solubility of the compound involved. This is called "salting out". Figure 7.5 illustrates the drop in solubility of some salts in aqueous solutions with an

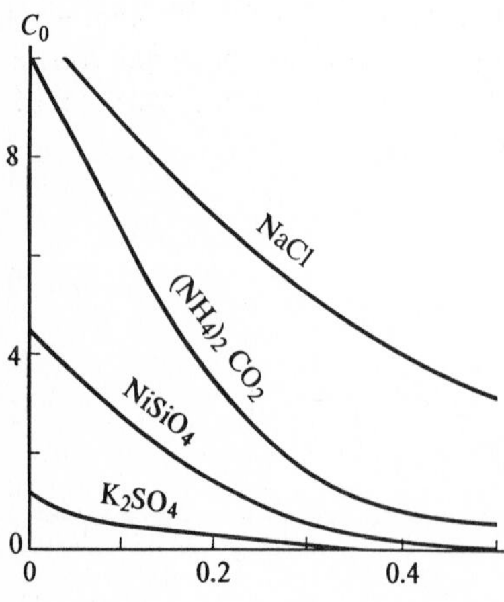

Fig.7.5. Dependence of the solubility of various salts on the amount of methanol added to the solution. The x axis indicates the methanol concentration in molar fractions in the methanol–water solvent, and the y axis the number of moles of the anhydrous salt per 100 g water [7.1]

increase in the content of methanol. Precipitating alum from aqueous solutions by adding ethanol to them yields an iron-free product. Conversely, in growing crystals of many organic substances from solutions in organic solvents, salting out is achieved by adding water.

The sizes of the small crystals obtained essentially depend on the ratio of the rates of crystal nucleation and growth. In industrial installations, crystallization takes place either spontaneously or on specially introduced tiny seeds. In the production of granulated sugar, for instance, the introduction of 500 g of $\sim 5\ \mu m$ crystal powder into a supersaturated solution yields 50 m^3 of crystallized product.

In spontaneous crystallization the effective nucleation rate increases with the intensity of stirring, a fact which is evidenced by the reduction in the average size of the crystals obtained [7.2]. For instance, an increase in the number of revolutions of an agitator from 0.06 to 18.9 rps reduces the average size of $NaNO_3$ crystals from 10^{-1} to 2.5×10^{-2} mm.

The increase in the effective nucleation rate may be partly due to so-called secondary nucleation, which is particularly prominent in mass crystallization with stirring. In secondary nucleation the probability that new nuclei will appear is higher in the presence of small crystals which have already formed than in a metastable solution free of such crystals. This is attributed to mechanical disintegration of crystals rubbing against each other, but it is not clear whether this is the only cause [7.21 – 23].

In tower crystallizers, a product of the required grain size distribution is obtained by quite a different method. Here, the initial (usually highly concentrated) solution or melt is sprayed in the upper part of the tower. The deposited droplets cool down and lose the solvent through evaporation (with solutions). This results in monocrystalline or polycrystalline granules. Their structure and size affect the future quality of the product, and in particular its tendency to cake, which can cause grave complications, for example in the use of fertilizers.

An example of a process in which the crystals obtained are not a recrystallization of the reagent, but have a new composition, is the production of $(NH_4)_2SO_4$ (ammonium sulfate). The ammonia formed in the coke oven is directed into a saturator with sulfuric acid, where the crystal form under conditions of vigorous agitations resulting from the reaction.

8. Growth from the Vapor Phase

The statistics in publications on crystal growth show that a large portion (~25% – 30%) of the papers in this field are devoted to the growth of crystals and films from the vapor phase. Together with crystallization from melts or solutions, crystallization from the vapor phase can be considered one of the most commonly used methods of crystal growth, particularly in semiconductor electronics. This chapter gives a systematic account of the foundations of the methods generally used to grow crystals from the vapor: physical vapor deposition (molecular-beam method, cathode sputtering, vapor phase crystallization in a closed system, gas flow crystallization); chemical vapor deposition (chemical transport, vapor decomposition, vapor synthesis); and crystallization from the vapor via a liquid zone. Growth mechanisms and the typical parameters of growth processes and of the grown crystals are also discussed for each technique.

8.1 Overview

Numerous methods of crystallization from the vapor phase have been developing intensively in recent years in connection with the requirements of modern technology and in particular electronics. A common feature of all of these methods is the need to deliver ("transport") material from some localized source; this is a consequence of the low concentration of the crystallizing substance in the medium. The location of the starting material is usually called the source zone, and the site at which it is deposited, the crystallization zone.

Crystallization from the vapor phase is widely used for growing bulk (usually isometric) crystals, epitaxial films, thin (polycrystalline or amorphous) coatings, and filamentary and platelet crystals. The choice of a concrete growing method depends on the material to be obtained. Some methods are specialized, and suitable only for growing a particular kind of crystal. Cathode sputtering, vacuum deposition from molecular beams, and close-spaced (or sandwich) methods are used for the preparation of films, whereas crystallization in a sealed ampule is more suitable for obtaining bulk crystals. Other methods offer more possibilities. Thus, the very widely used methods of chemical vapor deposition are applicable in the production not only of films, but also of bulk or filamentary crystals (or whiskers), but they are most frequently used in the preparation of films and whiskers.

Growing bulk crystals from the vapor phase is attractive because of the versatility of the methods: for practically any substance, at least one process (and sometimes several different ones) can be chosen which will ensure the growth of a single crystal.[1] The low growth rate resulting from the low concentration of the substance in the medium is naturally a drawback from the standpoint of economic efficiency, but the above-mentioned versatility, which is often combined with high purity, uniformity of composition, and perfection of structure in the resulting crystals, ensures the wide application of these methods in science and technology. Furthermore, the low growth rate typical of vapor phase methods proves an advantage in growing epitaxial films, particularly in producing multilayer structures. To obtain such structures, which consist of very thin layers (~ 100 Å $- 1$ μm) of differing composition, each layer must be grown separately. This can be done only at a very low rate of the process. As a result, even in the crystallization stage, devices can be produced which are almost ready for use. These advantages are responsible for the revolution in semiconductor technology that took place in the sixties.

The different requirements imposed on the growth rate in the production of bulk crystals and films mainly affects equipment design; in principle, the processes for growing differently shaped crystals have much in common. Indeed, so-called autoepitaxial (or homoepitaxial) growth of films does not differ essentially from ordinary growth of bulk crystals on a seed, while heteroepitaxial growth of films differs from autoepitaxial and ordinary growth only in some features of the initial stages (Sect. 8.2.4). Accordingly, there are many common modes in different techniques of growing variously shaped crystals from the vapor phase. But the most comprehensive and reliable information is obtained from investigations of atomic phenomena during the growth of epitaxial and in particular autoepitaxial films; this information helps to establish the crystallization mechanism and suggest process control methods. Such growth usually occurs on a definitely oriented single-crystal substrate, and the micromorphology of the growing surface is a highly sensitive indicator of atomic phenomena. Therefore, further on in this chapter the basic concepts of crystal growth from the vapor phase are illustrated with examples taken predominantly from the growth of epitaxial films.

Because the methods of crystallization from the vapor phase are so abundant, it is important to find a basis for classifying them. We distinguish two principal groups. In one, the methods are based on purely physical condensation, and in the other, a chemical reaction is involved which produces the crystallizing substance. This division is valid only as a first approximation, and does not exclude combined procedures. For example, the molecular-beam method (which is, in principle, a "physical" method) is used in different versions, includ-

[1] Note, for the sake of comparison, that by no means all substances can crystallize from the melt, because some sublimate, while others decompose at temperatures much below the melting point.

ing gas etching of the growing crystal (Sect. 8.3.1) or decomposition of molecules, emitted by a source, on the crystal surface [8.1]. The method by which the substance is supplied to the growing surface may serve as a basis for a more detailed classification within each of the two basic groups of methods (Sects. 8.3, 4).

8.2 The Physicochemical Bases of Crystallization from the Vapor Phase

A feature specific to vapor growth methods is that there is an adsorption layer at the crystal surface that (Sect. 1.3) differs in composition and properties from both the crystal and its environment. Accordingly, two stages of surface processes are distinguished: transition of the substance from the vapor to the adsorption layer, and incorporation directly into the lattice.

The phenomena occurring in the adsorption layer are not yet completely understood. It is usually assumed that migration and collisions of atoms and molecules take place there, and that associations of them are formed ("complexes" or "clusters"), as well as two-dimensional nuclei (with a thickness equal to one lattice spacing) and three-dimensional (thicker) nuclei.

The nature of surface processes depends on the properties of the adsorption layer and ultimately on its composition and density. The composition is obviously different for the above two basic groups: in physical condensation in an ideally pure medium, the adsorption layer has a composition identical to that of the crystallizing substance, while in chemical processes it contains other components. The layer density, or the amount of substance per unit area, depends on the surface activity and the density of the incident flux, or on the concentration of the substance in the medium.

8.2.1 Surface Activity and the Preparation of Substrates and Seeds

When crystals are grown from the vapor phase on a crystal surface of the same or a foreign substance, on some surfaces nucleation of new crystals occurs at many points even on small deviations from equilibrium, and the crystals formed are quite perfect. On other surfaces, however, growth begins only at high supersaturations and results in imperfect crystals. In the former case one speaks of active, and in the latter, of passive surfaces. The activity is evidently determined by the density of free chemical bonds. It depends on the structure (equilibrium or nonequilibrium) of the crystal faces and on the action of the various impurities forming the liquid phase or the adsorption coatings; it can, in addition, depend on the concentration of point defects in the layer underneath the surface of the crystal.

The structure of the crystal faces is determined by the density of the active elements — steps and kinks. In the ideal crystal, the equilibrium structure of a face depends on its orientation and on the temperature; on a surface in contact with its own vapor, the step density on a smooth face does not change substantially (i.e., no roughness develops) up to the melting point (Sect. 1.3). In a real crystal, the activity is higher because of the emerging defects (dislocations, twins, etc.) High supersaturations can result in kinetic roughness (Sect. 3.3.3), which also increases surface activity.

Impurities may change (reduce or increase) the activity considerably. For instance, in crystals of elementary substances — semiconductors and metals — the surface activity drops abruptly under the effect of oxidizing impurities, which primarily poison steps and kinks (Sect. 4.1). On the other hand, activization of the surface of the substrate or seed is achieved in practice by introducing foreign substances into the crystallization medium or by coating the growing surface with such substances. For instance, the surface of oxides (Al_2O_3, MgO, etc.) is rather inert at typical temperatures of film deposition ($\sim 1000°C$), but its activity can be raised by a thin layer of metal. Thus, if we evaporate a ~ 50 Å film of tantalum on sapphire (such a film is not continuous) through a mask, we can choose, during subsequent evaporation of silicon, a density of the incident flux such that crystallization will occur only on the portions with the previously deposited metal, that is to say, we can produce a definite "pattern" on the substrate. Another method of surface activization is by deposition of microscopic droplets ($\lesssim 500$ Å) of liquid metal; metals are usually chosen which do not substantially affect the electrophysical properties of the growing crystals, e.g., tin for germanium, gallium for gallium arsenide, etc. The activization mechanism is evidently attributable to the fact that the surface of the liquid phase, being ideally rough, has a large number of free chemical bonds, and therefore vigorously adsorbs the crystallizing substance. It is quite possible, moreover, that the atomic particles of the activizer, which "spread" over the crystal surface in the vicinity of the droplet, exert a catalytic effect.

The activity of the substrate can also be increased by raising the concentration of point defects in it, for instance by irradiation with x-rays or bombardment with fast ions (the latter naturally accompanies the process of ion implantation, which has recently found application in microelectronic technology).

It should be pointed out that these procedures for activizing inert surfaces are of particular importance in chemical as opposed to physical vapor deposition (Sect. 8.4), because the decomposing compounds are usually adsorbed much less than the vapor of the crystallizing substance.

Surface activity is highly temperature dependent. As the temperature decreases, the ratio of the substance crystallizing on the surface to the substance delivered (i.e., the condensation coefficient) increases as a result of the increase in the lifetime τ_s of the adatoms on the surface. The time τ_s depends exponentially on the temperature:

$$\tau_s = \tau_0 \exp\left(\frac{\varepsilon_s}{kT}\right). \tag{8.1}$$

Here, τ_0 is some constant, ε_s is the heat of desorption, and k and T have the usual meaning. The decrease in lifetime with increasing temperature explains the results of experiments by *Khariton* and *Shal'nikov* [8.2] in which they achieved condensation from the molecular beam only at a sufficiently low substrate temperature.

The temperature dependence of the activity (and of the condensation coefficient in particular) is very often related to the change with temperature in the amount and state of the impurity on the growing surface. The impurity effect may be responsible for the practically observed situation where during the condensation of atoms on the metallic surface, the condensation coefficient first decreases, and then increases, with rising temperature. This is attributed to the presence, on the metallic surface, of a deactivating oxide film, which is removed at a sufficiently high temperature. In other words, the surface is "refreshed" and becomes more active.

The preparation of substrates is one of the decisive factors in obtaining perfect crystals from the vapor phase. Substrate preparation usually amounts to the activization of those surfaces which should per se be active, but are poisoned by impurities.

A standard activization technique is high-temperature annealing. Its efficiency is higher, the purer the atmosphere is. Annealing is usually carried out in a high, or better, ultrahigh vacuum ($\sim 10^{-8}$–10^{-10} torr), in an atmosphere of purified (especially oxygen-free) inert gases, or in a reducing medium, e.g., hydrogen. The temperatures at which it is conducted are often higher than the crystallization temperature.

Two procedures for surface activation which are more effective and widespread are gas etching (usually with the aid of an $HCl + H_2$ mixture) and ion etching. In these, a surface layer of the substrate, ~ 1 μm thick, is usually removed thus "refreshing" the substrate. The etching is usually done in situ, i.e., in the crystallization chamber immediately before deposition.

8.2.2 Particle Flux Density in a Molecular Beam. Concentration of the Substance in the Medium

In growth from the vapor phase, it is usually the atoms or molecules constituting the adsorption layer that form the crystal, and hence growth depends on the supersaturation in this layer. This supersaturation differs, however, from that in the crystallization medium just above the growing surface, and that supersaturation may, in turn, differ from the average supersaturation in the bulk of the medium. Therefore growth conditions are usually chosen empirically. In such cases the controlled parameter determining the concentration of the substance in the adlayer is thedens ity of the substance flux in the molecular beam incident on the surface, or the concentration of the crystallizing substance in the medium. Various characteristics of the crystals and films to be obtained usually serve as the criteria for choosing the parameters of the crystallization process. Structural perfection of the crystals is the most objective and widely used criterion. It answers the requirements of practical utilization: as a rule, single-crystal, perfect films with a smooth surface are required, although fine-grained dense layers are sometimes necessary. The crystallization parameters may be chosen by evaluating the quality of the films on the basis of physical properties such as the mobility and concentration of charge carriers in semiconductors; usually high mobilities and low carrier concentrations are desired. In some cases the decisive criterion is a high deposition rate, even at the expense of the quality of the crystals.

8.2.3 Structural Perfection of Crystals. Minimal, Maximal, and Optimal Supersaturations. Epitaxial Temperature

Three types are distinguished in the structural state of solids. These are the single crystal, the polycrystal, and the amorphous phase. They are characterized by three kinds of electron diffraction patterns: Kikuchi lines or symmetric point reflexes, point reflexes imposed randomly in rings, and diffuse rings.

The conditions under which crystals form may substantially affect the perfection of the growing layers. Suppose that we deposit a film on a single-crystal, ideally perfect substrate. Under optimum crystallization conditions, we can hope to reproduce the perfection of the substrate. Under nonoptimum conditions, dislocations, stacking faults, microtwins, and other structural imperfections arise, but the layer grown is still considered to be a single-crystal layer: the electron diffraction patterns are composed of Kikuchi lines (additional reflexes can appear only at a high density of microtwins).

The perfection of such layers is usually characterized based on the dislocation density; one distinguishes between dislocation-free crystals, low-dislocation crystals (density of dislocation $\sim 10^2$–10^3 cm^{-2}), and crystals with a high dislocation density ($\sim 10^6$ cm^{-2}). In the last case the dislocation etch pits, observable in an optical microscope, usually overlap each other.

As the crystallization conditions deteriorate, crystals with a high dislocation density may be transformed into mosaic crystals (the Kikuchi lines on the electron diffraction patterns disappear, and symmetric point reflexes appear), and then into polycrystals. A texturized polycrystal (with a preferential grain orientation) may be formed as an intermediate version: the points forming the rings in the electron diffraction patterns gather into groups.

Together with an increase in dislocation density or independently of it, the density of stacking faults and microtwins may also increase. Microtwins are especially typical of high-symmetry (e.g., cubic) structures and of relatively high crystallization temperatures at which a plastic flow becomes possible in the crystals. As the density of stacking faults and microtwins increases with deteriorating crystallization conditions, polycrystals are also formed.

On the basis of the criterion of structural perfection, we shall point out the ranges of supersaturations and temperatures in which crystals with certain structural characteristics can be grown.

Let us first consider minimum supersaturations. Two cases should be distinguished here: (a) growth on a fresh surface, and (b) growth on a poisoned surface.

In the former, minimum supersaturations are determined by nucleation, and depend on the face structure (smooth or rough), the presence of imperfections (emergence of screw dislocations or twin boundaries) and the concentration of point defects. For different substances, media and temperatures, these minimum supersaturations vary within a comparatively small range, from fractions of one to several per cent.

In the latter case, minimum supersaturations may be considerably higher due to the necessity of suppressing the effect of impurities that poison crystal growth. For instance, during the crystallization of germanium from molecular beams in vacuum, residual gases, and especially active components of them such as oxygen, present a grave hazard. In order for residual gases not to hinder ordered (oriented) growth, their density in the adsorption layer must be at least one order less than that of the crystallizing substance itself. If we assume that the condensation coefficients for both oxygen and germanium on germanium are close to unity, then in a vacuum of $\sim 10^{-5}$ torr the flux of germanium atoms must be equivalent to a vapor pressure of $\gtrsim 10^{-4}$ torr. If the substrate is heated even to temperatures close to the melting point, the equilibrium pressure of the germanium vapor is only $\sim 10^{-7}$ torr. Since the supersaturation is defined here as the ratio of the incident and evaporating fluxes, I_{inc} and I_{ev}, and these are proportional to the respective pressures, the supersaturation is equal to $\sim 10^{-4}/10^{-7} = 10^3$, which is a rather high value.

The maximum supersaturations at which oriented (i.e., single-crystal) growth is still possible are higher, the more intensive the migration of adatoms and the higher the surface activity. The high surface activity enables a large number of atoms to join the crystal, and the intensive migration permits atoms incorrectly built into the lattice to shift rapidly into the required position. Both the migration of atoms and the creation of active sites on the surface are activated processes, and hence strongly depend on the temperature. Therefore as the temperature rises, the maximum permissible supersaturations must increase abruptly. The maximum permissible growth rates must increase to the same extent, since at high supersaturations they are proportional to the supersaturations.

The conclusion that the range of permissible supersaturations widens considerably with temperature is confirmed by experiment, but the numerical values of supersaturations differ substantially for the physical and the chemical methods.

Let us first consider the methods based on physical condensation.

If we suppose that the condensation coefficient is unity and define supersaturation as I_{inc}/I_{ev}, then at sufficiently high temperatures the permissible supersaturations may reach a high value for a number of substances. For instance, again in the case of germanium, single-crystal layers can, at $\sim 800°C$, be grown in a vacuum at the rate of ~1 μm/min (and this is by no means the limit), which corresponds to supersaturations of $\sim 10^6$. Moreover, single-crystal germanium layers were grown under similar conditions at a rate of ~1 μm/s, i.e., almost two orders higher. (There is, however, no assurance that the mechanism of growth itself will remain unaltered at such flux densities, since a quasi-liquid layer may form on the surface.)

The ranges of permissible maximum supersaturations strongly depend on the purity of the medium and the substrate. We mention in this connection the existence of the *epitaxial temperature*, which is the lowest temperature at which oriented growth is possible for a given degree of purity (Sect. 2.3). For instance, in silicon crystallization by evaporation in a vacuum of $\sim 10^{-5}$ torr, the epitaxial temperature is ~1000°C, and in a vacuum of $\sim 10^{-7}$ torr it is about 600°C. In an ultrahigh vacuum of $\sim 10^{-10}$ torr, epitaxial Si films were obtained on face (111) at 400°C, and on face (100) even at 350°C [8.3].

A set of experimental data on crystal growth from molecular beams is shown in Fig. 8.1. The curves denote different purity levels in the system, with the dashed lines characterizing the minimum supersaturations and the solid lines the maximum supersaturations. The corner points correspond to the epitaxial temperatures.

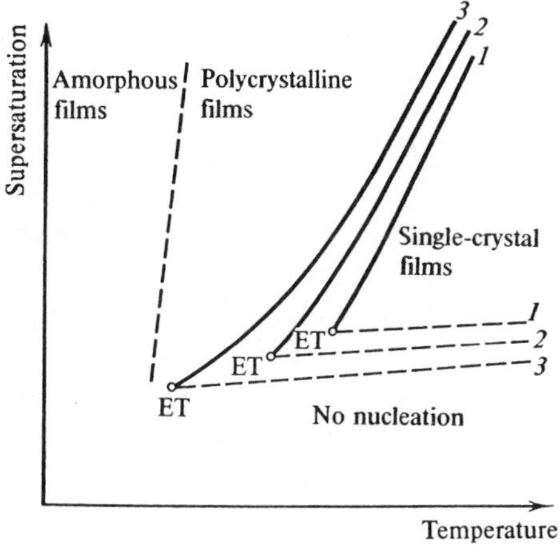

Fig.8.1. Supersaturation and temperature ranges in which layers with different structural characteristics are obtained. *ET* denotes the epitaxial temperature. The *solid curves* (*1–3*), which separate the regions of polycrystalline and monocrystalline growth, and the *dashed lines* (*1–3*), which separate the region of monocrystalline growth from the region where nucleation does not occur, correspond to different levels of purity in the crystallization medium and/or substrate. The purity improves from (*1*) to (*3*)

Note that at supersaturations exceeding the maximum for epitaxial growth, either polycrystalline (at high temperatures) or amorphous (at relatively low temperatures) films can be formed. As the temperature rises, amorphous films lose their stability and become crystalline. The transition temperatures differ depending on the impurities, supersaturation, and the method of deposition (molecular beams, cathode sputtering, etc.). According to various authors, they equal ~350°C for Si, from 200°C to 350°C for Ge, from 220°C to 400°C for GaAs, etc. [8.3].

The situation is somewhat different for chemical vapor deposition. Here, supersaturation is determined via the reaction equilibrium constants, but at a given supersaturation the growth rate also depends on the absolute concentrations of the reactants. Therefore, even at supersaturations which are not too high, the density of the adsorption layer may be so high that a polycrystal will grow rather than a single crystal, especially since the layer contains other components in addition to the crystallizing substance.

Another feature of the chemical methods is that with a rise in temperature the chemical reaction, which was previously kinetically retarded, may be accelerated to such an extent that the growing surface is unable to consume the entire substance, and a polycrystal is formed.

For these reasons, the relation between the extreme supersaturations and the temperature is not so simple in chemical methods as in physical vapor deposition. Nevertheless, the main trends in these two groups of methods are the same:

a) an abrupt extension in the range of permissible supersaturations (for oriented growth) with a rise in temperature;
b) the existence of an epitaxial temperature; the purer the growth conditions, the lower the temperature;
c) the important role of impurities, especially oxide impurities, which are capable of poisoning the active points of the surface.

In view of the ambiguity of supersaturation as a criterion for choosing the optimum crystallization conditions for perfect growth, it is advisable to proceed from typical growth rates, which change little from one material to another but depend substantially on the growth temperature and on the crystallization method used (Sects. 8.3–5).

8.2.4 Heteroepitaxial Growth

The foregoing pertained largely to growth on substrates of a material identical to the crystallizing substance, that is, autoepitaxial (or homoepitaxial) growth. In practice, one often has to deposit a substance on a foreign, sometimes single-crystal ("orienting") substrate. Disregarding deposition on an amorphous or polycrystalline substrate, a case where ordered growth is less probable, we shall briefly consider heteroepitaxial or oriented growth on a foreign single-crystal substrate [8.4].

At least three new factors enter here as compared with autoepitaxial growth: differences in the lattice parameters, in the types of chemical bonds, and in the thermal expansion coefficients of the substrate and film. The first two factors manifest themselves in the initial stages of growth, hindering nucleation in the first layer, and the third factor is in evidence in the cooling stage, causing imperfections, primarily microtwins and stacking faults.

A typical example of heteroepitaxy is the deposition of silicon on sapphire. Silicon is a typically covalent crystal, and sapphire a typically ionic one; therefore the formation of bonds between the substrate and the film is hindered.

It is believed that growth is facilitated by replacing the aluminum atoms on the surface by silicon. Three types of optimum heteroepitaxial relations have been established which ensure single-crystal growth: (111) Si‖(0001)Al_2O_3, (100)Si‖(10$\bar{1}$2)Al_2O_3, and (110) Si‖(11$\bar{2}$0) Al_2O_3. Heteroepitaxial silicon films always contain a large number of imperfections, mainly microtwins, and they cannot, in principle, be eliminated completely, although their number can be reduced to a certain level.

Numerous investigations, including electron-microscopic ones, have shown that in the early stages of heteroepitaxy, island-like films develop, evidently through the formation of three-dimensional nuclei (a similar picture is also cal of autoepitaxial growth if the substrate surface is not clean enough). In typisucceeding stages the islands coalesce into a continuous film, and the subsequent crystallization process does not differ essentially from the autoepitaxial process.

8.2.5 Oriented Crystallization on Amorphous Substrates

Many materials have been prepared successfully as thin films by means of both homo- and heteroepitaxial methods. However, many problems remain unsolved. First, by no means every material is available as a single-crystal substrate for epitaxial growth. Second, single-crystal substrates are, as a rule, expensive. Finally, there are a number of applications (e.g., three-dimensional integrated circuits, optoelectronics, integrated optics, etc.) in which multiple-layer structures with amorphous layers alternating with single-crystal intermediates are necessary or, at least, desirable.

Many attempts have been made and many approaches developed to solve these problems, with differing degrees of success. Fundamental phenomena in the nucleation and growth of crystals, and technical improvements in crystallization processes have proved important in this regard [8.5a].

Artificial Epitaxy, or Graphoepitaxy. In this approach, a surface pattern of micron or submicron scale is used to orient solid films on amorphous substrates [8.5b]. The pattern is most effective if its symmetry corresponds to the symmetry of a close-packed plane of the material to be crystallized. Single-crystal silicon films have been grown in this way on fused quartz by the vapor–liquid–solid mechanism in the hydrogen-tetrachloride process, and oriented films of some other materials have been deposited on a variety of substrates by various methods [8.5c]. In another version, single-crystal silicon films were prepared on fused quartz or on oxidized silicon substrates by a two-stage process: first an amorphous or polycrystalline silicon film was deposited on the patterned substrate, and then it was transformed into a single-crystal film by means of laser or strip-heater-oven recrystallization [8.5 d, e]. Silicon graphoepitaxial films have also been prepared on fused quartz by means of ion-beam deposition, both in a direct (one-stage) process and after beam annealing [8.5f]. Oriented deposition by graphoepitaxy has been reported for a number of metals and alloys deposited onto 0.3 μm period relief gratings in thin amorphous carbon foils [8.5g], and for electrodeposition of tin [8.5h].

Lateral Epitaxy. This new method makes it possible to grow single-crystal silicon films over insulating layers on single-crystal silicon substrates, as well as other materials on the corresponding single-crystal substrates. An amorphous or polycrystalline silicon film is deposited on the insulating layer, in which narrow strip have been opened to expose the substrate. By means of two graphite strip heaters, one of which is movable, the silicon film is melted and frozen in such a manner that solidification begins within the strip openings, where it is seeded by the substrate. The resulting single-crystal regions in turn seed lateral single-crystal growth over the adjacent insulating layer. Continuous single-crystal silicon films were prepared over SiO_2 insulating layers with areas as large as several square centimeters by this method [8.5i, j]. In another version, single-crystal GaAs and InP films were grown through strips in a masking SiO_2 layer on GaAs and InP substrates, and were cleaved off so that the substrates could be reused [8.5k, l].

Other approaches to oriented crystallization on amorphous (and, generally, arbitrary) substrates include laser recrystallization with shaped beams, micro-zone recrystallization of encapsulated films, rheotaxy, impurity-induced recrystallization of films, and the so-called printing technique [8.5a]. Generally, these methods open up new prospects in materials science and technology.

8.3 Physical Vapor Deposition

The feature common to this group of methods is that the substance is delivered to the growing crystal in the form of its own vapor, which consists of atoms and molecules as well as association of them—dimers, trimers, and polymers in general.

Four principal methods of physical vapor deposition are distinguished based on the method by which the substance is supplied to the crystallization zone. These are the molecular-beam method, cathode sputtering, vapor phase crystallization in a closed system, and crystallization in inert gas flow. The typical designs of these methods are presented schematically in Fig. 8.2. Figure 8.3 shows the general relationship of the methods to one another, together with their modifications.

8.3.1 Molecular-Beam Method

In the molecular-beam method, a compact source is heated to a high temperature in a vacuum. The source emits atoms or molecules, which propagate according to the laws of geometric optics and reach the substrate, where they are condensed (Fig. 8.2a). The temperature of the source is chosen based on the required beam intensity; it may be above or below the melting point of

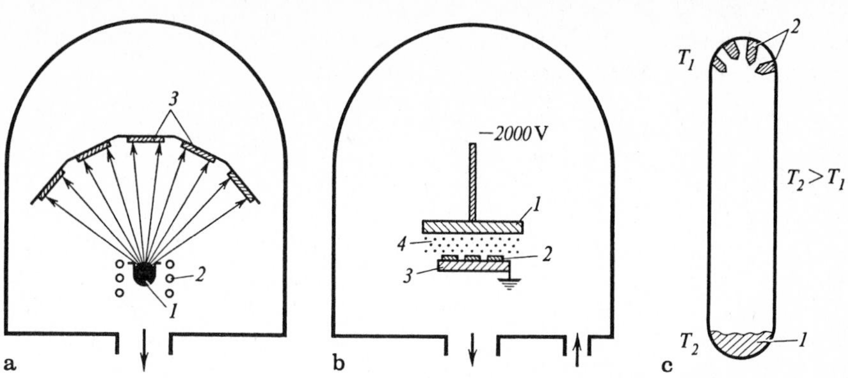

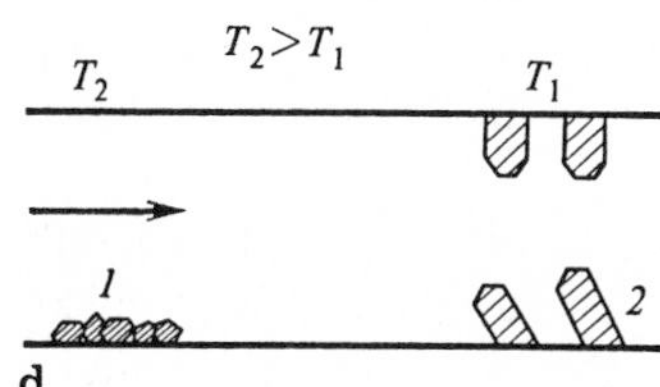

Fig.8.2a–d. The principal methods of physical vapor deposition. (**a**) Molecular beam method; (*1*) source, (*2*) heater, (*3*) substrates. The *arrow* indicates evacuation of the vessel. (**b**) Cathode sputtering: (*1*) cathode, (**2**) substrate, (*3*) anode, (*4*) Ar^+ plasma. The *arrows* indicate the inflow and evacuation of gas. (**c**) Vapor growth in a closed system: (*1*) source, (*2*) growing crystals. (**d**) Gas flow crystallization: (*1*) source, (*2*) growing crystals. The *arrow* indicates the inflow of gas

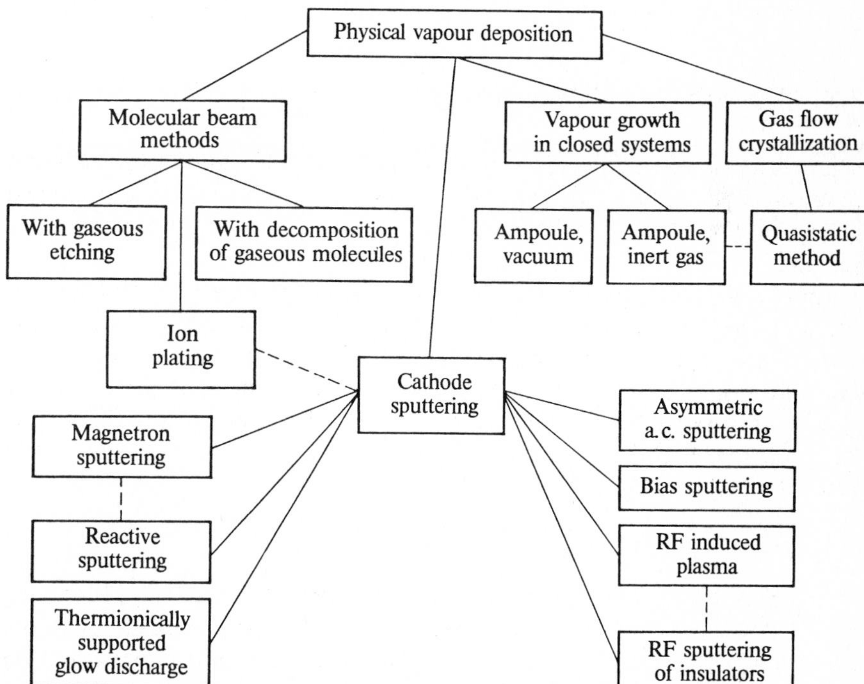

Fig.8.3. Classification of physical vapor deposition methods

the substance. For instance germanium, which is characterized by a relatively low equilibrium vapor pressure ($\sim 10^{-7}$ torr) at the melting point, has to be heated to a temperature much above the melting point. Silicon has a rather high vapor pressure ($\sim 10^{-1}$ torr) at the melting point; therefore, it can be evaporated (more precisely, sublimated) from a solid source. Since the source is compact, the distribution of the evaporated particles in space is, to a first approximation, the same as that for a point source, i.e., the beam intensity decreases with the distance r from the source by the law $\sim 1/r^2$.

The substance to be evaporated is placed in a crucible or boat of a refractory, chemically inert material, or in a wire basket of tungsten, molybdenum, tantalum, etc [8.6]. The source may be heated by resistance, induction, radiation or electron-beam heating.

At present, electron-beam heating is considered the best method, because it provides a high local temperature. Furthermore, it allows one to dispense with a crucible, which would otherwise inevitably introduce contaminations. A focused electron beam heats a small area on a comparatively bulky ingot of the substance to be evaporated, sometimes bringing the area to the melting point (Fig. 8.4a). A still higher energy concentration is achieved in a special design with magnetic focusing (Fig. 8.4b). The evaporator and substrate are placed in a chamber with cold walls in which a high vacuum is maintained, usually 10^{-5}–10^{-6} torr or better. Here the mean free path is far longer than the size of the chamber, and hence the particles of the crystallizing substance propagate without collisions. If a perforated screen (or "mask") is placed in their path, it cuts out individual particle beams. Thus the crystallization can be localized on selected portions, and herein lies the fundamental difference between the molecular-beam method and all the others. This feature is also a great advantage in solving a number of technological problems [8.7]. Another important advantage of the method is that it permits fine adjustment of the beam intensity, which means that the crystal growth rate can be controlled.

At the same time, the molecular-beam method does have some shortcomings which limit the utilization of the method in epitaxial technology.

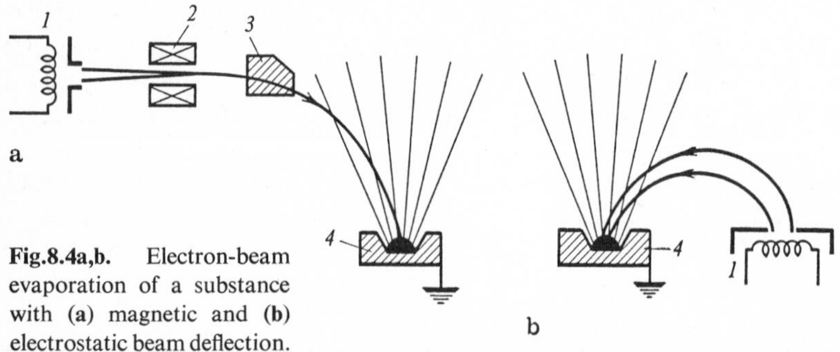

Fig.8.4a,b. Electron-beam evaporation of a substance with **(a)** magnetic and **(b)** electrostatic beam deflection. (*1*) Cathode, (*2*) magnetic lens, (*3*) magnetic prism, (*4*) water-cooled crucible containing the substance. The *arrows* indicate the electron beam

a) The molecular-beam method becomes greatly complicated in the crystallization of $A^{III}B^{V}$ compounds, whose components have drastically differing vapor pressures. To avoid difficulties, one of the following techniques is generally used: separate evaporation of components from different sources; three-temperature evaporation technique with the use of the "stoichiometry range", in which the compound itself is stable at the crystallization temperature, while the more volatile element of group V experiences reevaporation [8.3, 8, 9]; and the "flash evaporation" method, in which a compound is evaporated particle by particle when delivered into the source zone [8.3].

b) Another fundamental drawback of the molecular-beam method is related to doping, a problem typical of semiconductors. The simplest form of doping is evaporation of a doped source. However, some impurities, such as phosphorus, arsenic, or antimony (these are typical donors in silicon and germanium) are characterized by a high vapor pressure and/or a low condensation coefficient at growth temperatures, and therefore it is rather difficult to obtain n-type semiconductor layers. To solve this problem the crystallization temperature must be reduced (which requires a better vacuum), an additional impurity source introduced, or ionized impurities used [8.10].

c) Semiconductor or other films sensitive to structure or composition obtained by the molecular-beam method without special efforts have comparatively poor electrical parameters. An example is the low mobility of current carriers, which is due to the large number of defects, mainly of point type. This is a consequence of the strong nonequilibrium conditions for crystal growth: it was noted in Sect. 8.2.3 that supersaturations here may be as high as 10^3–10^6. And although electron diffraction patterns from such films usually show Kikuchi lines (which would seem to be an indication of high structural perfection), detailed investigation of etched specimens with the aid of an optical microscope, and of thin foils by means of transmission electron microscopy, reveals a great number of defects in films: point defects, line defects (dislocations), and two-dimensional ones (stacking faults, microtwins). The cause is evidently "haste" in building up the crystal at high supersaturations. To increase the perfection of films, a combination of the molecular-beam method and chemical deposition methods has been proposed. For this purpose, an additional flux of molecules, for instance of iodine, is directed at the growing surface. The iodine molecules etch the surface slightly, apparently acting for the most part on the imperfect positions of the atoms or on the impurity aggregates. In fact, the addition of a gaseous etchant results in a decrease in supersaturation, i.e., in an approach to equilibrium. In another version of the molecular-beam method coupled with chemical methods, a flow of the compound to be decomposed, rather than of the crystallizing substance itself, is directed toward a heated substrate. In silicon crystallization, molecules of monosilane SiH_4 are used [8.1], in chromium crystallization, CrI_3 [8.11], etc. The decomposition products of the compounds obviously remain on the surface for some time, and because of the fundamental reversibility of the chemical reactions, reduce the actual supersaturation in the adsorption layer, thus promoting the formation of more perfect crystals.

Due to the above-mentioned shortcomings, the molecular-beam method has, until recently, been used mainly on the laboratory scale to grow epitaxial films of germanium, silicon, $A^{III}B^{V}$, $A^{II}B^{VI}$, $A^{IV}B^{VI}$, and othei compounds both on substrates of the same substance and on foreign substrates (e.g., germanium on GaAs, CaF_2, NaCl, etc.; silicon on Ge, GaP, SiC, Al_2O_3, etc.; and lead telluride on BaF_2, etc.). In industry the method was commonly used in cases where the fine structure of films is not of primary importance, i.e., for optical and conducting coatings and for passive elements in microelectronics (resistors, capacitors, interconnections, etc.).

In recent years, an evolution of the molecular-beam method could be observed which was motivated by the desire to use this technique to grow device-quality single-crystal films (primarily semiconductor films) [8.12]. Two principal improvements were important for the progress in this field. The first was the creation of a new generation of molecular-beam apparatuses which permit crystallization at pressures as low as 10^{-9}–10^{-10} torr. This has made it possible to overcome a number of shortcomings by decreasing the crystallization temperature. The second is the use of in situ diagnostic techniques for studying crystal surfaces. Pilot equipment has already been developed in which the structure and composition of the growing films is controlled by means of low-energy electron diffraction (LEED), reflection high-energy electron diffraction (RHEED), secondary ion mass spectroscopy (SIMS), Auger and quadrupole mass spectroscopy, etc.

The principal aim of these developments is to provide industrial techniques for the production of epitaxial thin films from molecular beams; hence the term "molecular-beam epitaxy", or MBE, which is widely used for this process. Both elemental semiconductors such as silicon [8.13] and compound semiconductors such as $A^{III}B^{V}$ [8.14] and $A^{II}B^{VI}$ compounds [8.15] have been prepared by the MBE technique. Of special importance are the possibilities of growing ultrathin epitaxial layers (100–1000 Å) and producing "superlattices" (periodic structures of different materials 50–100 Å in thickness) and alternate-atomic-layer compositions. Commercial continuous-action apparatuses are being developed which feature "in-line" production by MBE of all integrated circuit components, both active and passive.

Below are some typical examples and conditions of crystallization.

Example 1. Crystallization of autoepitaxial germanium films is carried out under a water-cooled glass or metal bell 300 mm in diameter and 300–500 mm in height, in which a vacuum of 10^{-5}–10^{-6} torr is maintained. The source — a piece of germanium — is melted by heating to 1500–1800°C in a tungsten wire basket or a boat of tungsten, tantalum, graphite, or glassy carbon. The substrate is placed on a tungsten or tantalum strip, through which a current is passed to heat the substrate. The distance from the source to the substrate is

50–100 mm (this distance, as well as the source temperature, determines the density of the incident flux). Prior to crystallization, the substrate is shielded from the source by a chopper and the substrate surface is cleaned by heating at ~850°C for 10 to 20 min. Then the substrate is cooled to a temperature in the range of 500–800°C, and the chopper is removed. In measuring the temperatures, some additional heating of the substrate by the source radiation and the heat of condensation must be taken into account. Typical deposition rates at 800°C are 1 μm/min; typical film thicknesses are 5–10 μm; the dislocation densities in films are ~10^5 cm^{-2}; the conductivity is of the p type; the resistivity (with an extremely pure source) is ~1 Ω cm; and the carrier mobility is ~1000–1500 cm^2/V s. A typical epitaxial film surface is shown in Fig. 8.5 [8.16a].

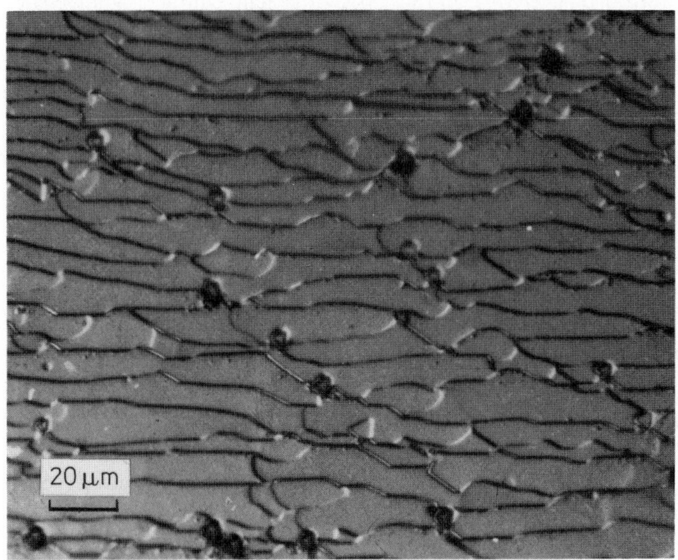

Fig.8.5. Micromorphology of an autoepitaxial Ge film obtained by the molecular-beam method. The photo was taken in an optical microscope [8.16]

Example 2. In silicon crystallization by sublimation, a source in the form of a cylindrical ingot is heated to ~1300°C with high-frequency currents. A substrate is placed parallel to the flat end face of the source at a distance of ~2 mm, and the substrate is then heated to ~1000°C by source radiation. Prior to crystallization, the substrate surface is usually cleaned by heating in an atmosphere of hydrogen, which is then pumped out. The growth rate here is low (~4 μm/h), and depends on the density of the molecular flux from the solid source [8.16b].

Example 3. For the preparation of GaAs and GaAlAs layers, molecular-beam epitaxy in an ultrahigh vacuum chamber is used [8.12a]. The structure and composition of the growing layers are monitored in situ by reflected high-energy electron diffraction (RHEED) and Auger and mass spectrometry. Prior to crystallization, the (100)-GaAs substrate is cleaned in situ by ion etching. Three effusion cells containing Ga, Al, and As serve as sources for synthesis of the compounds. The cells are heated by resistance furnaces, and their temperatures are monitored separately. Typical cell temperatures for gallium, aluminum, and arsenic are about 980, 1080, and 900°C, respectively (arsenic vapor is supplied by a cell containing polycrystalline GaAs). For doping the growing layers with p and n impurities, two auxiliary cells are used, with tin and magnesium maintained at 600–750°C and 420°C, respectively. The GaAs substrates are heated to 500–600°C. The distance from the effusion cells to the substrate is about 5 cm. Molecular beams of gallium and aluminium are alternatively shut off by choppers, so that a periodic structure of GaAs and AlAs layers is formed, a solid solution of GaAlAs, in fact, with a special, quite regular structure. Typical deposition rates are ~1 monolayer per second. The alternation periods for the gallium and aluminium beams are about 1 s. Up to 10^4 monolayers of GaAs and AlAs are generally formed, so that the overall thickness of the grown film is about 3 μm. The film has a single-crystal structure and contains an insignificant amount of imperfections [8.12a, b, f].

8.3.2 Cathode Sputtering

This versatile, readily controllable process is widely used for the deposition of films, primarily polycrystalline films (in particular fine-grained ones) and, more recently, single-crystal films [8.17].

The most developed among various such processes is diode cathode sputtering with the use of a self-sustained glow discharge.

The method consists essentially in the following (Fig. 8.6): A discharge is ignited between cathode 1, which usually has the shape of a plate, and earthed flat anode 2, which is approximately parallel to it and carries substrates 3. In an inert gas (argon, krypton, etc.) at pressures of 10^{-3}–10^{-1} torr and at cathode–anode distances of 2–4 cm, the discharge is ignited at a potential difference of 500 to 5000 V. The steady state of the discharge is maintained due to the

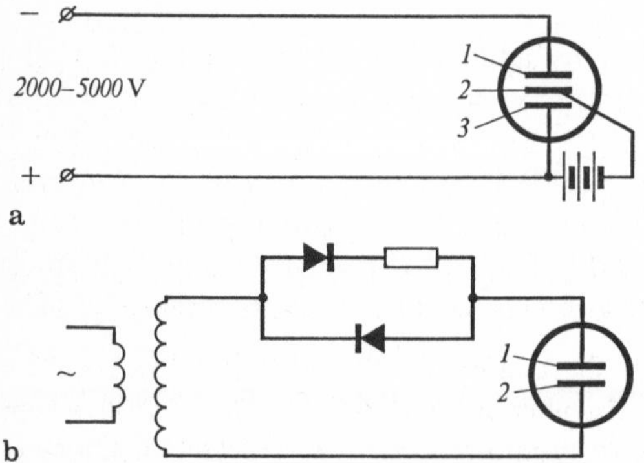

Fig.8.6a,b. Methods of cathode sputtering. **(a)** Sputtering with a bias: (*1*) cathode, (*2*) substrate, (*3*) anode. **(b)** Sputtering on ac current: (*1*) evaporated target, (*2*) substrate

dynamic equilibrium between the number of ions neutralized at the cathode and the number of new ions generated in the glow discharge plasma. The ions impinging the cathode knock atoms out of it by momentum transfer. The vast majority of these atoms are electrically neutral and reach the anode virtually without colliding with gas molecules. The energy of the sputtered atoms is usually much higher than that of atoms thermally evaporated in the molecular-beam method; therefore the condensation coefficient is close to unity. Sometimes atoms even become incorporated in the surface layer of the substrate, thus ensuring a high adhesion of the grown films to it.

In cathode sputtering, the mass transport rate, i.e., the flux density, is limited because of the physical phenomena in the gas discharge: when the pressure of the inert gas drops excessively, the discharge current rapidly diminishes, and the sputtering rate decreases despite the increase in cathode voltage (i.e., in ion energy). On the other hand, when the pressure increases, the probability that the sputtered atoms will return to the cathode also increases because of the collisions with gas molecules. The optimum pressures are 2×10^{-2} to 7×10^{-2} torr, and the deposition rate may be as large as $\sim$1000 Å/min.

In deposition on a single-crystal substrate that is sufficiently active (cleaned, heated, etc.), epitaxial growth can be achieved. Single-crystal films of Ge, GaAs, and other substances have been grown in this way; the epitaxial temperature was comparable to (and sometimes lower than) that in the thermal evaporation method, this being attributed to the cleaning action of the atoms colliding with the substrate.

Determining the supersaturation in cathode sputtering is difficult; therefore we shall give, as an example, some values of the principal parameters of the process which enable quality films to be obtained. Thus, epitaxial films of Ge can be grown on Ge at 220°C at rates of less than 15 Å/min; these rates can be achieved at a discharge current of 2.5 mA/cm^2 and a voltage of 3000 V. At a high current density (10 mA/cm^2) and the same growth rate, epitaxy is achieved only at 350°C. This is because the voltage has to be reduced if a low growth rate is to be maintained, and at a lower voltage the cleaning action of the atomic beam is not so effective. If the growth rate increases to 40 Å/min, the substrate temperature must be increased to $\sim$450°C, etc.

It should be noted that although in cathode sputtering the gas pressure in the system is rather high, the requirements imposed on the vacuum conditions before introducing the inert gas are even higher than in the molecular-beam method. This is due to the high chemical activity of the ionized components of the impurities in inert gases.

The most important advantage of the cathode sputtering method is that the source of the material used here is an unheated solid, and hence its interaction with the containers is negligible. This substantially facilitates the deposition of refractory metals (W, Mo, Ta, V, etc.) and, what is particularly important, of compounds such as $A^{III}B^{V}$, $A^{IV}N^{VI}$, etc. In the latter case the composition of the grown crystal reproduces that of the source, provided an additional source of more volatile elements is used [8.18]. In addition to the described principal method of cathode sputtering, numerous modifications of it have been developed to increase the quality of the deposited layers and the rate of deposition, or to extend the range of materials sputtered. The most common of these are listed below.

a) Thermionically Supported Glow Discharge. To eliminate one of the main drawbacks of cathode sputtering — the capture of discharge gases and particularly of their active impurities by the growing film — it is advisable to operate at as low gas pressures as possible. If this is done, however, the discharge is extinguished. To maintain the discharge at pressures of $\lesssim 10^{-3}$ torr, an additional thermal cathode and anode are introduced into the system and placed so that the electrons from the thermal cathode fly across the discharge gap, ionizing the gas molecules on their path.

b) RF-induced Plasma. A gas discharge at a low pressure can also be excited by an electromagnetic field. No electrons are needed within the discharge chamber if sufficiently high frequencies are used. At a frequency of several MHz, free electrons in the gas make a few oscillations in the applied field between each collision with a gas atom, absorbing sufficient energy and causing ionization. For instance, at a pressure of 10^{-3} torr, frequencies of 1–10 MHz, a power output of 200 W in the high-frequency circuit, and a target bias of 500 V, deposition rates of up to 100 Å/min were obtained.

c) Bias Sputtering. If we apply to the substrate a small potential which is negative with respect to the anode, the substrate will, in the course of deposition, be continuously bombarded by ions and cleaned effectively of adsorbed gases which would otherwise have remained in the film as impurities. A diagram of such sputtering is shown in Fig. 8.6a. The typical bias voltage is -200 V. This version is more effective than the frequently used technique in which preliminary etching of substrates is achieved by reversing the polarity of electrode voltage, because here "make-up etching" is ensured throughout the process of film growth.

d) Asymmetric ac Sputtering. This method is in essence similar to the preceding one. Alternating rather than direct voltage is applied between the "cathode" and the substrate, so that the electrodes are consecutively bombarded by ions in alternate half cycles. The electric circuit is so designed that a larger current flows during the half cycle when the "cathode" is negative (Fig. 8.6b), and therefore a net transfer of material from the "cathode" to the substrate is achieved. Bombarding the substrate during the half cycle when the "cathode" is negative removes the adsorbed gases and thus yields a purer film.

e) Reactive Cathode Sputtering. It was indicated above that the ionization of gas impurities makes them chemically more active than their neutral counterparts. This phenomenon can be utilized to form compound films, primarily oxides, nitrides, and carbides, by reacting the sputtering gas with the cathode material. Such "reactive sputtering" is achieved either by reaction at the cathode, followed by transport of the resulting compound to the substrate, or by reaction of the background gas with the growing film (a homogeneous reaction is unlikely to occur in the vapor phase). The working gas is usually a dilute mixture of an active gas and an inert gas. Thus by sputtering Si or Al in a medium with O_2, it is possible to obtain SiO_2 or Al_3O_2; by sputtering Si or Ta in medium of N_2 to obtain the compounds Si_3N_4 or TaN; by sputtering Ta in medium of CO or CH_4, the compound TaC, etc.

f) RF Sputtering of Insulators. It is often desirable to be able to sputter directly from an insulator cathode, particularly since the reactive sputtering process is limited to relatively low deposition rates. To neutralize the positive charge accumulating at the insulator target, radio-frequency sputtering is used, in which the insulator is alternately subjected to ion and electron bombardment. The frequency must be sufficiently high ($\gtrsim 10$ MHz), because the voltage applied must be switched in a period that is short compared with the time positive ions require to travel from the edge of the ion sheath to the surface of the insulator. The sputtering rates here are quite high; in sputtering SiO_2, rates of ~ 1000 Å/min are easily achieved.

g) Magnetron Sputtering. This method can be defined as a diode version in which crossed electric and magnetic fields are used. A permanent magnet is mounted in the cathode body so that the magnetic field lines emerge from the cathode and pass along the sputtering surface. Owing to this configuration, the electrons in the plasma move along a helical trajectory (as in a magnetron), and the discharge intensity increases accordingly by about two orders of magnitude, so that deposition rates can be as high as 1–1.5 μm/min [8.19a]. The increase in deposition rate not only raises the process productivity; what is more important, it improves both the structural and electrical properties of the film, because the relative amounts of residue gases captured by the film are decreased at the high deposition rates. Both dc and rf modes of deposition can be used [8.19b, c]. In the rf mode, insulators can be sputtered using the magnetron technique [8.19d]. Magnetron sputtering with additional electron-beam ionization is also useful [8.19e].

The above-described versions of cathode sputtering differ in the complexity of the equipment used and in the level of the problems solved. Some of them, such as thermionically supported glow discharge, bias sputtering, and asymmetric ac sputtering, make it possible to grow epitaxial films of semiconductors (including compounds) that are of different conductivity types and have electrical properties approaching those of bulk crystals. We note that cathode sputtering is, as a rule, inferior to molecular-beam methods as regards the mean growth rates (by about an order of magnitude) and requires more sophisticated equipment, but exceeds these methods in versatility, especially when making films of semiconductor compounds and refractory metals.

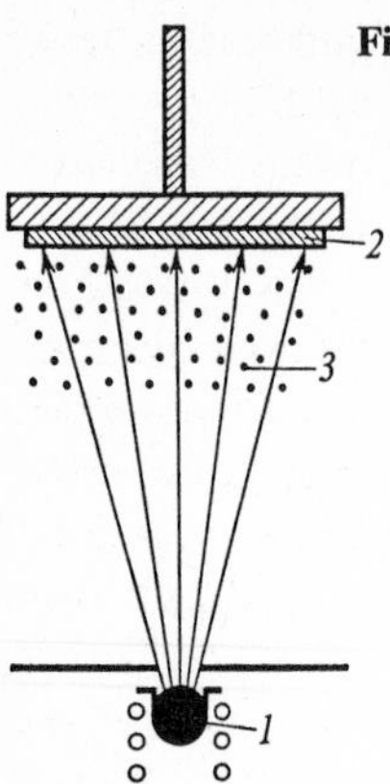

Fig.8.7. The ion plating method [8.20]. (*1*) Source, (*2*) substrate, (*3*) plasma

To conclude this section we shall consider one last method of film deposition. A combination of the evaporation (molecular-beam) and cathode sputtering methods, it accordingly combines the advantages of both. It is called ion-plating and can be described simply as evaporation in a glow discharge [8.20]. A schematic representation of this method is given in Fig. 8.7. The source material *1* is evaporated either by resistance or electron-beam heating. Substrate *2* serves as the negative electrode ("cathode"), and is situated in the zone of plasma *3*, which is maintained by a direct current or a high-frequency (diode or electrodeless) discharge. Thermally evaporated atoms or molecules pass through the plasma space and are ionized, accelerated to high velocities, and deposited on the cathode substrate. At the same time positive ions of the inert gas bombard the substrate, cleaning it continuously before and during growing. The effective growth rate is determined by the difference between the rates of arrival of film-forming ions and of the substrate sputtering. If resistance-heated evaporation is used, the rates may be as high as 3000 Å/min, and with electron-beam heating 20,000 Å/min. It should be remembered that an appreciable difference in pressures must be maintained in the deposition chamber: $\lesssim 5 \times 10^{-5}$ torr near the source (this is particularly important in electron-beam heating in order to avoid parasitic discharges), and $\sim 10^{-2}$ torr near the substrate; this is necessary to maintain the plasma. Finally, it should be noted that the deposition of material from ionized beams has been gaining in importance lately. This technique makes it possible to prepare epitaxial films at relatively low temperatures [8.20b] and to dope them with volatile impurities [8.10]. Some new versions of this technique make use of reactive media for the preparation of oxides or other compounds [8.21].

8.3.3 Vapor Phase Crystallization in a Closed System

In this method, a crystallization vessel is filled with vapor of the substance to be grown, with the different sections of the vessel heated to different temperatures. In such a system supersaturations are developed relative to the vapor as a whole and/or to its constituents ("partial supersaturations").

Two versions of the method may be distinguished:

a) the closed ampule method,

b) the demountable chamber method.

The ampule method is described in detail in the literature [8.22] and is widely used in the preparation of crystals in laboratories. It is as follows (Fig. 8.2c): A sealed or otherwise closed ampule (usually quartz) containing the crystallizing substance is placed in the temperature gradient. The charge material is placed in the hotter part of the ampule. Owing to the temperature dependence of the equilibrium vapor pressure, a concentration gradient is established in the ampule, the concentration being higher in the hotter part. Under the effect of this gradient the substance is transported to the colder end (crystallization zone), where the vapor is supersaturated and therefore can be condensed.

Two cases are distinguished:

a) A vacuum is set up in the ampule before the substance is loaded into it.

b) The ampule is filled with some "inert" gas[2] apart from the crystallizing substance.

Depending on the total and partial pressures in the ampule, different mass transfer mechanisms can be involved in this process: molecular flux, diffusion, or convection (the latter only in a vertical ampule when its lower part is hotter).

At the lowest total pressures ($\lesssim 0.1$ torr) the mean free path of the particles is greater than or comparable to the sizes of typical ampules. The transport rates of individual particles along the ampule are high, but the concentration of the substance in the ampule is low, and therefore the overall rate of crystal growth is also low, but increases in proportion to the vapor pressure. The crystal in fact grows from the molecular beam.

At higher pressures the transport mechanisms and rates are different, depending on whether or not an inert gas is present in the ampule.

In the absence of an inert gas, at vapor pressures of the substance from ~ 0.1 torr (i.e., $\sim 10^{-4}$ kp/cm^2) to ~ 1 kp/cm^2 (this is in practice the most important range for crystallization), the principal role is played by molecular diffusion, and the transport rate is, to a first approximation, proportional to the difference in the pressures in the source and crystallization zones. In this pressure range, the growth rate is limited by the transport rate ("diffusion regime of crystallization"). At still higher pressures, the effect of convection (if it is possible) becomes noticeable, and the convection transport rate is proportional to the mean pressure in the ampule. In this pressure range the

[2] By an inert gas we mean here a gas such as hydrogen, which does not interact with the crystallizing substance. The advantage of hydrogen lies in its ability to remove oxides from the substrate and the growing crystal. But in some cases, for instance when chalcogenides of group II (CdS, ZnS, ZnSe, and others) are crystallized, the possibility that they are partially reduced should be considered, because such a reduction can change the stoichiometry of the compounds.

transport rate may be so high that the growth rate will be limited by the surface processes, primarily in the crystallization zone ("kinetic regime of crystallization").

In the presence of an inert gas at pressures of $\gtrsim 0.1$ torr, the main role in the transport is played by diffusion of the crystallizing substance or its constituents. The diffusion coefficient is inversely proportional to the total pressure, and therefore, at a given concentration gradient, the transport rate is reduced with increasing pressure. In the practically important temperature range of 500–1000°C, diffusion is usually the limiting stage of the crystallization process. At pressures of $\gtrsim 1$ kp/cm^2, an appreciable role may be played by convection, which increases the transport rate.

In addition to affecting the mass transport rate, inert gases can influence heat removal from the growing crystal. This influence is manifested in changes in crystal morphology, for instance in the formation of dendritic and needle crystals at increased gas pressures.

Supersaturation in the ampule method is determined in the most straightforward manner as the ratio of the equilibrium vapor pressures in the source and crystallization zones (if one assumes that crystallization equilibria are established in both zones). Because of this simplicity the method is often used as a model for studying the mechanism and kinetics of crystal growth from the vapor phase.

In practice, the ampule method is usually used for growing bulk single crystals, since the preparatory operations (evacuating, loading, sealing, etc.) are rather laborious and it is inexpedient to obtain thin films in this way. In growing a large crystal it is desirable to have only one seed in the deposition zone. Technically, however, it is rather difficult to introduce a seed into an ampule, because it is sealed at high temperatures at which the seed could melt or evaporate. Therefore spontaneously formed seeds are generally used.

When growing large single crystals without any seed, it is necessary to preclude mass-scale spontaneous nucleation. Heterogeneous nucleation on the chamber walls usually plays the principal role. In order to deactivate the walls, they are first heated much above the crystallization temperature. Impurity atoms and their aggregates serve as nucleation centers (they evaporate on heating, spreading throughout the ampule, or become incorporated in the ampule walls), or areas with a disturbed structure may perform this function

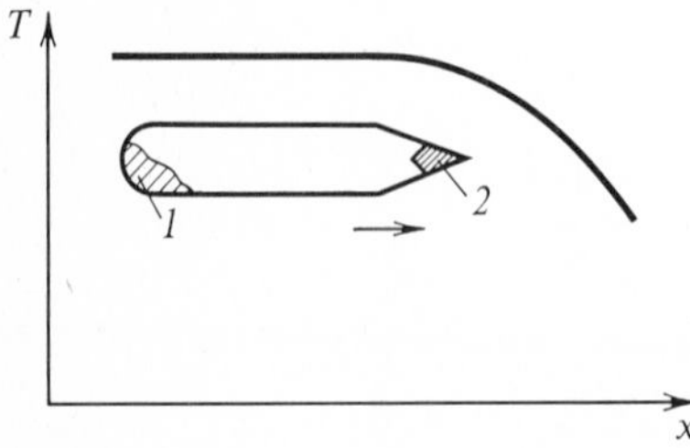

Fig.8.8. Crystal growth in an ampule during motion of the ampule along the abscissa (indicated by the *arrow*) in the temperature-gradient zone. (*1*) Source, (*2*) growing crystal

(they are then healed thermally). Afterwards the temperature of the crystallization zone is slowly reduced to several degrees below that of the source zone. The temperature is sometimes lowered locally by means of a heat-dissipating rod attached to the ampule tip, or a fine jet of a cold gas. With a local supersaturation of this kind the probability that a large number of nuclei will form is low, and sometimes a single crystallization center can be obtained.

In one of the most common versions of this method, a sealed ampule with a tapered tip is slowly moved through a temperature gradient so that the tapered tip is first to reach the region of comparatively low temperature (Fig. 8.8). A single crystal may form in the tip, and if several crystals appear, only one of them will survive, because of geometric selection. It should be pointed out that the problem of geometric selection (i.e., the "survival" of an individual crystal) is more complex here than in crystallization from the melt, for instance by the Bridgman-Stockbarger method. In vapor crystallization the delivery of the material to the necked part of the ampule is impeded, and there is therefore a high probability of nucleation where the necking just begins (Fig. 8.9a). Later on, these nuclei intercept the diffusing material in the vapor, and the supersaturation in the tapered tip of the ampule becomes insufficient for nucleation. Therefore ampules with rounded tips are now used more frequently, and a local increase in the supersaturation in them is achieved by placing the heat-dissipating rod in the proper place (Fig. 8.9b).

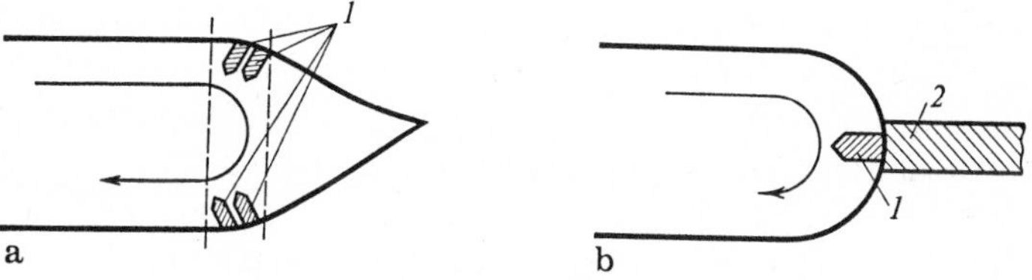

Fig.8.9a,b. Spontaneous nucleation in ampules of different shapes. (*1*) Growing crystals, (*2*) heat-dissipating rod. The *arrows* indicate the mass flow

An ampule method specifically for growing bulk crystals was recently developed on the basis of the traveling heater technique [8.23a–c]. Some empty, rather limited space is formed in the very tip of the ampule, while the other part of it is filled with a source material. The ampule is moved through a heated zone with a predetermined velocity, typically about 5 mm/d. A sublimation zone is formed between the growing crystal and the source. Such a configuration is typical of the close-spaced, or "sandwich" method. A similar approach has also been used for the growing of films [8.23d, e]. Below, a typical example is given of crystal growing by the ampule method.

Example [8.22d]. A vacuum of $\sim 10^{-5}$ torr is set up in a vertical quartz ampule 20 mm in diameter and 80 mm in height, with a heat-dissipating rod fused to the upper end. A source — lead selenide of near-stoichiometric composition — is placed in the bottom part of the ampule. The ampule is inserted into a furnace with a temperature gradient, and the bottom part of the ampule is heated to ~1100°C, i.e., a temperature above the melting point of PbSe (1080°C). The upper part is heated to 1000–1020°C, which means that the temperature difference determining the supersaturation in the crystallization zone is 80–100°C, according to thermocouple measurements. (The actual temperature at the growing surface is evidently appreciably higher because of the radiation from the source and the release of latent crystallization heat). The ampule is withdrawn from the furnace at the rate of 0.5–1.5 mm/h, and a single-crystal or large-grain ingot having a total weight of ~30 g is formed in its upper part (Fig. 8.10).

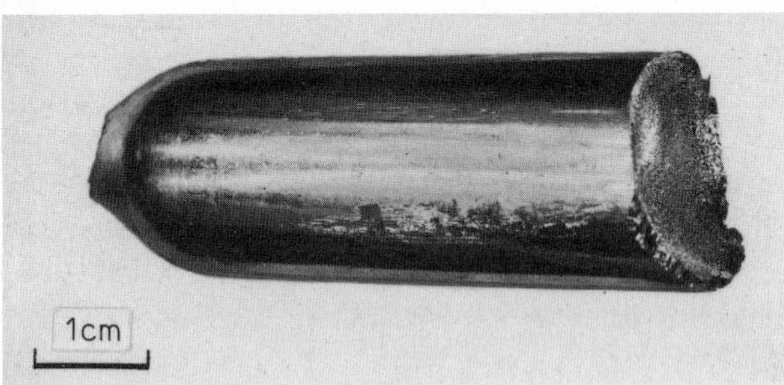

Fig.8.10. Single crystals of PbSe grown from the vapor in an ampule [8.23]

The demountable chamber method is widely used in growing bulk single crystals and epitaxial films from the vapor phase. It is sometimes called "hot wall epitaxy" [8.24]. Compared with the closed ampule method, it offers more possibilities of controlling the crystallization process. On the other hand, the demountable chamber method has much in common with the molecular-beam method, with the difference, however, that in the demountable chamber method, growth takes place under conditions far closer to thermodynamic equilibrium. Accordingly, crystals grown by this method are of high quality, both in structure and in purity.

The typical apparatus for this method consists in a vertical chamber with a sealed bottom and an open top. The source material is placed at the bottom, and the substrate or seed is mounted on a special holder on the open end. Thus the substrate or seed holder serves as a lid closing the tube. In growing epitaxial films, a shutter is usually mounted between source and substrate. The crystallization apparatus usually contains three independently controlled resistance furnaces. One of them heats the central (cylindrical) part of the chamber, while the other two are used to heat source and substrate. The entire chamber is placed in a vacuum vessel. In some versions of the method, the substrate or seed holder is placed at some distance from the tube assembly. This chamber cannot, of course, be regarded any longer as closed; however, with the proper design a well-collimated molecular beam can still be obtained.

In more sophisticated constructions, several sources are provided for different constituents and/or dopants.

The demountable chamber method is preferred for growing epitaxial films of $A^{II}B^{VI}$ and $A^{IV}B^{VI}$ compounds (CdS, CdSe, ZnS, ZnSe, PbS, PbSe, PbTe and their solid solutions) as well as low-melting $A^{III}B^{V}$ compounds (InSb, GaSb, InAs). Sometimes the method is also employed in growing bulk single crystals of such compounds.

8.3.4 Gas Flow Crystallization

This method is very similar to the ampule method with an inert gas. Dynamic, quasi-stationary, and stationary methods are the principal ones distinguished.

In the dynamic method the crystallizing substance is transported from one zone to another by means of an inert-gas flow in a tube (Fib. 8.2d). The "hotter" zone with the source to be evaporated or sublimated is placed in the path of the flow before the "colder" zone where crystallization proceeds on a special substrate or on the tube wall. In the crystallization of compounds (e.g., ZnS and CdSe), not the compound itself, but its volatile components are sometimes used as the sources, with each component being delivered through a special channel, and synthesis and crystallization taking place simultaneously in the mixture zone. In this case, the temperatures may be lower in the source zones than in the crystallization zone [8.25].

The total pressure in the zones does not usually exceed atmospheric pressure by more than several torr. This pressure drop determines the rate of flow, and must be sufficiently high to ensure a sufficient transport rate and exclude counterdiffusion of air. With a typical gas flow of ~ 1 l/min in a tube ~ 2 cm in diameter, the flow rate in a zone heated to $\sim 700°C$ is ~ 15 cm/s. At such a flow rate equilibria are not always established in the zones, and a calculation of supersaturations based on the temperature difference in the zones gives overestimated values. In practice, the temperature discrepancy between the source zone and crystallization zone should usually exceed 20–30°C in order to ensure supersaturation sufficient for growth.

The dynamic method is generally used to grow bulk, platelet and needle crystals, as well as epitaxial films of substances with sufficiently high vapor pressures such as $A^{II}B^{VI}$ compounds.

The quasi-stationary gas flow method is carried out in a demountable apparatus. First the main part of the apparatus, the crystallization chamber, is evacuated and its parts outgassed by annealing, and then a gas (usually inert) is fed into the chamber until the desired pressure is attained. A moderate gas flow is then maintained by means of an outlet valve, and crystallization is carried out in this state of back-up pressure. The main advantage of this method is that it makes it possible to intervene in the crystallization process at different stages in order to introduce impurities, control the transport rate (by changing the total gas pressure), switch from the nucleation to the growth stage, etc.

The stationary method is similar in design to the quasi-stationary method, although it could, in principle, be classified with the closed ampule method (Sect. 8.3.3). The stationary method makes it possible to carry out crystallization at increased inert-gas pressures, usually of up to 30–50 kp/cm². This is, for example, necessary in order to suppress the decomposition of compounds or the evaporation of volatile components. It is used to grow, among other things, single crystals of silicon carbide [8.26].

Example 1. Two zones with approximately flat temperature profiles of 850° and 750°C, respectively, are set up in a quartz tube 20 mm in diameter with the aid of a resistance furnace. A quartz boat containing ZnSe powder is placed in the first zone, and a quartz plate (substrate) in the other. Purified argon is passed through the tube at rate of 1 l/min. Platelet crystals of ZnSe having a size of 5 × 0.2 × 15 mm grow within 1 h on the substrate.

Example 2. A source — CdS powder bounded on two sides by perforated diaphragms — is placed in the central part of a quartz tube ~30 mm in diameter, over a length of 70 mm (Fig. 8.11). Quartz sleeves in which axial heat-dissipating rods are attached, are placed symmetrically on both sides of the source with spacings of ~50 mm, and serve as substrates for crystal growth. The outer diameter of the sleeves is chosen so that the clearance between them and the tube wall is 0.5–1 mm. The system is first evacuated and then filled with argon to ~2 kp/cm², after which a moderate gas flow of ~1 l/h is maintained. The surface of the sleeves is deactivated by setting their temperature above that of the source in the first stage (dashed line in Fig. 8.11) and then reversing the gradient (solid line). From that moment on, crystals begin to

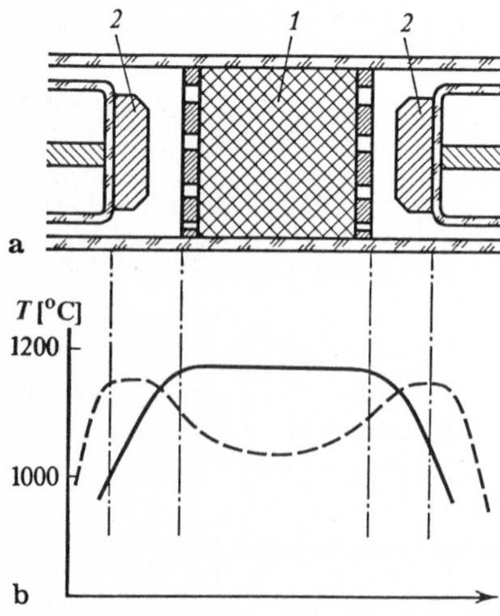

Fig.8.11. (**a**) The sublimation method in a quasistatic system; (**b**) the temperature distribution inside the tube in the preliminary stage (*dashed line*) and during growth (*solid line*) [8.27]. (*1*) Source, (*2*) growing crystals

grow from the central part of the substrates. The clearance and the form of the gradient preclude contact between the growing crystals and the tube walls, and hence any stresses orignating in the crystals during cooling are excluded or minimized, and very perfect crystals are formed. Within several hours, a crystal of 3–5 blocks, and sometimes a single crystal weighing up to 250 g, is formed. The dislocation densities in these crystals are as low as 10^3 cm^{-3}.

Example 3. A graphite sleeve 50 mm in diameter and 100 mm in height is placed, surrounded by numerous screens, in the center of a graphite furnace. The inside wall of the sleeve is lined with polycrystalline granules of SiC. The temperature in the sleeve is raised to $\sim$2500°C (the power consumed is $\sim$30 kW $= 10$ V $\times$ 3000 A), and axial and radial gradients arise. In the course of heating, the granules of SiC become sintered, and SiC vapor is formed inside the carbide void as a result of sublimation. In some zones the vapor is supersaturated, and platelet SiC crystals grow from it spontaneously. If nucleation is handled properly, the number of crystals is small, and hexagonal platelets $\sim$1–5 cm^2 in area and 0.1–0.5 mm in thickness can be grown within a few hours.

8.4 Chemical Vapor Deposition (CVD)

The feature common to this group of methods is that they involve a chemical reaction. This reaction serves not only to supply material for crystallization, but also (and especially) has an active influence on the crystallization process.

This involvement of a chemical reaction is a very favorable factor for a number of reasons.

1) The range of substances which can be crystallized from the vapor phase becomes considerably larger, because the requirement calling for sufficiently high vapor pressure of the crystallizing substance, inherent in physical vapor deposition methods, is eliminated.

2) Crystallization conditions are greatly improved. Owing to the fundamental reversibility of the chemical reaction, the process occurs near equilibrium, i.e., at low supersaturations, while the absolute concentrations of the substances (and hence the crystallization rates) may be quite high.

3) The chemical reaction ensures a stoichiometric ratio of the components forming the crystallizing substance.

Owing to these features, crystallization methods involving chemical reactions have proven highly efficient in obtaining perfect crystals and films, and have rapidly gained a firm foothold in the "technological market".

These methods can be roughly divided into three large groups:

a) Chemical Transport Methods. The crystallizing substance interacts in solid or liquid form with another substance in the source zone and transforms into gaseous compounds. These are then transported to a zone with a different temperature, and decomposing by the reverse reaction, they release the nutrient [8.28, 29].

b) Vapor Decomposition Methods. A volatile compound is introduced into the crystallization zone. Under the effect of a gaseous reducing agent and/or a high temperature or other factors, it decomposes, releasing the crystallizing substance.

c) Vapor-phase Synthesis Methods. The crystallizing substance (compound) is formed as a result of a reaction between the gaseous components directly in the crystallization zone.

As in the preceding section, no sharp distinction can be drawn between the groups; many methods can be classified as belonging equally to several groups. Nevertheless, each group has specific features of its own which affect the overall design of the crystallization process.

Before describing the separate groups, let us review briefly some general regularities characterizing heterogeneous chemical reactions and having a direct bearing on crystallization.

In studying a complex process it is convenient to divide it into separate stages and consider them consecutively. Here, the most general division includes two stages: mass transfer (the diffusion stage) and processes on the growing surface (the kinetic stage). The transfer stage encompasses the delivery of the reactants to the growing crystal and the removal of the reaction products into the gas phase. The surface processes consist in the adsorption of the reactants, chemical interaction between them on the surface, and desorption of the reaction products.

A more detailed gradation of the surface-process stage could be made by taking into account the surface migration towards a step or kink (see Sects. 3.1, 4.1). The crystallization process could also be divided up differently by classifying the mass transfer from the bulk of the medium to the surface, and also the adsorption, as the "material delivery" stage; the desorption and removal of the reaction products into the gas phase would then become the "reaction-products-removal" stage. A delay in the delivery of materials is as capable of retarding the chemical reaction proper as a delay in the removal of the reaction products.

Let us denote the concentration of the reactant in the bulk as C and its concentration at the surface as C_{surf}. Suppose that the flow of the substance consumed for crystallization as a result of surface processes isequal to $\beta(C_{surf} - C_0)$, where β is the constant of the rate of the crystallization reaction on the surface, i.e., the kinetic coefficient of surface crystallization in a given process (Sect. 3.1), and C_0 is the equilibrium concentration. On the other hand, an amount of material $\beta_{tr}(C - C_{surf})$, where β_{tr} characterizes the total rate of the indicated transport processes, is delivered to the surface by diffusion, convection, or molecular flow. In the steady-state mode, the above flows are equal:

$$\beta(C_{\text{surf}} - C_0) = \beta_{\text{tr}}(C - C_{\text{surf}}), \tag{8.2}$$

hence

$$C_{\text{surf}} = \frac{\beta C_0 + \beta_{\text{tr}} C}{\beta + \beta_{\text{tr}}}. \tag{8.3}$$

Consequently, the mass flow to the crystalline phase, which is proportional to the growth rate, is equal to

$$\beta(C_{\text{surf}} - C_0) = \frac{\beta\beta_{\text{tr}}}{\beta + \beta_{\text{tr}}}(C - C_0) = \beta_{\text{eff}}(C - C_0), \tag{8.4}$$

where

$$\beta_{\text{eff}} = \frac{\beta\beta_{\text{tr}}}{\beta + \beta_{\text{tr}}} \tag{8.5}$$

is the effective kinetic coefficient of crystallization, which simultaneously takes the surface reactions and the transport in the bulk into account.

This relation becomes particularly graphic if we consider not the reaction and diffusion rate constants, but their reciprocals:

$$\frac{1}{\beta_{\text{eff}}} = \frac{1}{\beta} + \frac{1}{\beta_{\text{tr}}}. \tag{8.6}$$

Hence in this simplest case the reciprocal reaction rates and diffusion rates—so-called diffusional and kinetic resistances—are added together.

This equation takes on a particularly simple form in two extreme cases, where one of the values—β or β_{tr}—is much larger than the other.

At $\beta \gg \beta_{\text{tr}}$ we get $\beta_{\text{eff}} \simeq \beta_{\text{tr}}$. In this case $C_{\text{surf}} = (\beta_{\text{tr}}/\beta)\, C \ll C$, and the rate of the total process entirely depends on the diffusion rate.

At $\beta \ll \beta_{\text{tr}}$ we obtain $\beta_{\text{eff}} \simeq \beta$ and $C_{\text{surf}} \simeq C$. Here, the rate of the summary process is wholly determined by the kinetics of the chemical reaction on the crystal surface, and is independent of the diffusion conditions.

The values appearing in the expression for β_{eff} depend on various parameters, of which temperature is the most important. The constant of the reaction rate depends exponentially on the temperature according to the Arrhenius law:

$$\beta \sim \exp(-E/RT), \tag{8.7}$$

where R is the gas constant.

The activation energy E for heterogeneous reactions proceeding at an appreciable rate in the temperature range of 500 to 1000°C, which is the most important range in crystal growing, is quite high, ~20 kcal/mol. This means that with a rise in temperature, for instance by 100°C, the reaction rate increases by a factor of 3 to 4.

On the other hand, although the diffusion coefficient in the gas phase increases with temperature, it does so only slightly (proportionally to the particle velocity, i.e., $D \sim \sqrt{T}$).

Therefore at relatively low temperatures the reaction generally proceeds in the kinetic mode, and an increase in temperature causes a transition from the kinetic to the diffusion mode.

These conclusions on the relationship between the kinetic and diffusion modes of crystallization have different implications for each of the indicated three groups of methods.

In analyzing chemical transport (which is commonly carried out in a sealed ampule), one usually proceeds on the basis of the concept advanced by *Schäfer* [8.28]: equilibria in the hot and cold zones of the ampule are established rapidly, and the gaseous substances move from zone to zone by diffusion, which actually limits the total crystallization rate. In other words, the conditions of temperature, as well as gas-dynamic and other experimental conditions in a sealed ampule are such that the diffusion mode is usually realized.

The vapor decomposition and vapor synthesis methods, which are usually implemented in open systems, offer much more freedom in the choice of crystallization conditions — temperature, the concentrations of gas mixtures (i.e., supersaturations), and flow rate. At the same time the requirements imposed on the growing crystals are rather high; the crystals are usually epitaxial semiconducting films which must be extremely homogeneous not only in structure (the most favorable are dislocation-free films and films with a minimum concentration of point defects), but also in composition (doped with impurities to a certain level) and in the state of the surface (atomically smooth, etc). Experience shows that these requirements are best met by crystals grown in the diffusion mode, since they show less anisotropy of surface processes and fewer related inhomogeneities. It should be noted that the transition from the kinetic to the diffusion mode and back is affected by all three principal parameters of the process: temperature, concentration, and flow rate, with temperature playing the decisive role. The effect of temperature was discussed above. An excessive increase in concentration may shift the process from the diffusion to the kinetic mode, since the reactants are adsorbed by the growing surface and may, like impurities, inhibit the motion of the steps. With an increase in flow rate the stagnant ("diffusion") gas layer over the substrate becomes thinner and thinner, and as a result the diffusion mode may be replaced by the kinetic one. But typically this layer remains sufficiently thick, of the order of several tenths of a millimeter or even several millimeters.

In practice, it is desirable to minimize the crystallization temperature in order to simplify the equipment, prevent the diffusion of impurities in the solid phase (which is important for semiconductors), combine different technological processes, etc. But often all these advantages must be forgone in order to shift the process to the diffusion mode.

Germanium offers a characteristic example. In the chloride process ($GeCl_4$ + H_2), single-crystal layers can be grown even at ~600°C, but commercial-quality films only at 800–850°C (the melting point of germanium is 939°C). The iodide transport process would appear to be even more attractive, as single-crystal growth of Ge can be ensured even at ~300°C. However, the quality of such films is unsatisfactory, and therefore this method has no practical application. For most substances, high-quality epitaxial films can be obtained from the vapor phase only at temperatures 0.75–0.90 T_0, T_0 being the melting point.

8.4.1 Chemical Transport

Imagine some substance A (solid or liquid), in which the equilibrium vapor pressure in the working temperature range is low. Suppose that when interacting with some gaseous component B according to a reversible reaction, this substance yields only gaseous products C and D. In other words, the reaction

$$\nu_A A_{s,l} + \nu_B B_{gas} \rightleftarrows \nu_C C_{gas} + \nu_D D_{gas} \tag{8.8}$$

proceeds from left to right (ν_A, ν_B, ν_C, and ν_D are the stoichiometric coefficients). If reaction (8.8) is reversed after C and D are transferred to a zone with a different temperature, the initial substances A and B will be evolved and A will be transferred to and separated at a new site. This process is called chemical transport [8.28].

Substance B is called the transporting agent. The most commonly used transporting agents are halogens (usually iodine, but also bromine and chlorine), hydrogen halides (generally HCl), elements of group VI B (sulfur, selenium, tellurium) and of group V B (arsenic, phosphorus), and various chemical compounds (H_2O, H_2S, $SiCl_4$, $AlCl_3$, etc.). In principle, any element or compound capable of entering into a reversible reaction with a given substance to form volatile products can be used for the transport of that substance.

The question then arises whether it is possible to choose beforehand, on the basis of the thermodynamic data, a transporting agent (and hence a reaction) which would ensure a maximum crystal growth and sufficiently high quality at the same time.

Basic information on the transport process can be obtained by studying the variation in the thermodynamic potential in the reaction, $\Delta\Phi^0$, and the reaction heat ΔH, which are related to one another and to the equilibrium constant $\mathcal{K}_0$, expressed in terms of the partial pressures of the vapor-forming component, by the following equations:

$$\Delta\Phi^0 = \Delta H - T\Delta S\,, \tag{8.9}$$

$$\Delta\Phi^0 = -RT \ln \mathscr{K}_0, \tag{8.10}$$

$$\frac{d(\ln \mathscr{K}_0)}{dT} = \frac{\Delta H}{RT^2}, \tag{8.11}$$

$$\mathscr{K}_0 = \frac{P_C^{\nu C} P_D^{\nu D}}{P_B^{\nu B}}. \tag{8.12}$$

Here, ΔS is the entropy variation in the reaction, and P_C, P_B, and P_D are the partial pressures of the gaseous components C, B, and D.

In order for crystal growth to proceed at the lowest supersaturations possible and at an acceptably high rate (i.e., at high concentrations of the reactants), the reaction must be sufficiently reversible, which corresponds to the values $\mathscr{K}_0 \sim 1$ or $|\Delta\Phi^0/RT| \ll 1$.

The direction of the transfer is determined by the sign of ΔH. In contrast to ordinary sublimation or evaporation, not all reactions transport the substance from the "hot" to the "cold" zone. It is only when $\Delta H > 0$, i.e., for an endothermic reaction, that the substance is transported towards the cold zone, whereas at $\Delta H < 0$, it is transported in the opposite direction. (It should be noted that the former case is more common.)

The quantity ΔH determines the temperature dependence of the equilibrium constant $\mathscr{K}_0$ and hence the necessary temperature drop ΔT between the zones. If ΔH is too large, high supersaturations, causing spontaneous nucleation, may arise even at low ΔT, in which case the temperature is difficult to control. If, on the other hand, ΔH is too small, such excessive ΔT may be required that experimental difficulties of a different kind arise: the temperature in the hot zone must not, as a rule, exceed $\sim$1100°C (this is the softening range for quartz, which is the typical ampule material), and it is not advisable to reduce the temperature in the other zone, because the surface processes may be retarded (especially if it is the crystallization zone).

The values of ΔH and ΔS have been tabulated for most substances. Therefore by calculating $\Delta\Phi^0$ one can choose a suitable transport reaction for practically any substance. Thus the method of growing crystals with the aid of chemical transport is quite versatile.

Below are some typical examples of the transport reactions used in crystal growing.

1) Transport of elements by means of the disproportionation reaction:

$$\mathrm{Ge} + \mathrm{GeI_4} \underset{350^\circ\mathrm{C}}{\overset{500^\circ\mathrm{C}}{\rightleftarrows}} 2\mathrm{GeI_2}, \qquad 2\mathrm{Al} + \mathrm{AlCl_3} \underset{650^\circ\mathrm{C}}{\overset{1000^\circ\mathrm{C}}{\rightleftarrows}} 3\mathrm{AlCl}.$$

2) Transport of oxides by hydrogen chloride or hydrogen:

$$\mathrm{Fe_2O_3} + 6\mathrm{HCl} \underset{800^\circ\mathrm{C}}{\overset{1000^\circ\mathrm{C}}{\rightleftarrows}} 2\mathrm{FeCl_3} + 3\mathrm{H_2O},$$

$$Al_2O_3 + 2H_2 \underset{1500^\circ C}{\overset{1800^\circ C}{\rightleftarrows}} Al_2O + 2H_2O\,.$$

3) Transport of chalcogenides by iodine:

$$CdS + I_2 \underset{400^\circ C}{\overset{1000^\circ C}{\rightleftarrows}} CdI_2 + \tfrac{1}{2}S_2\,,$$

$$ZnIn_2Se_4 + 4I_2 \underset{700^\circ C}{\overset{900^\circ C}{\rightleftarrows}} ZnI_2 + 2InI_3 + 2Se_2\,.$$

4) Transport in water vapors:

$$2GaAs + H_2O \underset{750^\circ C}{\overset{850^\circ C}{\rightleftarrows}} Ga_2O + H_2 + \tfrac{1}{2}As_4\,.$$

5) Iodide transport to obtain pure metals (Van-Arkel – de Boer process):

$$Zr + 2I_2 \underset{1450^\circ C}{\overset{280^\circ C}{\rightleftarrows}} ZrI_4\,.$$

In all but the last example, the substance is transported from the hot to the cold zone. There are also systems (such as W–Cl–CO) where the direction of the transport may be reversed, depending on the temperatures and pressures.

Let us now consider the various versions of crystallization systems. They may be closed, open, or quasi-closed (or close-spaced).

Closed Method. The simplest experimental device is a sealed ampule. The material to be transported is placed at one end of it, in the source zone, and is deposited at the other end, in the crystallization zone. A certain amount of transporting agent is introduced into the ampule, which is then placed in the temperature-gradient region. The crystallization is much the same here as in the closed ampule method in physical vapor deposition (Sect. 8.3.3): the same mass transport mechanisms depending on the total pressure in the ampule are in evidence, as well as the same problems with seeding and spontaneous nucleation. A fundamental difference is that in chemical deposition the concentraton of the substance transported in the vapor phase increases sharply, owing to the chemical reaction, and therefore it is possible to crystallize substances which have a negligible vapor pressure at crystallization temperatures. A remarkable feature of the chemical ampule method is that even minute amounts of transporting agent suffice to transport a virtually unlimited amount of the substance. The agent liberated in the crystallization zone diffuses into the source zone, recaptures the substance, transports it, and so on; in other words, the process is cyclic. The repeated participation in it of one and the same transporting agent ensures a high purity of the grown crystals. Another advantage of the closed system is its suitability for fundamental investigations of crystallization mechanisms and kinetics.

An ampule may be either evacuated or filled with a gas (argon, hydrogen, helium), thus permitting a certain degree of mass-transfer control. Tubular ampules of quartz glass are usually employed, making it possible to grow perfect and quite pure crystals at temperatures below ~ 1100°C. One such process is described at the end of this section (Example 1).

Two unconventional ampule designs are worth mentioning:

a) To overcome the difficulties involved in abundant spontaneous nucleation (Sect. 8.3.3), an ampule of the type depicted in Fig. 8.12 is used [8.30]. The source material is placed on the periphery of the ampule around the central area, where the temperature is lower because of the heat-dissipating rod attached to it; in this area the nucleation and growth of crystals occur. This design not only ensures a uniform (all-round) supply of material from the source to the crystallization zone, but also facilitates the introduction of a seed into the after its upper part is cut off. In addition, periodic temperature oscillations with an amplitude of 10–20°C and a cycle of 5–20 min are superimopsed on the system to control the nucleation. If a large number of crystallites nucleate during the period of positive supercooling, most of them can be reevaporated (or dissolved) by applying reverse supercooling for a short period. Periodic repetition of this process leads to the growth of just one crystal, because the growth and dissolution rate of a nucleus depends on its size, and the smaller crystallites have a higher vapor pressure according to the Gibbs-Thomson relation. Furthermore, when a single crystal grows, imperfect sites dissolve preferentially during the reversal period. By this method, quite large single crystals of Fe_3O_4 and HgI_2 were grown using spontaneous nucleation [8.30, 31], and isometric single crystals of SbSI, a substance with a filamentary structure, were grown for the

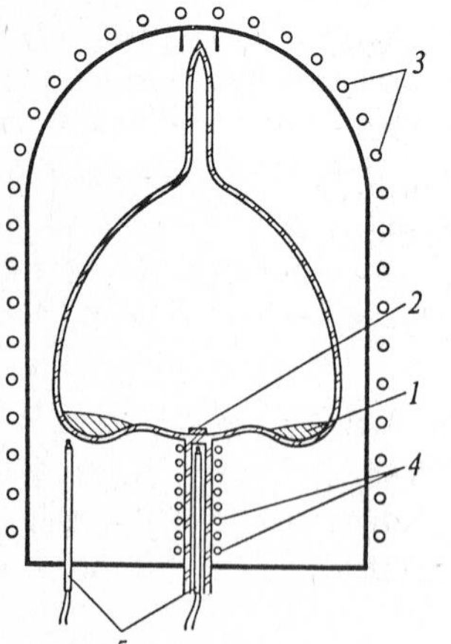

Fig.8.12. Crystallization ampule with a radial gradient and periodic temperature fluctuations [8.30, 31]. (*1*) Source material, (*2*) seed, (*3*) main heater, (*4*) auxiliary heater, (*5*) thermocouples

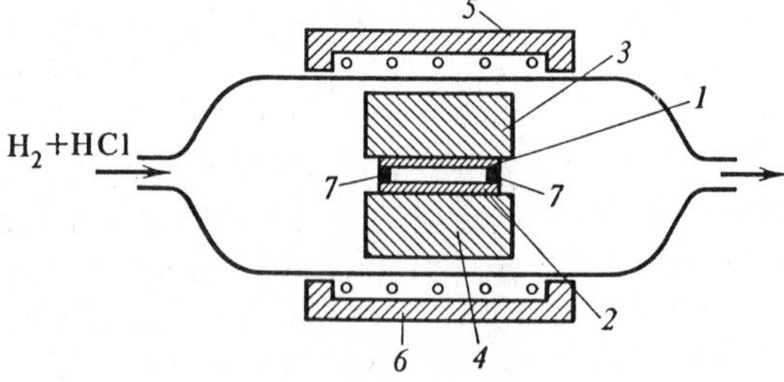

Fig.8.13. The sandwich process. (*1*) Plate substrate, (*2*) plate source, (*3, 4*) graphite blocks, (*5, 6*) infrared lamps, (*7*) quartz spacers

first time on a seed [8.32]. All these materials are very difficult to grow as perfect crystals.

b) Molybdenum or tungsten ampules are used for superhigh-temperature chemical transport ($\gtrsim$ 2000°C). The idea underlying this method, which is due to *Kaldis*, is that at relatively high temperatures (e.g., $\gtrsim$ 1700°C for molybdenum) any compound of the ampule material and iodine — a virtually universal transporting agent — is unstable, and hence the transport of the ampule material is prevented. This method has been used to grow single crystals of many rare-earth chalcogenides and europium oxides, as well as of other high-melting compounds [8.22a].

The Open Method. This technique of chemical transport is quite similar in its equipment to that used for physical vapor deposition (see Sect. 8.3.4). The fundamental difference is that here the high-temperature zone is not necessarily the first in the parth of the flow (although for most of the transport reactions the source must be at a higher temperature). The open method is more versatile than the closed method, because it permits wide changes of the growth conditions from stage to stage, the introduction of impurities, etc. Of particular importance is its versatility in the growing of films [8.33].

The Close-spaced Method (or "sandwich" method, due to *Nicoll* [8.34]). This method combines all the principal advantages of crystallization in closed and open systems. It consists essentially in the following: Two plates are placed parallel to one another at a distance which is small compared to the size of the plates, and a temperature difference is established between them (Fig. 8.13). Because of the small distance between the plates (tenths of one millimeter), the transport rate is much higher than in ampules, which are usually several centimeters long. This is one of the chief advantages of the method. Furthermore, since the exchange with the environment is reduced, the same amount of transporting agents repeatedly takes part in the transport, and thus the danger that the crystal will be contaminated by impurities is diminished (in this respect the close-spaced method is similar to the ampule method). Finally, the close-spaced method is most commonly used in a system with a small gas flow (although it can, in principle, be implemented in a sealed ampule as well); hence it is possible, as in the open method, to "intervene" in the process, thus making it quite easy to control.

Two shortcomings of the close-spaced method should be mentioned:

a) Cleaning the substrate surfaces of oxides and other impurities is rather difficult. Therefore the resulting films are less perfect than, say, in the open methods, where the substrates are cleaned in the chamber before crystallization.

b) The growth rate strongly depends on the spacing between the plates (especially at minimum spacings), and even a small misalignment of the "sandwich" results in appreciable inhomogeneities in film thickness.

For these two reasons the sandwich method has not found application in technology, although it is used quite often in growth experiments.

Below are some examples of chemical transport procedures.

Example 1. A thoroughly cleaned quartz ampule 18–20 mm in diameter and 150–200 mm in length, with hemispherical endfaces and a rod, as shown in Fig. 8.9b, is loaded with ZnSe powder and evacuated to ~ 10^{-5} torr. Using a special branch pipe, ~ 5 mg/cm^3 of iodine is distilled into the ampule, which is then sealed. It is held for ~ 10 h at an inverted temperature drop (700°C in the source zone and 800°C in the crystallization zone). The substance is distilled into the colder end, and any undesirable nuclei are removed from the crystallization zone. Then a normal temperature drop is established: 800°C in the source zone and 780°C in the crystallization zone. Within 10 days, 3 to 5 single crystals of zinc selenide, ~ 3 × 5 × 10 mm^3 in size, grow in the region of the rod (crystallization zone).

Example 2 (see Fig. 8.13). Single crystal plates *1* (substrate) and *2* (source) of gallium arsenide, 18–20 mm in diameter, are placed parallel to one another, and a spacing of 0.2 mm is established between them with the aid of three quartz inserts *7*. The plates are brought into contact with two bulky graphite blocks *3* and *4*, and the whole assembly is placed in a quartz chamber, through which an H_2 flow is passed at the rate of ~ 2 l/h. Infrared lamps *5* and *6* (~ 1 kW each) are mounted opposite the graphite blocks. In the first stage, lamp *5* is switched on and heats block *3* and plate *1* to ~ 750°C, while plate *2* attains a temperature of ~ 650°C by heat transfer. About 1% HCl is added to the H_2 flow, and within 5 minutes the surface of plate *1* is cleaned of impurities. In the second stage, lamp *5* is switched off and *6* is switched on; due to the inversion of the temperature gradient, an epitaxial film of GaAs about 30 μm in thickness grows on plate *1* within ~ 30 min.

8.4.2 Vapor Decomposition Methods

In vapor decomposition methods, a flow of a gas mixture containing the compound to be decomposed is introduced into a high-temperature zone. In this zone, a reaction takes place and the substance is evolved and deposited (usually on a single-crystal substrate), while the reaction products are carried away by the flow. It is usually assumed that the reaction proceeds heterogeneously, on the substrate. Two things should be remembered, however. Firstly, reactions in the vapor phase which result in the formation of intermediate compounds or radicals often play an important part, and secondly, release of the substance itself in the vapor phase near the growing surface cannot be ruled out, although it is not easy to verify this by experiment [8.35].

The two principal methods are reduction and thermal decomposition (pyrolysis) of compounds.

The typical reducing agent (which also serves as a carrier gas) is hydrogen; sometimes carbon monoxide (CO) or vapors of metals such as zinc are used. Typical compounds reduced are chlorides, and halides in general. Usually compounds having sufficient vapor pressures at room temperature are chosen in order to prevent condensation in communication pipes.

A typical example of such a process is the growing of silicon films by reduction of silicon tetrachloride with hydrogen [8.36–38]:

$$SiCl_4 + 2H_2 \rightleftarrows Si + 4HCl\,. \tag{8.13}$$

Other reactions are also possible, for instance with trichlorosilane ($SiHCl_3$) or dichlorosilane (SiH_2Cl_2). The process of obtaining germanium from $GeCl_4$ is similar.

A diagram of the chloride process is shown in Fig. 8.14. The hydrogen from cylinder *1* passes through gas purification system *2* and enters, at a pressure slightly above atmospheric pressure, a gas-handling system containing flowmeter *3*, valves *4*, evaporator with liquid chloride *5*, and condensor *6*. The gas-handling system produces a mixture containing $\sim 1\%$ gaseous compound and flowing at a rate of ~ 1 l/min. The mixture enters a reactor, where the substance is deposited at a high temperature. Diagrams of the most common reactors used in production are presented in Fig. 8.15. These are a vertical (a) and a horizontal (b) reactor with induction heating, and a "barrel"-type reactor (c) with internal resistance heating.

If the reactor contains single-crystal substrates with sufficiently clean and perfect surfaces, oriented autoepitaxial or heteroepitaxial layers will grow. By adding vapors of doping element compounds (such as $AsCl_3$ or BBr_3) to the gas mixture, n- or p-type films with different properties are obtained. To reduce the diffusion of the doping impurities in the solid state and simplify the equip-

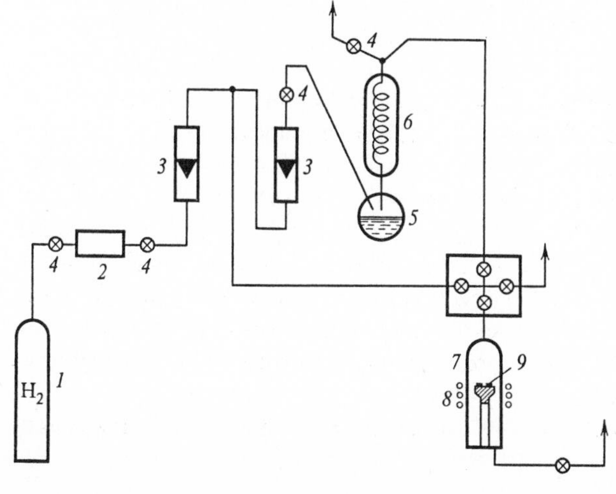

Fig.8.14. A crystallization setup for growing films in the chloride process. (*1*) Hydrogen cylinder, (*2*) gas purification system, (*3*) flowmeters, (*4*) valves, (*5*) evaporator, (*6*) condenser, (*7*) reactor, (*8*) inductor, (*9*) substrates. The *arrows* indicate gas venting

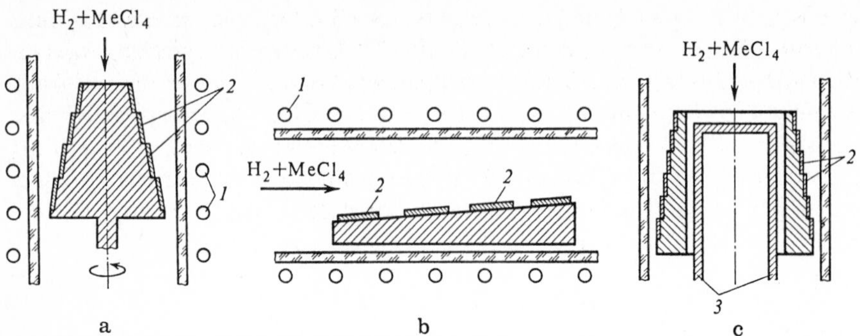

Fig.8.15a–c. Reactors for the deposition of epitaxial silicon or germanium films (Me = Si, Ge.) (*1*) Inductor, (*2*) substrates, (*3*) graphite resistance heater

ment, an attempt is usually made to minimize the growth termperature, but at relatively low temperatures ($\lesssim$ 1100°C for silicon) the density of imperfections in the crystals increases sharply, and therefore intermediate temperatures are generally chosen. For instance, silicon films are usually crystallized from tetrachloride-hydrogen mixtures at 1150–1200°C, although in principle, single-crystal layers can be grown even at ~900°C. It is worth noting that the decrease in the epitaxial temperature is directly related to the purity of the hydrogen with respect to oxygen, and that at equal partial pressures of the residual oxygen in the hydrogen and in a vacuum for physical vapor deposition (Sect. 8.2.4) we will have approximatedly the same epitaxial temperatures. This means that the poisoning mechanisms of contaminating impurities are common to both types of processes.

The dependence of the growth rate on the temperature, the concentration of the compound to be decomposed, and the gas flow rate offer important information for an understanding of growth mechanisms.

The temperature dependence here is typical of an activation process. Thus, plotting the dependence in Arrhenius coordinates, i.e., the logarithm of the growth rate versus reciprocal temperature in K^{-1}, gives, for reaction (8.13), an activation energy of 10 to 40 kcal/mol. The activation energy varies from one reactor to another, because the mixture is heated to different degrees before it approaches the substrates. One can assume that intermediate reactions, for instance reactions in which $SiCl_2$ is formed may play an important part here:

$$SiCl_4 + H_2 \rightleftarrows SiCl_2 + 2HCl\,.$$

This is also indicated by thermodynamic calculations and the experimental dependence of the growth rate on the concentration in the initial mixture [8.38–40]. Calculations show that in the practically important temperature range of 1000–1300°C the main components of the gas phase for reaction (8.13) are H_2, $SiCl_2$, HCl, $SiCl_4$, SiH_2Cl_2, and $SiHCl_3$, while the concentrations of other products (SiCl, SiH_3Cl, etc.) are negligible [8.41]. This conclusion was recently confirmed by mass spectrometric studies [8.42]. Furthermore, it follows from the calculations that the growth rate versus the initial molar concentration of $SiCl_4/H_2$ must pass a maximum, and at high concentrations become negative, i.e., the substrate must undergo etching. This is confirmed by experiment (Fig. 8.16). Consequently, in a given process the supersaturation may decrease with increasing concentration, while the quality of the crystals increases accordingly, as is proved in direct experiments.

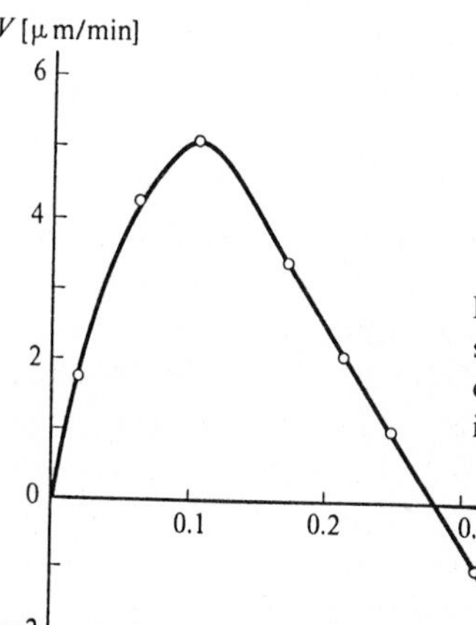

Fig.8.16. Typical dependence of the growth rate V of silicon films in the chloride process on the initial molar concentration of $SiCl_4/H_2$. The deposition temperature is 1270°C, the gas flow rate 1 l/min [8.37]

The dependence of the growth rate on the rate of flow (the growth rate increases as saturation is approached) points to the existence of a stagnant gas layer over the substrate. Model experiments show that in a horizontal reactor (Fig. 8.15b) this layer may reach several millimeters in thickness [8.43].

For practical purposes it is important to ensure a high output (typically up to 40 silicon wafers, each 76 mm in diameter) with a sufficiently uniform thickness of the grown layer (the spread from one wafer to another and within a given wafer should not exceed $\pm 2\%$). To do this, the substrate wafers are set up at an angle of 5°–15° to the flow (Fig. 8.15). Typical flow rates are 5 to 10 m^3/h and the concentration of $SiCl_4/H_2$ is about 2 mol. %. Then the depletion of the gas mixture along the path of the flow is offset by the inflow of the substance from flow sites far removed from the substrates. Typical growth rates of silicon film in the tetrachloride process are 1–2 μm/min at ~1150–1200°C.

The structural perfection of epitaxial films is extremely important in microelectronic applications of the process. Unfortunately, it is rather difficult to grow epitaxial films that are as perfect as the underlying substrate or even better, because apart from imperfections in the substrate (usually dislocations propagating into the film), new imperfections—in both number and type—are developed from the substrate–film interface. Typical imperfactions of epitaxial layers are dislocations (including aggregations, Fig. 8.17a), stacking faults

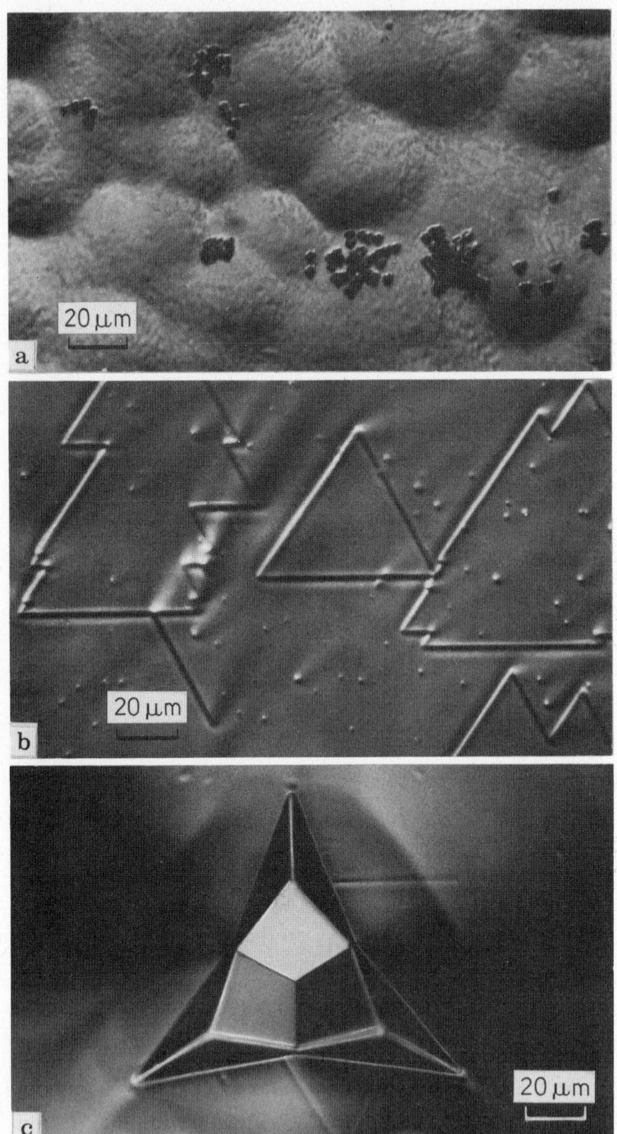

Fig.8.17a–c. Typical imperfections in autoepitaxial films of germanium. **(a)** Accumulations of etch pits corresponding to dislocation bundles (dislocations are usually formed at comparatively low crystallization temperatures); **(b)** stacking faults on the (111) surface; **(c)** "tripyramid" forming as a result of multiple twinning in (111) layers [8.44]. The photos were taken in an optical microscope

(Fig. 8.17b), and microtwins (Fig. 8.17c). It is generally accepted that they are caused by impurity particles (mainly oxides) on the substrate, although the formation mechanisms of imperfections in films are not completely understood.

The only way to eliminate them is by thoroughly cleaning the substrates. The most effective and widespread cleaning method is gas etching, which removes a layer approximately 5 μm thick from the substrate. A typical gas etchant is an $H_2 + HCl$ mixture, but the gas mixtures $H_2 + H_2O$, $H_2 + Cl_2$, $H_2 + SF_6$, and $He + Cl_2$ are also used in the semiconductor industry.

Let us now consider methods of thermal decomposition, or pyrolysis. A typical example is the silane method growing silicon films [8.45, 46]. Monosilane SiH_4 (normally a gaseous substance) decomposes at as little as ~700°C, but in practice the process is conducted at ~1000–1050°C. In terms of crystallization apparatus, this process is simpler than the chloride one (there is no need for bulky saturators, the gas flow control can be automated, etc.), but it has been found that the films are less perfect here, probably because the reaction

$$SiH_4 \rightarrow Si + 2H_2$$

is virtually irreversible. The quality of the films is improved by adding gaseous HCl to the $SiH_4 + H_2$ mixture and hence returning to the reversible chloride process; this combined method has found more extensive application in the semiconductor industry for silicon film preparation than the simple silane method.

Over the past few years, vapor decomposition methods conducted at low gas pressures have been developed and widely used. Two cases may be distinguished.

a) In the monosilane process with hydrogen as a carrier gas, silicon deposition rates are limited by the desorption of hydrogen from the substrate. Therefore, under reduced hydrogen pressure (10–100 torr) or when the hydrogen is replaced by helium, the deposition rates increase and/or it is possible to obtain epitaxial layers at relatively low temperatures [8.47]. The same is true for the silicon chloride–hydrogen process [8.48] and for the GaAs chloride process [8.49].

b) At lower pressures (<1 torr), mass transport is faster than surface kinetics. Under such conditions, very perfect, homogeneous epitaxial and polycrystalline Si films can be prepared [8.50].

In addition to the typical semiconductors (silicon and germanium), many other elements (tungsten, molybdenum, tantalum, etc.) can be deposited by hydrogen reduction or pyrolysis of halides or other volatile compounds [8.51]. Especially important is crystallization of diamond from the vapor. Auto-epitaxial diamond films have been grown on substrates of natural diamond by thermal decomposition of volatile carbon-containing compounds at about 1 kp/cm^2 or lower (i.e., at reduced pressures as against the diamond thermodynamic stability range) and at ~ 1000°C [8.52].

Finally, crystals and films of various compounds may be obtained by hydrogen reduction or pyrolysis of volatile compounds, Single crystals, films, and whiskers of silicon carbide have been grown by decomposition of methyltrichlorosilane (CH_3SiCl_3), dimethyldichlorosilane [$(CH_3)_2SiCl_2$], and trimethylchlorosilane [$(CH_3)_3$ SiCl] [8.53].

Owing to a number of advantages such as relative simplicity of the apparatus, high efficiency combined with good reproducibility of results, and ease of adaptibility to mass production, chemical vapor deposition methods have become classical in the modern epitaxial technology of semiconductors [8.54].

8.4.3 Vapor-Synthesis Methods

Vapor-synthesis methods are used to obtain crystals of compounds. Vapors of two, three, or more elements or their compounds are reacted in the high-temperature zone, and the synthesis reaction occurs, as a rule, on the surface of the crystal substrate. If the initial compounds are volatile, a single high-temperature zone is sufficient, and the equipment used in these methods is similar to that described in the preceding section. On the other hand, if at least one of the reactants is nonvolatile, furnaces having two, three, or more zones are used, usually with resistance heating.

Let us consider concrete methods for crystallizing the most important classes of compounds.

a) Silicon Carbide. Volatile or gaseous compounds such as $SiCl_4$ and CCl_4, $SiCl_4$ and CH_4, SiH_4 and CCl_4, and $SiCl_4$ and C_3H_8 are usually employed. The reaction is conducted in a hydrogen flow at 1200–1500°C, and the cubic β modification of SiC is formed. Plates of hexagonal α-SiC, and also silicon, sapphire, graphite, etc., are used as substrates. Sometimes the source of silicon is its own melt, because the vapor pressure over it at 1500°C is already considerable.

b) Semiconductor Compounds $A^{III}B^{V}$. Chemical vapor deposition is particularly attractive in the deposition of epitaxial layers of the $A^{III}B^{V}$ semiconductor compounds because of the difficulties that arise if they are decomposed by thermal evaporation in vacuum (Sect. 8.3.1). Let us consider the vapor synthesis methods for gallium arsenide, a typical representative of this class of substances. Laboratory and industrial methods have been developed to grow crystals of it, and these methods serve as a basis for obtaining other $A^{III}B^{V}$ compounds as well. Of the various approaches in vapor synthesis and epitaxial growth of gallium arsenide, the chloride process in the system $H_2 + AsCl_3 + Ga$ is at present the most typical and widely used in laboratories and industry.

The chloride process is presented in Fig. 8.18 [8.55, 56]. A quartz tube is placed in a three-zone furnace. The gas mixture entering the tube consists of H_2 and $AsCl_3$. In the first zone, which is heated to 400–600°C, $AsCl_3$ is reduced according to the reaction $2AsCl_3 + 3H_2 \rightarrow 6HCl + \frac{1}{2}As_4$. This arsenic vapor is initially absorbed by the gallium source, contained in a boat in the second zone, at about 800°C, until saturation is reached at 2.25 at. % (in accordance with the phase diagram of Ga–As). After this period, the arsenic vapor is trans-

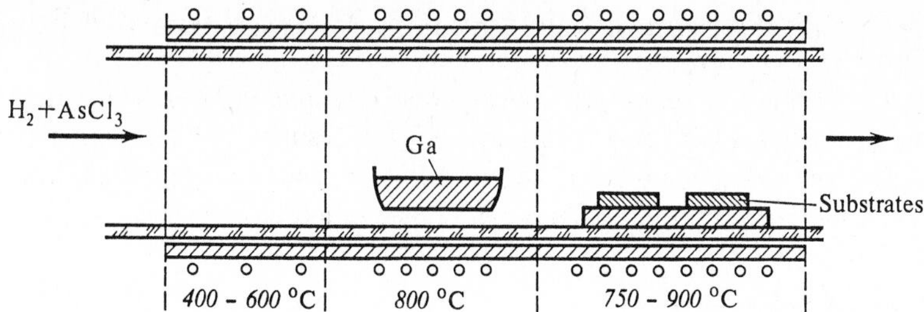

Fig.8.18. Diagram of the setup for growing epitaxial GaAs films in the chloride system H_2 + $AsCl_3$ + Ga

ported to the third zone together with GaCl, the product of the reaction of HCl and gallium. In this zone, at temperatures of 750–900°C, synthesis of gallium arsenide takes place on the surfaces of a single-crystal GaAs substrate, so that epitaxial films are formed. This process thus combines decomposition, transport, and synthesis. Epitaxial films of GaP, InP, InAs, etc., can be prepared similarly. In recent years, much attention has been directed to the deposition of quaternary compounds such as InGaAsP on InP substrates by this method [8.56]. The main advantage of the method is that the principal source materials—gallium, arsenic chloride, and hydrogen—are readily purifiable and hence ensure the preparation of high-purity crystals.

There are also other methods of synthesizing $A^{III}B^{V}$ compounds in which the different compounds of group-III and group-V elements, or chemicals generated as a result of different transport processes serve as sources of these elements.

In the hydride method, the source of the elements of group V is hydride, and of group III, chloride, which is formed by interaction of a metal with hydrogen chloride. For instance, the gas mixture H_2 + AsH_3 is reacted at 850°C with another gas mixture, which is formed when H_2 + HCl is passed over gallium at 750°C. Quaternary compounds can also be prepared by the hydride process [8.57].

Of special interest is the method of obtaining $A^{III}B^{V}$ compounds with the aid of metallo-organic sources. The outstanding characteristics of this method are that the sources of the elements of group III are its metallo-organic compounds (at room temperature, many of them are liquids with sufficiently high vapor pressures), and that a single heated zone is sufficient for crystal growth. In crystallizing GaAs use is made of a mixture of trimethylgallium $(CH_3)_3Ga$ and arsine AsH_3 in hydrogen; the compound GaAs is synthesized, and a film is crystallized at 650–750°C. The principal advantage of this method is that owing to the absence in the medium of chlorides capable of forming side products with the substrate, epitaxial deposition of GaAs films on sapphire, spinel, etc., is facilitated. This method has been used to obtain virtually all compounds of the type $A^{III}B^{V}$, including mixed crystals of GaAsP, GaInAs, InAsSb, etc. [8.58].

In connection with the requirements of optoelectronics, researchers have recently turned their attention to semiconductor compounds with a forbidden zone of ~2–3 eV, such as GaN, AlP, BP, BN, and AlN. A number of new processes are in use in the preparation of single crystals and epitaxial films of these compounds. For instance, to grow crystals and films of GaN, two flows are reacted in the high-temperature zone (1000–1100°C)—a mixture of NH_3 and H_2, and a Ga vapor carried by HCl gas. In growing crystals and films of boron nitride, $BBr_3 + H_2$ and $NH_3 + H_2$ mixtures are used, and for boron phosphide, $BBr_3 + PCl_3 + H_2$ or $B_2H_6 + PH_3 + H_2$, etc. [8.54, 59].

c) Semiconductor $A^{II}B^{VI}$ Compounds. The simplest method of synthesizing such compounds is to react vapors of elements (e.g., cadmium and sulfur for CdS), carried by an inert gas (i.e., argon) [8.60]. In another technique, Cd and H_2S vapors are used. In one of the latest methods the above-mentioned metallo-organic process is employed [8.4]. For instance, in the crystallization of zinc and cadmium sulfides and selenides, use is made of H_2S or H_2Se and diethylzinc or dimethylcadmium. To crystallize tellurides, the second element is also obtained by decomposition of dimethyltellurium, a metallo-organic compound. The crystal growth temperatures are 500–750°C.

To conclude this section, we note that although each of the above-considered methods can be used in growing both bulk crystals and films, there is a certain specialization, as in the methods using physical condensation. Chemical transport methods, both in the closed (ampule) and open (flow) systems, are used primarily to obtain bulk crystals, whereas the close-spaced ("sandwich") transport method, and methods involving the decomposition of compounds are used to obtain films. Synthesis methods are employed for both purposes.

8.5 Externally Assisted Vapor Growth

Many attempts have been undertaken in the past to stimulate crystal growth processes by external factors such as light, an electrical field, etc. Recently, advances have been made with various ion and/or plasma actions and with laser stimulation. In particular, ion-beam deposition facilitates epitaxial overgrowing [8.61], and plasma-assisted CVD processes can be implemented at temperatures 100–300°C lower than usual. Photostimulation improves the perfection of epitaxial layers of $A^{II}B^{VI}$ and $A^{IV}B^{VI}$ compounds [8.62], while the oxidation of silicon can be induced by laser illumination [8.63].

Of special importance are investigations on laser photodeposition of semiconductors and metals [8.64], which have opened up new prospects in microelectronics and materials science.

8.6 Crystallization from the Vapor via a Liquid Zone

8.6.1 A General Description of the Vapor–Liquid–Solid Growth Mechanism (VLS)

In 1964 *Wagner* and *Ellis* discovered a new growth mechanism, known as the "vapor–liquid–solid mechanism" (VLS) [8.65, 66], which proceeds essentially as follows: A gold particle is placed on a single-crystal silicon substrate of (111) orientation. On heating, the particle alloys with the substrate in accordance with the phase diagram of Au–Si of the eutectic type (with eutectic temperature of ~370°C), forming a droplet of a solution of Si in Au (Fig. 8.19). When a mixture, for example H_2 + $SiCl_4$, is introduced into the deposition zone, the reduction process occurs at the surface of the liquid droplet. The droplet becomes supersaturated with silicon to a value critical for growth at the solid–liquid interface, the excess silicon is deposited on the {111} planes which form the faceted interface, and the liquid droplet rises over the original substrate surface on top of the growing crystal. A prismatic column of Si grows under the droplet, thus continuing the substrate epitaxially, and the diameter of the column depends on that of the droplet.

The physical essence of the VLS growth mechanism is associated with the structure and properties of the vapor–liquid and liquid–solid interfaces. A liquid surface can be considered ideally rough, i.e., characterized by a high condensation coefficient, whereas a close-packed crystal face, such as (111) for a silicon crystal, has a relatively small condensation coefficient. The absolute values of

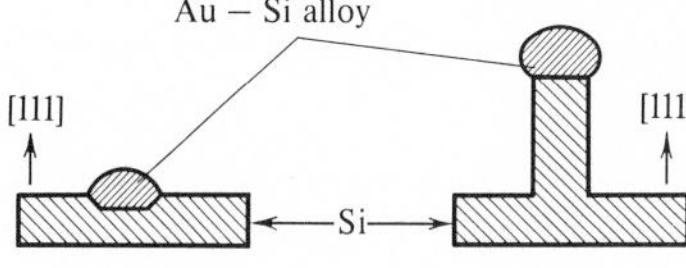

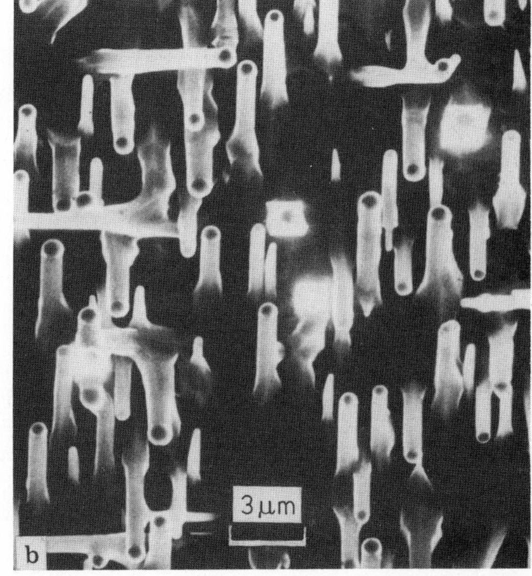

Fig.8.19. (**a**) Schematic diagram of growth of whiskers by the vapor–liquid–solid mechanism [8.60]. (**b**) GaAs whiskers on the (001) substrate of GaAs, plan view (the angle of incidence of the beam in the scanning electron microscope is 0°). The whiskers grow principally in four inclined ⟨111⟩ directions, with only some of them growing along the normal to the (001) face

these coefficients may differ considerably for crystallization from the vapor proper and from a chemical compound, but what is important for the action of the VLS mechanism is that these coefficients are much higher for a liquid surface than for a solid one. For example, during the growth of silicon whiskers in the silane and chloride processes, the evolution coefficient[3] of Si atoms from the corresponding compound is, under typical conditions, $\sim 10^{-3}$ for a liquid and $\sim 10^{-5}$ for a crystalline surface. Hence the efficiency of the liquid phase for a heterogeneous chemical reaction is about 10^2 times higher than that of the crystalline phase [8.67, 68].

Another important factor affecting the growth rate in the VLS mechanism is the nucleation of new crystal layers at the liquid–solid interface. Experiments show [8.68] that it is the processes at this interface that constitute the slowest of the four successive stages comprising the VLS mechanism. These are: 1) mass transport in the gas phase; 2) condensation and chemical reaction at the vapor–liquid interface; 3) diffusion in the liquid phase; and 4) the incorporation of atoms into the crystal lattice. The last stage, being the slowest, determines the overall rate of the process. Typically, the liquid–solid interfacial energy is 5–10 times lower than the corresponding energy of the vapor–solid interface. Consequently, the two-dimensional nucleation rate is many times higher at the liquid–solid interface than at the vapor–solid interface. This circumstance, coupled with the increased condensation coefficient on the liquid surface, the low diffusion resistance in micron-sized droplets, and the protective action (against unwanted impurities) of the liquid phase, sharply increases the growth rate by the VLS mechanism as compared with the vapor–solid mechanism.[4] The difference in rates is greater, the lower the temperature is; therefore whiskers often form at temperatures 100–200°C below typical temperatures for the growth of bulk crystals and films from the vapor phase.

Let us consider the case where the liquid–solid interface is formed by two different faces, one of which is close-packed. The free energy of this face is lower than that of the other; thus the barrier to two-dimensional nucleation is higher on it, and the growth rate is correspondingly lower than on the other face. Therefore the surface area of the close-packed face will increase, while that of the other, faster-growing face will reduce until the face ultimately disappears. At a growing surface which is formed of a number of faces, all but the close-packed faces will disappear. For this reason, on substrates (110)

[3] By evolution coefficient we mean the ratio between the number of atoms incorporated into the crystal or liquid surface and the number of molecules of the decomposing compound incident on the surface. This is analogous to the condensation coefficient in a chemical reaction.

[4] It should be noted that a comparatively low energy barrier to nucleation as well as protective action of the liquid phase, are also characteristic of ordinary high-temperature solution crystallization, i.e., of liquid phase epitaxy (Sect. 9.4). In the latter, however, the typical rates of single-crystal growth are much lower because of the high diffiusional resistance of the comparatively thick liquid layer.

of (100), which are not close packed in crystals with the structure of a diamond lattice (Si, Ge) or of sphalerite (GaAs, ZnSe), the vast majority of whiskers grow in inclined directions ⟨111⟩; in a few cases where the growth direction differs from the preferential one, which corresponds to the close-packed face, the crystallization front nevertheless consists of several such faces, which form a "roof." These experimental facts, taken together, indicate that in VLS crystallization the layer-by-layer mechanism is operative.

8.6.2 Growth Kinetics by the VLS Process

Whisker growth by the VLS mechanism offers a unique possibility of studying the kinetics and mechanism of crystallization, because conditions at the growing surface (such as the dimensions of the growing area, supersaturation, and temperature) remain virtually unchanged for a long time. Especially interesting in this respect are submicron filamentary crystals, in which the surface energy begins to play a substantial role [8.69, 70]. Experiments show that the growth rate of whiskers depends on their diameter (Fig. 8.20) and reduces to zero at some critical value that depends on the supersaturation (Fig. 8.21a). From this, and also from the fact that a drastic reduction in growth rate begins in all

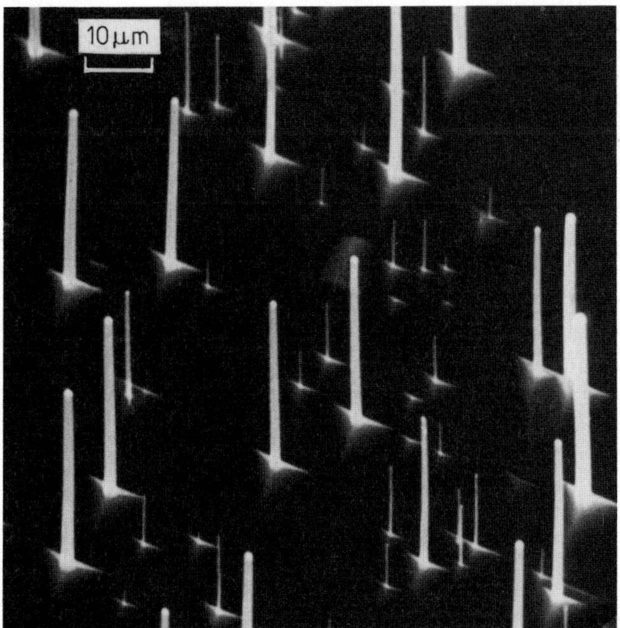

Fig.8.20. Silicon whiskers of different diameters. In the time elapsed since the beginning of the experiment, thick crystals ($d \simeq 1$ μm) elongated much more than thin ones ($d \simeq 0.1$ μm). The photo was taken in a scanning electron microscope [8.63]

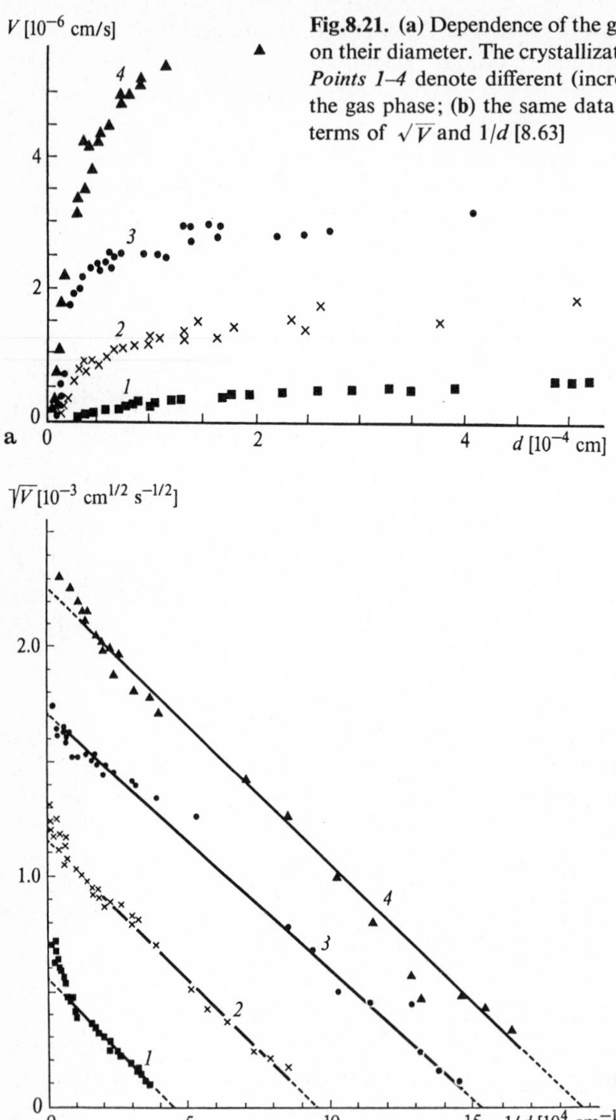

Fig.8.21. **(a)** Dependence of the growth rate V of Si whiskers on their diameter. The crystallization temperature is 1000°C. *Points 1–4* denote different (increasing) supersaturations in the gas phase; **(b)** the same data as in **(a)**, but expressed in terms of $\sqrt{V}$ and $1/d$ [8.63]

cases starting with diameters of $\sim 1\ \mu$m, we can conclude that the indicated dependence is due to the Gibbs-Thomson effect. In other words, the vapor pressure and solubility of the crystallizing substance (e.g., silicon) in the liquid increase as the whisker diameter becomes smaller, and therefore the supersaturation decreases. According to this effect, the supersaturation $\Delta\mu/kT$ is defined by the equation

$$\frac{\Delta\mu}{kT} = \frac{\overline{\Delta\mu}}{kT} - \frac{4\Omega\alpha}{kT}\frac{1}{d}, \tag{8.14}$$

where $\Delta\mu$ is the effective difference between the chemical potentials of silicon in the vapor phase and in the whisker; $\overline{\Delta\mu}$ is the same difference at a plane interface (i.e., when the diameter $d \to \infty$); α is the specific free energy of the whisker surface; Ω is the atomic volume of silicon; and k and T have the usual meaning.

On the other hand, the growth rate v (in this case whisker elongation) depends on the supersaturation. This dependence is not known a priori, but it can be determined from the experimental data. For singular faces the dependence is not linear, and in many cases is expressed in a power law:

$$V \sim \left(\frac{\Delta\mu}{kT}\right)^n. \tag{8.15}$$

If, instead of the proportion, we write an exact expression

$$V = \beta_n\left(\frac{\Delta\mu}{kT}\right)^n. \tag{8.16}$$

the proportionality factor β_n, which is called the kinetic coefficient, should not depend on supersaturation. This circumstance can be utilized in determining experimentally the values appearing in (8.16). Combining (8.14) and (8.16), we obtain

$$\sqrt[n]{V} = \frac{\overline{\Delta\mu}}{kT}\sqrt[n]{\beta_n} - \frac{4\Omega\alpha}{kT}\sqrt[n]{\beta_n}\frac{1}{d}. \tag{8.17}$$

Hence the dependence of $\sqrt[n]{V}$ on $1/d$ must be linear. If we rearrange the experimental points in Fig. 8.21a on the indicated coordinates for $n = 1, 2, 3, \ldots$, we find that at $n = 2$ they fit on straight lines with an equal slope, which corresponds to independence of the kinetic coefficient from supersaturation (Fig. 8.21b). If we assume that $n \neq 2$, the lines are not parallel, but converge ($n = 1$) or diverge ($n = 3, 4, \ldots$). This means that the growth of whiskers obeys the law $V = \beta_2(\Delta\mu/kT)^2$. The kinetic coefficient β_2 can be calculated from the slopes of the straight lines. The intercepts on the horizontal axis from the extrapolation of the straight lines determine the critical diameters d_{cr} (they also designate the diameters of the critical nucleus), and the intercepts on the vertical axis of the equation

$$\frac{\overline{\Delta\mu}}{kT} = \frac{4\Omega\alpha}{kT}\frac{1}{d_{cr}} \tag{8.18}$$

determine the effective supersaturations in the vapor phase at different concentrations of the gas mixture. Using (8.14), we can now determine the real supersaturations at the growing surface in relation to the diameter of the whisker.

The values corresponding to Fig. 8.21 for the growth of silicon whiskers in the chloride process are presented in Table 8.1. It can be seen that under typical conditions of crystallization, supersaturations are not large here, and the kinetic coefficient has values of $\sim 10^{-3}$–10^{-4} cm/s, which are typical for growth from concentrated solutions.

Table 8.1. Numerical data for the curves presented in Fig. 8.21 [8.69]

Experiment No.	$SiCl_4/H_2$ [mol %]	Kinetic coefficient β_2 [cm/s]	Critical diameter d_C [cm]	Effective supersaturation in vapour phase $\left(\frac{\Delta\mu_0}{kT}\right)$
1	0.3	4.6×10^{-4}	3.3×10^{-6}	0.20
2	0.75	4.5×10^{-4}	6.9×10^{-6}	0.11
3	1.5	3.8×10^{-4}	11×10^{-6}	0.07
4	3.0	4.2×10^{-4}	14×10^{-6}	0.05

The above dependence and values must be compared with the overall regularities of crystal growth. The quadratic dependence of the growth rate on supersaturation established here cannot be identified with the well-known parabolic law for the dislocation mechanism (Sect. 3.3.2), since experiments show that dislocations are absent in whiskers growing by the VLS mechanism. It has been shown that the quadratic dependence $V \sim (\Delta\mu)^2$ here can be regarded as empirical, and as approximating the weak exponential dependence; this corresponds to crystal growth by two-dimensional nucleation with a low energy barrier [8.68].

8.6.3 The VLS Mechanism and Basic Regularities in Whisker Growth

The VLS mechanism was put forward by *Wagner* and *Ellis*, primarily to explain the one-dimensional growth of whiskers from the vapor phase. The Sears model, which had previously been propounded and which related the growth of whiskers to the action of an axial screw dislocation, failed to explain the basic regularities, namely, 1) the unexpected, abrupt cessation of whisker gowth, which is easily explained in the VLS mechanism by the observed loss of the droplet as a result of instability; 2) the role of impurities stimulating whisker growth, which here is ascribable to the formation by the impurity of a liquid "cap," the active point on the growing surface; and 3) branching, bending, and periodic changes in the direction of growth and in the diameter of the whiskers, which can also be explained in the VLS mechanism. The branching is attributed to the breaking of the droplet, the bending and periodic changes

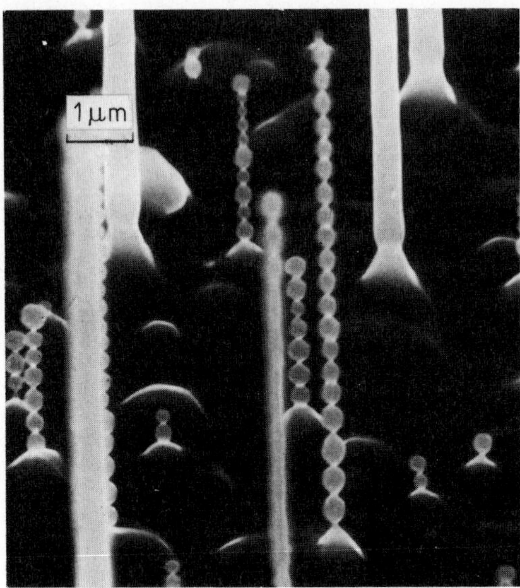

Fig.8.22. Radial periodical instability of Si whiskers. The photo was taken in a scanning electron microscope [8.62]

in growth directions to the climbing of the droplet from one tip face to another, and the changes in the whisker diameter (Fig. 8.22) to fluctuations in the contact angle of the droplet on the whisker tip.

The available experimental data show that the VLS mechanism is the principal one for whisker growth from the vapor phase. This must be interpreted as follows: A liquid phase alone is able to ensure unidirectional crystal growth, whereas other factors (such as axial dislocations, microtwins, poisoning impurities, mechanical stresses, etc.) play an auxiliary role, i.e., they can promote whisker formation only when combined with the action of the liquid phase.

The latest experimental results give direct evidence of the action of the VLS mechanism in the growth of metal whiskers by means of hydrogen reduction of halide salts, e.g., growth of copper whiskers by CuI reduction, with the salt being the liquid-forming impurity [8.71]. The VLS mechanism has also been shown to be operative in whisker growth by physical vapor deposition, for instance in the growth of Cd whiskers with Bi as the liquid-forming impurity [8.72].

8.6.4 Controlled Growth of Whiskers

As was indicated above, the diameter of a whisker growing by the VLS mechanism depends on the diameter of the drop, and therefore this mechanism can serve as a basis for controlled growth of whiskers. By creating a system of high-temperature solution droplets of the required composition on the substrate, one can grow an array of oriented whiskers. Figure 8.23a shows such an array on a substrate, part of which was coated prior to crystallization with

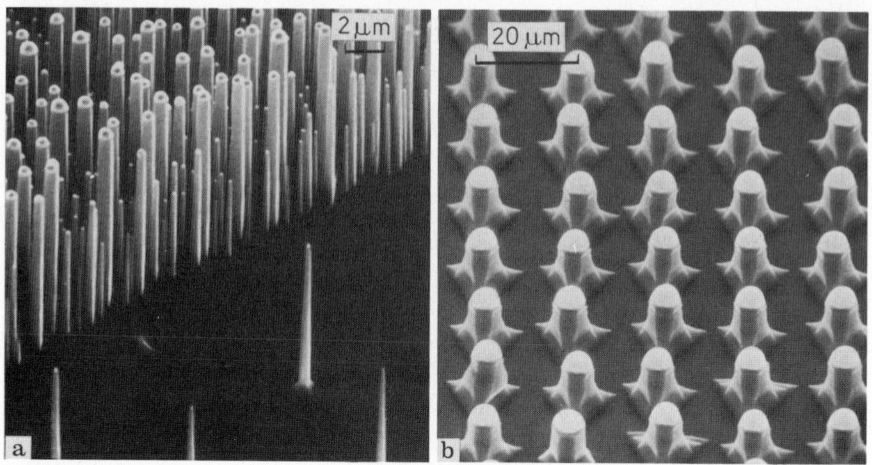

Fig.8.23. (**a**) Oriented array of Si whiskers on a (111)-Si substrate; (**b**) regular array of Si whiskers (photos taken in a scanning electron microscope [8.62])

Si–Au droplets, while the other was almost free of them. Figure 8.23b shows a regular whisker array obtained by depositing isolated metallic impurity dots by evaporation through a mask.

This method was also employed to grow whiskers of Ge, GaAs, GaP, SiC, ZnS, B, CdSe, etc. It is universal in the sense that it can be used to grow whiskers of any substance. *Wagner* and *Ellis* [8.73] established some requirements for liquid-forming impurities which must be satisfied in VLS whisker growth:

a) The distribution coefficients of the impurity must be far less than unity; otherwise it will be exhausted during growth.
b) The equilibrium vapor pressure of the impurity over the liquid alloy should be small.
c) The impurity must be inert to chemical reaction products.
d) The contact angle of the liquid alloy on the whisker tip must be large. (More precisely, to ensure stable growth it must exceed 90°, and optimally it should be about 95°–120°.)

It is possible to select, for any material, a solvent which meets these conditions and a chemical reaction which ensures the generation of the material in the required temperature range.

It should be pointed out that attempts to grow epitaxial films from a vapor through a liquid-phase layer failed, because a thin liquid layer is unstable and breaks into separate droplets under the effect of the surface tension, while a thicker layer shows a very large diffusion resistance.

Example. A chemically polished (111) silicon wafer is coated by vacuum evaporation with a gold film having a thickness of ~500 Å. The wafer is introduced into a chamber for epitaxial growth and heated in hydrogen to ~950°C. This results in the formation of a thin layer of liquid Au + Si solution, which breaks down spontaneously into separate droplets 100 Å to 10 μm in size, with a density of up to 10^7 cm^{-2}. Upon the delivery of H_2 + $SiCl_4$ mixture with a concentration of ~2 mol %, whiskers begin to grow, each droplet producing one whisker. Silicon whiskers ~25 μm high grow within five minutes [8.68].

8.6.5 The Role of the VLS Mechanism in the Growth of Platelets, Epitaxial Films, and Bulk Crystals and in Crystal Vaporization

Experiments show that in many cases platelet[5] crystals grow by the VLS mechanism. It is significant that whiskers and platelets can be genetically related so that they transform into one another depending on the crystallization conditions (temperature, supersaturations, etc.) [8.68]. Figure 8.24 shows an example of such a transformation. At first the CdSe whisker grew at a relatively low temperature, and then, at a higher temperature, which promotes disturbance of the stoichiometry, it was transformed into a platelet.

Recent investigations show that in many cases bulk crystals can also grow from the vapor with participation of the liquid phase. Owing to impurities (uncontrolled or intentionally introduced into the crystallization medium), a liquid phase can be formed on the step rises of the growing surface, in accordance with the crystal–impurity phase diagram. In some sections of the

Fig.8.24. Transformation of a CdSe whisker (indicated by the single *arrow*) into a platelet (indicated by two *arrows*) during growth by the VLS mechanism. The photo was taken in a scanning electron microscope [8.62]

[5] Platelike growth is generally typical of substances having a hexagonal structure with a clearly defined basal plane (such as crystals with wurtzite structure) and of compounds having a tendency towards disturbed stoichiometry.

steps the liquid phase forms droplets distinguishable even in a light microscope. These droplets serve as sites for the preferential deposition of atoms, and as a result so-called "protuberances," or spikes, are formed, which entrain the other, less active parts of the steps. This phenomenon can be termed "two-dimensional VLS growth" [8.74, 75]. An example is the growth of paratoluidine crystals in the presence of a water vapor (Sect. 3.4.1).

Another possible cause for the formation of the liquid phase on the crystal surface in growth from the vapor is a disturbance of the stoichiometry. The phase diagrams of many substances are such that when one of the components (such as Pb) is in excess, it can serve as the solvent of the compound forming on its basis (in this case PbS, PbSe, or PbTe). Given suitable conditions, a liquid alloy is formed on the surface of the crystal, and growth actually occurs from the high-temperature solution [8.22d, 24]. Other crystals that probably grow by a similar mechanism are single crystal platelets of SiC (from solution in Si), crystals of GdP (from solution in Gd), etc.

Finally, vapor–solid transformations via a liquid zone can also occur in the reverse direction. Specifically, vaporization of crystals via the liquid phase has been observed [8.76, 77]. As in growth by the VLS mechanism, the liquid phase accelerates the transformation here also, owing to the lowering of the barrier to two-dimensional nucleation. Under suitable conditions, elongated etch figures under crystal surfaces, or "negative whiskers," can be formed by the solid–liquid–vapor mechanism [8.78].

To conclude this section, it should be noted that crystal growth from the vapor via a liquid zone is a rather common phenomenon [8.79]. It reflects Ostwald's "step rule," according to which in phase transformations a substance must pass through all the intermediate stages. Further experiments will undoubtedly produce new examples of the operation of this mechanism.

9. Growth from Solutions

Growth from solutions is the crystallization of a compound whose chemical composition differs markedly from that of the initial liquid phase. Commonly used solvents are water, multicomponent aqueous or nonaqueous solutions, and melts of some chemical compounds. Differentiations are made based on growth temperatures and on the chemical nature of the solvent. In growth from low-temperature aqueous solutions, the temperatures generally do not exceed 80–90°C; in growth from superheated aqueous solutions (hydrothermal method), they reach 800°C, and in growth from salt melts (molten-salt, flux, or fluxed melt growth), they usually do not exceed 1200–1300°C, but sometimes reach 1500°C.

Growth from solutions is the most widespread method of crystal growing. It is always used for substances that melt incongruently, decompose below the melting point, or have several high-temperature polymorphous modifications and is also often efficient in the absence of such restrictions. The equipment is relatively simple, the crystals exhibit a high degree of perfection, and the conditions of growth–temperature, composition of the medium, types of impurities can be widely varied. It is also significant that the crystals generally grow at temperatures much below their melting point, and are therefore free of many defects inherent in specimens grown from the melt. On the other hand, the crystals are not grown in a one-component system, and the presence of other components (a solvent) materially affects the kinetics and mechanism of growth. The migration of the nutrient to the crystal faces is hindered, and thus diffusion plays an important part. The heterogeneous reactions at the crystal–solution interface are complicated by adsorption of the solvent on the growing surface and by the interaction between the particles of the crystallizing substance and the solvent (hydration in aqueous solutions and solvation in nonaqueous ones).

A theoretical analysis of the effects of the above factors on the mechanism of growth, the morphology, and the deficiencies of crystals is generally rather complicated. Therefore predicting processes and analyzing the data is generally more difficult here than when the crystal grows from its own vapor or melt. For this reason we have given more weight to empirical investigations.

9.1 The Physicochemical Basis of Growth from Solutions

9.1.1 Thermodynamic Conditions and Classification of the Methods

All methods of growing crystals from solutions are based on the dependence of the solubility of a substance on the thermodynamic parameters of the process: temperature, pressure, and solvent concentration. In the majority of cases the temperature dependence of solubility is used. Let us consider in more detail the effect of the indicated parameters on the solubility of various substances in water.

Most compounds are ascribed to one of two types, depending on their water solubility at increased temperatures [9.1]. The first includes many low-melting salts with a high solubility, such as nitrates, halides of alkali metals (except NaF and LiF), etc., and most alkalis and acids. The most characteristic compounds of the second type are H_2O and Na_2CO_3, NaF, K_2SO_4, SiO_2, and $CaMoO_4$ [9.1, 2].

The phase diagram of a system of the first type is shown in Fig. 9.1 [9.3] in projections onto the P–T, P–C, and T–C planes (C is the solute concentration). The solubility of a substance, i.e., its equilibrium concentration in solution, increases continuously with temperature up to the melting point T_0 of the poorly volatile compound (projection *T–C*, curve L of the compositions of saturated solutions). An increase in the concentration of the component less volatile than H_2O (NaCl in Fig. 9.1) increases the critical temperature T_{cr} of the solution (for pure water, T_{cr} = 374°C). The higher the water solubility of a

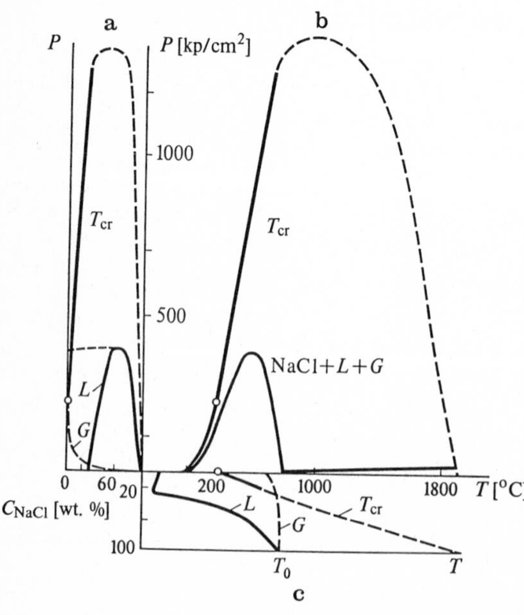

Fig. 9.1a–c. Example of a phase diagram of a first-type system [9.3]. Projections of the three-dimensional phase diagram of the NaCl–H_2O system onto coordinate planes *P–C* (a), *P–T* (b), and *T–C* (c). (**a**) Compositions and pressure of the coexisting liquid (L) and vapor (G) phases, (**b**) NaCl + *L* + *G* is the curve of the vapor pressure of saturated solutions; (**c**) *L* is the curve of the solubility of NaCl in H_2O vs the temperature; and *G* is the curve of the compositions of the vapor phases coexisting with the solutions. T_{cr} designates the curves of critical temperatures and the circles denote the projections of the critical temperature for water

given compound, the higher is T_{cr}. Consequently, if the poorly volatile component has a sufficiently high solubility, critical phenomena appear at very high temperatures. The critical curve intersects the T axis on the T–C plot at a temperature higher than T_0. The curves for the critical temperatures of solutions and for the three-phase equilibrium (solid–saturated solution–vapor, (curve NaCl + L + G) do not intersect in the projection onto the P–T plane (Fig. 9.1b), and hence no critical phenomena arise in saturated solutions. Such systems are characterized by a continuous three-phase equilibrium surface and a continuous critical curve, extending from critical T and P for one component (H_2O) to critical T and P for the other. The three-phase equilibrium curve (P–T projection, Fig. 9.1b) of first-type systems has a maximum; the temperature at the point of maximum, T_{max}, is related to the melting point T_0 by the empirical equation $1/T_{max} = 1/T_0 + 0.0021$, where the temperatures are expressed in kelvins.

The water solubility of compounds of the second type decreases with increasing temperature (Fig. 9.2c, T–C projection) and is low near the critical temperature of water. Compounds of this type include many substances which dissolve well in water at room temperature, but have a negative temperature coefficient of solubility (Na_2CO_3, Na_2SO_4; see below), and substances poorly soluble in water both at room and at higher temperatures (oxides and salts of high-melting metals, sulfides, silicates, etc.). For solutions of these salts the critical temperature of the solution rises only slightly because of their low solubility, and the critical phenomena arise in the saturated solution. In the phase

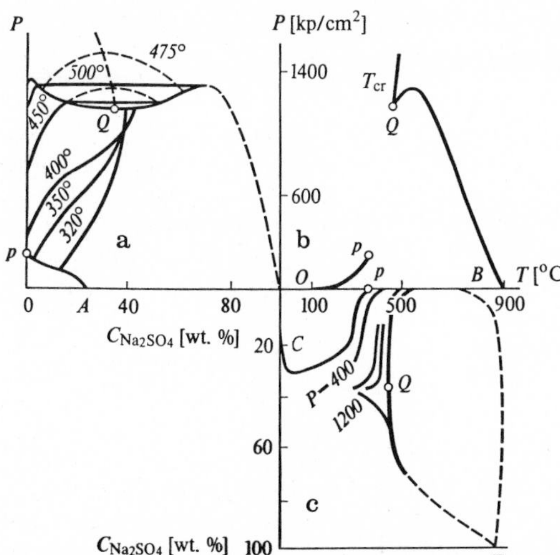

Fig.9.2a–c. Example of a phase diagram of a second-type system [9.2]. Projections of the space model for the phase diagram of a Na_2SO_4–H_2O system onto coordinate planes P–C **(a)**, P–T **(b)**, and T–C **(c)**. **(a)** Isotherms of the solubility of sodium sulfate. The 450°C isotherm corresponds to stratification of the solution; p is the lower, and Q the upper critical point; Ap is the curve of the vapor pressure of saturated solutions in the lower three-phase region. **(b)** Op is the curve of the vapor pressure of saturated solutions in the lower three-phase region, QB that of the vapor pressure of saturated solutions in the upper three-phase region and T_{cr} that of critical temperatures in the upper three-phase region. **(c)** The curve of solubility in the presence of the vapor phase is designated by pC. The *dashed lines* indicate the boundaries of the upper three-phase regions

diagram of such a system, the critical curve T_{cr} intersects at points p and Q with the curve for the three-phase equilibrium (solid–saturated solution–vapor), which consists of sections Op and QB (*P–T* projection, Fig. 9.2b; the critical curve is shown only in the upper three-phase range). At temperatures higher than that of the lower critical point p, the so-called fluid range is located —the divariant equilibrium range, where the supercritical fluid is in equilibrium with the solid phase. Above the second critical point Q, the system behaves like those of the first type.

Single crystals of compounds of the first type can be grown from low-temperature aqueous solutions. Compounds grown by the hydrothermal method are usually allocated to the second type in this classification; they will be treated in more detail in the section on growth from hydrothermal solutions (Sect. 8.3).

We shall now examine the principal prerequisites for growth from aqueous solutions.

Let us consider section *T–C* of the diagram for the most typical case, where the solubility of the test substance increases with the temperature (Fig. 9.3). The solubility curve (solid line) divides the diagram into two main regions: that of unsaturated solutions (below the saturation curve) and that of supersaturated solutions (above the curve). The region of supersaturated solutions is divided into a metastable and a labile zone. The metastable region is formed because energy must be expended on the formation of a critical-size crystal nucleus. Labile solutions are unstable, strongly supersaturated solutions in which crystal nuclei form readily owing to spontaneous fluctuations in the concentration of the substance (Sect. 2.1.2). The boundary separating the metastable and labile zones indicates the maximum supersaturation at which excess solute (relative to equilibrium solubility) does not crystallize spontaneously.

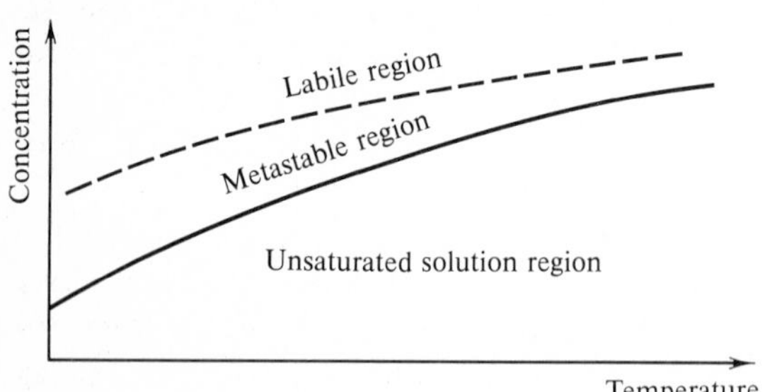

Fig.9.3. Schematic diagram of solubility for a substance whose solubility increases with temperature

Solutions are relatively stable in the metastable state (Sect. 2.1.2). This is due to the considerable work of formation of a critical-size crystal nucleus; the energy barrier cannot be overcome by means of natural fluctuations in the concentration of the substance. The simplest way of initiating growth from supersaturated solutions is to introduce seeds (small crystals of solute or some mechanical impurity), which serve as nucleation centers. Some metastable solutions lose their stability under the influence of factors which cause a chemical reaction in them (light and mechanical effects, an electric discharge, etc.). Since the stability of supersaturated solutions depends on many facotrs, the boundary between the labile and metastable zones usually cannot be clearly defined. It depends not only on the nature of the solutions and of the processes occurring in them, but also on the degree of purity of the starting materials. Solutions generally withstand supersaturations caused by supercooling only up to several degrees. Specially purified solutions tolerate supersaturations of several tens of degrees.

Controlled crystal growth is possible only from metastable solutions. The driving force of the process is the deviation of the system from equilibrium (Sect. 1.1.2). This can be conveniently characterized either by the supersaturaation ΔC or by the value of "supercooling," ΔT, i.e., the difference between the temperature of saturation of the solution and that of growth. The supercooling ΔT is related to the supersaturation ΔC by the expression $\Delta C = (\partial C_0/\partial T)\Delta T$, where $\partial C_0/\partial T$ is the temperature coefficient of solubility, i.e., the change in solubility of the substance per 1°C change in the temperature of solution. If the solubility of the substance is not known, the supercooling ΔT serves as a rough estimate of the deviation from equilibrium.

The methods of growing crystals from solutions are classified into several groups according to the principle by which supersaturation is achieved.

1) Crystallization by *changing the solution temperature*. This includes methods in which the solution temperature differs in different parts of the crystallization vessel (temperature-difference methods), as well as isothermal crystallization, in which the entire volume of the solution is cooled or heated.

2) Crystallization by *changing the composition* of the solution (solvent evaporation).

3) Crystallization by *chemical reaction*.

The choice of method depends on the solubility of the substance and the temperature solubility coefficient $\partial C_0/\partial T$. A few rules can be formulated to serve as guidelines in selecting a method.

1) If the temperature coefficient of solubility differs appreciably from zero, crystallization by changing the solution temperature may be used. Several versions should be distinguished here.

a) If the temperature solubility coefficient is relatively low (0.01–0.1 g/l deg), temperature-difference methods are preferable, irrespective of the absolute value of solubility. Such methods ensure prolonged, continuous growth of crystals in one part of the crystallization vessel as a result of continuous dissolution of the substance in the other part. Cooling methods are hardly suitable here, because cooling over a wide temperature range is required to extract appreciable amounts of substance from solution.

b) When the temperature coefficient of solubility is high (over 1 g/l deg), but the absolute solubility of the substance is low (several weight percent), temperature-difference methods are also preferable, because solution cooling, even over a wide temperature range, results in the extraction of only scant amounts of substance.

c) With high solubility and a high temperature solubility coefficient it is best to use solution cooling (or heating, if the solubility of the substance decreases with rising temperature). The higher the temperature coefficient of solubility is, the less suitable temperature-difference methods are here, because it is difficult to control the vigorous process of spontaneous nucleation.
2) If the temperature solubility coefficient is very low, crystallization can be achieved by solvent evaporation or by chemical reaction. The absolute solubility is not particularly important here, but it must not be too low in evaporation methods.
3) When dealing with poorly soluble substances, crystallization by chemical reaction is expedient.
4) Growth at a constant temperature and at a constant supersaturation in the growth zone is best ensured by feed-up methods, where the substance is fed into the crystallizer continuously as the crystal grows. The temperature-difference method can be considered a version of the feed-up method. In other versions a supersaturated solution is added to the crystallizer making use of various techniques and forced convection of the solution is realized.

The choice of method is often affected by factors other than solubility and the temperature coefficient of solubility. For instance, temperature difference methods are hardly suitable for viscous solutions, because solution-convection is impeded there. Therefore these methods are used comparatively rarely in growing crystals from solution in molten salts, as will be shown below. Under hydrothermal conditions, crystallization by reducing the overall temperature of the solution is inefficient because of the limited volume of the crystallizer and the relatively low solubility of the substances, and methods using convection under the effect of the temperature difference are generally used. Crystallization in gels is possible only with low-temperature solutions.

9.1.2 Mechanisms of Growth from Solutions

Crystal growth from solutions always occurs under conditions in which the solvent and the crystallizing substance interact. This interaction manifests itself in the formation of complexes of the solvent molecules (or ions) with those of the solute in the bulk of the liquid phase, as well as in the solvation (hydration in aqueous solutions) of the crystal faces.[1] In the latter case it is important to remember that the solvent is, as a rule, the predominant component of the solution, and therefore its effect depends on the heat and selectivity of adsorption, rather than on concentration, as in the case of trace impurities. At high adsorption energies a layer of a new chemical compound actually forms on the crystal surface, and the crystal ceases to grow.

Here, adsorption selectivity means preferred adsorption from solutions of structural associates whose short-range order corresponds to the arrangement of

[1] Crystals growing from solution usually have faces with low indices.

the adsorption sites on the crystal face. Because of the different crystallographic structure of the faces, selective adsorption on them occurs with differing intensity, and its effect on crystal growth is also manifested selectively.

An example of the influence of selective adsorption on crystal morphology is crystallization of corundum (Al_2O_3) from solutions in melts of tungstates of alkaline or alkaline-earth elements [9.4] at temperatures of 1100–1250°C. Corundum crystals are usually of platy habit with well-developed (0001) pinacoid faces or, in much rarer cases, of rhombohedral habit. In tungstate melts, the faces of the dipyramidal growth form are stable; the faces of a dipyramid $\{22\bar{4}3\}$ and a large rhombohedron are the most developed. Under the growth conditions used, tungstate melts usually polymerize, and chain linear tungsten–oxygen anions form. The change in the morphology of corundum crystals can be attributed to the selective adsorption of these anions on the faces. Analysis of the adsorption of polymer anions, taking into account the crystallochemical requirements and the geometric similarity in the structure of these anions and the arrangement of the adsorption sites on different faces of corundum, shows that chain tungsten–oxygen anions on $\{22\bar{4}3\}$ faces are preferentially adsorbed. As a result, an epitaxially adsorbed solvent layer is formed which completely covers the $\{22\bar{4}3\}$ faces and hinders their growth. If the polymer chains of tungsten–oxygen anions in the melt are destroyed, for instance by introducing fluorine ions, the adsorption conditions change. This is because for isolated tetrahedra of $[WO_4]^{2-}$, epitaxial adsorption on faces $\{0001\}$ is optimal, as demonstrated by analysis. Accordingly, corundum crystals grown from solutions in tungstate melts in the presence of fluorine have a platy habit.

The above example is limited to chemical adsorption of a solvent film having virtually monomolecular thickness. In actual fact the solution layer ordered by the orienting action of the crystal surface extends for a certain distance from the crystalline "substrate." It is not improbable that in aqueous solutions the thickness of the ordered liquid layer may reach tens and even hundreds of angstroms, although the available experimental data on layers of such thickness are not convincing enough. As in the above-discussed case of tungstate melts, the thickness of a partially ordered layer of an aqueous solution depends on the crystallochemical structure of the different faces of the crystal, and is at a maximum when the arrangement of the adsorption sites on the face corresponds most closely to the short-range order in solution. It is precisely the differing thickness of the adsorbed solution layer that is sometimes seen as responsible for changes in crystal morphology in aqueous solutions, sectorial entrapment of solution by the crystal, and a number of other defects.

The second "function" of the solvent is that it changes the manner in which the substance is transported to the growing face. In a motionless solution the delivery of the substance is ensured by diffusion, and growing crystals exhibit all the defects of growth in the diffusion regime (Sect. 5.1). In a pure diffusion regime the supersaturation differs over different areas of the faces (Sect. 5.1). To reduce this nonuniformity of the supersaturation and nutrition of different areas of the faces, motion of the crystal and solution relative to one another must be ensured. This is achieved either by free convection of the solution, resulting, for instance, from the temperature difference in different zones of the crystallizer, or by stirring the solution.

Even free convection of the solution markedly improves the nutrition of the growing crystal. But it is often insufficient, because in the steady state of free convection the flow of the solution near the crystal may be nonuniform: faces perpendicular to the solution flow, for instance, are in the most favorable position, while other faces may suffer from malnutrition. Furthermore, the solution velocity here can only be increased by increasing the temperature difference between the zone where the growing crystal is placed and the remaining volume of the solution, and this inevitably changes the supersaturation. Such poorly controlled convective stirring occurs in the hydrothermal method and also in growing from solution in the melt, where a different method of stirring the solution involves technical difficulties. In low-temperature aqueous solutions, rotation of the crystal in solution or stirring is usually applied. Stirring ensures motion of the solution relative to the crystal at velocities ranging from units to several tens of cm/s. The supersaturation is then regulated independently. The methods of stirring are treated in more detail in the following section.

9.2 Growth from Low-Temperature Aqueous Solutions

Crystallization from low-temperature aqueous solutions is extremely popular in the production of chemical reagents, fertilizers, and other crystal products. The required compounds are formed either by spontaneous crystallization or by the special introduction of small seeds (20–100 μm). The term "growth from solutions" emphasizes the more specific task of obtaining large single crystals, and it is this problem that we consider in the present section.

Before turning to the different methods of growing crystals from low-temperature aqueous solutions, we shall continue our consideration of the principal methods of stirring solutions used in crystal growth at low temperatures.

The simplest method is unidirectional central rotation of the crystal on the holder. It is used only rarely, because the solution flow which arises leads to the formation of cavities in the crystal and consequently later promotes the trapping of inclusions (Fig. 9.4). The faces receiving more nutrition grow more rapidly, while the others, which suffer from malnutrition, grow slowly and acquire various defects. Eccentric reversive crystal rotation is used more often (Fig. 9.5f). In reversive rotation the reversal period may last several minutes. The assigned rotation speed depends on the length and shape of the crystal support, the viscosity and amount of solution, and the crystal size, and may vary from several tens to several hundred rpm.

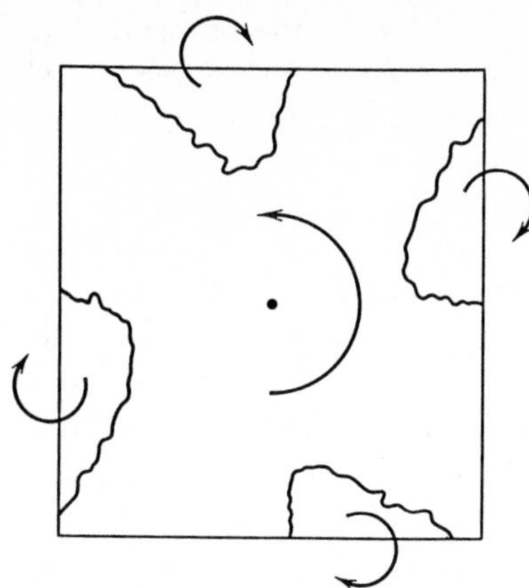

Fig.9.4. The formation of caverns in a crystal during unidirectional rotation [9.5]. The *arrows* indicate the rotation of the crystal and stationary eddies of the solution in the caverns

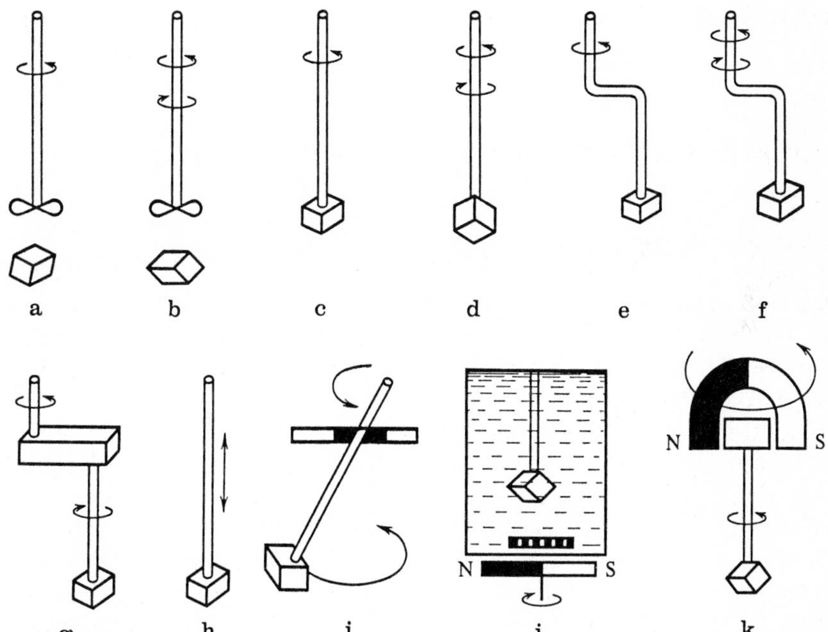

Fig.9.5a–k. Methods of solution agitation during crystal growth [9.5]. Unidirectional **(a)** and reversive **(b)** central stirring of the solution over the crystal; unidirectional **(c)** and reversive **(d)** central rotation of the crystal; eccentric unidirectional **(e)** and eccentric reversive **(f)** rotation of the crystal; planetary rotation of the crystal **(g)**; reciprocating motion of the crystal **(h)**; circular motion of the crystal **(i)**; stirring with a magnetic agitator **(j)**; rotation of the crystal holder with the aid of a magnet **(k)**

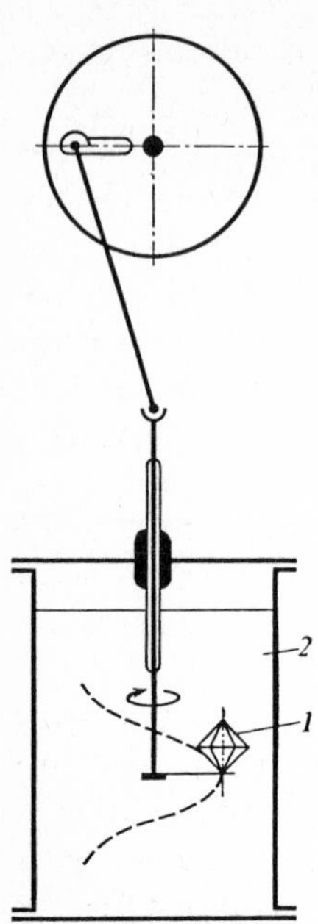

Fig.9.6. Spiral motion of a crystal (indicated by the *dashed line*) [9.6]. (*1*) Crystal, (*2*) solution. The period of rotation indicated by the *arrow* is half the period of the reciprocating motion

Reciprocating motion of the crystal compares favorably with rotation as regards stirring efficiency (Fig. 9.5h). If the crystal is fixed in a wide ring of large diameter, the ring acts as a piston, agitating the solution vigorously.

To grow crystals of perfect shape, spiral motion of the crystal has been used. It ensures uniform nutrition of all the faces (Fig. 9.6).

The introduction of agitators or rotating crystal holders into the crystallizer somewhat complicates its design, because special seals have to be used at the places where the rotating holder or stirrer is inserted. Various methods of hermetic sealing are described in [9.5]. Magnetic stirrers (Fig. 9.5j,k) do not have this drawback; they are extensively employed in crystal growing from low-temperature aqueous solutions.

9.2.1 Methods of Growing Crystals from Low-Temperature Aqueous Solutions

a) Crystallization by Changing the Solution Temperature

In this method two techniques are used to achieve supersaturation by reducing the temperature in the zone of the growing crystal:

a) Gradual reduction of the temperature *throughout* the crystallizer. As a rule, the reduction is continued for the entirety of the growing cycle.

b) Setting up *two zones with different temperatures* in the crystallizer. The substance is dissolved in one of the zones, and the crystal grows in the other. The mass transfer between the zones is maintained by natural or forced convection. Such methods are united under the general term "temperature-difference methods".

Crystallization by temperature reduction results from the cooling of the solution according to a preassigned program. In order for constant supersaturation to be maintained throughout the growth cycle it is important that the composition-temperature point on the phase diagram moves parallel to the solution saturation line in the metastable region. A shift to the unstable region must be avoided; otherwise there is danger of mass nucleation of parasitic crystals. The rate of temperature reduction varies from case to case and depends on the slope of the solubility curve and the growth rate at given supersaturations, i.e., on the condition of mass balance in solution. One advantage of the method is that the process can be calculated before hand. By using the solubility data, one can compute the rate at which the solution temperature must be changed in order for the required amount of substance to be extracted within a given time. The amount of substance extracted per unit time in a given temperature regime can also be determined.

The simplest illustration of the method is crystallization in a closed vessel (Fig. 9.7). A solution saturated at an above-room temperature is poured into a crystallizer, which is then hermetically sealed. Prior to this the solution is heated to a temperature slightly above the saturation point in order to avoid spontaneous crystallization at the moment it is filled into the crystallizer. A seeding crystal is suspended in the solution, and the crystallizer is placed in a water thermostat, whose temperature is reduced according to a preassigned plan. This method often produces sufficiently large crystals. Often, however, crystals acquire defects characteristic of immobile crystals grown in the diffusion regime. Therefore, more sophisticated crystallizers are usually used to grow large crystals; they ensure crystal rotation or intensive solution stirring. Heating elements are mounted directly in the crystallizer, and contact thermometers

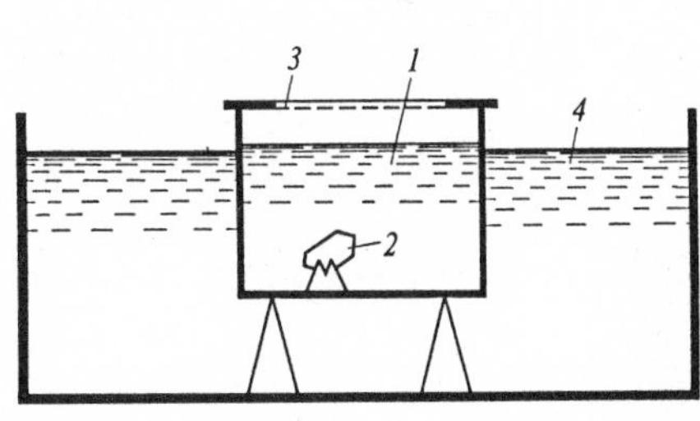

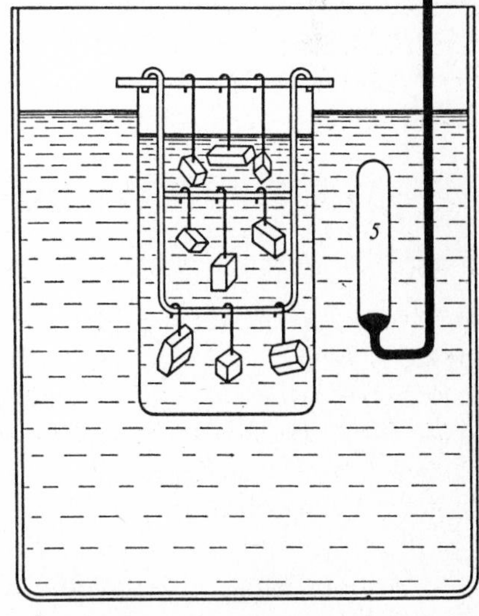

Fig.9.7a,b. Crystal growth in a closed vessel **(a)** [9.5] and growth of Rochelle salt crystals by slow cooling of a saturated solution in a Moore gas-heated crystallizer **(b)** [9.7]. (*1*) solution, (*2*) growing crystal, (*3*) glass lid, (*4*) thermostated liquid, (*5*) temperature control

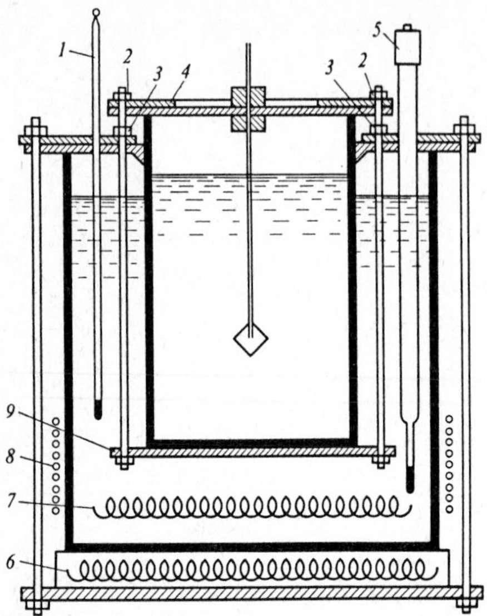

Fig.9.8. Setup for growing crystals by changing the solution temperature [9.5]. (*1*) Thermometer, (*2*) screws for fastening the lid, (*3*) screws for adjusting the position of the support in the thermostat, (*4*) lid, (*5*) temperature control, (*6–8*) various positionings of heater, (*9*) crystallizer support

are introduced. In another version the crystallizer is placed in a thermostat (Fig. 9.8), the growth temperature usually being maintained with an accuracy of ± 0.05°C.

A great number of crystals have been grown by the temperature-reduction method: Rochelle salt, triglycinesulfate, alums, etc. Figure 9.9 shows a crystal of triglycinesulfate, weighing about 1 kg, that was grown by the solution-cooling method in the temperature range of 55–20°C in a 4.5-l crystallizer [9.8]. The crystallizer was placed in a 20-l thermostat. Prior to crystallization the thermostat and solution were held 30–40 min at a temperature 5°C above the saturation point. The temperature was reduced according to a special plan so that the rate of crystal growth was several mm/d (up to 5.6 mm/d along the *c* axis). The solution was agitated vigorously.

Fig.9.9. Triglycinesulfate crystal weighing about 1 kg. Courtesy of I.V. Gavrilova

b) Temperature-Difference Methods

Temperature-difference methods are based on the formation of two regions with different temperatures in the crystallizer. In one of them, the substance which is always in excess in the form of the solid phase is dissolved, and in the other, crystal growth takes place. In the simplest version a tall vessel is used, with its lower part containing the nutrient and its upper part the suspended seed. The temperature in the lower part is kept higher than in the upper part. This results in convection of the solution, which ensures continuous upward movement of the substance into the growth zone. This arrangement is used in hydrothermal growing of crystals, and is discussed in more detail in Sect. 9.3. In growth from low-temperature aqueous solutions, two vessels connected by tubes are usually employed. The substance is dissolved in the vessel with the higher temperature, and the crystal is grown in the other (Fig. 9.10). The exchange between the vessels is achieved both by natural convection of the solution and by stirring with a mechanical agitator. At the beginning of the experiment, an equal temperature is set up in both vessels and maintained until the solution is completely saturated. Then the solution in the growth vessel is heated by a few degrees, and a seed is introduced. The vessel is held at this temperature for a short period and then cooled until the required temperature difference is established between the two volumes.

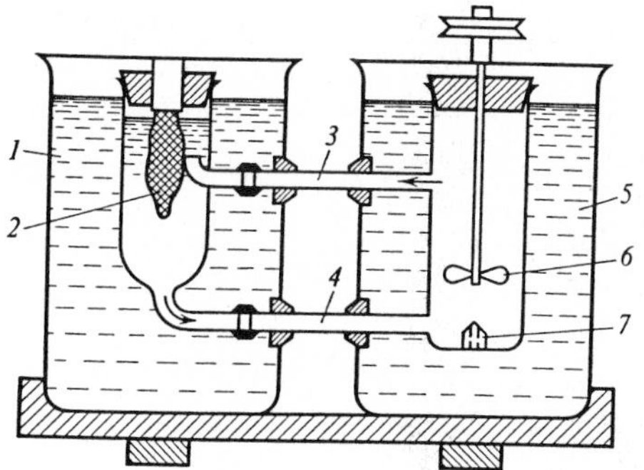

Fig.9.10. Growth from solution by the temperature-gradient method in a two-tank setup. (*1*) Thermostat for dissolution with a temperature T_1, (*2*) nutrient, (*3, 4*) connecting tubes, (*5*) thermostat for growth with a temperature T_2, $T_2 < T_1$, (*6*) vane-type agitator, (*7*) growing crystal

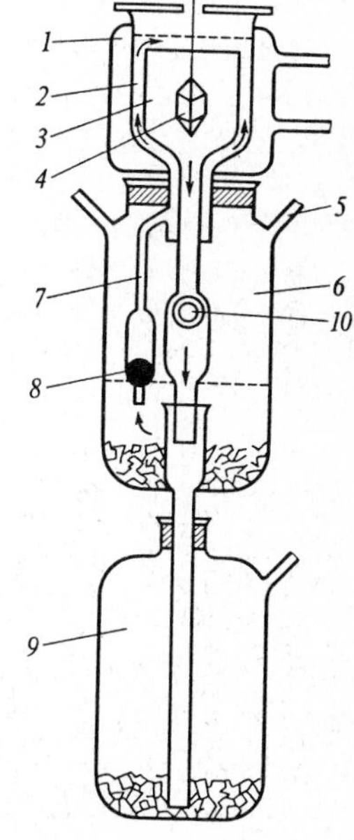

Fig.9.11. Setup for growing crystals by the temperature-difference method (see [9.9]). (*1*) Cooler, (*2*) tank containing the crystallization vessel, (*3*) crystallization chamber, (*4*) growing crystal, (*5*) branch pipe connected to a rubber bulb, (*6*) dissolution vessel, (*7*) branch pipe of the vessel containing the crystallizer, (*8*) ball walve, (*9*) standby vessel in the same thermostat as vessel (*6*), (*10*) valve. The *arrows* indicate the direction in which the solution moves

Due to natural convection in the above-described setup, the solution from the dissolution chamber moves along the upper tube into the growth chamber, and along the lower tube in the reverse direction. To avoid crystallization in the narrow connecting tubes, thermal insulation or additional heating of these tubes can be used. With the help of the agitator, the solution can be made to flow into the growth chamber along the lower tube, and the forced-convection mode then becomes established.

A number of setups have been developed with the crystallization and the dissolution vessels placed one over the other. One of them is illustrated in Fig. 9.11. As a result of periodic compression of the rubber bulb on tube 5, the solution, saturated at temperature T_1, arrives from vessel *6* via tube *7* at reservoir *2* with crystallization chamber *3*, where a temperature T_2 is maintained ($T_2 < T_1$). The depleted solution reenters vessel *6*. Any parasitic crystals which may form in crystallizer *3* are easily separated from the walls and removed to vessel *9*.

Walker and *Kohman* [9.10] designed a three-stage setup (Fig. 9.12). Growth of the crystal, which is fastened to a rotating crystallizer, takes place in vessel *I*, the nutrient is dissolved in vessel *II*, and the solution is heated above the saturation point in vessel *III*. This last operation dissolves any possible minute crystal nuclei. The setup has been used for growing ammonium dihydrogen phosphate (ADP).

c) Crystallization by Concentration-Induced Convection

Here, in contrast to the above-described method, the solution exchange between the dissolution and growth zones is ensured by the difference in density

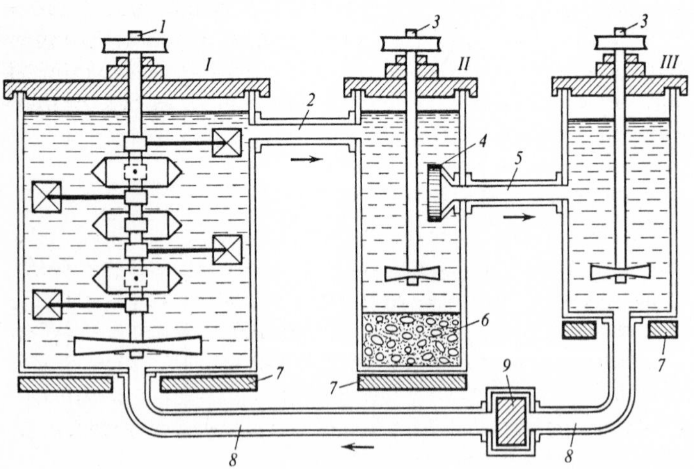

Fig.9.12. Three-tank setup for crystal growth [9.10]. (*I*) Crystallizer, (*II*) saturator, (*III*) superheater. (*1*) Rotating holder with crystals and stirrer, (*2, 5, 8*) connecting tubes, (*3*) stirrers, (*4*) filter, (*6*) nutrient, (*7*) support, (*9*) pump. The *arrows* indicate the direction in which the solution moves

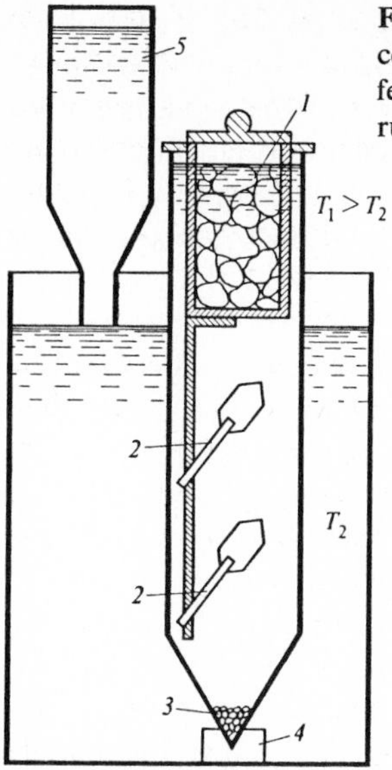

Fig.9.13. Setup for crystal growth by concentration-induced convection of solution [9.5]. (*1*) Dissolution chamber with the feeding substance, (*2*) crystal holders, (*3*) parasitic crystals, (*4*) rubber gasket, (*5*) vessel for replenishing thermostat water

between the saturated and unsaturated solution. The nutrient is placed in the upper part of the crystallizer, while the seeds are suspended below. The temperature is kept higher in the upper zone than in the lower one, and thus thermal convection is completely suppressed. One of the setups for this method is depicted in Fig. 9.13. The process is conducted in a glass tube 40–50 mm in diameter, the lower part of which is narrowed at the end to prevent further growth of the falling parasitic crystals. The saturated denser solution descends from the upper to the lower chamber, where it becomes supersaturated, which results in crystal growth. The chamber with the nutrient is usually a beaker made of glass or plastic. Crucibles with porous walls yield good results.

d) Crystallization by Solvent Evaporation

Supersaturation is achieved in this method by evaporating the solvent: as it evaporates, the solute concentration increases to above-equilibrium value. The process is carried out at a constant temperature under strictly isothermic conditions. Preferential evaporation of the solvent occurs "spontaneously" if contact is provided between the solution and the atmosphere. The rate of evaporation can be readily controlled by changing the solution temperature. To accelerate and control evaporation at a constant temperature, a stream of air or gas is often passed over the solution.

In crystal growing by solvent evaporation ali unfavorable factors resulting from temperature changes are eliminated. On the other hand, as the solvent evaporates, the solution is enriched in impurities whose distribution coefficient is below unity. Their concentration in the crystal changes accordingly. Another drawback of the method is that the supersaturation changes during the growth run. The change depends on the evaporation rate, which in turn is determined by the geometry of the setup, above all the surface area of the solution. In the absence of a seed the rate of change in supersaturation is proportional to the ratio of the area of the evaporating surface to the solution volume. In a cylindrical vessel this ratio is inversely proportional to the hight of the solution column. In other words, when equal volumes of solution are evaporated, the supersaturations are higher, the lower the column is.

In the presence of seeding crystals the increase in supersaturation is controlled by the growing crystal; the larger the surface area of the crystal, the more intensively it "takes up" the excess substance. Thus the most favorable situation, from the standpoint of keeping the supersaturation constant, obtains when many parasitic crystals form.

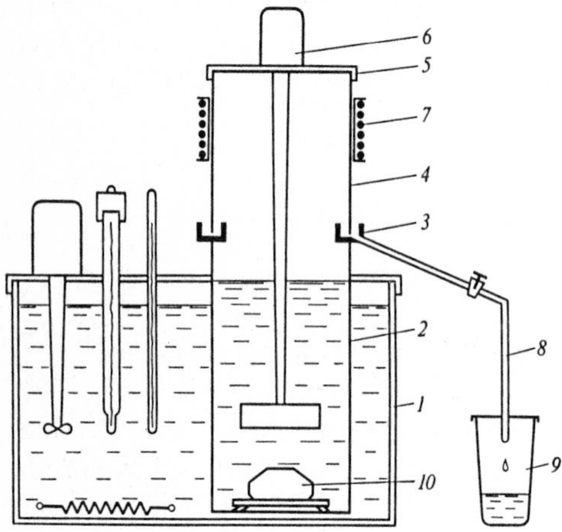

Fig.9.14. Apparatus for crystal growth by solvent evaporation [9.11]. (*1*) Water thermostat, (*2*) dismountable glass crystallizer, (*3*) organic glass ring with gutter, (*4*) glass cylinder (upper part of the crystallizer), (*5*) organic glass lid, (*6*) motor with stirrer, (*7*) heater, (*8*) drain tube, (*9*) beaker, (*10*) growing crystal

Figure 9.14 shows one apparatus design for growing crystals by solvent evaporation [9.11]. The glass crystallizer *2* is dismountable. Its upper part is a glass cylinder with a lid, on which a motor with a stirrer is mounted. Ring *3* (with gutter) serves to join the two parts of the crystallizer and simultaneously to drain the condensate precipitating on the upper part. The condensate flows down into a beaker through tube *8*. The rate of evaporation is controlled by changing the position of heater *7* on the cylinder (the higher the heater is raised, the more solvent is evaporated per unit time) and by opening and shutting the cock of the drain tube at the right time. This can be done according to a preassigned plan. The excess condensate again flows down the crystallizer walls into the solution. At an air temperature of 23°C, a solution temperature of 40°C, and with the annular heater switched off, the maximum rate of evaporation in this setup was 100 cm^3/d. Potassium alum crystals weighing up to 550 g grew in the apparatus within 52 days.

e) Growth from Aqueous Solutions at a Constant Temperature and a Constant Supersaturation

In the methods described above, growth proceeds either at changing temperature or at variable supersaturation. These two parameters — temperature and supersaturation — determine the growth rate and the impurity entrapment. The growth rate and the distribution coefficient, in turn, affect the purity and perfection of the crystal; therefore in order to obtain quality crystals crystallization should be carried out at a constant temperature and supersaturation. This is achieved by growing crystals under isothermic conditions and by adjusting the solution concentration in the course of the experiment. The adjustment can be made by diffusing the saturated solution into the crystallizer, introducing it in separate batches, or by forced circulation of the nutrient past the crystal.

One setup is presented in Fig. 9.15. It consists of crystallization vessel *1*, vessel *2* containing a solution saturated at a higher temperature, and thermostats enclosing both vessels. In vessel *2*, where an additional amount of substance equal to the weight of the growing crystal is dissolved, the solution is heated

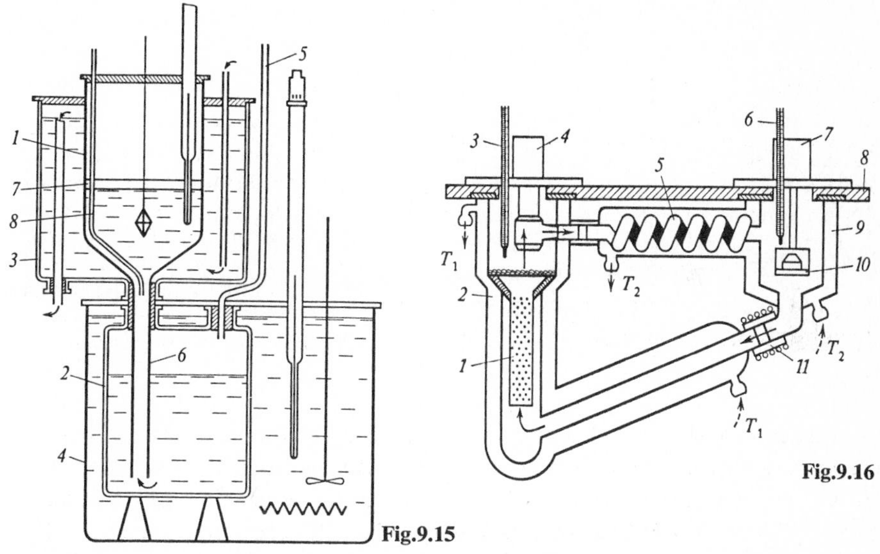

Fig.9.15. A setup for crystal growth at a constant temperature and supersaturation which permits the concentration in the crystallizer to be adjusted by adding superheated solution [9.12]. (*1*) Crystallizer, (*2*) vessel with adjusting solution, (*3*) thermostating jacket (the arrows indicate the direction in which the thermostating liquid moves), (*4*) ultrathermostat, (*5*) tube for vaseline oil supply, (*6*) tube for transporting the adjusting solution to the crystallizer, (*7*) layer of vaseline oil

Fig.9.16. Crystallizer for growing crystals at a constant solution supersaturation [9.13a]. (*1*) Quartz vessel with nutrient, (*2*) saturation chamber, (*3, 6*) control thermometers, (*4*) pump, (*5*) coil, (*7*) electric motor, (*8*) support, (*9*) growth chamber, (*10*) crystal holder, (*11*) additional heater. The *solid arrows* indicate the direction of motion of the solution, the *dashed arrows* that of the thermostating liquid

8–10°C above the saturation temperature. According to a preassigned schedule, vaseline oil is fed through tube *5* into vessel *2*, and pushes the solution into the crystallizer through tube *6*.

The setup for growing crystals at constant supersaturation shown in Fig. 9.16 has a growth chamber and a special feeding chamber with an excess of crystallizing substance. Each chamber has a thermostated jacket of its own; a temperature T_2 is maintained in the growth chamber, and T_1 in the feeding chamber ($T_1 < T_1$). The solution flows through the nutrient in the feeding chamber, becomes saturated, and is pumped to the growth chamber through a thermostated coil with the aid of a centrifugal pump. The depleted solution reenters the saturation chamber by an inclined chute. The inclined position of the chute and the conical bottom of the growth chamber promote the rapid removal, by the solution, of all parasitic crystals to the saturation chamber. This cyrstallizer was used to grow Rochelle salt crystals. At a growth rate of 0.2 mm/h, a solution pumping rate of 0.1 l/min is sufficient to saturate completely the solution in the feeding chamber. This crystallizer is convenient for growing large crystals in a small volume.

f) Crystallization by Chemical Reaction

This process is based on the extraction of solid products in the course of interaction between dissolved components. For instance, in the reaction $AC_{in} + BD_{in} \rightarrow AB_{sol} + CD_{in}$, AC_{in} and BD_{in} are the initial components, and AB_{sol} is the solid reaction product which can be extracted in the form of crystals.

This method is obviously possible only when the solubility of the resulting crystal is lower than that of the initial components. The absolute supersaturation is determined by the difference between the solubility of substance AB, which is equal (to a first approximation) to the product $[A^+][B^-]$, and the real product of the concentrations $[A^+]^*[B^-]^*$, which results from the dissolution of the initial components. Here $[A^+]^*$ and $[B^+]^*$ denote actual (non-equilibrium) concentrations of the ions in solution.

Chemical reactions in solution usually proceed at quite a high rate, leading to very high supersaturations and the precipitation of small crystals. Hence it is of the utmost importance that the rate at which the crystallizing substance forms be controlled. The reaction rate can be limited sufficiently to produce quality crystals either by using weakly soluble nutrient or by controlling the rate at which the substance is supplied to the reaction zone.

One version of this method is shown in Fig. 9.17a. Solutions of compounds AC and DB are poured into a crystallizer partitioned into three zones by a semipermeable membrane. The crystal seed AB is suspended in the middle zone. The crystal grows as a result of diffusion of A^+ and B^- ions into the middle zone through the semipermeable membranes. In another version, solutions of AC and DB are slowly added, for instance from droppers, to a vigorously agitated solution containing the seeding crystal. The rate at which these components are added is then used to control.

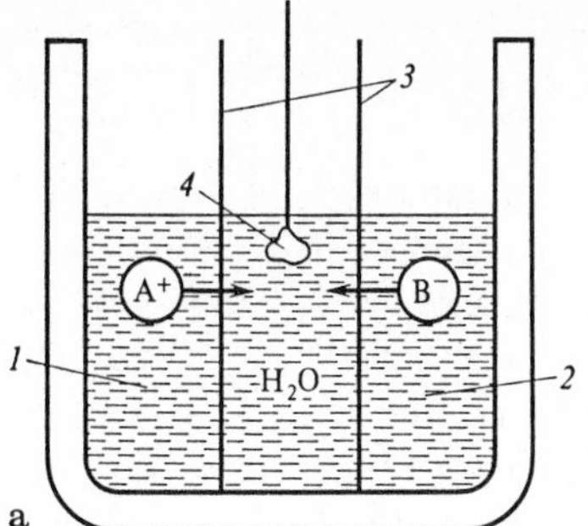

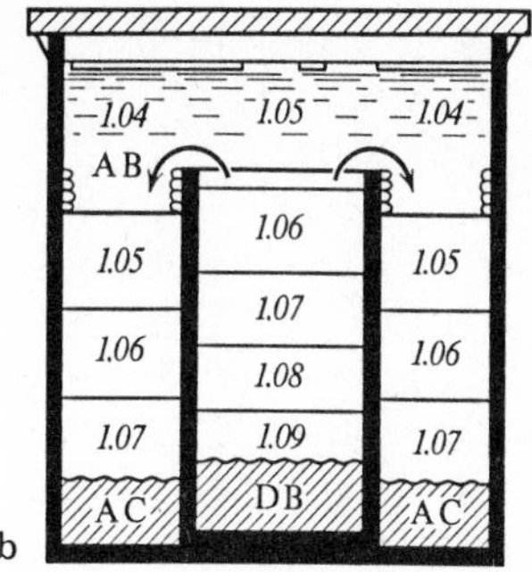

Fig.9.17a,b. Crystal growth involving a chemical reaction. **(a)** Setup with semipermeable membranes [9.13b]. (*1*) Solution of AC, (*2*) solution of DB, (*3*) semipermeable membranes, (*4*) seed. **(b)** Growth by the diffusion method [9.5]. The *numbers* give the solution densities at different depths, and the *arrows* indicate the substance diffusion from the inner to the outer beaker

If the substance enters the crystallization zone by diffusion, the reaction rate can be regulated by increasing the diffusion path or reducing the diameter of the crystallizer. One version of this method proceeds as follows (Fig. 9.17b): AC powder is poured onto the bottom of the crystallizer, and a beaker of a smaller diameter containing substance DB in powder form is placed in the center of it. Water is poured into the crystallizer, covering the beaker completely. Substances AC and DB dissolve, and a solution density gradient directed from top to bottom is set up in both vessels (the solution density increases with nearness to the solid deposit). Under these conditions, no concentration convection is observed, and the substance is delivered upwards only by diffusion. When solutions of AC and DB are mixed, crystals of AB are formed at the level of the beaker brim.

One of the most efficient ways of regulating the diffusion rate is by increasing the viscosity of the medium. This is commonly achieved by using gels.

g) Growth in Gel Media

The following experiment may serve as the simplest illustration of this method. A viscous substance, usually gelatin, an agar–agar solution, or silica gel (water glass) is poured into a u-shaped tube. A solution of AC is poured over the gel into one branch of the tube, and a solution of BD into the other. A^+ and C^- ions, and B^+ and D^- ions slowly diffuse towards each other in the gel medium and react to form a weakly soluble compound, for instance AD. The result is growth of crystals of this compound. A schematic representation of the process of growing calcium tartrate crystals in gel media by this method is given in Fig. 9.18.

Sometimes a different procedure is used: a gel containing one of the reactants is poured into a flask, and another reactant is poured over the gel. Crystal growth results from diffusion of the latter reactant into the gel medium. Growth in gels has produced many crystals of various classes, including carbonates,

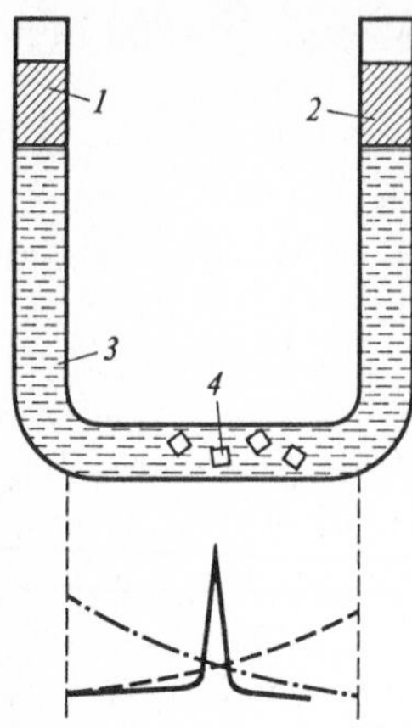

Fig.9.18. Growth of calcium tartrate crystals in gel [9.14]. (*1*) Concentrated solution of $CaCl_2$, (*2*) concentrated solution of tartaric acid, (*3*) gel containing concentrated tartaric acid, (*4*) growing crystals. The *lower part* of the figure gives values for the horizontal part of the tube. The concentration distribution of Ca^{2+} is shown by the *dot-dashed line*, the distribution of $C_4O_8^{2-}$ by the *dashed line*, and the product of the concentration of the ions $\mathcal{K}_i = [Ca^{2+}][C_4O_8^{2-}]$ by the *solid line*

tungstates, chlorides, silicates, etc. [9.14a–d]. As a rule, the crystals are well formed, stress free, and exhibit high optical perfection. But a substantial drawback of the method is the low growth rate; the crystals rarely attain a size of more than a few millimeters, even after growing for several months.

In addition to the above method of producing calcium tartrate crystals based on counterdiffusion of ions, the effects of dilution of the solution, decomposition of complex compounds, etc., are also implemented in crystallization in gel media. Some examples of the application of these methods are given below.

In growing gold crystals, an $AuCl_3$ solution with a concentration of 0.2n was mixed with a gel of sodium silicate having a density of 1.04 g/cm^3. The mixture was placed in a flask, and a 0.1–0.2n solution $(COOH)_2$ was poured over it. As a result of diffusion of the solution into the gel and of the reaction $2AuCl_3 + 3(COOH)_2 = 2Au + 6HCl + 3CO_2$, gold crystals up to 10^{-2} cm in size were grown within three days [9.15].

To grow the compounds HgS, AgI, and AgBr, which are poorly soluble in pure water, use was made of mineralizers, which considerably increase the water solubility of these compounds. Thus HgS was dissolved in an aqueous solution of Na_2S, and AgI and AgBr were dissolved in solutions of the corresponding potassium, sodium, ammonium, or calcium halides. The solutions thus obtained were poured carefully over a gel of sodium silicate with a density of 1.05 g/cm^3. The diffusion of the solvent into the gel medium changes the concentration of the solutions, and HgS, AgI, and AgBr are crystallized in them. Crystals as large as 3 and even 8 mm across were readily obtained. Crystals grew in the gel as well (as a result of partial diffusion of HgS, AgI, and AgBr into the gel), but these were much smaller in size [9.16].

Lead acetate was added to silica gel obtained by mixing a solution of sodium silicate and 3n hydrochloric acid, and a thioacetamide solution was carefully poured over the mixture. Hydrolysis of this solution was accompanied by the release of sulfur ions, which diffused into the gel and reacted with the lead ions. As a result, PbS crystals were formed [9.17].

One of the difficulties involved in growth from gels is the change in the process rate due to the steady reduction in the concentration of the reactants. In order to overcome this difficulty and achieve prolonged growth runs under constant conditions, a large volume of initial solutions must be available. The equipment proposed for this includes two large tanks connected by a horizontal tube of a relatively small diameter filled with gel medium. The large size of the tanks and the small volume of the reaction zone (the tube containing the gel medium) ensure a virtually constant concentration of the nutrient over a long period. For instance, crystals of CuCl were grown with the use of two 1-l vessels. One of them contained a solution of CuCl in hydrochloric acid, and the other distilled water. The connecting tube was filled with silica gel. Crystallization was achieved by diluting the CuCl–HCl solution with water in the gel bulk [9.18].

h) Crystallization by Electrochemical Reaction

Electrocrystallization can be regarded as a particular case of crystallization by chemical reaction involving electrons. A typical example is the extraction of metals in an electrolytic bath. To achieve electrocrystallization, two electrodes are placed in a solution, and an electric current is passed between them. The metal is deposited at the cathode. For instance

$$Zn^{2+} + 2e^{-} \rightarrow Zn_{sol} .$$

If a direct current is used, the metal is released at one of the electrodes, and with an alternating current, deposition proceeds at both electrodes.

Electrocrystallization has a number of advantages over the other types of growth in aqueous solutions. The process can be easily controlled by changing the current density; the substances can be easily separated and purified by selecting the appropriate voltage; and prolonged, continuous crystallization can be ensured by redeposition of the anode substance on the cathode.

On the other hand, the following requirements are necessary if successful electrocrystallization is to be achieved:

a) absence of strong interaction between the crystallizing substance and solvent,
b) limited voltage, not exceeding that of water decomposition (if an aqueous solution is used), and
c) high electrical conductivity of the solution or melt and of the substance deposited on the electrode.

Noncompliance with the third requirement leads to a rapid damping of the process as a result of electrode passivation.

It should be remembered that passivation of the electrodes (coating of their surface with a nonconducting substance) may also occur under the effect of impurities in the solution. These may be either cations and anions of a foreign substance, or solvent dipoles.

The above requirements are met most efficiently by metals; therefore electrocrystallization is mainly used for metal deposition. But the method can also be employed to obtain a number of other crystals by electrolytic decomposition of melts. For instance, crystals of MoO_2 were deposited by electrochemical reduction of a melt of sodium molybdate at 670°C. Single crystals of UO_2 up to 3 mm in size were deposited on a platinum cathode in electrolysis of a uranium chloride solution (UO_2Cl_2) in a eutectic sodium chloride–potassium chloride melt under reducing conditions [9.19, 20].

In a further development of this method, the technique of pulling the crystal out of the melt is combined with simultaneous electrolysis (Czochralski's electrochemical method). The growing crystal becomes one of the electrodes, and must possess sufficiently high electrical conductivity at the growth temperature. The possibilities presented by the method can be illustrated by cubic crystals of sodium-tungsten bronzes (Na_xWO_3, $0.38 < x < 0.9$) grown from a melt of Na_2WO_4 containing 5–58 mol % WO_3 [9.21].

The process was conducted according to the general reaction:

$$\tfrac{1}{2}x\ NA_2WO_4 + (1 + \tfrac{1}{2}x)\ WO_3 = Na_xWO_3 + \tfrac{1}{4}xO_2$$

The oxygen is released at the anode, while Na_xWO_3 is crystallized at the cathode. At a process temperature of 750°C and crystal pulling and rotating speeds of 3–4 mm/h and 30 rpm, respectively, Na_xWO_3 crystals 2.5 cm in diameter and up to 11 cm in length were obtained.

A survey of papers on crystal growing by electrochemical reaction can be found in [9.8, 22a–c]. Electrocrystallization has also found wide application in studying the mechanism of crystal growth [9.22b]. It was this method that first revealed the formation of two-dimensional nuclei on the surface of dislocation-free silver crystals in an $AgNO_3$ solution and made it possible to measure the linear energy of the steps bounding them (Sect. 3.3.4).

9.2.2 Growth of KDP and ADP Crystals

Crystallization of potassium dihydrogen phosphate (KH_2PO_4, or KDP) and ammonium dihydrogen phosphate [$(NH_4)H_2PO_4$, or ADP] is a characteristic example of crystal growth from low-temperature aqueous solutions [9.23–30].

The water solubility of KDP and ADP increases with temperature, and therefore solution-cooling (or dynamic) methods are predominantly used to grow them. The cooling rate is hundredths of a degree per day. For instance, large KDP crystals were grown at a temperature-reduction rate of 0.05 deg/d at the beginning of a growth run and 0.03 deg/d at the end. Crystals weighing up to 400 g grew in 1.5–2 months in a 3-l crystallizer. The crystal rotation

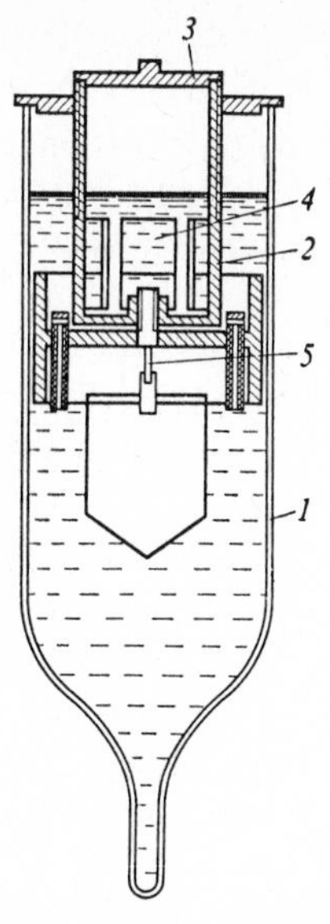

Fig.9.19. Crystallizer with an insert for crystal growth under static conditions [9.27]. (*1*) Crystallizer, (*2*) cylinder with nutrient, (*3*) lid, (*4*) windows covered with kapron textile, (*5*) crystal holder

speed was 80–100 rpm. The usual crystal growth rate is about 1 mm/d. ADP crystals are grown under similar conditions.

A static method has been developed for growing KDP crystals. In it, a temperature difference exists between the upper part of the crystallizer, where crystals (or a powder) are placed for dissolving, and the lower part, where the crystal is grown [9.27]. The crystallizer is a cylinder with a tapered bottom (Fig. 9.19). The nutrient is placed for dissolution at the top of the crystallizer in a special cylindrical insert. The growing crystal is affixed below it. From the insert, the saturated solution passes into the crystallizer through wide openings covered with kapron textile and descends to the bottom. The temperature difference can be set up in the crystallizer with the aid of a special heating element introduced into the dissolution zone.

Several such crystallizers are simultaneously placed in a thermostat in which a temperature difference is also established between the upper (T_1) and the lower (T_2) zone ($T_1 > T_2$). The thermostat is divided into two zones by a horizontal partition.

The growth rates are slightly lower in this method than under dynamic conditions, but it provides a constant growth temperature and nearly constant supersaturation in the growth zone, which relieves the stresses in the crystal.

The growth mode and the defect structure of ADP and KDP crystals depend on a number of factors, such as the presence of impurities in the solution, the growth rate, and the defect structure of the seeding crystal. Regularly faceted KDP crystals usually have the approximately equally developed faces of a tetragonal prism and dipyramid. The maximum growth rate is observed in the $\langle 001 \rangle$ direction, and therefore the crystals are elongated along the c axis.

Crystal surfaces normal to [001] begin to grow upon attaining a certain (critical) supersaturation value, which is equal to 0.5 g and 2.0 g per 100 g H_2O for ADP and KDP, respectively. The critical supersaturation for faces (100) and (010) is 3 and 5 g per 100 g H_2O for ADP and KDP. At higher supersaturations the growth rate is proportional to the square of the supersaturation. More accurate investigations have shown that the dependence of the growth rate of the $\{101\}$ faces of ADP crystals on the relative supersaturation σ in the range of $\sigma = 0.03$–0.06 is in fairly good agreement with that for the dislocation growth mechanism, and at $\sigma > 0.06$ can be described by a unified model of dislocation growth and growth by two-dimensional nucleation [9.31]. The experimenter often observes a decrease in the growth rate of $\{101\}$ faces of KDP crystals at constant supersaturation. This fact finds logical explanation within the framework of the unified growth model and can be attributed to the reduced activity of the dislocation groups leading the growth [9.32].

With a constant supersaturation the growth rates of the faces usually increase with temperature. For $\{101\}$ faces of KDP, an oscillating temperature dependence of the growth rate is observed, with anomalous peaks in the growth rates in the range of 50–53°C [9.33]. This dependence agrees with the changes in the viscosity of the KDP solution observed in this temperature range by *Wojciechowski* and *Karniewicz* [9.34]. The origin of this nonmonotonic behavior is not yet clear. It is discussed in [9.35].

The crystal habit transforms under the influence of impurities. When a KDP or ADP crystal is several millimeters across, its growth ceases almost completely, and increased supersaturation is required for it to be renewed. Further growth causes distortion of the crystal shape. The opposite faces of a prism become nonparallel, the faster-growing faces of a pyramid wedge out (Sect. 5.2.1), and the whole crystal acquires the shape of a rectangular bayonet (so-called wedging-out of a crystal). Then its growth becomes slower and slower, and finally stops altogether. With a subsequent increase in supersaturation, growth is not resumed.

If a seed plate is parallel to a prism face, the steps nucleating at the center of the growing surface coalesce and become higher under the influence of impurities, and their tangential motion in both directions parallel to the c axis slows down. As a result, the overgrown layer acquires the shape of a gable roof.

This kind of change in the growth mode is caused by the ions of multivalent metals such as Sn^{4+}, Cr^{3+}, Fe^{3+}, Al^{3+}, and Ti^{4+} contained in the solution.

The presence of even trace amounts of chlorides of trivalent metals may result in unstable growth, because impurities enter the adsorption layer. Accumulation of the impurity in the adsorption layer alternates with impurity trapping, and as a result the growth rate osillates. This is characteristic of $\{101\}$ faces. Consequently, the impurity enters the crystal nonuniformly and gives rise to striation [9.36, 37]. A similar, but still more pronounced pattern of impurity distribution is typical of the growth pyramid $\{100\}$, although no perceptible oscillations in the growth rate have been observed in this case. It can be assumed that the mechanism of interaction between impurities and faces {101} and {100} is differs slightly because of the difference in the structure of the atomic nets of these faces [9.28].

If the solutions are thoroughly cleansed of impurities, the distortion (wedging-out) of KDP and ADP crystals is less pronounced. For instance, in preparing solutions, the use of KH_2PO_4 that has been recrystallized 5–times markedly reduces the distortion of the crystal habit, but does not eliminate it completely. An increase in the initial temperature of growth to 78–83°C also produces favorable results. Another way of suppressing the impurity effect is to change the form of its presence in the solution. This form is an even more important parameter than the fact that an impurity is present. This is indicated by the obvious dependence of the degree of distortion in the crystal shape on the solution pH. At pH exceeding 4.5 (for ADP) or 5.0–5.5 (for KDP) the crystals retain their habit even if the solution is contaminated by Fe^{3+}. These ions may enter the crystal, causing coloring. An increase in pH to 5.8–8.6 results in precipitation of the impurity Fe^{3+} in the form of colloid particles of iron phosphate or hydroxide, and causes a sharp increase in the growth rate of the prism faces of KDP crystals and a decrease in that of the pyramid faces. Under these conditions it is easy to obtain crystals with a regular habit, but they are generally of low quality; they grow with many inclusions, and with internal stresses and crevasses.

At pH below 4.5–5.0 the wedging-out of ADP and KDP crystals is very clearly manifested. Since no considerable numbers of impurity colloid particles are observed here, they cannot be regarded as the cause of wedging-out. The wedging-out effect is due rather to complex-ion impurities in the solution. For instance, at pH below ~ 5 the solution may consist predominantly of intricate complexes of trivalent cations M^{3+} whith OH^- and H_2O of the type $[M(H_2O)_4(OH)_2]^+$. Their specific interaction with the growing surface leads to the phenomenon observed [9.38]. The effect of iron impurities on the wedging-out of ADP at pH values of less than 4.5 is ascribed to positively charged complex ions $[FeHPO_4]^+$ in solution. Preferential adsorption of these ions on a prism face reflects that the atomic structure of its surface layer consists of like-charged ions. The increase in solution pH breaks up the indicated complexes, and their effect on the crystal morphology diminishes accordingly.

Large KDP and ADP crystals are grown on seeding plates. These can be differently oriented, but preference is given to plates parallel to planes (001) and (011). Corresponding to each orientation is a specific type of regeneration zone, i.e., a region of the built-up layer adjacent to the seed. On plates cut out parallel to plane (001) crystal growth begins with the "restoration" of faces {101}. When this process is completed, a highly defective regeneration zone with a large number of solution inclusions, cracks, and other defects is established under these planes. The presence of such a zone is useful for the perfection of the subsequent layers, because any possibility of their inheriting the defects of the seed is precluded. In other words, the perfection of the layer formed is subsequently independent of the defectiveness of the seeding plate, since after the regeneration period the crystal develops on "restored" surfaces (101), which are highly perfect. X-ray topography investigations show that dislocations do not continue into the overgrown layer from the seed, and that new dislocations are generated only at the apex of the pyramid, at the places where planes {101} meet, or at the center of faces (Fig. 9.20).

When seeding plates parallel to (011) are used, the regeneration zone is either absent or only very weakly developed. When the natural crystal face {011} is used, the boundary between the seed and the newly formed layer is often unnoticeable, and only a limited number of small inclusions of solution are found along it. In this case the newly formed layer inherits practically all of the defects of the seeding crystal (Fig. 9.21) Dislocations also arise on some of the small solution inclusions at the boundary between the seed and the new layer. If the seed dissolved slightly prior to beginning of growth, the number of solution

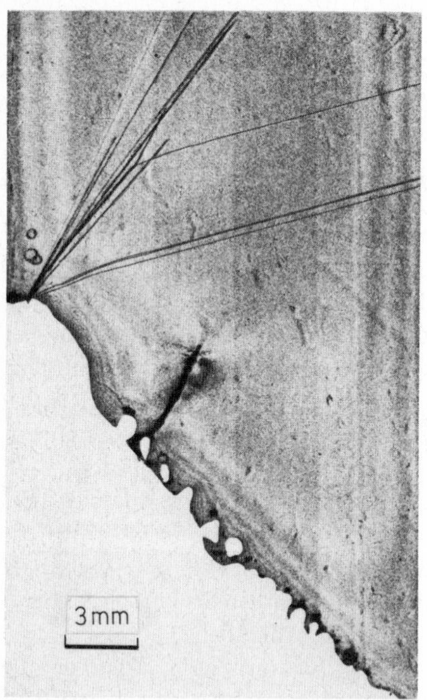

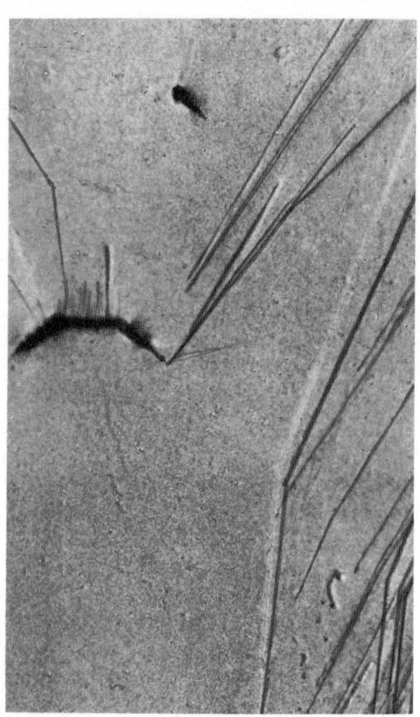

Fig.9.20 **Fig.9.21**

Fig.9.20. X-ray topographic pattern of a KDP crystal grown on a seed plate paralled to (001) [9.29]. Only the grown layer is shown. The seed separated during treatment of the crystal. The ragged lower edge is the boundary between the seed and the grown layer

Fig.9.21. X-ray topographic pattern of the X section of a KDP crystal grown on a seed plate paralled to (011). The position of the seed–crystal interface is determined from a slight change in the direction of the dislocations inherited by the crystal from the seed. There is no diffraction contrast at the interface; this is an indication that there are no noticeable streses in the region [9.29]

inclusions along the boundary between the seed and the fresh layer increases, and hence more dislocations are generated at the boundary. It is particularly undesirable to use seeds with a cut or polished (011) surface, as a maximum number of dislocations are then generated at the boundary.

In using seeds along (011), a favorable effect is achieved by artificially increasing the thickness of the regeneration zone. To do this, the seeding plate is cut out at an angle to surface (011). The growth begins with the restoration of this surface. This results in the formation of a wedge-shaped regeneration zone whose thickness depends on the degree of deflection of the plate from plane (011). The subsequent growth is similar in nature to the above-described growth on seeds (001): the regeneration zone prevents the fresh layer form inheriting the seed defects, and practically no new dislocations arise on the restored surface (011) (Fig. 9.22). This method can be used to grow sufficiently perfect crystals on seeding plates parallel to (011).

Dislocations in a crystal arise not only on the surface of the seeding plate, but also on defects in the bulk of the newly formed layer — on inclusions of

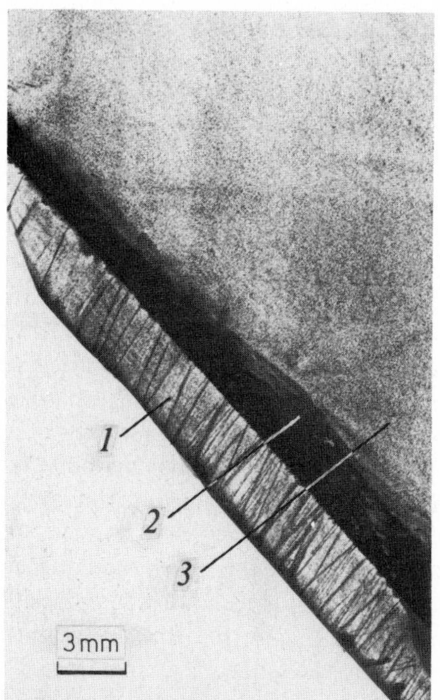

Fig.9.22

Fig.9.23

Fig. 9.22. X-ray topographic pattern of a KDP crystal grown on a seed plate parallel to (011) [9.29]. (*1*) Seed, (*2*) regeneration zone hindering the growth of dislocations from the seed into the crystal, (*3*) a newly grown, dislocation-free crystal

Fig. 9.23. Bundles of dislocations originating on inclusions in an ADP crystal. X-ray topograph, pattern [9.29]

impurities, of mechanical particles [9.37], and of solution (Fig. 9.23). The formation of these defects is, in turn, promoted by stepwise fluctuations in the growth temperature and the attendant sharp changes in supersaturation, by high growth rates, and by insufficient solution circulation. For instance, when the rate of ADP crystal growth along the c axis is 0.7 mm/d, stepwise temperature changes of 0.03°C, which result when the heating elements are periodically switched on and off, are sufficient to cause the formation of inclusions. Improved thermostating of the crystallizers decreases the probability of formation of solution inclusions and thus reduces the dislocation density in grown crystals.

9.3 Growth and Synthesis in Hydrothermal Solutions

In the preceding section we discussed crystallization from aqueous solutions at comparatively low temperatures of less than 100°C, at which the vapor pressure over the solution is appreciably less than atmospheric. For a large number of substances, the solubility is so low at such temperatures that growing crystals becomes impracticable. One way of increasing the solubility is to increase the solution temperature. The various techniques of crystallizing substances from high-temperature aqueous solutions at high vapor (solution) pressures are united under the term "hydrothermal method." This method is characterized by the presence of an aqueous medium, temperatures above 100°C, and higher-than-atmospheric pressures.

The term "hydrothermal" is of geologic origin. Minerals formed in the postmagmatic stage in the presence of water at elevated temperatures and pressures are said to be of hydrothermal origin. Early attempts to produce artificial crystals under hydrothermal conditions were undertaken by mineralogists with the aim of studying the conditions of natural mineral formation; the size of the crystals obtained usually did not exceed thousandths or hundredths of a millimeter [9.39]. The first steps in developing the hydrothermal method for single-crystal growth were made by *De Senarmon* and *Spezia* (see [9.40]), who grew crystals of α-quartz.

The hydrothermal method of crystal growing, like the other methods of growing from solution, is based on the dependence of the equilibrium concentration of the crystallizing substance in solution, C_A, on the thermodynamic parameters determining the state of the system: pressure P, temperature T, and solvent concentration C_B. An important feature of hydrothermal growth is the use of a mineralizer B, which is introduced into the A–H_2O system to improve the solubility of the poorly soluble component A. The mineralizer is often simply called a solvent, although strictly speaking, the mineralizer solution in water, B + H_2O, is the solvent. Growth systems are generally composed of at least three components, such as A–B–H_2O, where A is the compound to be crystallized, and B is a highly soluble compound, i.e., the mineralizer.

Thus, the hydrothermal method essentially consists in creating (by means of high temperatures, high pressures, and the introduction of a mineralizer) conditions that ensure the transformation of the crystallizing substance into a soluble state, the necessary supersaturation of the solution, and the crystallization of the test compound. The supersaturation value can be controlled by changing the system parameters determining the solubility C_A of the test substance (temperature, pressure, type and concentration of the mineralizer) and the temperature difference between the dissolution and growth zones. The hydrothermal method makes it possible to grow crystals of compounds with high melting points at much lower temperatures, and to obtain crystals which cannot be grown by any other method. Crystals of the cubic modification of ZnS (sphalerite) are a vivid example.

Sphalerite (ZnS) crystals cannot be obtained from the melt, because they undergo a polymorphous transformation at 1080°C and change to the hexagonal modification, wurtzite. Under hydrothermal conditions, however, sphalerite growth proceeds at a temperature (300–500°C) much lower than the indicated phase transition temperature, i.e., in the stability range of the cubic modification [9.41, 42].

The hydrothermal method has proved to be extremely efficient both in the search for new compounds with specific physical properties and in systematic physicochemical investigations of intricate multicomponent systems at elevated temperatures and pressures.

The choice of a particular procedure for growing crystals depends above all on which parameter is the decisive one in changing the solubility of a given compound.

9.3.1 Methods of Growing Crystals from Hydrothermal Solutions

a) Temperature-Difference Method

This method is most extensively applied in hydrothermal synthesis and crystal growing. The necessary supersaturation is achieved by reducing the temperature in the crystal growth zone. The starting material is placed in the lower part of an autoclave filled with a specific amount of solvent. The autoclave is heated so as to set up two temperature zones. The temperature T_1 in the lower part of the autoclave (the dissolution zone) exceeds that in the upper part (growth zone), T_2. We denote the concentrations of the crystallizing substance A in the solution as C_1 in the "hot" part and C_2 in the "cold" part of the autoclave. The solution density ρ depends on the solution temperature and concentration, and therefore ρ has different values along the autoclave axis. An increase in concentration C usually increases the density, and a rise in temperature reduces it. If $\rho(C_1, T_1) < \rho(C_2, T_2)$, i.e.,

$$\frac{\partial \rho}{\partial C}(C_1 - C_2) + \frac{\partial \rho}{\partial T}(T_1 - T_2) < 0\,, \tag{9.1}$$

the cooler and denser solution in the upper part of the autoclave descends while the counterflow of lighter solution ascends. Thus upon the attainment of a certain value of ΔT which is specific to each compound, the thermal expansion of the liquid in the dissolution zone reduces the solution density to a greater extent than the charge dissolution increases it, and a convective motion of the solution arises in the autoclave. The aqueous solution saturated with component A at temperature T_1 is transported upwards, the solution becomes supersaturated in the upper part as a result of the reduction in temperature to T_2, and crystallization sets in. The value of supersaturation ΔC_A is assigned by the difference in temperature between the growth and dissolution zones, ΔT.

The above method ensures continuous mass transfer from the lower to the upper zone of the autoclave until the nutrient completely dissolves. The amount of substance crystallized per unit time, $dm/d\tau$ (where m is the mass of the newly formed crystals, and τ is the time) largely depends on the convection velocity; the relationship between these two values is either linear or more complicated.

The temperature-difference technique is most commonly used and is practically the only technique in industrial growing from hydrothermal solutions. It has been used to grow large single crystals of quartz (SiO_2), ruby (Al_2O_3), calcite ($CaCO_3$), and zincite (ZnO), and has produced single crystals of almost all classes of compounds, from elements to complex silicates [9.40, 43–48]. The version of this method which implements a direct temperature difference ($T_1 > T_2$) can obviously be used for substances with a positive temperature coefficient of solubility. If the solubility of a substance reduces with an increase in temperature ($C_1 < C_2$ at $T_1 > T_2$) and $\rho(C_1, T_1) \geq \rho(C_2, T_2)$, the method of inverted temperature difference is used, in which the charge is placed in the cooler zone, and $T_1 < T_2$. Here it is also advisable to position the autoclave horizontally, as this permits multiple recrystallization of the nutrient by changing the temperature and alternating the growth and dissolution zones in the course of the experiment. Cyclic recrystallization is of considerable importance in purifying the starting material[2] and enlarging the nutrient particles. With complex compounds it also makes it possible to synthesize the individual components beforehand and grow the crystals afterwards.

The temperature-difference technique, in any of its versions, is applicable only when the solubility of the substance changes appreciably with the temperature. The higher the absolute value of the temperature coefficient of solubility $\partial C_A/\partial T$, the higher is the supersaturation $\Delta C_A = (\partial C_A/\partial T)\Delta T$ that can be achieved at the same temperature difference. For each substance there exists a minimum supersaturation which is necessary to ensure an appreciable growth rate. The values of relative supersaturation $\sigma = \Delta C_A/C_A$ that are typical of hydrothermal conditions lie in the range of 0.01–0.1.

[2] Crystals obtained hydrothermally usually contain fewer impurities than the initial material; "accidental" (nonstructural and mechanical) impurities pass into the solution.

b) Temperature-Reduction Technique

In this technique crystallization takes place without a temperature difference between the growth and dissolution zones; the supersaturation necessary for growth is achieved by gradually reducing the temperature of the solution in the reactor volume. Mass transfer to the growing crystal occurs mainly by diffusion; there is no forced convection. As the solution temperature decreases, a large number of crystals nucleate and grow spontaneously in the autoclave. The temperature-reduction technique was used by *Rooijmans* [9.49] to crystallize lead oxide (PbO) in NaOH solutions. The autoclave was held at 450°C until the solution was completely saturated, and then cooled to 250°C at the rate of 1–2 K/h. Platy PbO crystals having an area of up to 50 mm^2 grew as the temperature decreased.

The disadvantage of the temperature-reduction technique is the difficulty involved in controlling the growth process and introducing seed crystals. (The seed must be isolated from the solution until it is completely saturated and the temperature begins to decrease.) For these reasons the technique is rarely used in hydrothermal crystal growing.

The technique of temperature reduction with a constant temperature difference between the growth and dissolution zones was used to grow Na_2CoGeO_4 by spontaneous crystallization and by crystallization on a seed. In the former case a NaOH solution was saturated with sodium-cobalt germanate at a temperature of about 500°C, and then nucleation in the growth zone was initiated by setting up a temperature gradient. Simultaneous reduction of the temperature in the growth and dissolution zones ensured the growth of high-quality crystals at high growth rates of up to 8 mm/d along the [001] axis.

c) "Metastable-Phase" Technique

This technique is based on the difference in solubility between the phase to be grown and that serving as the starting material. The initial nutrient consists of compounds (or polymorphous modifications of the crystallizing substance) which are thermodynamically unstable under the conditions of the experiment. In the presence of polymorphous modifications the solubility of the metastable phase will always exceed that of the stable phase, and the latter will crystallize due to the dissolution of the metastable phase. This technique is usually combined with the temperature-difference or temperature-reduction technique. The growth of corundum α-Al_2O_3 may serve as an example of hydrothermal growth by the metastable-phase technique [9.50]. The nutrient consisted of gibbsite [$Al(OH)_3$], and the seeds were crystals of α-Al_2O_3. In this case the solubility of the metastable phase exceeds that of the stable phase, and the solution saturated with Al with respect to $Al(OH)_3$ is strongly supersaturated with respect to Al_2O_3, so that the excess Al ensures the growth of corundum crystals.

The metastable-phase technique is generally used for growing crystals of compounds with very low solubility.

Some variations of the techniques listed above are briefly discussed in the following.

The separated-nutrient techique is a variation of the temperature-difference technique and is used for crystallizing complicated compounds consisting of at least two components. The initial components are spatially separated and placed in different zones of the autoclave. The lower zone generally contains the more soluble and readily transportable component, and the upper zone the less soluble component. In the course of dissolution the first component is transferred by convection flows to the upper zone, where the substance to be crystallized is formed and a crystal grows as a result of interaction with the second component. This technique was used to obtain crystals of lead titanates and yttrium-iron garnet [9.51, 52].

The separation of the initial components and solvents is a comparatively new procedure in growing crystals [9.53]. It was developed to grow crystals of the compound $SbSbO_4$, which contains antimony ions of two degrees of oxidation (3+ and 5+). The nutrient is placed in two separate cells of an insert, and different solvents are used for each of the charge components: KF for Sb_2O_3 and (KHF_2 + H_2O_2) for Sb_2O_3. This technique is especially effective in growing compounds containing identical or similar ions of different valence.

The tilted-reactor technique is widely used in producing epitaxial films under hydrothermal conditions. The main aim is to shorten the time of contact between the seed (substrate) and the solution (until the desired temperature is achieved) and thus to reduce the etching of the substrate. Until the preassigned growth temperature is attained the seeds must remain in the vapor phase; after the solution reaches saturation the autoclave is tilted and the seed is brought into contact with the solutions [9.54].

Other techniques of crystal growing from hydrothermal solutions are, in principle, also possible. They make use of continuous passage of a flow of saturated solution over the growing crystal, evaporation of solvent (at pressures equal to those of the saturated vapor and at low temperatures), forced heat removal through the seed holder, or the reduction of pressure (for compounds whose solubility changes considerably with the pressure). Because of their complexity and low efficiency, however, these methods are practically never used for hydrothermal crystal growth.

9.3.2 Equipment for Hydrothermal Crystal Growth

The equipment design for the various techniques of hydrothermal synthesis depends on the specific objective of the experiment. The crystallization vessels used in all the methods are autoclaves. They must be able to withstand high temperatures and pressures simultaneously for prolonged periods, and be convenient and safe to operate and sufficiently simple in design. Furthermore, the autoclave material must be inert with respect to the solvents.

A conventional autoclave is a thick-walled steel cylinder with a hermetic seal (Fig. 9.24). The thickness of the autoclave walls is calculated from the required experimental parameters: temperature, pressure, inner diameter and length of the vessel. For research purposes, 20–1000 cm^3 autoclaves are generally used. In industrial growing, much larger autoclaves are necessary; the volume of autoclaves for growing quartz crystals may reach several cubic meters.

The autoclave shown in Fig. 9.24 is designed for a maximum working pressure of up to 3000 kp/cm^2 and a temperature of up to 600°C; the temperature range can be slightly increased by using special grades of steel. For high temperatures (> 600°C) multilayer autoclaves can be used.

The closure is the most important element of the autoclave. A cylindrical closure, one of the simplest types, is depicted in Fig. 9.24. It consists of plunger *1*, locknut *2*, tightening nut *3*, coupling ring *4* and sealing ring *5*. In closing the autoclave *2* exerts pressure on *4*, and the pressure is transmitted to *5*, which expands and presses against the autoclave walls. This ensures preliminary hermeticity of the autoclave interior. As the autoclave is heated, the pressure developing inside it tends to push the plunger out, upon which the sealing ring increasingly expands, thus ensuring that the reaction space of the autoclave is securely sealed. Closures based on this principle of an unsupported area are usually called "self-energized closures".

Gaskets are generally made of plastic materials, most frequently of annealed copper. At temperatures below 300°C, Teflon can be used. If the closure is provided with a cooling system, rubber gaskets are employed. Sealing such vessels does not require as much mechanical effort as with copper gaskets.

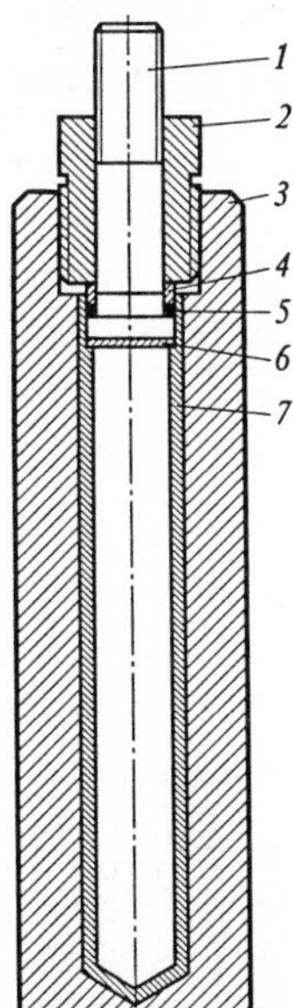

Fig.9.24. Standard hydrothermal autoclave lined with a protective material, with a cylindrical-type seal. (*1*) Plunger, (*2*) lock nut, (*3*) autoclave body, (*4*) steel ring, (*5*) copper ring, (*6*) titanium gasket, (*7*) titanium lining insert

Many designs have been developed for seals. Some of them are not provided with gaskets, and sealing is achieved by a strong pressure of the plunger against the autoclave body (e.g., knife-type closures).

In most cases steel-corroding solutions are used in hydrothermal experiments. To prevent contamination of the crystallization medium by corrosion products, special protective inserts are used. They may have a shape fitting into the internal cavity of the autoclave, in which case they are pressed into it (contact-type inserts). A "floating" type insert occupies only part of the autoclave interior, while the remaining space is filled with water; the degree of filling is chosen so as to equalize the pressure inside the insert and in the free volume of the autoclave. Floating-type inserts sometimes have special corrugated areas (Fig. 9.25), so that the volume of the insert may also change slightly, thereby offsetting a possible drop in pressure.

Inserts may be made of carbon-free iron, copper, silver (for alkaline media), titanium, platinum, various grades of glass, or molten quartz (for acid solutions), depending on the temperature and the solutions used, or of Teflon (for all media, but at temperatures below 300–350°C). In some cases the autoclave cavity need not be lined. Thus quartz or Na silicates can be grown directly in a steel autoclave, because a reaction between NaOH and Fe in the presence of silica yields a practically alkali-insoluble compound $Na_2O \cdot Fe_2O_3 \cdot 4SiO_2$ (acmite), which forms a protective film on the autoclave walls.

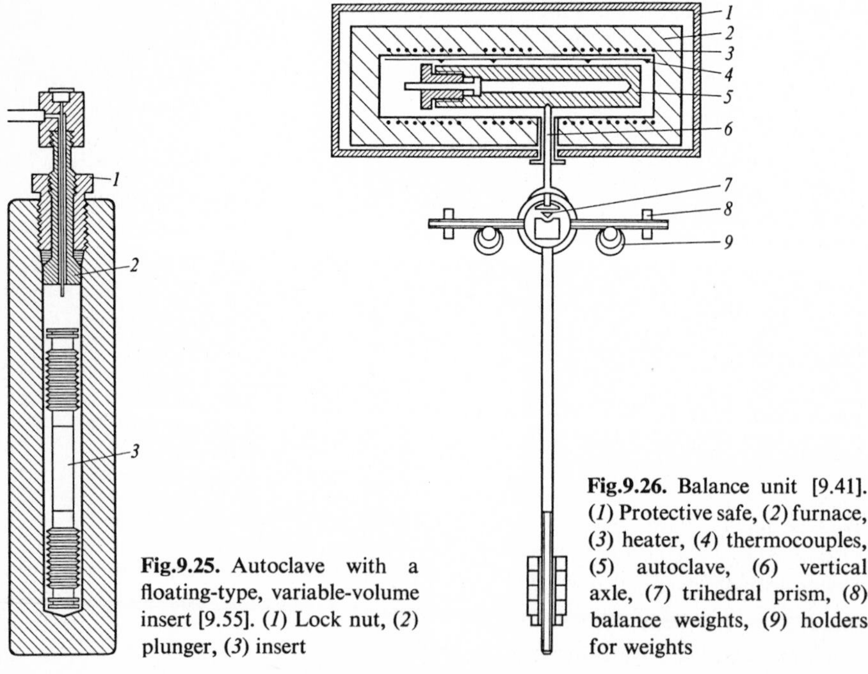

Fig.9.25. Autoclave with a floating-type, variable-volume insert [9.55]. (*1*) Lock nut, (*2*) plunger, (*3*) insert

Fig.9.26. Balance unit [9.41]. (*1*) Protective safe, (*2*) furnace, (*3*) heater, (*4*) thermocouples, (*5*) autoclave, (*6*) vertical axle, (*7*) trihedral prism, (*8*) balance weights, (*9*) holders for weights

Various types of special-purpose equipment are available. They include units for investigating PVTC ratios, i.e., the relationships between pressure P, solution volume v, temperature T, and mineralizer concentration C_B in aqueous solutions, and units for solubility measurements, visual observation of growth, and quantitative control of mass transfer. Without dwelling on the relevant designs,[3] we shall outline the operating principle of the most interesting units.

A weighing unit (Fig. 9.26) has been designed for quantitative control of mass transfer inside the autoclave. The main part of the unit is a horizontal autoclave mounted at the center of gravity on the vertical axle. The autoclave swings freely in the furnace, and the deflection of the axle from the vertical indicates a shift of the mass inside the autoclave to one side or the other. Special weighing scales make it possible to estimate the amount of displaced material, and additional weights help to return the autoclave to horizontal position.

Various types of autoclaves with windows of melted quartz or sapphire allow observation of the processes inside the autoclave during the experiment (Fig. 9.27).

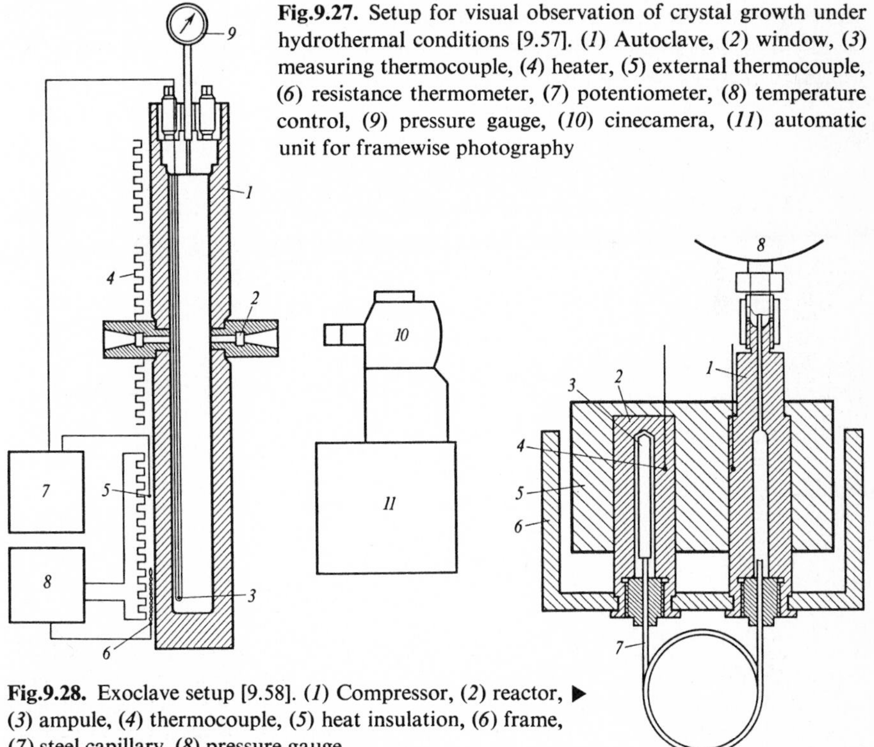

Fig.9.27. Setup for visual observation of crystal growth under hydrothermal conditions [9.57]. (*1*) Autoclave, (*2*) window, (*3*) measuring thermocouple, (*4*) heater, (*5*) external thermocouple, (*6*) resistance thermometer, (*7*) potentiometer, (*8*) temperature control, (*9*) pressure gauge, (*10*) cinecamera, (*11*) automatic unit for framewise photography

Fig.9.28. Exoclave setup [9.58]. (*1*) Compressor, (*2*) reactor, ▶ (*3*) ampule, (*4*) thermocouple, (*5*) heat insulation, (*6*) frame, (*7*) steel capillary, (*8*) pressure gauge

[3] They are described in more detail in [9.47, 56].

The unit presented in Fig. 9.28 (an exoclave) permits independent changes in temperature and pressure, in contrast to the devices described above. It consists of crystallization vessel *2* and compressor *1*, which communicate through steel capillary *7*. The closure is cooled by running water, and therefore rubber gaskets can be used. The exoclave is designed for use at temperatures of up to 700°C and a pressure of 3000 kp/cm^2.

9.3.3 Hydrothermal Solutions. Solvent Characteristics

Under conditions of hydrothermal synthesis the media used in crystal growing are aqueous solutions of salts, alkalis, and acids, which are the simplest two-component, B-H_2O systems at high temperatures and pressures. Such systems consist of a highly volatile component (H_2O) and a salt or base poorly volatile compared to water.

Compounds grown by the hydrothermal method usually belong to the second type in the classification given in Sect. 9.1.1; therefore they cannot be obtained in the form of sufficiently large crystals in the simplest A-H_2O system because of their low water solubility. For this reason a mineralizer is introduced into the hydrothermal system to increase its solubility. The mineralizer is usually a compound of the first type. Introducing it not only increases the solubility of the test compound, but also changes the temperature dependence of solubility. For instance, the solubility of $CaMoO_4$ in pure water, C_A, reduces with an increase in temperature from 100 to 400°C, but increases by almost an order and changes the sign of its temperature coefficient of solubility ($\partial C_A/\partial T$) to positive if a highly soluble salt (KCl, NaCl is introduced. A change in the sign of $\partial C_A/\partial T$ may also occur with one and the same mineralizer, depending on its concentration. Thus, the solubility of Na_2ZnGeO_4 in the NaOH solution decreases with the rise of temperature at NaOH concentrations below ~20 wt. % and increases in more concentrated solutions (Fig. 9.29).

The use of compounds belonging to systems of the second type (Sect. 9.1.1) as mineralizers is expedient only at high pressures, when the solubility of the

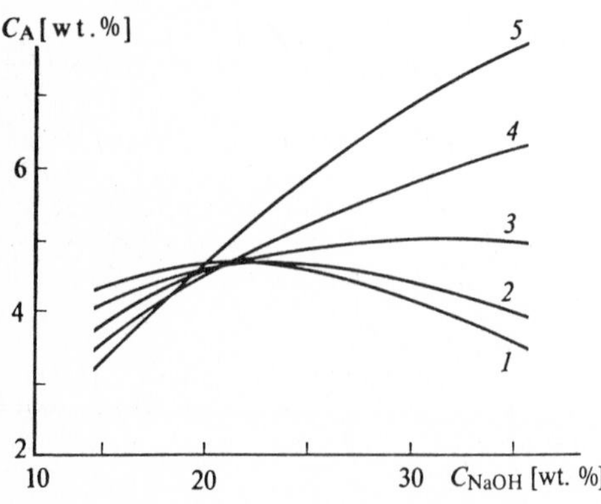

Fig.9.29. Solubility of sodium zinc germanate in aqueous solutions of sodium hydroxide vs solvent concentration [9.59]. (*1*) 200°C, (*2*) 220°C, (*3*) 250°C, (*4*) 300°C, (*5*) 350°C

mineralizer salt reaches a high value. At low pressures near the critical temperature of water, the low solubility of the salt limits its mineralizing effect on the dissolution of compound A. Thus the solubility of $CaWO_4$ in K_2SO_4 solutions (the K_2SO_4–H_2O system belongs to the second type) at elevated temperatures and low pressures is low because of the low solubility of K_2SO_4. An example of the application of a water–salt system with two critical points for crystal growing is the use of the Na_2CO_3–HO_2 system to obtain crystals of Al_2O_3 [9.50, 60].

A complete characterization of a mineralizer must include not only the temperature dependence of its solubility in water, but also the relationship between all the thermodynamic parameters determining the hydrothermal process in a wide range of temperatures, pressures, and concentrations (the PVTC relations).

In an enclosed space, a change in temperature leads to a corresponding change in pressure and in the volumes of the liquid and vapor phases in the system. As the temperature rises, the volume of the liquid phase and the vapor pressure over the solution increase. The volume ratio of the liquid and vapor phases at different temperatures can be calculated from data on the densities of aqueous solutions at high temperatures. By way of example we give the results of such a calculation for aqueous solutions of the alkali NaOH with a concentration of 20 and 30 wt. % (Fig. 9.30). The temperature at which the liquid-vapor interface vanishes (the homogenization temperature) is higher, the lower the fill factor F of the autoclave is.[4] After this temperature is achieved, any slight change in temperature results in an abrupt change in pressure.

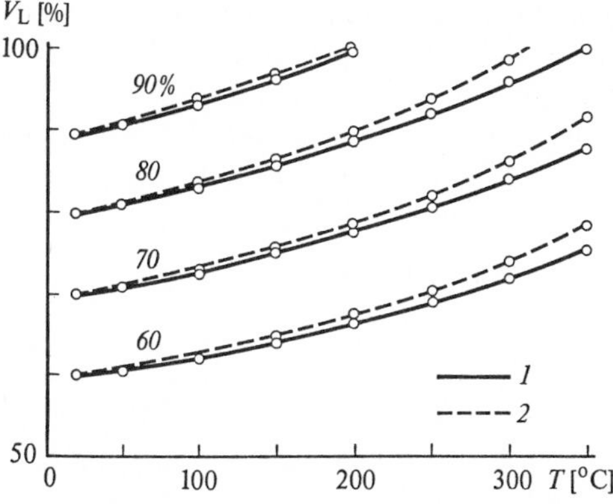

Fig.9.30. The volume V_L occupied by the liquid phase in an autoclave at temperatures of 25°–350°C and NaOH concentrations of 30 wt. % (*1*) and 20 wt. % (*2*) [9.43]. The *figures* above the curves give the fill coefficients at room temperature

[4] The fill factor is defined by the following expression relating the volume of the liquid phase, V_L, the volume of the solid phase, V_S, and the working volume of the autoclave, V: $F = V_L/(V - V_S)$. The fill factor is determined at room temperature and is always less than unity. The term "fill coefficient" is also used for this concept.

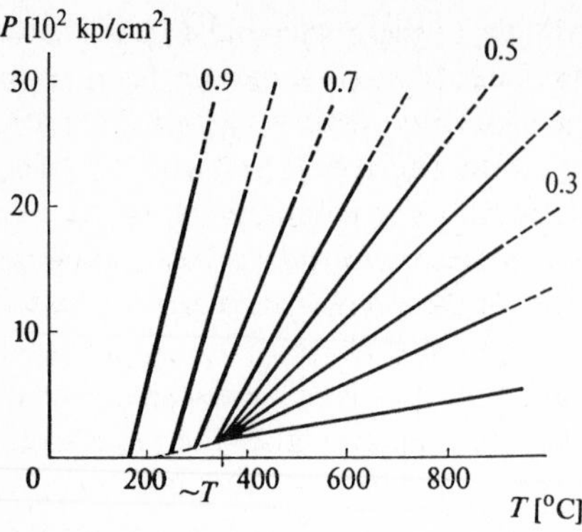

Fig.9.31. PVT diagram for water [9.61]. The *figures* above the curves give the fill coefficients at room temperature

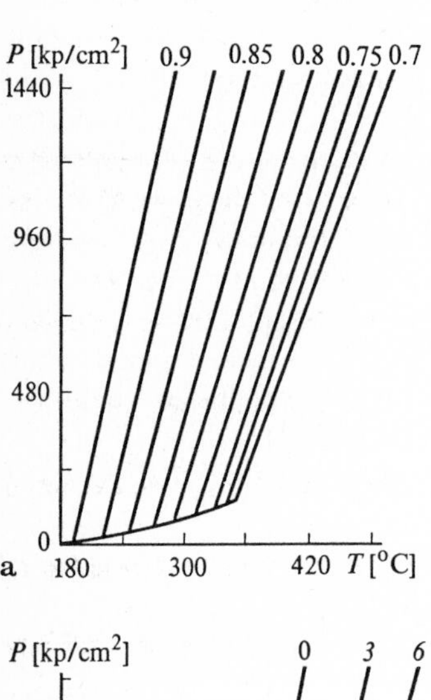

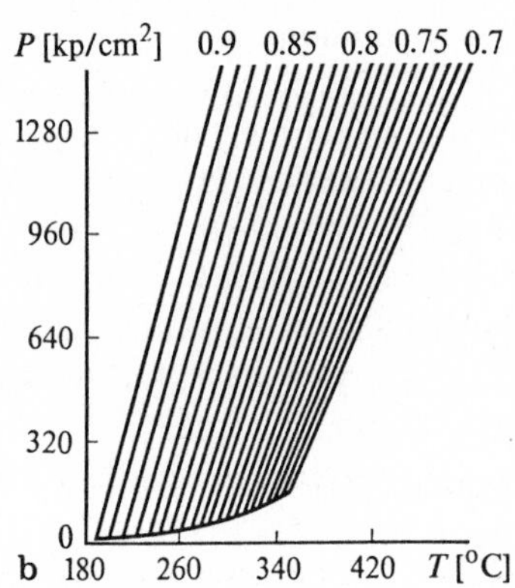

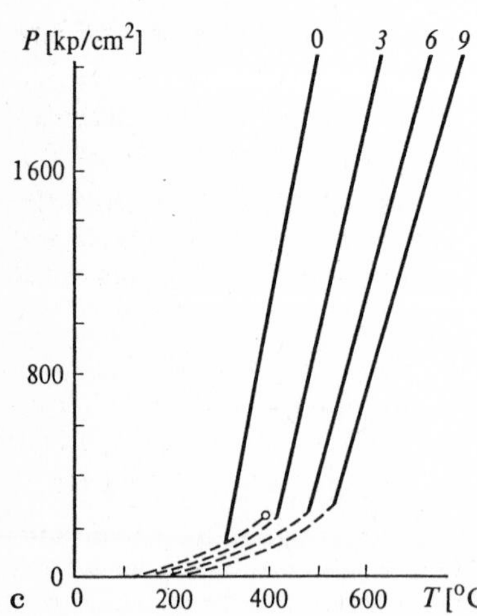

Fig.9.32. PFTC diagrams for hydrothermal solutions. **(a)** NaOH solution with a concentration of 10 wt. % [9.63], **(b)** K_2CO_3 solution, 10 wt. % [9.63]. The *figures* above the curves correspond to the fill coefficient at room temperature. **(c)** LiCl solutions, fill coefficient 0.7 [9.62]. The *figures* above the curves denote the molal concentration of the solution

The most comprehensively studied PVT diagram is that of pure water (Fig. 9.31). The introduction of a salt (base, acid) reduces the pressure in the system while the temperature remains the same (at least at pressures below several thousand kp/cm^2). Therefore the PVT diagram of water helps in estimating the maximum pressure possible in the system at a given degree of filling of the autoclave and at the temperature at which the experiment is conducted. The PVTC ratios of aqueous solutions of acids, alkalis, chlorides of alkali metals and of a number of other compounds [9.62–64] have been investigated. The decrease in pressure in these solutions at T = const as compared with pure water is illustrated by the PFTC diagrams shown in Fig. 9.32, where F is the fill coefficient of the autoclave.

Knowing F, the temperature of the experiment, and the solvent concentration, it is possible to estimate, from the data on the PFTC ratios, the pressure in autoclaves which have no pressure gauges or other pressure indicators.

9.3.4 Interaction of the Crystallizing Substance with the Solvent

It is possible to control the growth of crystals from high-temperature solutions if the following conditions are ensured during the interaction of the crystallizing substance A with the solvent [9.43]:

1) Congruence of dissolution of the crystallizing substance.
2) Sufficiently high solubility of the crystallizing substance. This is necessary if the crystals are to grow at an appreciable rate.
3) Sufficiently sharp changes in the solubility of the crystallizing substance with changes in temperature or pressure.
4) Formation, in the solution, of mobile complexes readily decomposed upon changes in temperature.
5) Establishment of the redox potential necessary to ensure the existence of ions of the required valency.

In growing single crystals by recrystallization, the congruence of solubility is a necessary condition for growth. If the solubility is incongruent, a solution saturated with one of the components is poor in the other. This precludes crystallization of the compound in its initial composition, because one of the components is removed from the reaction sphere. Note that in some cases different compounds can be synthesized under conditions of incongruent solubility as a result of partial or complete decomposition of the initial compound and the formation of others.

The high solubility of the crystallizing substance ensures a sufficiently high concentration of it in the initial solution, as well as an acceptable rate of crystallization. The solubility of inorganic compounds in water (A–B–H_2O system) and the variation of this solubility with temperature and pressure are determined by the nature of the compound (type of chemical bond), the degree of electrolytic dissociation of the compounds in solution, and the energy of interaction between the solute and the solvent particles; this energy defines the deviation of the solution from the ideal. In the vast majority of cases the solubility of substances under crystal growing conditions ranges from 1 to 5 wt. %. The solubility of quartz in NaOH solutions is 2–4 wt. %, that of corundum in Na_2CO_3 solutions 3.4 wt. % at 500°C, and that of zincite (ZnO) in NaOH about 4 wt. % at 200°C. Only in rare cases is the content of the crystallizing component less than 1 wt. % under growing conditions. An example is potassium niobate-tantalate [K (Ta, Nb)O_3], whose solubility is ~0.4 wt. %, and yttrium orthoferrite [$YFeO_3$], where it is 0.3–0.4 wt. %. This value is close to the limit below which the substance virtually does not crystallize on a seed under hydrothermal conditions. It is for this reason that pure water is not used in crystal growing; the solubility of most compounds grown hydrothermally in it does not usually exceed 0.1–0.2 wt. % even at high temperatures.

The relationship between the solubility of the crystallizing substance and the temperature is the most important characteristic of a system. It determines the choice of growing procedure and permits control of crystallization. The pressure and temperature dependence of the solubility of a substance must be investigated as a preliminary stage to growing any new crystal.

Most of the compounds crystallized under hydrothermal conditions form systems of the second type (discussed in Sect. 9.1.1) with water. The temperature dependence $C_A(T)$ of these compounds may be direct, i.e., the solubility increases with the temperature and $\partial C_A/\partial T > 0$; it may be reversed ($\partial C_A/\partial T < 0$); or it may change sign (when the solubility decreases in a certain temperature range and increases in another).

The concrete values for the solubility of a crystallizing substance at a given temperature and pressure are usually obtained experimentally (see, e.g. [9.65]). The simplest and most widely used method of determining the solubility under hydrothermal conditions is the weight-loss method. The solubility is determined from the weight loss of a crystal plate immersed into a solution which is then heated in an autoclave to the required temperature, held until equilibrium is established, and cooled rapidly. Other procedures for determining solubility are the sampling method, which requires specially designed autoclaves, the radioactive isotope method, the method of taking *P–C* and *T–C* curves, etc. [9.66, 67].

In determining the quantitative characteristics of crystal growing from experimental data on the temperature dependence of the solubility of compounds, the following Arrhenius-type equation is used:

$$(\partial \ln \mathscr{K}_0/\partial T)_P = -\Delta H/RT^2 . \tag{9.2}$$

Here, $\mathscr{K}_0$ is the equilibrium constant of the reaction (in this case, the equilibrium solubility of the compound), and ΔH is the heat of solution, which is the characteristic used in selecting the heat conditions for growing.

As a rule, experiments aimed at precise determination of the equilibrium solubility at elevated temperatures and pressures are extremely laborious and time consuming. For instance, at $T = 250°C$, it takes from two to seven days, depending on the solution concentration, for equilibrium to be established in an unstirred set-up. Therefore various investigators have tried repeatedly to find the values of solubility under hydrothermal conditions by calculation, proceeding from the thermodynamic characteristics of the system under standard conditions. But most such calculations produce reliable results only for the simplest compounds [9.68].

Variations in solubility with temperature and pressure can theoretically be calculated more accurately by considering the dissolution process as a combination of two equilibria: the equilibrium between the solid phase and the neutral particles of the corresponding composition in solution, and the equilibrium between these neutral particles and the ions into which they dissociate in solution [9.69]. Then the solubility of the compound, C_A, is composed of the equilibrium concentration C_n of the neutral particles of the dissolved material and the equilibrium concentration C_i of the ions making up the dissolved compound; $C_A = C_n + \sum C_i$. The equilibrium concentrations C_n and C_i are expressed via the solution characteristics, and can be completely defined by the following set of parameters: the product of the solubility and the dissociation constant of the dissolving electrolyte, and the coefficients of activity of the ions formed. These parameters are, in turn, related to the temperature and pressure by expressions that are known from solution theory and that contain such characteristics of the substances as ion mass and charge, solution dielectric constant and density, and the standard changes in enthalpy and entropy on dissolution.

The dependences $C_A(P, T)$ thus obtained show agreement with experiment and explain such properties of systems as their division into two classes (Sect. 9.1.1), stratification, the appearance of the second critical point (Fig. 9.2), and the change of sign in the temperature dependence of a poorly soluble electrolyte in a solution of another electrolyte. The described approach helps to determine the type of the solute–water system on the basis of low-temperature data on the temperature dependence of the solubility.

The solubility of most compounds is usually less dependent on pressure than on temperature. In the general case the pressure dependence is described by the equation

$$\frac{\partial \ln X_\alpha}{\partial P} = \frac{V_{\alpha L} - V_{\alpha S}}{RT} - \frac{\partial \ln \gamma_\alpha}{\partial P}, \tag{9.3}$$

where X_α is the molar fraction of the given component in the system, $V_{\alpha L}$ is the partial molar volume of component α in a saturated solution, $V_{\alpha S}$ is the molar volume of the solid component α at the same temperature and pressure, and γ_α is the coefficient of activity of component α. The effect of the pressure on the solubility depends on the difference $V_{\alpha L} - V_{\alpha S}$. Since this difference is usually small, the solubility depends only slightly on the pressure. In most cases $V_{\alpha L} < V_{\alpha S}$. As the pressure rises, the partial molar volume of the salt in solution $V_{\alpha L}$ increases, while the molar volume of the solid salt $V_{\alpha S}$ falls off, and the sign of change in solubility with pressure may reverse. In the general case, if the energy of interaction between the salt and water particles is high at low temperatures, and the solubility of the salt is low, then an increase in pressure usually increases the solubility as well. One of the important conditions ensuring controlled crystal growth is the *reversibility of the dissolution reaction* of substance A. In other words, associates of substance A with the solvent, which form in the liquid phase on dissolution, must be readily decomposable. In this respect the hydrothermal method can be regarded as a specific case of chemical transport reactions. The maximum mass transfer is ensured when the equilibrium constant of the reaction by which the complex is formed does not differ substantially from unity. Otherwise the mass transfer decreases. The reason is either that the solubility of the substance is very low, or the complex is highly stable and does not decompose upon a change in temperature when the solution is transferred from the dissolution zone to the growth zone [9.70].

9.3.5 Hydrothermal Growth of Crystals

Under hydrothermal conditions crystals can be grown either by synthesis or by recrystallization. In either case crystals can be obtained as a result of spontaneous crystallization, recrystallization, or seeded crystallization. Crystal synthesis is used but rarely for growing large crystals, especially if the crystallizing compound has a complex composition. The synthesis method is generally employed for studying the phase diagrams of crystallization in multicomponent systems. These diagrams are not investigated in order to obtain large single crystals of a specific composition (the size of synthesized crystals may vary from 10^{-3} to 10^{-1} cm), but to determine the regions in which solid phases crystallize and the composition of steadily crystallizing compounds. Such investigations make it possible to search extensively for and produce crystals of new compounds, perhaps of previously unknown composition, and possessing valuable physical properties. Plotting such diagrams and investigating solubility form a preliminary stage in growing large single crystals, since they help to establish the region of crystallization and the composition of the accompanying phases, and to evaluate the effect of variable parameters on the crystallization of a given compound.

Phase diagrams of crystallization differ radically from the conventional phase diagrams plotted for equilibrium conditions. The former are based on experimental material on crystallization in the systems under investigation. The conditions may be such that there is either a temperature gradient along the vertical axis of the autoclave or a fixed temperature difference with isolation of the growth and dissolution zones. Crystallization diagrams are plotted in different coordinates: X–C_B (where X is the nutrient composition, C_B is the

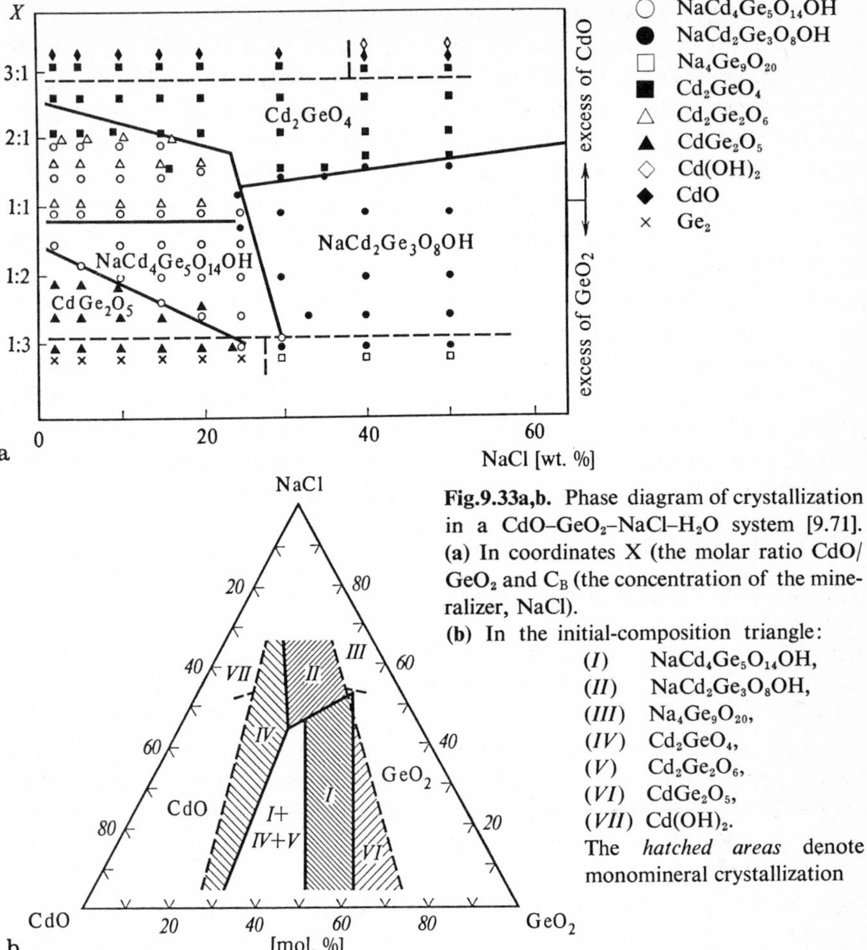

Fig.9.33a,b. Phase diagram of crystallization in a CdO–GeO_2–$NaCl$–H_2O system [9.71]. **(a)** In coordinates X (the molar ratio CdO/GeO_2 and C_B (the concentration of the mineralizer, NaCl). **(b)** In the initial-composition triangle: (*I*) $NaCd_4Ge_5O_{14}OH$, (*II*) $NaCd_2Ge_3O_8OH$, (*III*) $Na_4Ge_9O_{20}$, (*IV*) Cd_2GeO_4, (*V*) $Cd_2Ge_2O_6$, (*VI*) $CdGe_2O_5$, (*VII*) $Cd(OH)_2$. The *hatched areas* denote monomineral crystallization

concentration of the solvent mineralizer), X–T, C_B–T, etc. The phase relationships are represented less often on the initial-composition triangle. An example of a phase diagram of crystallization is given in Fig. 9.33a. The initial nutrient composition, i.e., the relative content of oxides, CdO/GeO_2, is plotted in molar fractions on the vertical axis, and the concentration of the solvent, NaCl, on the horizontal axis. The continuous lines indicate the regions of monomineral crystallization of germanates, and the dashed lines the beginning of crystallization of the excessive component (CdO or GeO_2) in the form of a separate compound. In Fig. 9.33b the same phase relationships are shown on the triangle of the initial compositions; each point on the triangle describes the composition of the initial oxides for a given crystallizing compound at fixed temperatures, solvent concentration, and initial liquid-to-solid volume ratio.

Neither spontaneous crystallization nor synthesis is generally used to grow single crystals because of the complexity involved in controlling the process. (Spontaneous crystallization means nucleation and growth on spontaneously arising crystallization centers.)

In recrystallization, growth occurs as a result of the dissolution and assimilation of neighboring grains (crystals). This process usually leads to the enlargement of grains or of separate charge crystals, and is not used either, because it is virtually impossible to control the supersaturation during the growth of separate crystallites.

Large single crystals are generally grown by direct recrystallization of a nutrient having the same composition as the desired crystal, under previously established PVTC conditions.

To obtain large single crystals of the required orientation, oriented growth on seeds is used. The seeds are placed in the upper zone of the autoclave, which is separated from the "hot" zone by special baffles with openings or with flow-guiding tubes. The baffles aid in regulating the direction and flow velocities of the solution. The role of a baffle in redistributing the temperature in an autoclave is shown in Fig. 9.34.

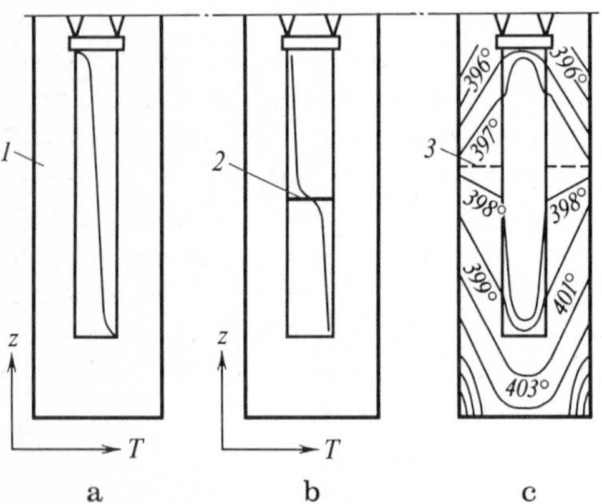

Fig.9.34a–c. Temperature distribution with natural convection in an autoclave (**a**) without internal fittings, and (**b**) with a baffle. (**c**) Temperature distribution with natural convection, pressure 500 kp/cm² [9.72]. (*1*) Autoclave body, (*2*) baffle, (*3*) horizontal isotherm

If the delivery of the feed solution to the growth zone is continuous and uniform, the crystal grows on seeds at a specific rate which is constant for given crystallization conditions. The principal parameters influencing the growth of a crystal on seeds are the crystallographic orientation of the seed, the values of supersaturation, temperature, pressure, and mineralizer concentration, the temperature coefficient of solubility, the chemical nature of the solvent, the ratio of the surface area of the nutrient, S, to that of the seed, S_1, and the value and arrangement of the openings in the internal baffle of the autoclave.

The last two parameters determine the amount of material arriving at the growth zone. The chosen ratio S/S_1 is based on the relationship between the growth and dissolution rates. If the conditions are the same, the rate of crystal dissolution usually considerably exceeds that of growth. Therefore the effect that dissolution of the initial charge has on the growth rate manifests itself only when S is less or slightly greater than S_1. It has been shown with quartz and corundum that at $S/S_1 \geq 5$ (SiO_2), and $S/S_1 \geq 20$(Al_2O_3) the growth rate remains constant [9.60, 73]. For this reason it is expedient to use a sufficiently large-grained porous nutrient. An amorphous and finely dispersed nutrient often cakes; the solution does not penetrate into it, and dissolution proceeds only on a small area.

The total area of the openings in the baffle, f, affects the velocity of solution convection and hence the mass transfer to the growth zone. When f is small (less than $\sim 1\%$ of the cross section of the autoclave cavity), two "zones" of convective motion of the solution may be set up upstream and downstream of the baffle, with a limited mass exchange between them. Under these conditions, either the crystal grows at a very low rate, or dissolution of seeds is even observed. With increasing f the delivery of the dissolved nutrient to the growth zone increases, as does the growth rate. In most cases the value of f is 3%–8% of the internal cross-sectional area of the autoclave; this is evidently close to the optimum value.

Thus, at low values of S/S_1, and at low dissolution rates and convection velocities, the amount of substance arriving at the growth zone is insufficient to ensure the maximum growth rate possible at the given temperature and temperature difference. The conditions of mass exchange are not intensive enough, and the crystals often show defects indicating inadequate nutrition. In the opposite case the growth rate is independent of the rate of nutrient dissolution and transfer of material to the growth zone (excessive mass exchange). Such a situation is most favorable for crystal growth.

For all crystals whose growth rates have thus far been studied under hydrothermal conditions, it has been found that the growth rates of faces increase linearly with the supersaturation:

$$V = \beta \, \Delta C_A . \tag{9.4}$$

The kinetic coefficient β was discussed in Sects. 3.1, 2.

In practice, it is more common to measure the dependences of the growth rates on the temperature difference between the growth and dissolution zones, ΔT. Supersaturation ΔC_A usually varies linearly with ΔT:

$$\Delta C_A \sim \Delta T , \tag{9.5}$$

and therefore the relationship between the growth rate and "supercooling" ΔT is also expressed by a straight line.

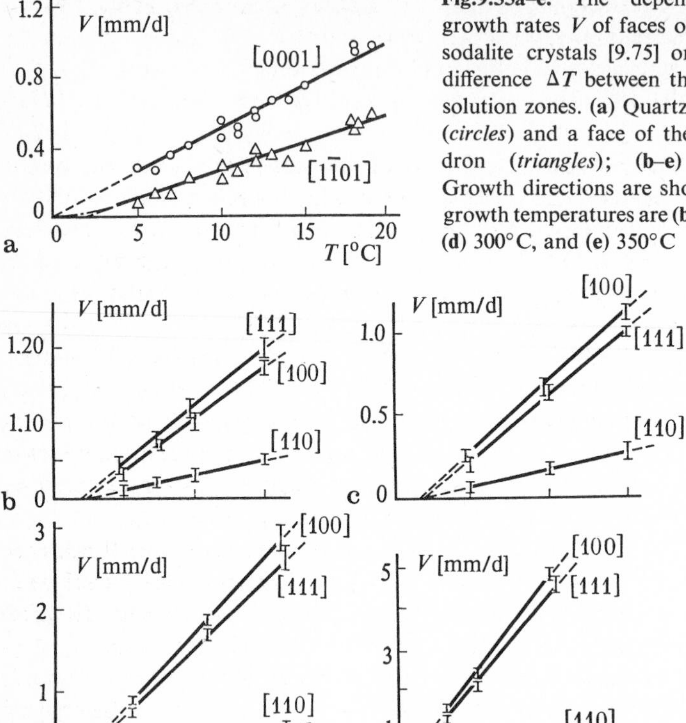

Fig.9.35a–e. The dependence of normal growth rates V of faces of quartz [9.74] and sodalite crystals [9.75] on the temperature difference ΔT between the growth and dissolution zones. (**a**) Quartz crystal, basal face (*circles*) and a face of the small rhombohedron (*triangles*); (**b–e**) sodalite crystal. Growth directions are shown on the *curves*; growth temperatures are (**b**) 200° C, (**c**) 250° C, (**d**) 300° C, and (**e**) 350° C

An example of the dependence of growth rates on ΔT for crystals of quartz [9.74] and sodalite $Na_8Al_6Si_6O_{24}(OH)_2 \cdot nH_2O$ [9.75] is given in Fig. 9.35. The slope of the lines differs for the various faces, i.e., the anisotropy of the growth rates (see Sect. 3.1.3) slightly increases with the supersaturation. It is also important that the straight lines $V(\Delta T)$ do not arrive at the origin on extrapolation to $\Delta T = 0$, but intersect the x axis at a certain value of ΔT. The supersaturation corresponding to this value of ΔT is termed critical. As a rule, it is different for the different faces, and may sometimes reach very high values. Such faces do not grow except under special conditions. The critical supersaturations reduce with an increase in temperature, and strongly depend on the composition of the solution. For instance, the growth rate of the (0001) face of a corundum crystal in a solution of sodium and potassium carbonates is practically zero, while in bicarbonate solutions it reaches an appreciable value even at relatively low ΔT, about 10°C.

At a constant supersaturation the growth rates of the faces increase with the temperature. This is reflected in the temperature dependence of the coefficient

β_V in (9.4). In many instances (or all, for rapidly growing faces) this change is such that the following Arrhenius equation holds:

$$\partial \ln \beta / \partial T = E / RT^2 \tag{9.6}$$

(E is the growth activation energy), and the temperature dependence of the growth rate in coordinates $\ln V - 1/T$ is expressed by a straight line. In some instances (usually for slowly growing faces), a deviation from linearity is observed. The slope of the lines $\ln V(1/T)$ relative to axis $1/T$ describes the value of E. This value has been measured for quartz, corundum, sodalite, zincite, etc., and lies between 10 (more often 15) and 40 kcal/mol for different faces. As has been confirmed for quartz and corundum, it does not change appreciably in solutions of different types (Na_2CO_3, K_2CO_3, $NaHCO_3$). The comparatively high values of the activation energy of face growth are an important indication that the growth process is not limited by the diffusion of the substance in solution (whose activation energy does not exceed 4–5 kcal/mol), but largely depends on the phenomena occurring directly on the surface of the growing crystal.

The pressure generally used in hydrothermal experiments evidently does not produce a direct effect on the growth rate, but acts through other parameters: mass exchange, solubility, etc. The pressure also has an effect on the dissociation of intricate complexes in solution; the effect is usually such that dissociation is prevented. Increasing pressure influences the mass exchange in the autoclave via nonuniform changes in the specific volume of the water. The velocity of solution convection can, to a first approximation, be considered proportional to the difference between the specific volumes of the solution at the temperatures of the upper and the lower parts of the autoclave. For water (and for low-concentration aqueous solutions) at a fixed temperature difference, this difference reduces as the pressure rises, and hence the solution velocity must also decrease. This results in a change in the temperature distribution inside the autoclave, a change in the supersaturation in the growth zone, and a corresponding change in the growth rate.

A mineralizer may similarly affect the growth rate via the solubility of the crystallizing substance (which is clearly related to the mineralizer concentration). On the other hand, if changes in mineralizer concentration also change the temperature coefficient of solubility, $\partial C_A / \partial T$ (see, e.g., [9.59]), as the mineralizer concentration at a given ΔT increases, the growth rate will increase (or decrease) because of the change in supersaturation in the growth zone. Experience in crystallizing corundum, zincite, and sphalerite shows that when the mineralizer concentration ensures a comparatively high solubility C_A(1.5 wt. % for ZnS, 2.5 wt. % for Al_2O_3 and ZnO), the growth rate is virtually independent of the mineralizer concentration [9.57, 40].

A mineralizer may also adsorb directly on crystal faces and inhibit their growth, in which case it must be regarded as an ordinary impurity. Finally, a mineralizer may interact with an impurity previously adsorbed on the surface, and cause the growth rate to increase. Thus the growth rate of corundum and zincite crystals in solutions containing potassium (KOH, K_2CO_3) is higher than in solutions containing sodium (NaOH, Na_2CO_3). This increase cannot be explained by changes in supersaturation in the growth zone, because supersaturation in potassium solutions must be lower than in sodium solutions. One possible explanation of this phenomenon is interaction between potassium and a surface-adsorbed impurity, which is assumed to be a chemisorbed layer of water. [9.76].

9.3.6 Defects and Methods for Their Elimination in Crystals Grown Hydrothermally

Since hydrothermal crystallization is carried out at relatively low temperatures, much below the melting point, the crystals show no strong thermal stresses, plastic deformations, or structure defects of various kinds (blocks, waviness, etc.) unless they are caused by defects in the seed crystal. In crystals of leucosapphire Al_2O_3 and orthoferrites of rare-earth elements and yttrium, the maximum disorientation of the blocks does not exceed 3′. In spontaneous crystallization it is possible to obtain crystals of rutile (TiO_2) with a low dislocation density, and in growth on seeds, dislocation-free quartz crystals having an area of up to 300 cm^2. At the same time, if stresses do for some reason arise, they can no longer be relieved by plastic deformations, and the crystal breaks down if the elasticity limit is exceeded. Cracks are one of the most common defects in hydrothermally grown crystals.

Defects in crystals grown by the hydrothermal method are associated both with the quality of the seed and with the growth conditions. One of the main causes of defects such as stresses, blocks, or cracks are defects in the seed. A mismatch in the crystallochemical characteristics of the seeding and newly formed crystals (a large difference in unit cell parameters, or a difference in chemical composition) also causes defects in a growing crystal. Quartz cracked when the difference in unit cell parameters was ~ 0.0005 Å, Al_2O_3 at ~ 0.001 Å when grown on sapphire, and ~ 0.0005 Å on growing ruby [9.40, 77].

The permissible difference between the unit cell parameters of a seeding and a growing crystal varies depending on the type of crystal and seed orientation. Cracking is promoted by crystal cleavage, the presence of impurities, and nonuniformity in their distribution.

The last factor is associated not only with the growth conditions (temperature fluctuations resulting in changes in the solubility of the substance and in the content of the main components in solution), but also with the differing capacity of the various faces to absorb impurities, and the difference in the coefficient of impurity entrapment for individual growth pyramids. Nonuniformity in impurity distribution manifests itself in the zonal and sectorial structure of crystals (Sect. 4.3), which is particularly important in growing doped crystals.

As in the terminology adopted for quartz, the impurities can conveniently be classified as nonstructural and structural. Nonstructural impurities encompass inclusions of the liquid phase and, in the case of quartz, of the colloid-disperse phase of sodium and potassium silicates. Such impurities do not cause strong stresses in the crystal, but noticeably affect physical characteristics such as optical properties. Structural impurities are ions replacing ions of the basic substance in the crystal lattice or interstitial ions. Structural impurities may include components of a solvent a dopant, or (when the starting materials are not pure enough) impurities from the nutrient and solvent. The presence of structural impurities changes the unit cell parameters and cause stresses. For instance, in impurity-free hydrothermal corundum the stresses do not exceed 1.4 kp/mm^2, whereas in crystals containing a chromium impurity they are five times as high.

Typical impurities in hydrothermally grown crystals are H_2O, H^+, and OH^-. They may exert a considerable effect on thermal conductivity, microhardness, electrical and optical characteristics, luminescence, etc. In quartz, entrapped water ions (molecules) adversely affect the Q factor, phase transition temperature, and density; in ruby, they affect the strength, hardness, and internal friction. To produce a high-quality crystal it is often necessary to reduce the amount of impurity water in it. This problem can be solved by purifying the starting reactants of impurities that increace the coefficient water entrapment selecting crystallization modes in which the capture of water is minimal, and annealing the grown crystals under special conditions. For instance, for single crystals of rare-earth ferrogarnets, the half-width of the ferromagnetic resonance line correlates with the content of H^+ protons, and the quality of the crystal improves appreciably on replacement of H_2O by D_2O in the initial solution [9.78]. During the growth of quartz a reduced concentration of Al^{3+}, Fe^{3+}, Fe^{2+}, and Na^+ in the nutrient leads to a decrease in the trapping of hydrogen ions [9.79]. By reducing the amounts of alkali-earth metals it is possible to reduce the content of H^+ in crystals of yttrium-iron garnet and rare-earth orthoferrites [9.78], thus improving the quality of the crystals.

Quartz and zincite are examples of crystals which are freed from impurity water by heat treatment. Thus, annealing quartz crystals in a hydrogen atmosphere may reduce the number of hydroxyl OH^- ions in crystals. Annealing zincite crystals in a lithium atmosphere improves their stoichiometry with respect to oxygen and increases their dark resistivity [9.80].

Typical defects caused by high crystal growth rates are liquid-phase inclusions, which are signs of dendritic skeletal growth. *Ikornikova* [9.45] showed for calcite that under conditions of high supersaturation the system $CaCO_3$–$NaCl$–CO_2–H_2O exhibits degeneration of the faces of the calcite rhombohedron with the formation of dendrites. The morphology of the antiskeleton (in the case of $CaCO_3$, characteristic clover leaves are formed) is attributed to the concentration distribution near the growing crystal; the concentration is at a maximum near the crystal surface and decreases with the distance from it (Sect. 5.3).

A decrease in growth rate below a certain critical value appreciably reduces the trapping of colloid inclusions by quartz crystals; the value of the critical trapping rate is specific to each face (Sect. 6.1.2). The increase in the growth rate of quartz at $T > 400°C$ in the growth zone augments the amount of structural Al additive in crystals [9.81]; at $T < 400°C$, the dependence reverses [9.82].

The nature of defects and their distribution in a given crystal may be associated with the type of solvent. For instance, in $Y_3Fe_3O_{12}$ grown in KOH solutions the content of H^+ impurity is much lower than in crystals grown in NaOH [9.52]. The relationship between defects and the type of solvent is clearly seen in the case of ZnS.

Crystals of sphalerite (ZnS) grown under hydrothermal conditions, while having a highly perfect structure on the whole, still show point defects, stacking faults, and inclusions [9.83]. Crystals obtained at low solution pH (mineralizer H_3PO_4) contain fewer planar structural defects and have practically no inclusions of a nonstructural impurity—the second phase of ZnO. However, the concentration of point defects in such crystals is high. Compared with them, crystals grown in alkaline media (KOH) have a low content of intrinsic point defects, but show an increased concentration of stacking faults, and contain potassium impurities and zincite (ZnO) inclusions. The increased concentration of stacking faults in ZnS is evidently caused by a shortage of sulfur in the crystals and by inclusions of zinc oxide. Crystals grown in solutions containing NH_4Cl have the most defective structure. It can be substantially improved by introducing an excess of sulfur into the solution, increasing the crystallization temperature, and heat treatment of the nutrient beforehand.

As has already been mentioned, seed defects are one of the main causes of defects in hydrothermal crystals. The structural defects of the seeding material are usually inherited by the growing layer. Therefore the availability of high-quality seeds is an important condition for obtaining perfect crystals. Crystals grown previously by the hydrothermal method at similar conditions can best be used for seeding.

When using low-quality seeds, more perfect crystals can be obtained by running successive building-up cycles. After each run, the as-grown layer is separated and used for seeding in the next experiment. If high-quality crystals are to be obtained, selection cycle of this kind may involve a dozen or more runs. Seed selection has been used in growing single crystals of sodalite (sodium aluminosilicate). The first experiments were performed with seeds prepared from natural crystals with a large number of impurities, macroinclusions, stress, and other structural defects.

Seed selection is particularly important in heteroepitaxial growth. To obtain single crystals, different in composition but similar in structure must often be used. (Examples are the growth of germanates on seeds of isostructural silicates, and the growth of ferrogarnets on seeds of yttrium-aluminum garnet.) Then, if a single crystal is desired that is without impurities of foreign elements contained in the seeding material (Si, Y, and Al for the above cases), a sufficiently long selection process is necessary.

Another method of obtaining quality crystals is growth from point seeds, usually in the form of small crystals obtained hydrothermally. This method has been used in growing sodium zincogermanate (Na_2ZnGeO_4), a number of tungstates ($CdWO_4$, Li_2WO_4), yttrium orthoferrite ($YFeO_3$), etc.

9.3.7 Some Crystals Grown by the Hydrothermal Method

A large number of compounds belonging to practically all classes have been synthesized under hydrothermal conditions: elements, simple and complex oxides, chalcogenides, tungstates, molybdates, carbonates, silicates, germanates, etc.

Hydrothermal crystallization of quartz has produced the best results. This method has proved to be the only efficient way of growing single crystals of quartz SiO_2.

The need for quartz crystals is a consequence of their application in radio engineering. The dramatic advances achieved in growing high-quality quartz crystals can be ascribed to the comparatively simple growth system (SiO_2–Na_2O–H_2O or SiO_2–Na_2CO_3–H_2O), and hence the simple process technology. Quartz crystals are obtained by crystallization in aqueous solution of NaOH or Na_2CO_3 at temperatures of about 300°C and pressures of 700 kp/cm^2. The crystallization vessels, several cubic meters in size, require no special lining.

Piezoelectric devices made of optically homogeneous single crystals of quartz operate successfully in all kinds of radioelectronic equipment and compare favorably with natural quartz products in terms of parameters (Q factor and temperature coefficient of frequency).

Zincite (ZnO) is the strongest piezoelectric among the semiconductor materials, which explains the interest in its production. The hydrothermal method has proven to be the best way to produce high-quality isometric crystals. Aqueous solutions of NaOH and KOH with a concentration of 4–15 m are used as solvents [9.42, 80]. The temperature of the dissolution zone is 300–450°C, that of the growth zone 250–380°C, and the fill coefficient is 0.7–0.9. It has been found that the mineralizing additives LiOH and LiF (0.1–2 wt. %) greatly improve the quality of the crystals. Zincite crystals grown under the indicated conditions have low resistivities (10–10^{-2} Ω cm), and annealing in air at 800°C in a melt of Li_2CO_3 increases the resistivity to 10^8–10^{13} Ω cm.

Sphalerite ZnS of cubic modification is interesting as a semiconducting, piezoelectric, electrooptical material. Its structure is similar to that of wurtzite, the hexagonal modification of ZnS (both modifications have identical packing of atoms in their layer), and therefore polytypic modifications of ZnS can arise.

Crystals grown from the gas phase or a melt when the growth temperature is close to that of the phase transition (1020°C) are characterized by the presence of interlayers of the hexagonal phase in the matrix of the cubic crystal. Under hydrothermal conditions the crystallization temperature is much lower than that of phase transformation, and therefore it is possible to obtain a perfect "pure" cubic phase. Single crystals of sphalerite can be obtained by using aqueous solutions of alkalis (KOH) or acids (H_3PO_4) as solvents [9.42, 83]. ZnS crystals 1.5 cm^3 in size have been grown in 30–40 wt. % solutions of KOH at temperatures of 355–365°C and $\Delta T \approx 12°C$. The growth rate V may be as high as 0.15 mm/d and is different for different crystallographic faces:

$$V_{(111)} > V_{(110)} > V_{(100)} \, .$$

In concentrated solutions of H_3PO_4, single crystals of ZnS have been obtained at temperatures of 360–400°C and pressures of about 1000 kp/cm^2.

Sodalite $Na_8Al_6Si_6O_{24}(OH)_2 \cdot nH_2O$. Single crystals of sodalite are of interest because of the piezoelectric and photochromic properties of the compound. Photochromic centers are activated in a field of penetrating electromagnetic radiation or fast nuclear particles, an absorption band of about 530 nm is induced in the crystals, and they acquire a crimson coloring. The most efficient ionizing radiation consists of γ-rays of ^{60}Co, which have high penetrating power. This induced absorption band can be deleted by visible light and restored by uv radiation with a wavelength of less than 350 nm.

Sodalite contains OH groups and water, and crystals of it are therefore difficult to obtain by other methods. Single crystals of sodalite several cubic centimeters in volume have been grown under hydrothermal conditions in highly concentrated solutions of NaOH (30–50 wt. %) at temperatures of 200–450°C, pressures below 1000 kp/cm^2, and temperature differences of 10–30°C [9.75]. Silver or Teflon inserts were used to protect the autoclaves from corrosion by alkaline solutions. The piezoelectric modulus, as measured on sodalite crystals, proved to be the highest among the known cubic crystals of class 43 (12.9×10^{-8} cgs). Converters made of sodalite [9.84] have the same shear parameters as on γ sections of quartz, but offer advantages over quartz insofar as they show a stronger trend towards a peaked frequency and experience a pure shear with polarization along (001).

With sodalite crystals, a new method of growing single crystals has been developed and implemented [9.85]. Amorphous layers of carbon or silicon dioxide, or polycrystalline layers of gold or silver were applied by thermal evaporation onto the surface of a single-crystal seed, the surface having been treated beforehand by mechanical and chemical polishing or ion bombardment. The thicknesses of the carbon and silicon dioxide layers were 70–100 Å, and those of gold and silver 500–5000 Å. Seeds coated with such layers were placed in conventional silver-lined autoclaves, and the regime and experimental conditions used in growing sodalite by the temperature-gradient method were maintained.

The growth of single sodalite crystals on seeds covered with intermediate layers was observed (Fig. 9.36). The maximum thicknesses of these layers which yet permit single crystals to grow were established, as well as the optimum thicknesses promotion growth of the most perfect crystals. For gold layers, the limiting thickness in sodalite growing is 3000 Å; for silver it is 2500 Å. These values differ for different crystallographic directions.

Fig.9.36. Single sodalite crystal grown on a seed coated with an intermediate layer of silver. Courtesy of O.K. Melnikov

Single sodalite crystals grown through intermediate layers exhibit greater structural perfection than those grown on seeds without such layers; specificially, they contain many fewer cracks than seeds. The layers "filter off" some of the outcropping structural defects. In growing on seeds coated with intermediate layers, the as-growh single-crystal layer does not adhere to the surface of this layer, and the built-up layers are readily separated from the intermediate layer and the seed.

Ferrogarnets. The growth of $A_3B_5O_{12}$ compounds (the simplest composition is $Y_3Fe_5O_{12}$) has been attracting much attention in connection with the valuable magnetic properties of these crystals, which permit their utilization in storage cells.

Ferrogarnets can be obtained either as bulk single crystals [9.54] or, more importantly, as films [9.86]. The growth conditions are similar for different garnets: temperature 330–530°C, $\Delta T = 30$–40°C, $F = 0.6$–0.8, solvent 10–20 m KOH, 4 m NaOH. When growing a bulk crystal, a seed of the same material is used, whereas in producing films, the growh is heteroepitaxial and takes place on nonmagnetic (or magnetic) substrates of a different composition. Films can be obtained either by synthesis from oxides and hydroxides or by recrystallization from a charge of the same composition. The substrates used in [9.87a] consisted of $Gd_3Ga_5O_{12}$, $Gd_3Sc_{5-x}Ga_xO_{12}$, and $Nd_xGd_{3-x}Ga_5O_{12}$. The hydrothermal heteroepitaxy method was used to obtain magnetic films of various compositions of the type $R'_xR''_{3-x}Ga_yFe_{5-y}O_{12}$, where R′ and R″ are rare-earth elements.

9.4 Growing from High-Temperature Solutions (Flux Growth)

High-temperature crystallization from solution (so-called flux growth) has gained wide acceptance in growing single crystals of complex multicomponent systems. This method takes advantage of the high solubility of high-melting compounds in liquid inorganic salts and oxides. One of its important features is that it occurs in air at a temperature much lower than the melting point of the crystallizing substance. A disadvantage is that a solvent is present which can be trapped by the growing single crystal.

Flux growth was one of the first methods employed in growing technically important crystals. Single crystals of corundum were obtained in this way at the end of the nineteenth century. At present, flux growth is used to grow single crystals of diamond, iron-yttrium garnet, barium titanate, etc. [9.87b, c].

Flux methods are most frequently applied in the search for new crystals, because they permit crystallization of complex multicomponent systems by simple techniques. These techniques take into account the specifics of the phase diagram of the system: crystallizing substance–solvent. In the ideal case the phase diagram should not contain chemically stable compounds or solid solutions; it must be the simplest phase diagram similar to that of two-component systems (Fig. 9.37). The most important parameter is the slope of the liquidus curves, on which the temperature–time crystallization regime depends. Growth from high-temperature solutions is also possible for systems with more complicated phase diagrams, for instance ones in which a chemically stable compound is formed. The compounds that form simultaneously with the crystallizing substance then affect primarily the solution supersaturation. The situation is more complicated when solid solutions of the crystallizing substance and the solvent form, because then the solvent enters the crystal.

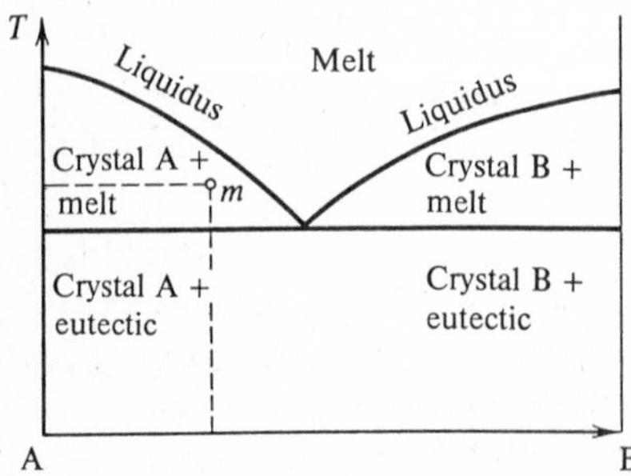

Fig.9.37. Phase diagram of a two-component system. Point *m* shows the growth temperature and the composition of the initial flux

Thus, successful use of flux growth depends above all on the choice of solvent, which must

a) dissolve the crystallizing substance sufficiently (between 10 and 50 wt. %);
b) provide a temperature coefficient of solubility (slope of the liquidus curve) not less than 1 wt. %/10°C;
c) have a low vapor pressure;
d) be inert in interaction with the container material and the crystallization atmosphere.

The necessary supersaturation of the solution is achieved either by reducing the temperature or by evaporating the solvent. The former method is the more popular, and served as a basis in developing growth procedures for technically important crystals such as yttrium-iron garnet, mica, diamond, etc.

A crystallization process occurring under conditions in which the composition of the liquid phase diverges considerably from the stoichiometric composition of the crystallizing substance can also be classified among the flux methods. Here the excess component plays the part of the solvent. For instance, in growing single crystals of the cubic modification of barium titanate $BaO \cdot TiO_2(BaTiO_3)$, TiO_2 is used as the solvent. The process is conducted in a platinum crucible with a 5% excess in titanium dioxide with respect to the stoichiometric composition (Fig. 9.38). A diagram of the crystallization setup is presented in Fig. 9.39. The growing method is similar to that of Czochralski

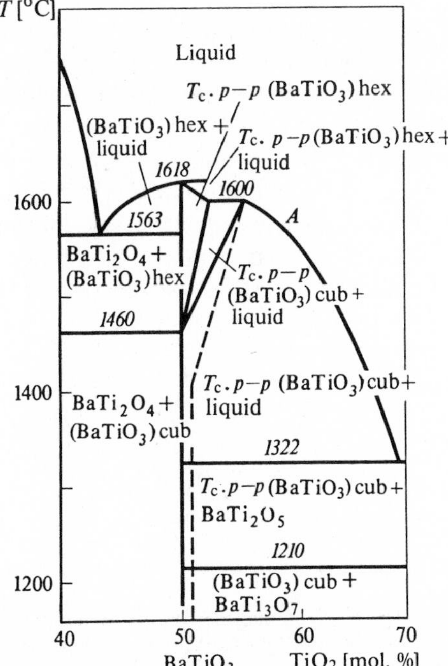

◀ **Fig.9.38.** Phase diagram of $BaO–TiO_2$

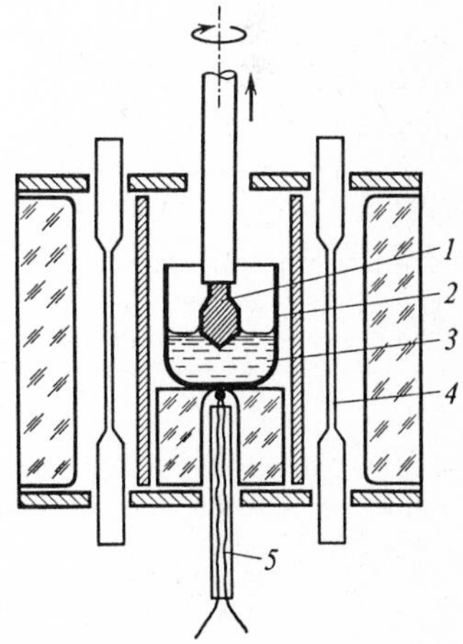

Fig.9.39. Setup for growing crystals from solution in the melt by pulling. (*1*) Growing crystal, (*2*) crucible, (*3*) solution, (*4*) heater, (*5*) thermocouple

(Sect. 10.2.1): the crystal is pulled from the solution at a very low rate of reduction in the solution temperature ($\lesssim 0.1$ deg/h) and a high accuracy of maintenance (± 0.1°C). With this technique, highly perfect crystals about $10 \times 10 \times 10$ mm^3 in size can be grown in crucibles having a diameter of 50 mm and a height of 60 mm. Using the excess of one of the components of the crystallizing substance as the solvent is convenient in that it does not complicate the solution composition. A special solvent, however, permits this composition to be varied over a wide range. A solvent such as PbF_2–PbO–B_2O_3 is used most commonly. In growing single crystals of barium titanate, use is also made of a KF solvent, which interacts only weakly with the platinum walls of the container. This solvent has a number of other advantages. No spontaneous nucleation of barium titanate crystals on the solution surface is observed, because the density of barium titanate exceeds that of potassium fluoride. The high water solubility of KF makes it easy to take the crystals out of the solution. The initial solution contains 70 wt. % barium titanate and 30 wt. % potassium fluoride. The following temperature–time regime is used: the crucible is heated in 8 h to a temperature of 1500°C and cooled to 900°C at a rate of 20–50 deg/h. Then, after the solution is decanted, the crystals in the crucible are slowly cooled in the annealing regime at a rate of 10–50 deg/h.

Another example of growing single crystals by temperature reduction is synthesis of mica ($KMg_3[AlSi_3O_{10}]F_2$). The solvent used is a mixture of fluorides of alkaline and alkaline-earth elements: LiF–MgF_2, LiF–BaF_2, and LiF–CaF_2. The mica concentration in solution is 20–50 wt. %. The temperature at the beginning of crystallization lies in the vicinity of 1000°C. The container is made of iron and platinum. This method can be used to grow faceted single crystals of mica with a volume of several cubic centimeters.

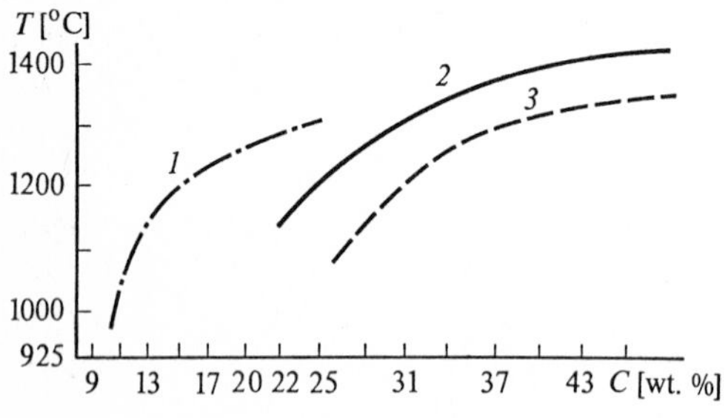

Fig.9.40. Curves of the solubility of $Y_2Fe_5O_{12}$ in solvents of different compositions [9.88]. (*1*) 0.8 PbO–0.3 B_2O_3–1 PbF_2, (*2*) 1.3PbO–0.3 B_2O_3–1PbF_2, (*3*) 1.1PbO–0.5B_2O_3–1PbF_2–1.5BaO (in molar fractions)

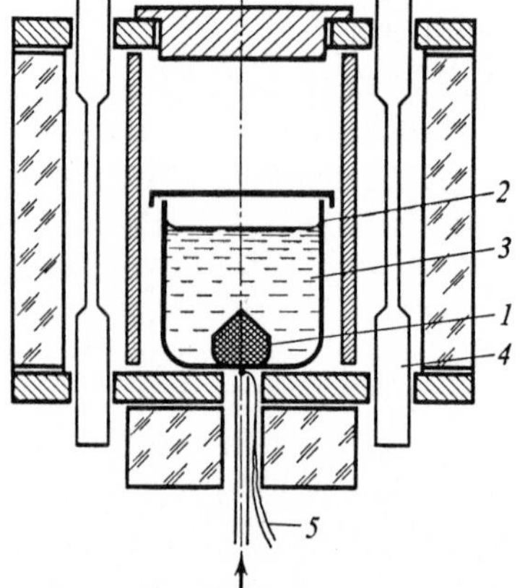

Fig.9.41. Setup for growing yttrium-iron garnet ($Y_3Fe_5O_{12}$). (*1*) Growing crystal, (*2*) crucible, (*3*) solution, (*4*) heater, (*5*) thermocouple. The *arrow* indicates the direction of a stream of cold gas (argon)

The method of reducing the solution temperature has gained the widest application in growing large single crystals of yttrium-iron garnet. Figure 9.40 shows the solubility curves of $Y_3Fe_5O_{12}$. A PbO–PbF_2–B_2O_3 mixture is used as the solvent, the crystallization temperature is 1300–950°C, and the cooling rate is about 0.5 deg/h. An important part of the procedure is the decantation of the solution from the crucible at 950°C through an opening made in the platinum foil welded into the crucible bottom. A diagram of the setup is given in Fig. 9.41. With it, single crystals of yttrium-iron garnet about ten cubic centimeters in volume can be grown.

The solution evaporation method can be used to grow quality single crystals of substances that are stable within a narrow temperature range. The solution evaporates incongruently, so that one phase may become stable and will begin to crystallize. This method has been used to grow such high-melting crystals as HfO_2, ThO_2, CeO_2, etc. The evaporation temperature is usually in the range of 1200–1500°C.

The flux methods are divided into two groups: spontaneous crystallization and crystallization on a seed.

Spontaneous crystallization is mainly used in research work. It is technically simple, but the sizes and quality of the crystals are generally rather low, because the crystallization conditions cannot be sufficiently controlled. The search for ways to diminish the number of crystallization centers led to the use of the local-supercooling method, which results in the formation of one or several crystal nuclei. In this method, which is shown in schematic form in Fig. 9.41, a jet of cool gas is directed at a small area at the center of the crucible bottom, where a seed crystal then appears. Since the volume of the solution is usually large (about 10–20 liters), the overall supersaturation of the solution changes only slightly during the growth run. As a result, crystallization proceeds at a nearly constant supersaturation. This method of seeding has been used to grow large single crystals of iron garnets and orthoferrites.

Single crystals of diamond were first synthesized by spontanoeus nucleation from a solution consisting of a carbon–metal system. The crystallization temperature was about 1500°C at a pressure in the chamber of 50,000–60,000 kp/cm^2. In this kind of synthesis, transition metals of variable valence such as chromium, manganese, cobalt, nickel, and palladium are used as the catalyst and solvent. Graphite usually serves as the source of carbon. On heating, a solution is formed in which diamond crystals appear if the indicated conditions are provided. Figure 9.42 is a schematic representation of superhigh-pressure equipment used to obtain diamonds. A small cylinder made of a mixture of graphite and a metal is compressed, and the mixture is melted by direct heating from an electric current. This method of spontaneous nucleation helped in solving the problem of synthesizing abrasive diamonds, because the production cycle of small crystals (not exceeding tenths of a millimeter) lasts about an hour.

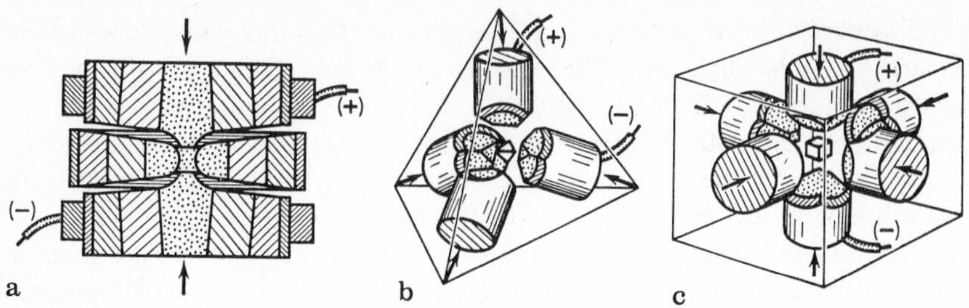

Fig.9.42a–c. Setups for growing diamond single crystals [9.89]. (**a**) "Belt" unit, (**b**) tetrahedral unit, (**c**) cubic unit. The *arrows* indicate the direction of compressive stresses; (±) are electric guides

Crystallization on a seed is widely used in growing large and perfect single crystals. Several seeding methods have been developed. The simplest consists in placing a seeding crystal in a solution. After preliminary partial dissolution of the crystal it is grown by reducing the solution temperature. An important factor here is the supersaturation of the solution near the crystal. Uniform delivery of the substance to the crystal is achieved by rotating it. Stirring is most effectively achieved by the accelerated crucible rotation technique (ACRT) due to *H.J. Scheel.* With this technique, the crucible is rotated in periodically alternating directions, with the speed varying according to a previously adjusted plan [9.90].

In another method of crystallization on a seed placed in solution, the substance is transferred from the lower, more saturated zone to the upper zone

Table 9.1. Flux growth conditions [9.91]

Compound	Formula	Solvent	Conditions
Yttrium-iron garnet	$Y_3Fe_5O_{12}$	PbO	Slow cooling of solution from 1300°C at a rate of 1–5 deg/h
Barium titanate	$BaTiO_3$	TiO_2	Pulling out of solution at 1200°C at a cooling rate of 0.1–0.5 deg/h
Yttrium-aluminium garnet	$Y_3Al_5O_{12}$	$PbO–PbF_2$	Slow cooling from 1500°C to 750°C at a rate of 4–5 deg/h
Beryl	$BeAl_2Si_6O_{16}$	$Li_2O–MoO_3$, B_2O_3, $PbO–PbF_2$	Slow cooling from 975°C at a rate of 6 deg/h
Magnesium ferrite	$MgFe_2O_4$	PbP_2O_7	Slow cooling from 1310°C to 900°C at a rate of 4.3 deg/h
Yttrium vanadate	YVO_4	V_2O_5	Slow cooling from 1200°C to 900°C at a rate of 3 deg/h
High-melting oxides	HfO_2, TiO_2 ThO_2, GeO_2 $YCrO_3$ Al_2O_3	PbF_2 or $BiF_3+B_2O_3$	Evaporation at 1300°C

(crystallization zone) as a result of a temperature difference. This method has been used to grow single crystals of potassium niobate and of ferrites and yttrium-iron garnet.

The method of crystallization on a seed has undergone considerable development as a result of the practice of pulling a crystal out of the solution. The seeding crystal is introduced from above until it comes into contact with the solution, and then the crystal growing on the seed is pulled out of the solution at a rate of about 0.1–0.5 mm/h. This method requires accurate maintenance of the temperature ($\pm 0.1^\circ$C at the level of 1500°C) and of the rate of pulling (± 1 μm/s). It was developed in connection with the growing of single crystals of barium and strontium titanates.

Table 9.1 lists the growing conditions for a number of technically important single crystals.

Temperature-gradient zone melting (TGZM) is similar to conventional zone melting; a narrow zone of solution moves along the specimen as a result of a temperature gradient [9.92]. The method is shown in schematic form in Fig. 9.43, which also presents the corresponding phase diagram. A thin layer (of about 1 mm) of a solid B, which is to serve as solvent, is placed at the boundary between the seeding crystal and a ploycrystalline ingot of substance A. During heating in a high-gradient temperature field, a temperature below the melting point of the crystallizing substance is maintained. Under these conditions partial dissolution of the crystallizing substance takes place, and the thickness of the solution layer exceeds that of the initial layer of the solvent. If $T_1 < T_2$ (see Fig. 9.43), the liquid zone will move upwards in the rod. When it reaches the region $T_1 - T_2$, dissolution of the substance occurs at the cooler liquid–solid boundary until the equilibrium concentration of the solvent is reached: $C_B = C_1$. The dissolution of the substance at the hotter liquid–solid boundary continues until a concentration $C_B = C_2 < C_1$ is established. The concentration gradient which arises in the solution cause the crystallizing substance to diffuse towards the cooler interface, where the solution becomes supersaturated and substance A crystallizes. A continuous solution–diffusion–crystallization process takes place.

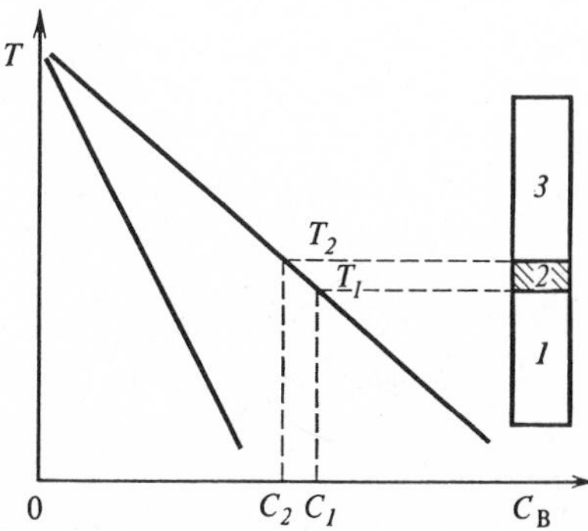

Fig.9.43. Part of the phase diagram for a system recrystallizing substance (A) and a solvent (B) [9.93]. On the *right* is a rod subjected to temperature gradient zore melting. (*1*) Newly grown crystal of substance A, (*2*) solution A + B, (*3*) solid nutrient A. The solvent concentration C_B is plotted on the *x* axis.

To obtain crystals with a low impurity content, a solvent with a low solubility in the solid crystallizing substance is used. The same result can be achieved by reducing the rate of motion of the solution zone by a corresponding reduction in the temperature gradient.

With the use of TGZM, single crystals of gallium arsenide, the α modification of silicon carbide (solvent: chromium), germanium (solvent: lead), and gallium phosphide (solvent: gallium arsenide) have been obtained.

Film growing from high-temperature solutions. Flux growth is used successfully in growing epitaxial films and filamentary crystals and is also known as liquid-phase epitaxy. It is most commonly used to obtain single-crystal films of semiconducting compounds of the elements of groups III–V, and also of garnets and orthoferrites. A large number of such methods have been developed. They can be divided into three groups [9.94, 95]. The first is comprised of methods in which solution is removed from the film surface by simple decantation, the second encompasses those involving forced removal of solution from the film surface, and in the third, the substance is completely crystallized without removal of solution from the film surface.

A diagram of the setup in which solution is decanted from the film surface (the rocking-container method) is given in Fig. 9.44. The substrate and the solution are in a container made of a material which does not react with the solution, usually graphite, quartz, or platinum. In the initial position the container is tilted in such a way that the solution is in the lower part, while the upper part contains the substrate on which the film is to be grown. The container is then turned (a technique of turning the furnace has also been developed), resulting in a complete flooding of the substrate by the solution. Following this the temperature is reduced to ensure supersaturation of the solution and growth of the film. The process is completed by returning the container to the initial position. This method has been used to obtain films of gallium phosphide.

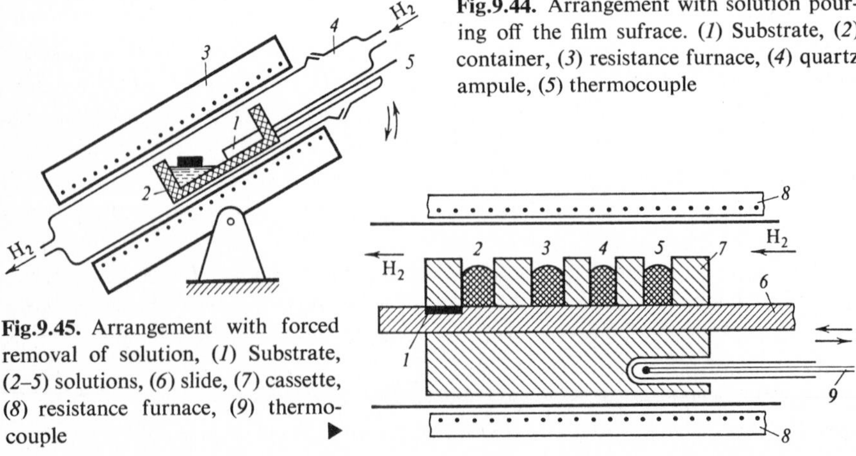

Fig.9.44. Arrangement with solution pouring off the film sufrace. (*1*) Substrate, (*2*) container, (*3*) resistance furnace, (*4*) quartz ampule, (*5*) thermocouple

Fig.9.45. Arrangement with forced removal of solution, (*1*) Substrate, (*2–5*) solutions, (*6*) slide, (*7*) cassette, (*8*) resistance furnace, (*9*) thermocouple ▶

Despite its simplicity it has a number of substantial shortcomings, which restrict its application. These include a high expenditure of solvent and difficulties in controlling epitaxy conditions and film thickness.

The method of liquid-phase epitaxy by rapid withdrawal of a substrate dipped into the solution has gained wide recognition. It is carried out in a resistance furnace enclosing a container with a saturated solution. The substrate holder rotates about its axis at a speed of 2–10 rpm so that the substrate is immersed in solution to a depth of 1–3 mm once a period. The temperature is first increased by about 5°C to ensure etching of the substrate, and then reduced to attain the necessary supersaturation and film formation. When this process is completed the substrate is withdrawn from the solution. With this method, the solution can be batchfed in the course of film growth. Films can also be grown using the liquid encapsulation technique. It has been used to grow films of gallium phosphide (GaP) and solid solutions of GaInP under conditions where the evaporation of the volatile components of the solution is cut down considerably by encapsulation.

One of the basic shortcomings of the method of rapid withdrawal of the substrate from solution is its low efficiency. Special cassettes with several substrates considerably increase the output of films, and are widely used in growing films of GaP, GaAs, GaAlAs, etc. When films of a specific configuration are to be produced, a special mask (obtained by depositing a 2500-Å film of SiO_2 on the substrate) is applied to the substrate to protect certain areas of it from film formation. This method is used to make light-emission diodes, contacts for Gunn oscillators, etc.

The second group of methods involves forced removal of solution from the film surface and is used for obtaining multilayered epitaxial structures. The films are built up as follows: A graphite container is divided into several isolated cells containing the initial solution. The container and the substrate are shifted with respect to one another in such a way that the substrate comes into contact with the solution in each cell successively (Fig. 9.45). After a certain amount of substance is built up from a given cell the solution is removed either with the aid of graphite "scrapers" or by rinsing the substrate, and then the substrate enters into contact with the solution of the next cell, and the growth of the next layer begins. Multilayered structures are obtained by using cassettes with cells containing the necessary impurity, which is introduced into the solution immediately before epitaxy. The growth rate of the films is about 1 μm/min. The grown layers show a high degree of perfection and are used for making semiconductor lasers and Gunn oscillators.

The third group of methods, in which the solution is not removed from the film surface until the film (or multilayer) growth process is completed, includes the so-called capillary film method. Two substrates are arranged parallel and superimposed on one another at a spacing of about 0.1 mm. They are heated to 800–900°C in a hydrogen atmosphere to free their surfaces of oxide films. A specific temperature gradient is set up between the plates. A drop of a substance, for instance GaAs (or GaP), is introduced into the gap between the substrates and drawn into it under the effect of capillary forces. Two versions of crystallization by substrate cooling are possible. In the first, cooling takes place immediately after the drop is drawn in, and in the second it begins only after a certain period of mass transfer from the hotter plate to the cooler one. In both versions films are formed on both substrates, but in the second case the film on the cooler substrate is much thicker. An important advantage of this method is that thin films can be obtained simultaneously on a considerable number of substrates; however, the films deteriorate at the end of the growth run because the entire volume of the solution is crystallized.

10. Growth from the Melt

Single crystals are most commonly grown from the melt, and it is by this method that more than half of the technically important crystals are currently obtained. Most of the substances crystallized from the melt have a simple composition; they include elemental semiconductors and metals, oxides, halides, chalcogenides, etc. This does not mean, however, that this method cannot be applied in growing single crystals of more complex substances such as tungstates, vanadates, and niobates. In many cases single crystals containing five or more components are grown from the melt.

The substances most suitable for growth by this method are those that melt without decomposition, have no polymorphous transitions, and exhibit low chemical activity. The main advantage of this kind of crystallization is that it permits higher growth rates than in crystallization from solution and, in part, from the vapor.

10.1 The Physicochemical Bases of Growing Single Crystals from the Melt

Crystallization from the melt is accompanied by a number of physical and chemical processes, which can be broadly divided into four groups:

1) Processes affecting the melt composition: thermal dissociation of the nutrient, its chemical interaction with the environment, and evaporation of dissociation products and impurities.
2) Processes at the crystallization front, which determine the phase transition kinetics.
3) Heat transfer processes determining the temperature distribution in the crystal and the melt.
4) Mass transfer processes, particularly impurity transfer due to convection and diffusion in the melt.

The processes at the crystallization front and the kinetics of growth are discussed in detail in Sects. 3.1–4.1. Crystals often grow from the melt according to the normal growth mechanism (in which case the crystal–melt interface is rough). The growth rate is then independent of the crystallographic direction, and the supercooling at the crystallization front is extremely low. As a result the front has a rounded shape and practically coincides with the isothermal surface, which is at melting temperature.[1]

Heat transfer processes determining both the shape and stability of the crystallization front during growth from the melt are described in Chap. 5, and the stresses in crystals are discussed in Chap. 6. An analysis of the distribution of impurities and their effect on growth from the melt is given in Sects. 4.2, 3.

Here we shall consider physicochemical processes due to (A) thermal dissociation of the crystallized substance, (B) the chemical interaction between the substance and the container material, and (C) the interaction between the substance and the crystallization atmosphere. Without taking these processes into account it is impossible to determine the temperature–time regime of crystallization and hence the conditions and method for growing single crystals.

10.1.1 State of the Melt

Thermal dissociation of a substance [10.1] and also chemical reactions in the melt can disturb its stoichiometric composition and promote the formation of inclusions, impurity striations, grain boundaries, dislocations, and other defects in the crystal [9.93, 10.2]. The intensity of these reactions depends on the temperature–time regimes, whose optimum characteristics can be determined from the phase diagrams of the principal components of the crystallizing substance with due account of the presence of stable compounds, the extent of stability regions of solid solutions, the presence of phase transitions, and a possible shift in composition because of evaporation, etc.

When dealing with problems of single-crystal preparation it is convenient to divide the phase diagrams into simple ones, which depict one-component systems described by pressure and temperature (PT diagrams), and complex ones, which depict systems of two or more components described by composition, in addition to pressure and temperature (PTC diagrams) [10.1].

For one-component systems (e.g., metals and semiconductors), the study of a PT diagram helps in establishing the growth range and the nature of the crystallization atmosphere and pressure. This is seen from the phase diagram of silicon (Fig. 10.1), whose hatched area corresponds to the practical growing range. According to the diagram, silicon can be crystallized in a vacuum and under elevated pressure. Figure 10.2 presents the phase diagram of carbon, according to which single crystals of either graphite or diamond can be obtained. Single crystals of graphite can be produced from the melt at a temperature of the order of 4000°C and a pressure of 1 kp/cm^2. Diamond single crystals can be grown at the same temperature, but at a pressure of about 2×10^5 kp/cm^2. The

[1] Flat areas, i.e., crystal faces growing layer by layer and requiring considerable supercooling for their growth, may often appear at a rounded growth front. The conditions under which this occurs, and the so-called facet effect are discussed in Sects. 5.2.5 and 10.3.2.

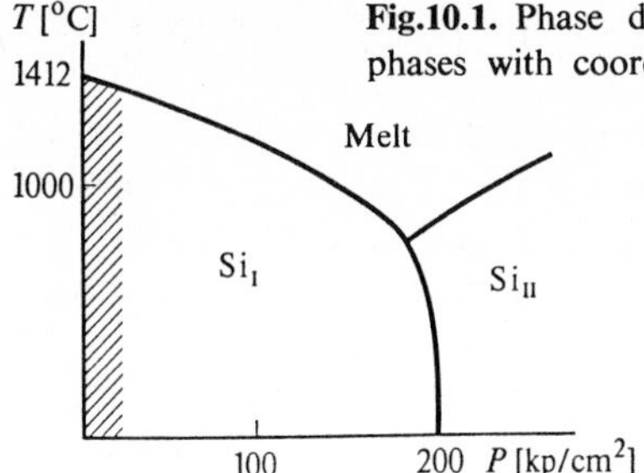

Fig.10.1. Phase diagram of silicon [10.3] Si_I and Si_{II} denote the phases with coordination numbers 4 and 6, respectively

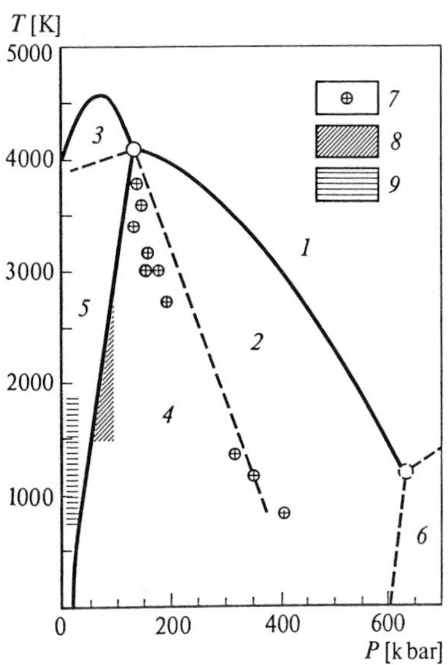

Fig.10.2. Phase diagram of carbon [10.4]. (*1*) Liquid, (*2*) stable diamond, (*3*) stable graphite, (*4*) stable diamond and metastable graphite, (*5*) stable graphite and metastable diamond, (*6*) hypothetic range of existence of other solid phases of carbon, (*7*) points corresponding to the conditions realized in experiments on the direct transformation of graphite into diamond, (*8*) P–T conditions under which diamond crystals are prepared with the use of metals, (*9*) experimental conditions of diamond formation at low pressure

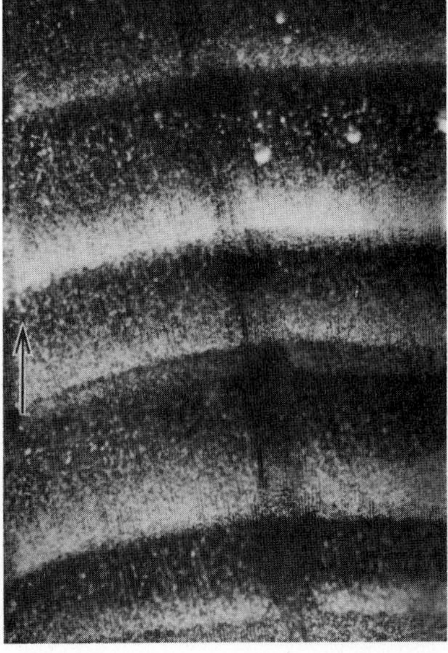

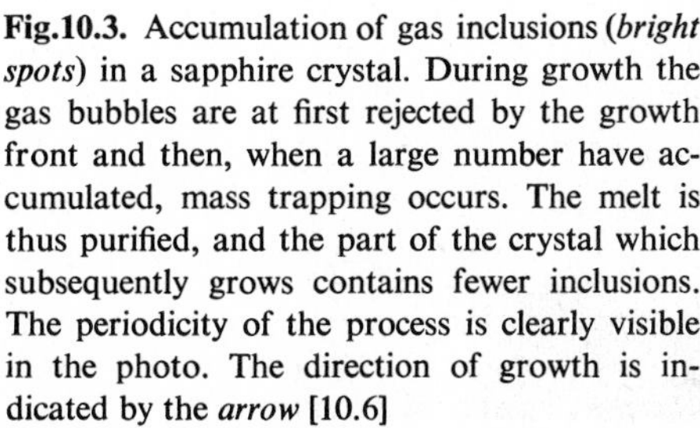

Fig.10.3. Accumulation of gas inclusions (*bright spots*) in a sapphire crystal. During growth the gas bubbles are at first rejected by the growth front and then, when a large number have accumulated, mass trapping occurs. The melt is thus purified, and the part of the crystal which subsequently grows contains fewer inclusions. The periodicity of the process is clearly visible in the photo. The direction of growth is indicated by the *arrow* [10.6]

melting of alumina at normal pressure may be accompanied by thermal dissociation and the formation of Al_2O and probably other gaseous species [10.5]. Because of the high vapor pressure of the dissociation products, the melt is saturated with gas bubbles that accumulate at the crystallization front (Fig. 10.3) and strongly affect the growth kinetics and quality of the crystals.

The difficulty involved in unambiguous determination of crystallization conditions increases in more complex systems, where the behavior of the individual components must be taken into consideration. For instance, accord-

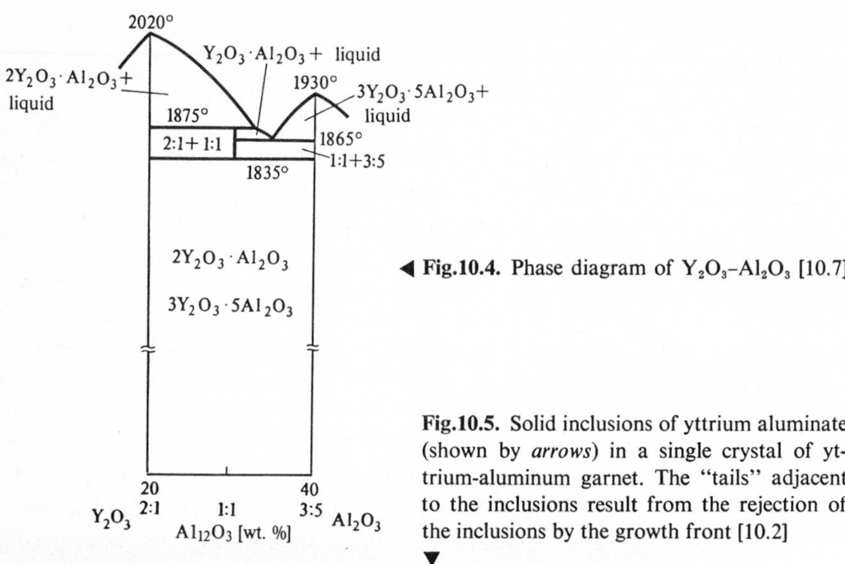

◀ **Fig.10.4.** Phase diagram of Y_2O_3–Al_2O_3 [10.7]

Fig.10.5. Solid inclusions of yttrium aluminate (shown by *arrows*) in a single crystal of yttrium-aluminum garnet. The "tails" adjacent to the inclusions result from the rejection of the inclusions by the growth front [10.2]
▼

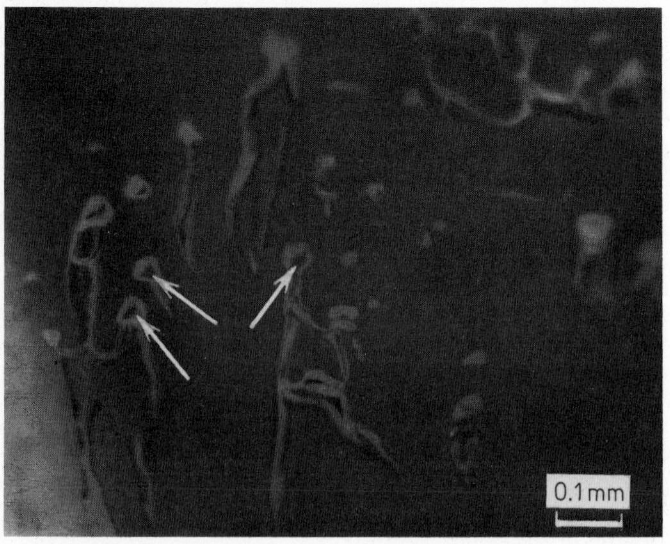

ing to the phase diagram of Y_2O_3–Al_2O_3 (Fig. 10.4), the chemical formula of yttrium-aluminum garnet can be represented as $3Y_2O_3 \cdot 5Al_2O_3$. If thermal dissociation of alumina is taken into account, one must expect depletion of the melt in this oxide and hence the formation of another chemically stable compound, yttrium aluminate ($Y_2O_3 \cdot Al_2O_3$), adjacent to the yttrium-aluminum garnet on the phase diagram.

Thus, a disturbance in the melt stoichiometry results in the formation of another chemically stable compound, which precipitates in the form of solid inclusions (Fig. 10.5).

The intensity of thermal dissociation can be weakened by making use of its dependence on the external pressure, the temperature, and the time the substance remains in the molten state [10.1]. Using increased pressures involves technical difficulties, and therefore researchers strive to find conditions permitting crystallization in a vacuum or at normal pressure. Thus they often use maximum allowable melt overheating temperatures (at which the intensity of thermal dissociation is still moderate), and permit the solution to remain in the molten state for only a short period. The upper limit of melt overheating depends on the intensity of dissociation and evaporation of the substance and on its chemical interaction with the container material and the crystallization atmosphere. The lower limit depends on the viscosity of the melt, which hinders convective mixing. For instance, the viscosity of an alumina melt changes about twofold in the range of 2150–2050°C. The average value for the overheating of the melt is in practice 1.01–1.03 T_m.

The time that the solution spends in the molten state is reduced if use is made of zone melting, in which only a small part of the substance is liquid.

Losses by evaporation are usually offset by delivering the lacking components in batches to the melt. The amount of the substance to be supplied is found experimentally. In growing single crystals of yttrium-aluminum garnet, the losses are compensated by increasing the content of alumina in the initial charge nutrient by 0.5–1 wt. % with respect to the stoichiometric composition. The maximum permissible excess of the component is detected by the appearance of constitutional supercooling resulting from the accumulation of this component at the crystallization front (Sect. 5.3.3).

10.1.2 Container Material

Interaction between the melt and the container material is the decisive factor in selecting a crystallization method. The basic requirement is mutual insolubility and the absence of chemical interaction. This is not enough, however, since a reaction between the melt and the container material may be caused by a third component, such as oxygen (and water), which arrives from the crystallization atmosphere or the initial charge and adsorbs on the walls of the crystallization chamber, the working elements of the furnace, and the container.

The general rule in choosing a material for the container is that the chemical bonding forces of the container material must differ drastically in nature from those of the crystallizing material. Dielectric crystals are grown in metal containers, organic crystals are grown in containers made of inorganic dielectrics, etc. In addition, the container material must have sufficient mechanical strength and be machinable; the coefficients of expansion and compression of the container material and the grown crystal must have dose values; high electrical conductivity is necessary in connection with high-frequency heating; and the container must be cleanable by chemical or other etching methods. All these problems vanish if the container is made of the crystallizing substance itself (Sect. 10.2).

Table 10.1 lists the principal container materials used in growing single crystals. The most common ones are glass, quartz, graphite, platinum, molybdenum, iridium, and tungsten. In the low-temperature range (below 800°C) preference is given to copper, nickel, iron, and also glass and organic materials; in the medium-temperature range (800–1800°C) to noble metals, graphite, and molten quartz, and in the high-temperature range (over 1800°C), to graphite, high-melting metals, and oxides.

Since it is not always possible to find a material that is neutral with respect to the melt, the container walls are sometimes coated with various linings which prevent reaction with the melt. These linings must have the necessary mechanical strength, and their coefficient of expansion must be close to that of the container material. For instance, if molybdenum is coated with silicon, molybdenum disilicide forms, and the molybdenum is protected from oxidation up to 1400–1500°C; coating platinum with iridium and molybdenum with tungsten prolongs the life of the container.

10.1.3 Crystallization Atmosphere

In assigning growth conditions one must bear in mind the possibility that the melt and the container material could react with the crystallization atmosphere. The main requirements imposed on the composition and pressure of this atmosphere are related to the vapor pressure and chemical activity of the crystallizing substance. The role of the crystallization atmosphere may be passive or active; in the latter case the gases react with the melt, inhibiting or promoting certain processes. For instance, in a neutral atmosphere chromium, introduced in the trivalent state into alumina, remains unaltered, while in a hydrogen atmosphere it changes its valency. The divalent state of samarium in fluorite, on the other hand, is preserved in a reducing atmosphere and changes to a trivalent state in a neutral atmosphere.

In choosing the crystallization atmosphere, preference must be given to an atmosphere containing volatile components of the crystallizing substance. For instance, an oxygen-containing atmosphere is used in growing oxides, a fluorine-containing atmosphere for fluorides, a sulfur-containing atmosphere

Table 10.1. Operating conditions for some container materials

	Container material	Max. T^{a}_{work} [°C]	Crystallizing substance	Crystallization atmosphere
	1	2	3	4
Metals	Iridium	2200	Simple oxides, molybdates, tungstates, tantalates, garnets, aluminates	Vacuum, inert, and reducing
	Molybdenum	2500	Elemental oxides, garnets, aluminates, nitrides	
	Tungsten	3000		
	Tungsten-molybdenum alloys	2500		
	Graphite	2500	Fluorides, sulfides, nitrides, phosphides	Fluorine-containing
Oxides	Aluminium	1800	Metals, arsenides, phosphides	Vacuum, inert, oxidizing, and reducing
	Zirconium	2500		
	Magnesium	2500		
	Carbides, Nitrides	2500		Inert
Metals	Platinum	1500	Fluorides, tungstates, molybdates, germanates, fluorine-containing oxides	Vacuum, inert
	Rhodium	1700		
	Platinum-rhodium alloys	1650		
	Molten quartz	1200	Metals, sulfides, arsenides, phosphides, nitrides	Vacuum, oxidizing
	Vycor	1000		
Metals	Iron	1300	Oxides, organic compounds, fluorides	Vacuum, reducing, inert
	Copper	800	Organic compounds	
	Aluminium	500		
	Pyrex	600	Organic compounds, metals	Oxidizing, inert, vacuum
	Teflon	200	Metals, organic compounds	Oxidizing, vacuum, neutral, reducing, fluorine-containing

[a] Maximum permissible working temperature

for sulfides, etc. Ultimately, two conditions must be met by the crystallization atmosphere: there must be an absence of reaction with the melt, the material of the container, and that of the furnace structures; and the removal of foreign impurities must be technically simple.

In growing single crystals, a distinction is made between crystallization in a vacuum, in a neutral atmosphere (helium, argon, nitrogen), and in a reducing atmosphere (air, oxygen). A vacuum is used to purify the melt of dissolved gases, volatile contaminants, and products of thermal dissociation. The depth of the vacuum depends on the vapor pressure of the crystallizing substance and the temperature of the melt. A vacuum of the order of 3×10^{-5} torr is most often used. The possibilities of utilizing a vacuum are limited by the fact that at temperatures exceeding 800°C and in a vacuum of less than 10^{-4} torr, reactions resulting in the destruction of metal heaters and containers are intensive. A vacuum is used for growing single crystals of metals (aluminum, copper, iron, etc.), semiconductors (silicon, germanium, etc.), and dielectrics (fluorite, corundum, yttrium-aluminum garnet, etc.).

A neutral atmosphere is very widely used to reduce the intensity of evaporation of a substance. It finds extensive application in growing single crystals of a number of high-melting oxides and their compounds, and also of sulfides, nitrides, fluorides, arsenides, etc. The gases predominantly used for the crystallization atmosphere are helium, argon, and nitrogen, the reason being that effective systems of chemical purification have been developed for them.

A reducing atmosphere is used to prevent oxidation reactions in the melt. For instance, in the crystallization of fluorite (CaF_2), a hydrogen fluoride atmosphere prevents the development of hydration reactions accompanied by the formation of $CaHCO_3$, and the crystallization of metals in a hydrogen atmosphere makes it possible to obtain oxygen-free single crystals.

An oxygen-containing atmosphere (air, oxygen) is used in growing single crystals whose composition is depleted in oxygen. The possibility of using an oxidizing as opposed to an inert atmosphere depends on the crystallization temperature, the material of the container, and the substance making up the structural elements of the furnace. If the concrete experimental conditions rule out an oxidizing atmosphere, the process is conducted in a neutral one, and the grown crystal is subsequently subjected to thermal annealing in an oxygen atmosphere at $\frac{1}{2}$–$\frac{2}{3}$ T_m. This operation is called oxygen annealing.

The pressure of the crystallization atmosphere depends on the vapor pressure of the crystallizing substance. A distinction is made between crystallization at normal pressure (1 kp/cm^2), medium pressures (up to 200 kp/cm^2), and high pressures (above 200 kp/cm^2). The vast majority of crystals are grown at normal gas pressure, since the use of medium and high pressures involves technical difficulties. But in growing some semiconducting crystals, such as sulfides, nitrides, arsenides, etc., medium gas pressures are often used.

The successful utilization of gas media is dependent upon the degree of their chemical purity, i.e., primarily upon the extent to which the media are purified of oxygen and water vapor. The purification methods are divided into three groups, the first of which is based on thermal diffusion of a gas through a membrane. For instance, deep purification of hydrogen is achieved by diffusion through a palladium plate at a temperature of 300–400°C. The second group is based on the adsoprtion of contaminants by molecular sieves, and has gained wide acceptance in connection with the removal of impurities such as oxygen, nitrogen, and water. The third group is based on chemical reactions between the contaminants contained in the gas, which result in the formation of chemically stable compounds such as oxides, nitrides, and fluorides. An overall system of gas purification generally uses several methods simultaneously.

10.2 Principal Methods of Growing Single Crystals from the Melt

The methods of growing single crystals from the melt are divided into two groups:

1) Methods with a large melt volume (the Kyropoulos, Czochralski, Stockbarger, and Bridgman methods)
2) Methods with a small melt volume (Verneuil and zone melting methods).

The volume of the melt affects the nature and intensity of a number of physicochemical processes occurring in it. The molten substance may experience thermal dissociation, and the dissociation products may evaporate into the atmosphere. As indicated in Sect. 10.1.1, the time during which the substances remain in the molten state should be limited; in other words, the crystals should be grown by methods with a small melt volume. The same condition must be fulfilled for substances whose melts react actively with the container material and the crystallization atmosphere; the smaller the volume of their melt, the less the crystal will be contaminated by products of reaction between the melt and the environment. Attention should also be given to the drastic difference in convection conditions for the two groups of methods. In a large melt volume the convective flows caused by the differences in temperature in the various parts of the melt develop freely, and the convective transfer of mass and impurities plays an important part. In a small melt volume, on the other hand, convection cannot develop in such a manner, and transport is effected by diffusion. The most striking example of the influence of the melt volume on the quality of the growing crystal is the distinct difference between the impurity distribution in cylindrical crystals grown by unidirectional crystallization and in those grown by zone melting. Unidirectional crystallization is one of the methods used with a large melt volume. A cylindrical container holding the melt moves into the cool region of the furnace, so that a crystal nucleating at the cooler end of the container gradually increases in size until it occupies the whole container. Zone melting is conducted with a small melt volume, and can be used to grow cylindrical crystals; they are obtained by moving a narrow zone of the melt

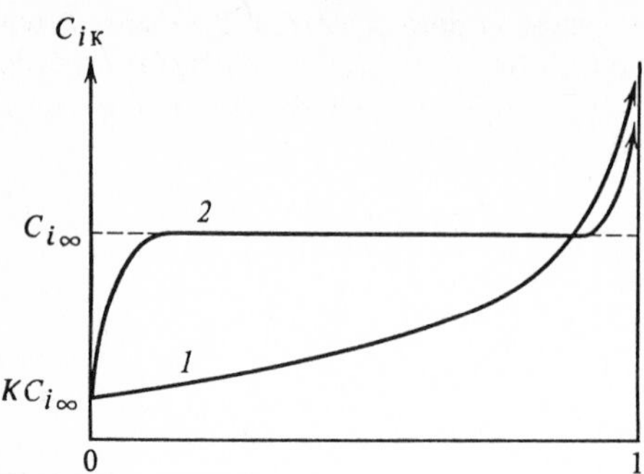

Fig.10.6. Concentration C_{ic} of impurity with a distribution coefficient $K < 1$ in a crystal grown (a) by unidirectional crystallization and (b) by zone melting. The impurity concentration in the nutrient is $C_{i\infty}$; the crystals grew from left to right

through the initial polycrystal or single crystal. The impurity distribution with a distribution coefficient $K < 1$ along the length of the grown crystal is presented for these two methods in Fig. 10.6. In unidirectional crystallization the amount of impurity in the middle portion of the crystal remains constant, and therefore this method is preferred for the impurity activation of crystals. In zone melting, the initial part of the crystal contains less impurity than the initial reactant, and hence this method is used successfully in the purification of crystals.

10.2.1 Kyropoulos and Czochralski Methods

In the Kyropoulos method, crystal growth is achieved by gradually reducing the melt temperature with the aid of a special cooler. Either spontaneous nucleation (Fig. 10.7a) or crystallization on a seed, which is lowered from above until it touches the melt (Fig. 10.7b), is possible. Spontaneous nucleation can be carried out by local cooling of the container bottom, for example with a jet of cool gas (Fig. 10.7a). In this method the entire volume of the melt can be crystallized at a decreased temperature gradient of ~ 1 deg/mm [10.8]. The Kyropoulos method has gained wide recognition in growing large crystals with

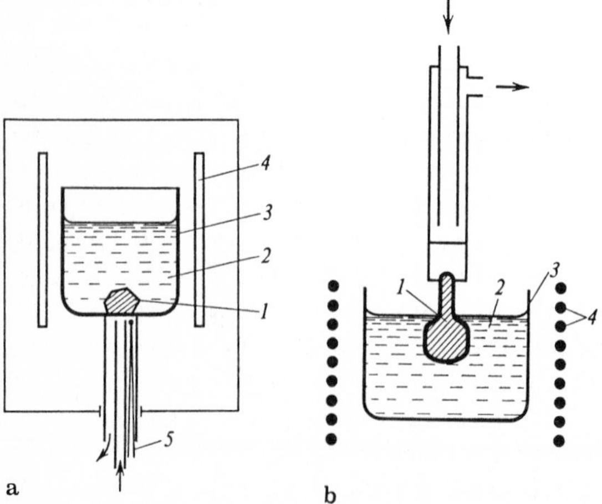

Fig.10.7a,b. Equipment for growing crystals by the Kyropoulos method, (**a**) using spontaneous crystallization and (**b**) on a seed. (*1*) growing crystal, (*2*) melt, (*3*) crucible, (*4*) heater, (*5*) thermocouple. The *arrows* indicate the direction of the cold gas jet

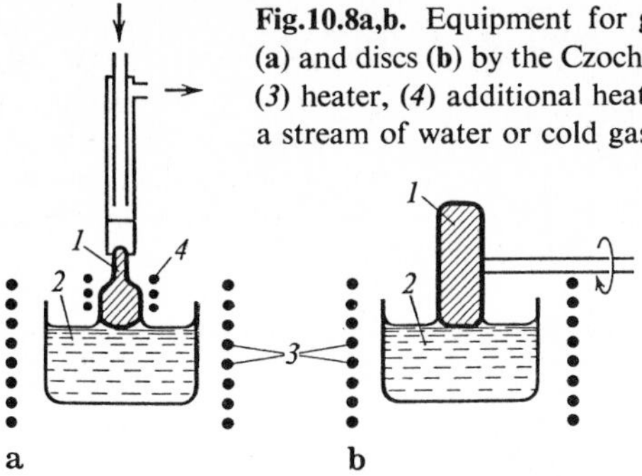

Fig.10.8a,b. Equipment for growing single crystals in the shape of rods (**a**) and discs (**b**) by the Czochralski method. (*1*) Growing crystal, (*2*) melt, (*3*) heater, (*4*) additional heater. The *arrows* indicate the flow direction of a stream of water or cold gas

weights of several kilograms. A serious disadvantage, however, is that the growing rate is not constant, because the heat exchange varies in a complex manner as the crystal increases in size. This method is used in growing crystals of fluorides, chlorides, etc., and single crystals of corundum.

The Czochralski method (Fig. 10.8) differs from the Kyropoulos method in that the melt temperature is kept constant and the crystal is slowly pulled out of the melt as it grows. This ensures a virtually constant crystallization rate. The pulling rate depends on the physicochemical characteristics of the crystallizing substance and the diameter of the crystal, and ranges from 1 to 80 mm/h. The melt temperature and the crystallization rate, which depends on the rate of heat removal, can be changed independently.

The substance is melted in a metal crucible, usually by high-frequency or resistance heating. Figure 10.9 presents a photo of the crystallization unit "Kinovar," which is supplied with high-frequency heating from a motor generator and has a system of programmed stabilized precision crucible heating. Figure 10.10 shows a silicon crystal grown by the Czochralski method.

The Czochralski method has a number of advantages which are particularly worth mentioning. Firstly, there is no direct contact between the crucible walls and the crystal, a circumstance which helps in producing unstressed single crystals. Secondly, the crystal can be extracted from the melt at any stage of growth; this is important in investigating the growing conditions. Thirdly, the geometric shape of the crystal can be changed by varying the melt temperature and growth rate. This is done to obtain crystals with few or no dislocations: by devising a special pulling plan one can reduce the crystal diameter, and then most of the dislocations meet the lateral sides of the crystal at the necking site, i.e., they leave the crystal (taper off). Then the crystal diameter increases again, and the dislocation density in it becomes very low, because no new dislocations arise. This is called the "pinch" method [10.9].

To avoid any possible disturbance in the stoichiometry of the melt as a result of evaporation of the volatile components of the feedstock, the free surface of the melt is insulated from the atmosphere by a liquid layer of a specially introduced substance which is immiscible with the melt and does not form chemically stable compounds with it (liquid encapsulation technique).

Fig.10.9. Crystallization unit "Kinovar"

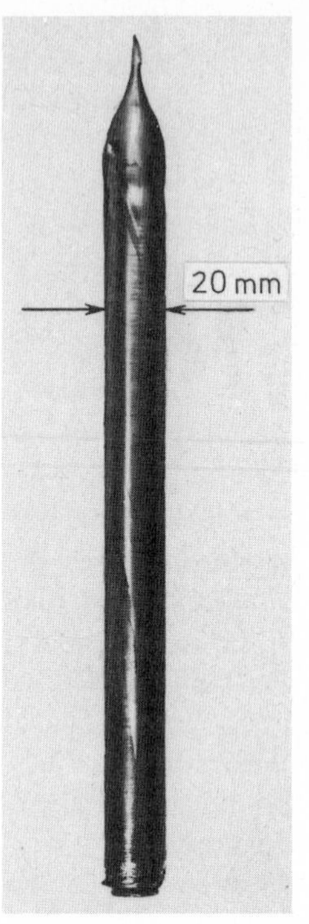

Fig.10.10. Single crystal of silicon grown by the Czochralski method

Due to the above advantages the Czochralski method has gained wide recognition, particularly in growing single crystals of silicon, germanium, corundum (Al_2O_3), yttrium-aluminum garnet ($Y_3Al_5O_{12}$), lithium niobate ($LiNbO_3$), gallium phosphide, gallium arsenide, etc. [10.6, 10, 11]. Dislocation-free crystals of silicon and germanium were first obtained by this method.

A grave shortcoming of the method, however, is formed by the presence of a heated container, which may become a source for contamination of the melt. Attempts to dispense with a container made of a foreign material have given rise to the "cold crucible" or skull method.

In this method [10.12, 13] a water-cooled container is used which consists of several segments insulated from one another (Figs. 10.11, 12). The substance is melted by induction currents of a frequency of 50–400 kHz. The crystallizing substance adjacent to the cold container remains unmelted and forms, together with the crystallized crust, a crucible containing the melt. The stable position of the melt–skull interface is maintained by controlling the power output of the heater. If the crystallized substance has a low electrical conductivity at room temperatures or even near the melting point, then prior to induction heating,

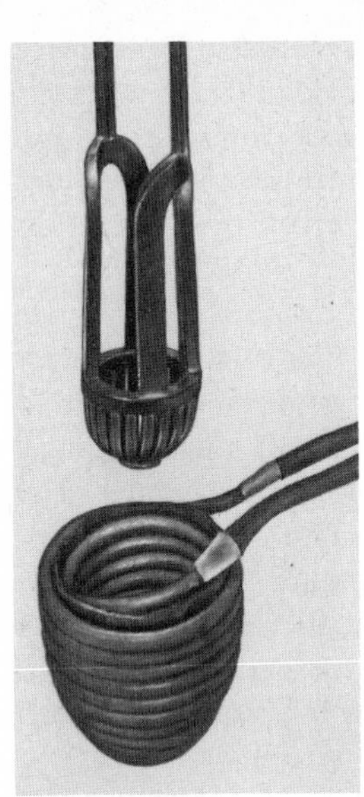

Fig.10.11. *Top*—"cold crucible," *bottom*—induction heater [10.12]

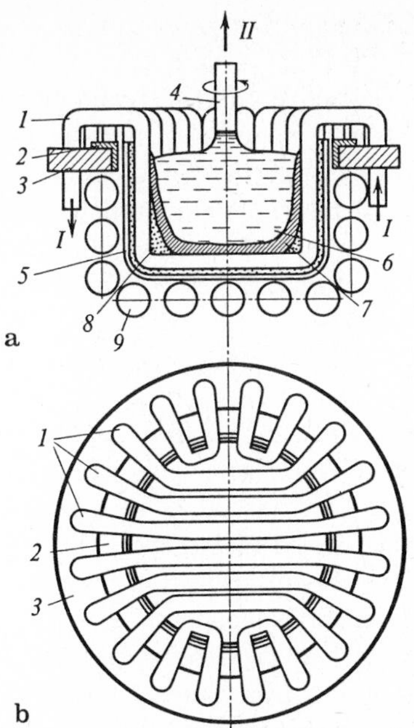

Fig.10.12a,b. Skull method [10.13]. **(a)** Cross section of the Apparsus; **(b)** plan view of the container. (*1*) Cotainer tubes, (*2*) protective quartz ring, (*3*) fluoroplast ring, (*4*) growing crystal, (*5*) quartz vessel, (*6*) melt, (*7*) layer of crystallized melt, (*8*) unmelted charge, (*9*) induction heater. The *arrows* indicate the motion of the cooler (*I*) and pulling of the crystal (*II*)

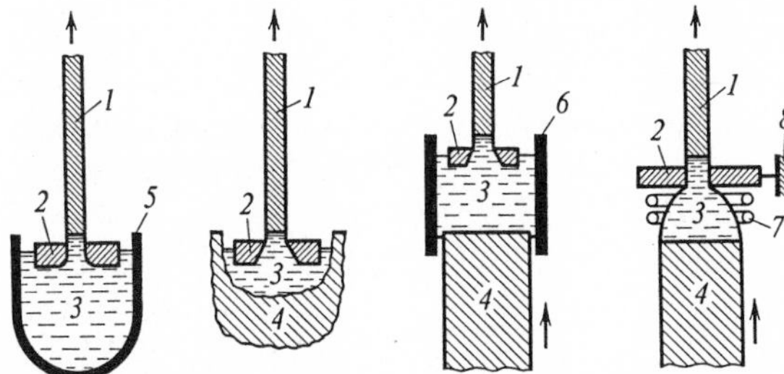

Fig.10.13. Versions of Stepanov's method [10.14]. (*1*) Growing crystal, (*2*) shaper, (*3*) melt, (*4*) solid material for melting, (*5*) crucible, (*6*) melt holder, (*7*) induction heater for melting, (*8*) support for the shaper. The *arrows* indicate the direction of crystal pulling and transport of the material

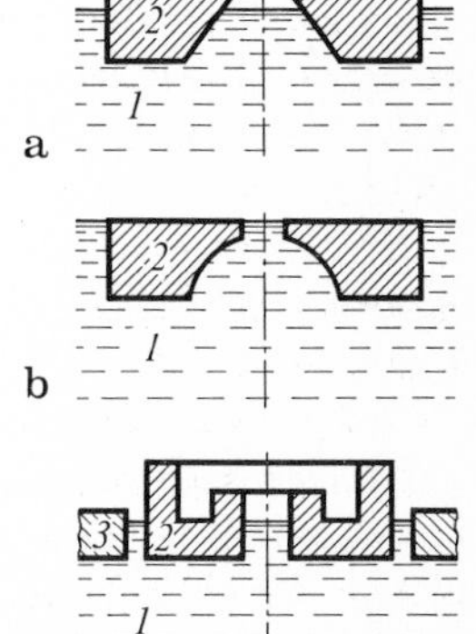

Fig.10.14. Various types of shapers [10.14]. (*1*) Melt, (*2*) shaper, (*3*) lid covering the melt surface

the substance is heated to a temperature at which its electrical conductivity increases abruptly (up to 2–10 Ω^{-1} cm^{-1}), and then induction heating with a 2–7 mHz current is switched on. The preliminary heating can also be achieved by means of induction currents: pieces of a metal contained in the crystallizing substance are introduced into the initial charge. For instance, in growing single crystals of corundum, pieces of aluminum introduced into the charge oxidize on heating in air, forming the crystallizing substance. This method is used to obtain crystals having diameters of ~ 20 mm and lengths of ~ 100 mm.

The Czochralski method was further developed by *Stepanov* [10.14]. He proposed pulling the crystal through a die float placed on the surface of the melt (Fig. 10.13). The die imparts the necessary shape to the crystal and is referred to as a shaper. Figure 10.14 shows various types of shapers for producing single crystals of complicated geometric shapes such as rods, tubes, and plates. The substance is first shaped in the liquid state (mainly as a result of the capillary effect) and then transformed to the solid state. The parameters of the process are the temperature in the crystallization zone, the pulling rate, and the intensity with which the growing crystal cools.

Mlavski and co-workers proposed a method of pulling out filaments, plates, and single crystals with intricate cross sections [10.15] which was actually a further development of Stepanov's technique. A melt flowing along the capillaries of a shaper of preassigned profile accumulates against its end face (Fig. 10.15). When the crystal is pulled, the melt film is replenished continuously due to the capillary forces, which move the melt from the crucible to the end face of the shaper. Corundum filaments 0.1 mm in diameter are pulled out at a rate of 500 mm/min.

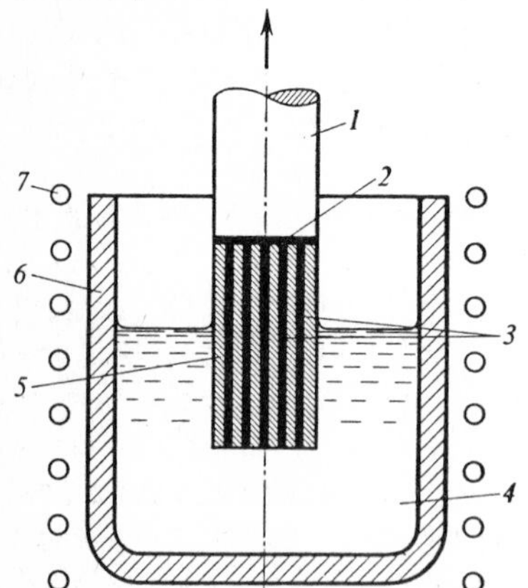

Fig.10.15 Schematic diagram of crystal pulling (see *arrow*) with the aid of a capillary shaper [10.15]. (*1*) Growing crystal, (*2*) film of the melt, (*3*) capillaries, (*4*) melt, (*5*) shaper, (*6*) crucible, (*7*) in duction heater

The Czochralski method and its modifications were among the first methods to serve as a basis for a fully automated system with computer-controlled growing of single crystals [10.16]. The main element in such systems is a control pickup, which must be not only highly sensitive and accurate, but also reliable when used for prolonged periods.

The automated systems developed for the Czochralski method are based on precise and continuous weighing of the crucible containing the melt when growing single crystals. By setting a specific law for change in the weight of the crucible (with melt) in time, one also assigns a law for change in the weight of the crystal, since these characteristics are interdependent.

10.2.2 Stockbarger-Bridgman Method

This method, which is also called the unidirectional crystallization method, differs from the Czochralski and Kyropoulos methods in that the entire volume of the melt, which is usually placed in a cylindrical container, is crystallized. The method is technically simple, and crystals of a preassigned diameter are produced by selecting the appropriate container. But here the container effect is not limited to possible contamination of the melt; the elastic interaction of the container walls with the crystal during cooling may cause stresses in the crystal.

In papers by *Tamman*, *Bridgman*, *Stockbarger*, *Obreimov*, and *Shubnikov* [10.17], two modes of crystallization in a container are discussed. In one the container is moved through the melting zone, and in the other, the temperature is reduced gradually at a constant temperature gradient. These two modes have been realized for two versions: crystallization in the vertical and crystallization in the horizontal direction.

The more common of these is crystallization in which the container holding the substance is displaced vertically (downward) through the melting zone [10.2]. To set up the necessary temperature gradients in the crystallization zone, Stockbarger employed thermal screens (Fig. 10.16). Figure 10.17 is a photo of the crystallization unit "Granat-2," which operates by Stockbarger's method using resistance heating. Figure 10.18 gives schematic representations of the main types of resistance heaters.

With the Stockbarger-Bridgman method, single crystals can be grown either by spontaneous nucleation or on a seed. In the former case the container has a conical shape. As it is lowered into the cold zone, a few crystallization centers appear at the top of the cone. The small crystals appeared that way undergo geometric selection (Sect. 7.2), and one crystal remains, which increases in size until it fully occupies the cross section of the container.

The Stockbarger-Bridgman method is applied in growing a wide range of substances, but is most commonly used to obtain metallic, organic, and a number of dielectric single crystals such as oxides, fluorides, sulfides, and halides. Figure 10.19 shows a single crystal of sapphire (Al_2O_3) grown by this method. In recent years horizontal unidirectional crystallization (HUC) in a boat-shaped container (the "boat" method) has gained a firm foothold. This technique, by which single crystals can be grown in the shape of plates, has been used successfully in growing large single crystals of corundum (Al_2O_3) and yttrium-aluminum garnet ($Y_3Al_5O_{12}$) (Fig. 10.20) [10.2, 6]. A schematic representation of it

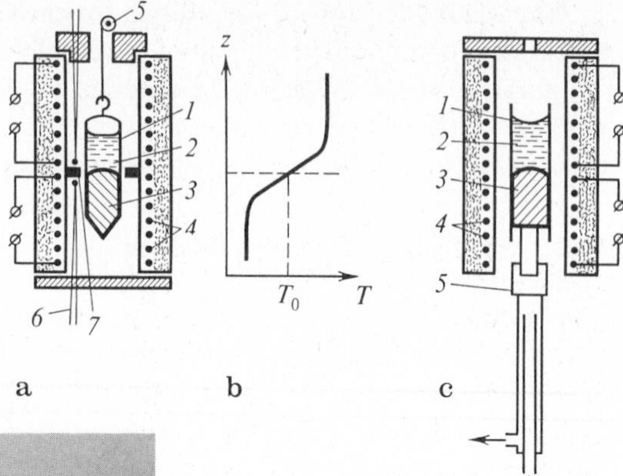

Fig.10.16a–c. The Stockbarger method (**a**, **c**) and temperature distribution along the furnace (**b**), T_0 is the melting point of the crystallizing substance. (*1*) Container, (*2*) melt, (*3*) growing crystal, (*4*) heater, (*5*) arrangement for lowering the container, (*6*) thermocouple, (*7*) thermal screen. The *arrows* indicate the flow direction of a stream of water or cold gas

Fig.10.17. Crystallization unit "Granat-2" (*1*) Crystallization chamber, (*2*) mechanism for lowering the container, (*3*) control assembly

Fig.10.18a–d. Types of resistance heaters. (**a**) Helical, (**b**) split type, (**c**) rod type (the rods are placed around the container holding the molten substance), (**d**) coaxial

▼

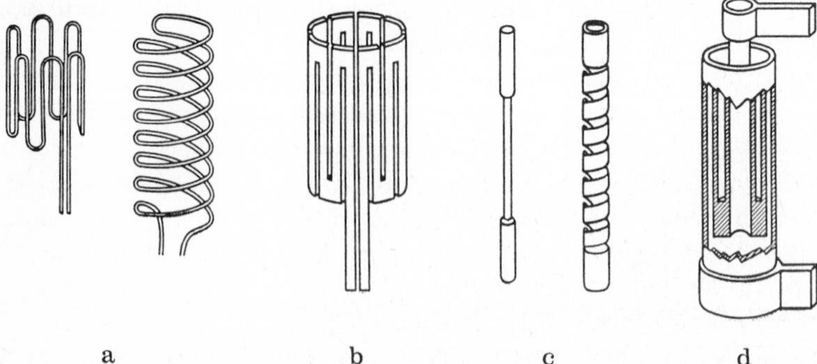

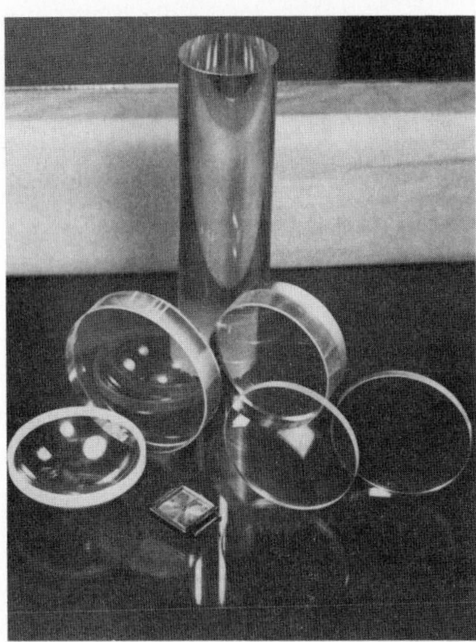

Fig.10.19. Corundum single crystal grown by the Stockbarger method, and products made from it

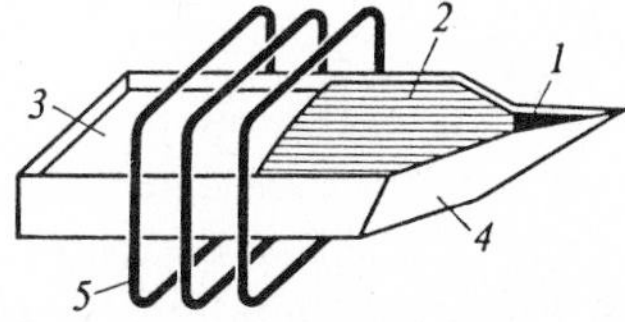

Fig.10.21. Horizontal unidirectional crystallization. (*1*) Seed, (*2*) crystal, (*3*) melt, (*4*) container, (*5*) heater

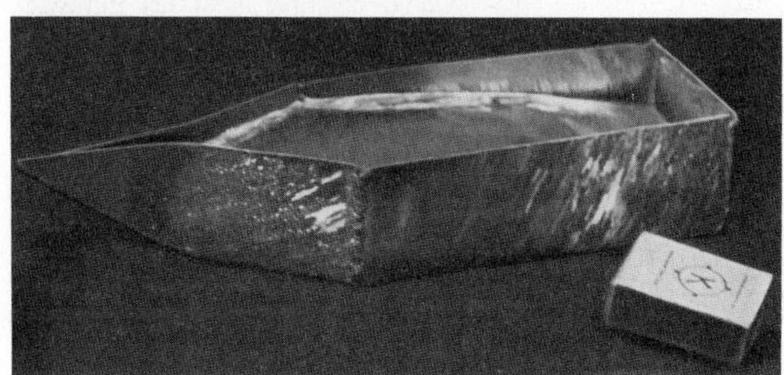

Fig.10.20. Corundum single-crystal plate in a boat Courtesy of Kh. S. Bagdasarov

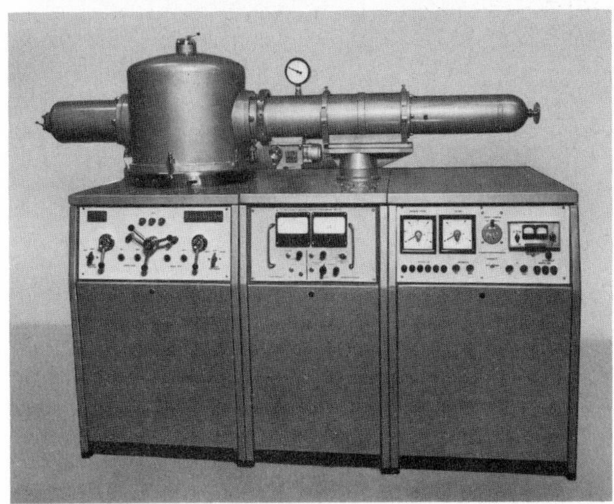

Fig.10.22. Crystallization unit "Sapfir-1"

is given in Fig. 10.21. The principal difference between horizontal and vertical crystallization is that in the former the melt depth remains nearly constant during the growth run, which improves the stability of the process. Furthermore, the considerable surface area of the melt, which is typical of HUC, ensures the evaporation of foreign impurities during crystallization. The crystals have a platelike shape suitable for technical applications. The HUC method makes it possible to grow crystals on a seed placed at the front of the container (see Fig. 10.21), and produces flat crystals of sizes so large as to be difficult to obtain by other methods.

Figure 10.22 presents a photo of a horizontal "Sapphire-1" crystallization unit with resistance heating. The block diagram of the unit is shown in Fig. 10.23, and Fig. 10.24 shows the position of the tungsten heater and molybdenum screens. This system is highly reliable and permits crystallization in the vicinity of 2000°C without major repairs for 3–5 months.

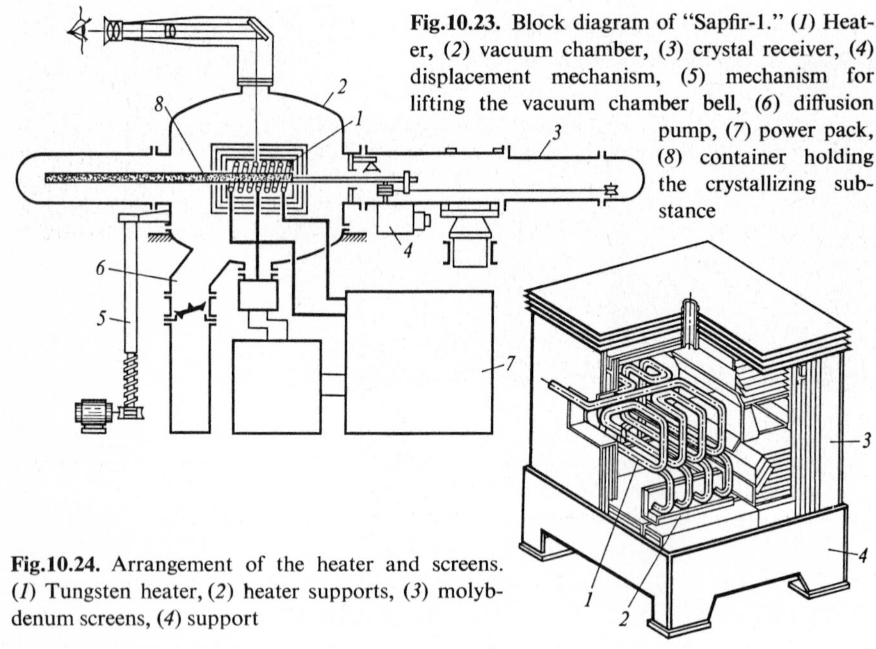

Fig.10.23. Block diagram of "Sapfir-1." (*1*) Heater, (*2*) vacuum chamber, (*3*) crystal receiver, (*4*) displacement mechanism, (*5*) mechanism for lifting the vacuum chamber bell, (*6*) diffusion pump, (*7*) power pack, (*8*) container holding the crystallizing substance

Fig.10.24. Arrangement of the heater and screens. (*1*) Tungsten heater, (*2*) heater supports, (*3*) molybdenum screens, (*4*) support

10.2.3 Verneuil Method

This method, which is also called the flame fusion method, is one of the techniques in which a limited melt zone is used (Fig. 10.25). A powdered substance (particle size 2–100 μm) falls from a bin, passes through a gas burner, and reaches the upper fused end face of a single-crystal seed, which is slowly lowered with the aid of a special mechanism. Dropping through a hydrogen–oxygen

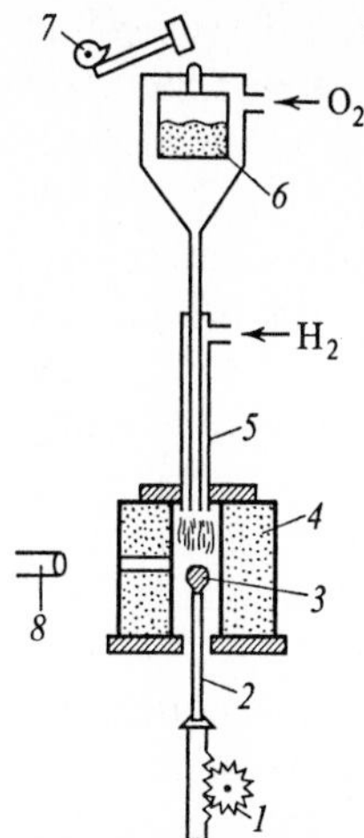

Fig.10.25. Unit for crystal growing by the Verneuil method. (*1*) Mechanism for lowering the crystal, (*2*) crystal holder, (*3*) growing crystal, (*4*) muffle, (*5*) burner, (*6*) bin, (*7*) shaking mechanism (*8*) cathetometer

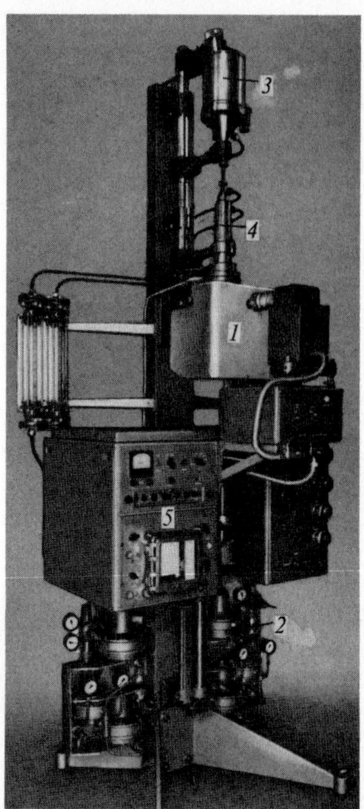

Fig.10.26. Crystallization unit "KAU-1." (*1*) Crystallization chamber, (*2*) lowering mechanism, (*3*) bin, (*4*) burne, (*5*) control assembly

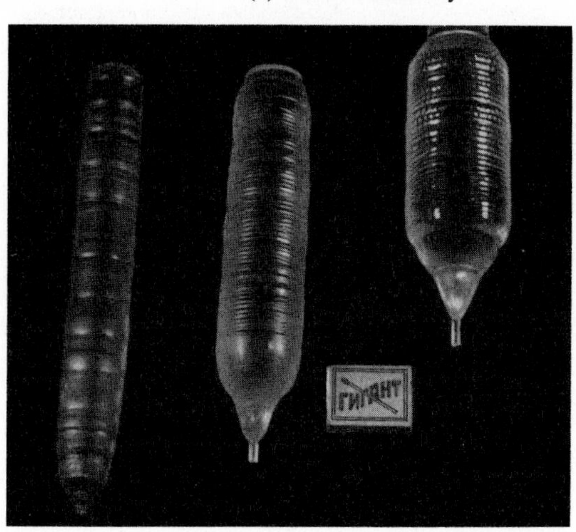

Fig.10.27. Single crystals of ruby and sapphire obtained by the Verneuil method

flame, the charge particles partially fuse and fall into a thin film of the melt ($\sim$ 0.1 mm). Since the seed is lowered slowly, the melt film crystallizes at the required rate, being continuously replenished from above. If the consumption of the charge and of hydrogen and oxygen is coordinated along with the rate at which the seed is lowered, the film thickness is kept practically constant.

Figure 10.26 shows the unit "KAU-1," used for crystallization by the Verneuil method. In it, rod-shaped crystals with diameters of up to 20 mm and lengths of up to 500 mm can be grown. To reduce the residual stresses in the crystals, various gas burners have been designed which provide the crystals with supplementary heating during growth. The resulting crystals may reach 40 mm in diameter (Fig. 10.27).

The popularity of the Verneuil method is the result of successful application of it in growing single crystals of ruby, sapphire, aluminum-magnesium spinel ($MgAl_2O_4$), and rutile (TiO_2).

The main advantages of this method are the following:

1) There is no container. This eliminates problems of physicochemical interaction between the melt and the container material, and also the problem of residual stresses resulting from the elastic effect of container walls.
2) Crystallization can be conducted at about 2000°C in air, with the oxidation-reduction potential of the crystallization atmosphere being regulated by changing the relative content of oxygen and hydrogen in the flame.
3) It is technically simple, and the growth of the crystals can be observed.

The disadvantages of the method include the following:

1) It is difficult to choose the optimum ratio between the rate at which the seed is lowered, the charge delivery, and the consumption of the working gas.
2) Impurities can enter the melt from the working gases, since these are expended in considerable amounts (0.7 m^3/h of O_2, 1.5–2 m^3/h of H_2), and also from the air and furnace ceramics.
3) High temperature gradients develop in the crystallization zone (30–50 deg/mm), causing high internal stresses in the crystals (up to 10–15 kp/mm^2).

In the Verneuil method the shape of the growing crystal can be changed by means of simple techniques. For instance, tubular crystals can be grown by arranging the axis of the rotating crystal holder and the axis of the burner so that they do not coincide (Fig. 10.28a). Ceramic tubes, which are used to manufacture muffles for the furnace of the Verneuil apparatus, can also be produced.

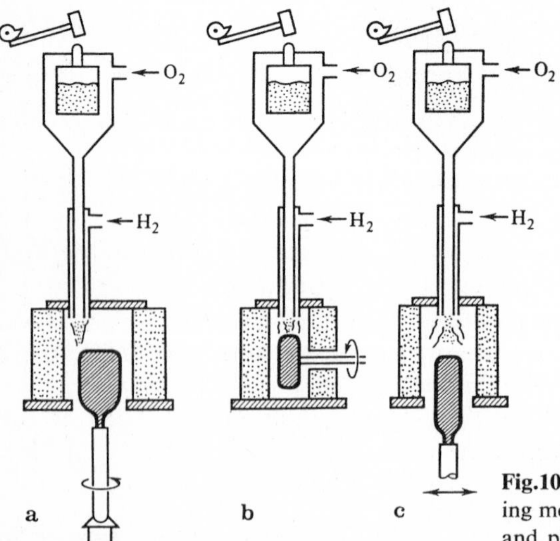

Fig.10.28a–c. Modified units for growing monocrystalline tubes (**a**), discs, (**b**), and plates (**c**) by the Verneuil method. The *arrow* in (**c**) indicates the direction of motion of the crystal

Figure 10.28b, c presents various diagrams of this apparatus, which can be used to grow single crystals in the shape of plates, discs, hemispheres, and cones.

The gas flame heating in the Verneuil apparatus is based on heat release in the reaction

$$H_2 + \tfrac{1}{2}O_2 = H_2O + 57.8\ \mathrm{kcal/mol}\,.$$

The maximum possible temperature in the oxygen–hydrogen flame is about 2500°C. This limitation is imposed by the fact that at higher temperatures the combustion products dissociate with subsequent heat absorption.

To increase the temperature in the apparatus, plasma, electron-beam, radiation, electric-arc, and other heat sources are employed.

Plasma heating is due to reactions of ionization and recombination of monoatomic and diatomic gases such as argon, helium, nitrogen, and oxygen, and also mixtures of them. Ionization of these gases is achieved either by an electric-arc discharge or by induction at a frequency of 4–8 MHz. Figure 10.29 presents a diagram of an apparatus with plasma heating. Although it is possible to obtain superhigh temperatures ($\sim$ 16,000°C) by this method, its use is limited by technical difficulties involved in powder transport to the melt film. The charge flow strongly affects the temperature distribution in the plasma and hence its stability.

Also of interest is radiation (light) heating, in which the radiation of a 5–10-kW tungsten lamp is focused on the end face of a seed, and growth by the

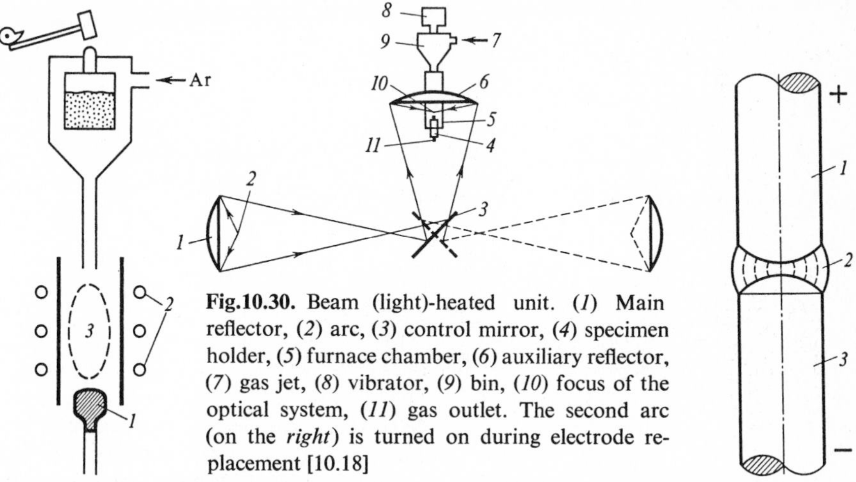

Fig.10.30. Beam (light)-heated unit. (*1*) Main reflector, (*2*) arc, (*3*) control mirror, (*4*) specimen holder, (*5*) furnace chamber, (*6*) auxiliary reflector, (*7*) gas jet, (*8*) vibrator, (*9*) bin, (*10*) focus of the optical system, (*11*) gas outlet. The second arc (on the *right*) is turned on during electrode replacement [10.18]

Fig.10.29. Plasma-heated unit. (*1*) Growing crystal, (*2*) induction heater, (*3*) plasma

Fig.10.31. Crystallization in an electric arc. (*1*) Initial polycrystal, (*2*) arc discharge, (*3*) single crystal

Verneuil method then ensues (Fig. 10.30). This system is particularly suitable for exploratory investigations, because the heat source can be insulated from the crystallization chamber. With this system it is technically easy to create a controlled atmosphere and ensure the required purity, but the crystals grow stressed. The unit shown schematically in Fig. 10.30 was used to obtain high-melting crystals of corundum, magnesium-aluminum spinel ($MgAl_2O_4$), rutile (TiO_2), yttrium oxide (Y_2O_3), etc.

A similar method is crystallization in an electric arc, the difference being that along with a high temperature, a directed flow of an electrically charged substance is set up (Fig. 10.31). This method has gained wide acceptance in growing single crystals of metals and semiconductors, as well as of dielectrics which show a considerable electrical conductivity near the melting point.

10.2.4 Zone Melting

This method consists in successive fusing of a feedstock ingot (Fig. 10.32). It was developed in *Pfann*'s investigations on the removal of contaminants from a substance [9.93]. First a narrow melt zone is created, and then recrystallization of the specimen is achieved by displacing the ingot or the heater.

An advantage of zone melting is that multiple recrystallization of the specimen is possible, which permits chemical purification of the substance and also ensures uniform distribution of the activator along the length of the crystal. In addition, when growing thermally unstable substances one can considerably reduce the disturbance in their stoichiometry by keeping the width of the melt zone to a minimum.

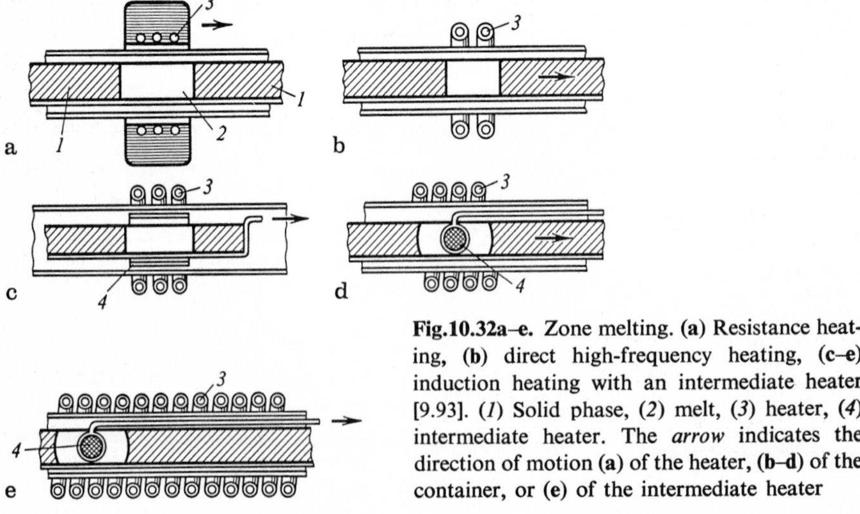

Fig.10.32a–e. Zone melting. (**a**) Resistance heating, (**b**) direct high-frequency heating, (**c–e**) induction heating with an intermediate heater [9.93]. (*1*) Solid phase, (*2*) melt, (*3*) heater, (*4*) intermediate heater. The *arrow* indicates the direction of motion (**a**) of the heater, (**b–d**) of the container, or (**e**) of the intermediate heater

A distinction is made between vertical and horizontal zone melting. Vertical displacement of the zone is especially convenient in crystallization without a container; because of the surface tension, the melt does not spread out. The maximum height of the zone is proportional to the coefficient of surface tension α:

$$h_{max} \simeq 2.8\sqrt{\alpha/\rho g}\,, \tag{10.1}$$

where α is the surface tension, ρ is the melt density, and g is the gravity acceleration.

Low-frequency electromagnetic fields coupled with an appropriately designed high-frequency inductor establish an upwardly directed force that acts continuously on the melt, and thus they prevent the molten zone from spreading. This method of confining a material with the aid of a high-frequency field without a mechanical contact is called levitation. It is brought about by the interaction between the external high-frequency field and that induced by eddy currents in the melt [9.93]. Levitation force

$$F \sim HI\,, \tag{10.2}$$

where H is the magnetic field induced in the melt by the high-frequency field, and I is the current in the melt. The levitation force is used in crucibleless vertical and horizontal zone melting of substances with a low α/ρ ratio.

Zone melting without a crucible is remarkable in that it creates precision-clean crystallization conditions due to the absence of a container. This method is used most commonly in growing single crystals of metals and semiconductors, whose melts have a high surface tension. It has been used successfully in growing large, dislocation-free single crystals of silicon (up to 100 mm in diameter).

For dielectric crystals, the container method of zone melting is used more often, especially with horizontal zone displacement.

Attempts to minimize the effect of the container walls gave rise to the "cold-crucible" method of horizontal zone melting [10.12]. Figure 10.33 shows a diagram of the method, by which single crystals of titanium and silicon were obtained.

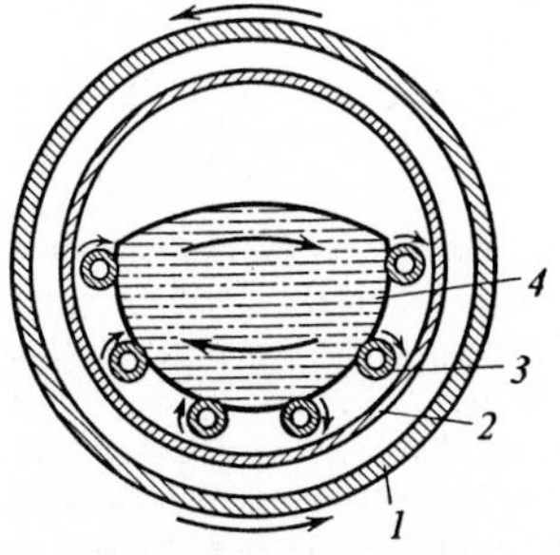

Fig.10.33. "Cold crucible" for horizontal zone melting [10.12]. (*1*) Induction heater, (*2*) quartz tube, (*3*) water-cooled silver tubes, (*4*) melt. The *arrows* indicate the direction of the electric currents. The melt is retained by surface tension

In setting up a melt zone, high-frequency heating (with 50–400 kHz currents) is widely used. For substances with high electrical conductivity, the specimen is heated directly (Fig. 10.32b), and for low-conductivity substances use is made of an intermediate heating element (Fig. 10.32c–e) made of a high-melting metal (usually platinum, a platinum-rhodium alloy, iridium, molybdenum, or tungsten).

In addition to induction heating, electron-beam and radiation (light) heating are commonly used. In recent years, the laser heating method has undergone rapid development.

Electron-beam heating (Fig. 10.34) is employed in growing high-melting metals by zone melting without a crucible. But this method requires a mandatory vacuum of at least 10^{-6} torr because of intensive electron scattering in a gas atmosphere. An advantage worth mentioning is that energy of a considerably higher density than in induction heating can be delivered to the specimen, which is helpful in melting refractory metals (Mo, W, etc.). When working with low-conductivity materials, the accumulating electric charges must be released, which is done by placing conducting nets on the specimen. Electron-beam zone melting is used in growing single crystals of tungsten, rhenium, molybdenum, and their alloys. Rods with a density similar to that of the single crystal serve as the feedstock. If the densities differ substantially, the mechanical stability of the liquid zone is disturbed.

The "cold cathode" method is based on an electrical discharge set up inside a ring-shaped hollow cathode surrounding the melting zone (Fig. 10.35). When

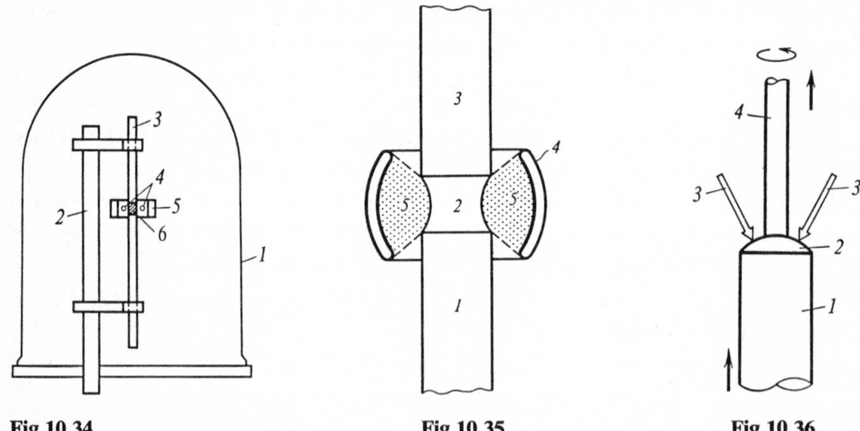

Fig.10.34 **Fig.10.35** **Fig.10.36**

Fig.10.34. Electron-beam heating [9.93]. (*1*) Vacuum bell, (*2*) support, (*3*) specimen, (*4*) cathode (electron emitter), (*5*) reflector, (*6*) melt

Fig.10.35. Heating with a "cold cathode." (*1*) Growing crystal, (*2*) melt zone, (*3*) initial material, (*4*) cathode, (*5*) plasma

Fig.10.36. Differential pulling method [9.93]. (*1*) Charge, (*2*) melt, (*3*) power supply to ensure melting, (*4*) growing crystal. The *arrows* indicate the direction of motion of the material, and the pulling and rotation of the crystal

the pressure of Ar, O_2, and other gases equals several millimeters of mercury and the cathode voltage is several kilovolts, the cathode electrons ionize the gas and form a plasma. By imparting an appropriate shape to the inner side of the cathode one can focus an electron or ion flux on the specimen. This method of heating has been employed in growing single crystals of sapphire, yttrium-aluminum garnet, etc.

Of interest among the modifications of zone melting are methods which incorporate elements of other growing techniques. An example is the pedestal technique (Fig. 10.36), in which a melt zone is set up by focusing the energy beam on the end face of the initial specimen and pulling a crystal of a smaller diameter than that of the initial rod.

The development of quantum electronics has made it possible to devise new methods of crystal growth using laser heating. A laser heat source presents substantial advantages: (1) The small angular divergency of the laser beam makes it possible to move the heat source out of the crystallization chamber. (2) The energy is transmitted to the melt zone by a highly coherent electromagnetic radiation field, and therefore standard optics can be used in the formation of the temperature field. (3) In a single-mode radiation regime a uniform energy distribution is achieved across the beam, and the highest temperature gradients exist at distances comparable to the radiation wavelength. (4) A laser heat source shows a high response. (5) It is simple in design. Furthermore, due to the high specific power of laser radiation, the volume of the substance melted can be rigorously limited, which is particularly necessary in zone melting without a crucible.

Laser sources operating in a continuous regime are most suitable for crystal growing, because they ensure highly stable growth conditions.

In contrast to other heat sources, a laser source radiates within a rather narrow range of wavelengths, and this imposes certain limitations on its application. Control of the temperature at the growth front is a prerequisite for growing more perfect crystals. However, the necessary temperature distribution can be provided by a laser source only when the crystal is transparent (and the melt opaque) to laser radiation. Hence, the choice of a laser source depends on the optical characteristics of the crystallizing substance.

The most popular lasers are those operating on CO_2 and on a CO_2–N_2–He mixture in a continuous regime and having a high radiation energy and efficiency (up to 20%). Using them, however, greatly restricts the possibilities of obtaining high-quality crystals, because the laser radiation wavelength lies in the vicinity of 10 μm, where most of the refractory substances are opaque in both the liquid and the solid phases. Greater possibilities are opened up by shorter-wave lasers, such as solid-state lasers which are based on crystals of yttrium-aluminum garnet activated by ions of trivalent neodymium and radiating in the range of 1 μm. In this range, crystals of most refractory dielectrics are highly transparent, whereas their melts are opaque. Therefore it is possible to set up thermal fields with the necessary values of axial and radial temperature gradients at the growth front.

10.2.5 Heat Transfer in Crystal and Melt

To create optimum temperature conditions for growth, the transport processes must be rigorously controlled, because otherwise they may vary substantially in intensity, depending on the growing method. Heat exchange in crystallization can occur by thermal conduction, radiation, and convection in the liquid phase. If these different mechanisms are taken into account, three cases must be considered: (1) $kl \gg 1$, (2) $kl \sim 1$, and (3) $kl \ll 1$, where k is the absorption coefficient, and l is the characteristic dimension of the object along the direction of the thermal flux.

The first case applies to opaque media, in which heat transfer is achieved only by molecular thermal conduction. Here, the temperature distribution obeys the Fourier law:

$$q = -\kappa_{\mathrm{m}} \operatorname{grad} T \tag{10.3}$$

(where q is the heat flux, κ_{m} is thermal conductivity, and T is the temperature) and the equation of heat transfer.

The second case applies to translucent media; the radiation from the heat source damps out inside the crystal. Heat transfer is achieved by reradiation. Here it is still possible to represent the heat flux as in (10.3). But the thermal conductivity must be represented by the sum

$$\kappa = \kappa_{\mathrm{m}} + \kappa_r , \tag{10.4}$$

$$\kappa_r = 16 n^2 \sigma T^3 / 3k . \tag{10.5}$$

Here n is the refractive index, and $1/\alpha$ is the "Rosseland mean" of the absorption coefficient [10.19]. In the general form the fraction of the molecular and radiation components of thermal conductivity is determined by the position of the maximum of the Planck function relative to the transmission band of the crystal. In sufficiently transparent crystals, at high temperatures the heat transfer may be completely determined by the optical properties. This is confirmed by the crystallization of dysprosium-aluminum garnet in the example given in Sect. 10.3.2, where faceting is described.

The third case relates to transparent media. As a transparent crystal grows, the radiation in it does not damp out, and the heat transfer is markedly affected by radiation from the heater, the container walls, etc. Here, relation (10.3) is not fulfilled, and the equation of heat transfer cannot be used. It must be replaced by the integral equations of radiant energy transfer. The heat transfer coefficient for transparent media, λ_{eff}, is an effective value, because it depends on the shape and state of the surfaces on which the reflection and refraction of radiation occur:

$$\lambda_{\mathrm{eff}} = 4\pi k T^3 n \phi . \tag{10.6}$$

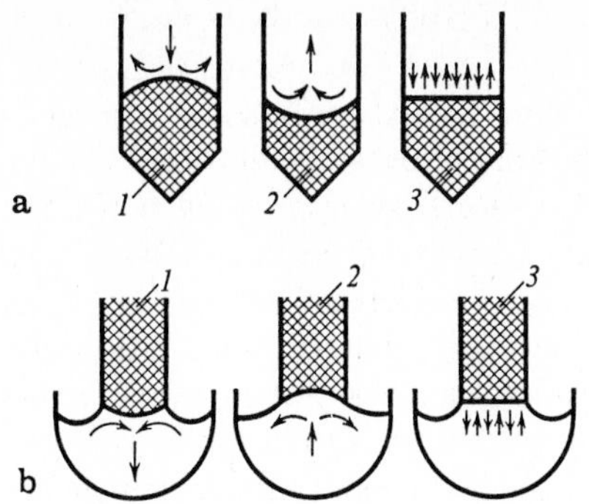

Fig.10.37a,b. Convective flows in the melt during crystal growth by (**a**) the Stockbarger method and (**b**) the Czochralski method; (*1*) near a convex, (*2*) a concave, and (*3*) a flat growth front. The *hatched areas* denote the crystals

Here, ψ is a factor depending on the optical properties and configuration of the system.

Taking the transparency of the substance and the reflective power of the walls into account aids in evaluating a given crystallization method as regards the possibility of attaining constancy of temperature across the crystal and constancy of the temperature gradient along the crystal axis. An increase in the transparency of the substance makes it more complicated to control the temperature gradients and, more importantly, to keep them at a constant level throughout the growth of single crystals. An important role is played in this connection by the convective transport in the melt, which may differ depending on the growing method. Figure 10.37 shows the directions of the convective flows in the melt during crystallization by the Stockbarger and Czochralski methods. Experience shows that with a convex isotherm (and hence a convex crystallization front) the flows in the Stockbarger method are directed from the melt to the center of the crystallization front, whereas in the Czochralski method they move in the opposite direction. The intensity of convective stirring in the melt depends on the temperature conditions; the larger the temperature gradients in the system, the higher it is. In some cases, forced stirring of the melt is introduced. In the Czochralski method this is achieved by intensive rotation of the crystal or the crucible ($\sim$50–100 rpm), or by rotating them simultaneously in opposite directions. In the Stockbarger method the melt is stirred with a special mechanical agitator.

Melt hydrodynamics have been intensively investigated in recent years (see the sections on fluid dynamics in [10.20, 21]).

10.2.6 Temperature Control and Stabilization Systems

A crucial condition in growing high-quality, homogeneous crystals is that the temperature and the temperature gradients in the crystallization zone be kept constant. Instability in the thermal conditions results in defects, primarily nonuniform distribution of uncontrollable or specially introduced impurities.

(The defects associated with temperature fluctuations in the crystallization zone are discussed in Sect. 10.3)

The stability of the temperature field depends on the nature of the heat source, the design of the furnace, and the reliability of temperature control.

Systems of temperature-field stabilization are developed along two lines. The first aims at elaborating low-response heating systems with a time constant $\tau > 10$ s, and the second at designing high-response systems with $\tau < 1$ s. The former is realized with the use of so-called passive control systems in which the controlled parameter is not the temperature, but either the power delivered to the control device, the current, or the voltage. High-response, or active temperature control can be realized if high-accuracy temperature pickups are available that can operate for long periods at working temperatures. The design of active temperature control systems depends on the working temperature range. Resistance thermometers and copper–constantan thermocouples are widely used for low temperatures (up to 600°C). Chromel–alumel and platinum–platinum-rhodium thermocouples are used for medium temperatures (600–1600°C), and tungsten–iridium, tungsten–molybdenum, and tungsten–rhenium thermocouples for high temperatures (> 1600°C). But the accuracy of control by thermocouples decreases considerably with increasing working temperatures (from ±1°C to 10°C). Furthermore, control systems based on thermocouples are contact devices, i.e., the pickup is embedded directly in the melt. In many cases direct contact between the control elements and the melt is inadmissible, and this restricts the use of such systems, especially in the high-temperature range.

Of interest are so-called contactless control systems which implement the laws of the radiation of bodies and their optical characteristics. The most common instruments are:

a) the *radiation pyrometer*, which uses the relationship between the temperature of a body and the overall energy flux emitted by this body over a wide range of wavelengths;
b) the *brightness pyrometer*, which uses the dependence of body brightness on temperature at a specific wavelength;
c) the *color pyrometer*, which is based on the temperature dependence of the energy distribution within a given region in the radiation spectrum of the body.

The contactless control systems have a number of advantages over the contact devices. They are highly responsive, sensitive, and accurate (±0.1–0.5 K), and their pickups are placed outside the crystallization chamber. The operating accuracy of such systems depends, however, on the transparency of the windows of the crystallization chamber, which may become veiled in the course of growth.

The low-response crystallization systems, on the other hand, eliminate all problems associated with the precision system of control and stabilization, because low thermal response damps out temperature disturbances and keeps the crystallization system in the steady state. The technical realization of such systems is straightforward, but they necessitate construction of bulky crystallization units. This is expedient only in growing large crystals and in mass production.

10.2.7 The Automatic Control System for Growing Single Crystals

The early computer-controlled automatic regulation systems were based on the Czochralski method [10.16], because with it the weight of the container holding the melt provides a convenient control signal. By assigning a law for the change in weight the process can be kept within the required temperature–time regime. Figure 10.38 shows how such a process can be controlled; the principal regulating element is continuous weighing of the container holding the melt during the growth of a single crystal. Figure 10.39 is a photo of a crystal of lithium tantalate ($LiTaO_4$) obtained by automatic control. With the use of a computer, hundreds of crystallization units can be controlled simultaneously.

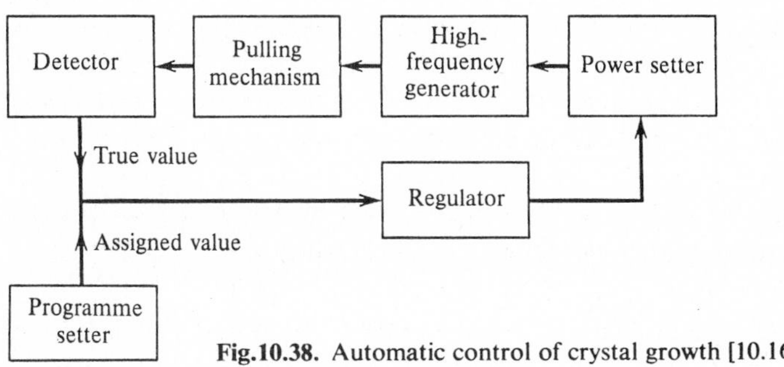

Fig.10.38. Automatic control of crystal growth [10.16]

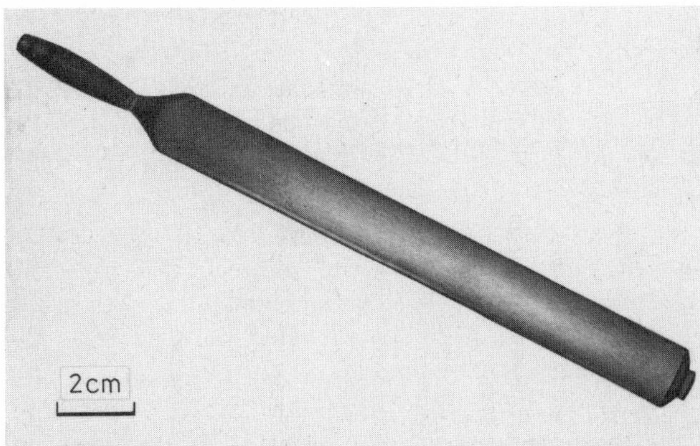

Fig.10.39. Single crystal of lithium tantalate grown using automatic control [10.16]

Table 10.2. Typical growth conditions for preparing single crystals

Item No.	Single crystal	Chemical formula	Growing method	Basic crystallization conditions		
				Growth rate [mm/h]	Container material	Crystallization atmosphere
1	Naphthalene	$C_{10}H_8$	Stockbarger	1–3	Glass	Air
2	Copper	Cu	Stockbarger	30–50	Graphite	Inert
3	Germanium	Ge	Czochralski	10–20	Graphite	Vacuum
4	Silicon	Si	Czochralski	10	Graphite	Vacuum
5	Fluorite	CaF_2	Stockbarger	10	Graphite	Fluorine-containing
6	Lithium niobate	$LiNbO_3$	Czochralski	2	Platinum	Oxygen-containing
7	Gallium-gadolinium garnet	$Gd_3Ga_5O_{12}$	Czochralski	2	Iridium	Oxygen-containing
8	Yttrium-aluminium garnets	$Y_3Al_5O_{12}$	Stockbarger	2	Tungsten, Molybdenum	Inert
9	Sapphire, ruby	Al_2O_3 Al_2O_3 (Cr^{3+})	Verneuil, Stockbarger, Czochralski	10–16	Tungsten, Molybdenum	Inert
10	Spinel	$MgAlO_4$	Verneuil, Stockbarger, Czochralski	4	Tungsten, Molybdenum	Inert
11	Tungsten	W	Czochralski, without crucible	30-50	Tungsten, Molybdenum	Vacuum 5×10^{-5} torr

In the Verneuil method, the process is monitored by measuring the position of the crystallization front (see Fig. 10.25) with the aid of a cathetometer. The growth is controlled by regulating the charge delivery and the consumption of the working gases. A system of this kind has been developed for growing single crystals of ruby.

10.2.8 Choosing a Method for Crystal Growth

The procedure and conditions for growing single crystals are selected on the basis of the physical and chemical characteristics of the crystallizing substance. Single crystals of metals with a melting point of up to 1800°C are grown primarily by unidirectional crystallization (the Stockbarger method), and those with a melting point above 1800°C by zone melting. Semiconducting crystals are grown chiefly by pulling out of the melt (the Czochralski method) and by zone melting. Single crystals of dielectrics with a melting point of up to 1800°C are usually grown by the Stockbarger or Czochralski methods, while higher-melting materials are produced by flame fusion (the Verneuil method). If the physicochemical processes involved in crystallization are taken into account, the optimum conditions can be established. These conditions pertain to the phase composition of the feedstock, its chemical purity and form (powder, pellets, ingot), the nature of the crystallization atmosphere, the material and shape of the container, the growth rate, the temperature gradients and shape of the crystallization front, the degree of stabilization of the growing conditions, and the way in which crystallization begins (spontaneous nucleation or crystallization on a seed).

Table 10.2 gives basic information on the methods and conditions for growing a number of technically important crystals. Metallic and semiconducting crystals are grown at rates averaging five times as high as those for the dielectrics. This is caused among other things by differences in the mechanisms of heat transfer and in the physicochemical stability of the crystallizing systems.

10.3 Defects in Crystals Grown from the Melt and Ways to Control the Real Structure of Grown Crystals

In single crystals grown from the melt, all the basic types of defects may appear: three-dimensional defects (inclusions, nonuniform impurity distribution, stresses) two- and one-dimensional defects (dislocations and stacking faults), and point defect (vacancies, interstitial and substitutional defects). We shall consider the three most widespread growth defects: foreign inclusions, nonuniform impurity distribution, and residual stresses. These defects may arise depending on the structure and shape of the phase boundary and the processes of heat and mass transfer.

10.3.1 Foreign Inclusions

Foreign inclusions are associated with a number of defects, such as scattering centers (Fig. 10.40), stresses (Fig. 10.41), dislocations (Fig. 10.42), and blocks (Fig. 10.43). Macroscopic inclusions may be formed in crystals either through the trapping of foreign particles by the growth front or through the precipitation of excessive components and impurities in the solid state upon cooling or annealing of the crystal.

The origin of inclusions in crystals can be established on the basis of their average size and the nature of their distribution in the crystal. Inclusions resulting from thermal annealing are much smaller and are distributed statistically uniformly. Their arrangement generally correlates with dislocation lines and grain boundaries. Conversely, foreign particles trapped by the crystallization front are usually distributed nonuniformly in the crystal, so that crystal layers containing many trapped inclusions alternate with layers almost free of them (see Fig. 10.3).

The formation of foreign particles that are trapped by the crystallization front may be due to a number of factors: thermal dissociation of the melt, release of a soluble gas in crystallization, reaction of the melt with the container material and the ambient atmosphere, or the presence of contaminants (see also Sect. 6.1). Such inclusions range from 10 to 1000 μm in size. In optically

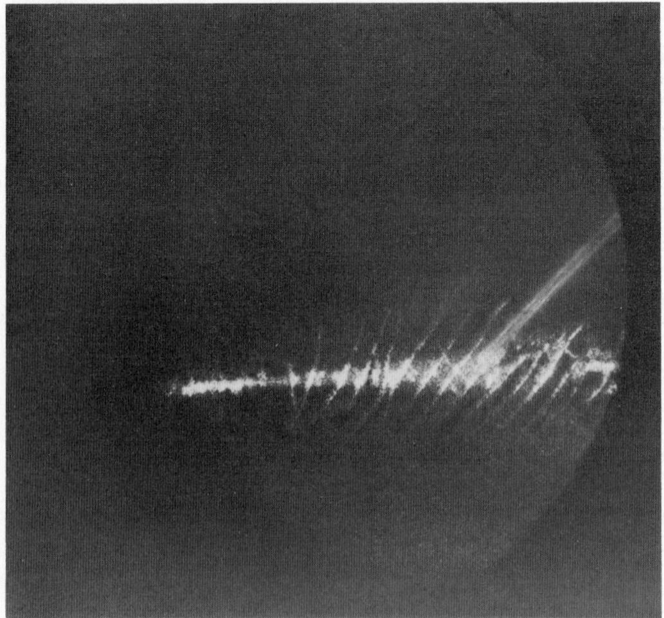

Fig.10.40. Scattering of helium–neon laser beam in a ruby crystal due to the presence of foreign particles. Magnification × ca. 10. Courtesy of V.Ya. Khaimov-Mal'kov

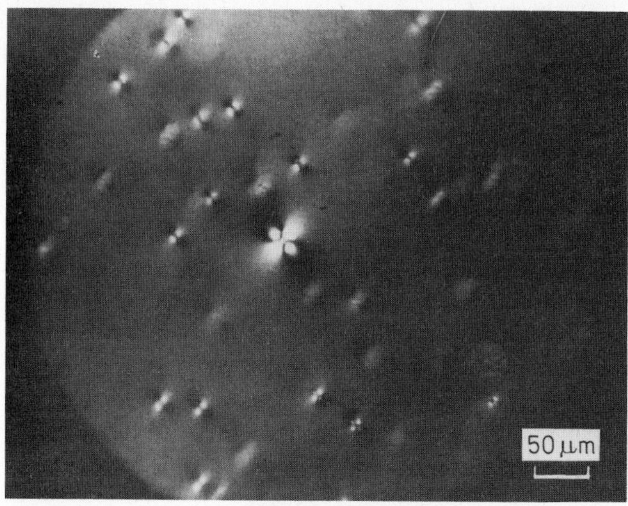

Fig.10.41. Local stress fields induced by foreign particles of yttrium aluminate in a single crystal of yttrium-aluminum garnet. Polarized light. Courtesy of I.A. Zhizheiko

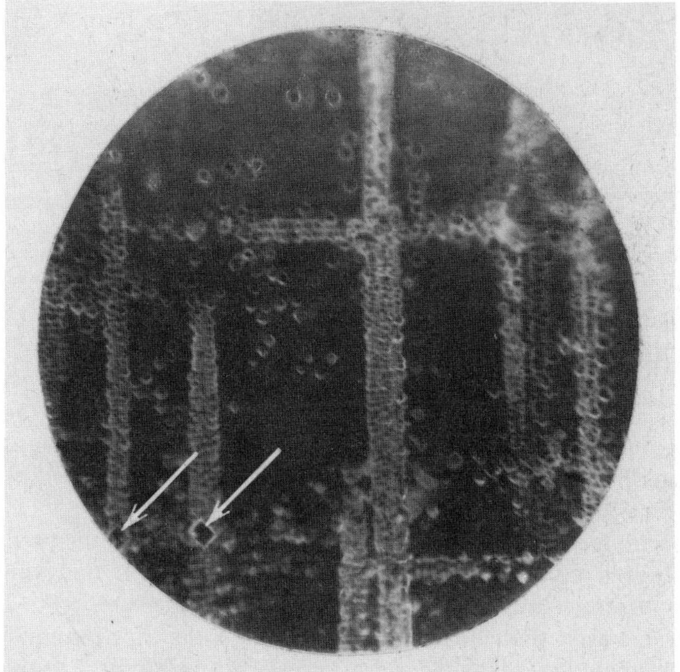

Fig.10.42. Etch figures on dislocations in a crystal of yttrium-aluminum garnet. The dislocations are caused by yttrium aluminate particles. Magnification × ca. 50. Courtesy of V.A. Meleshina

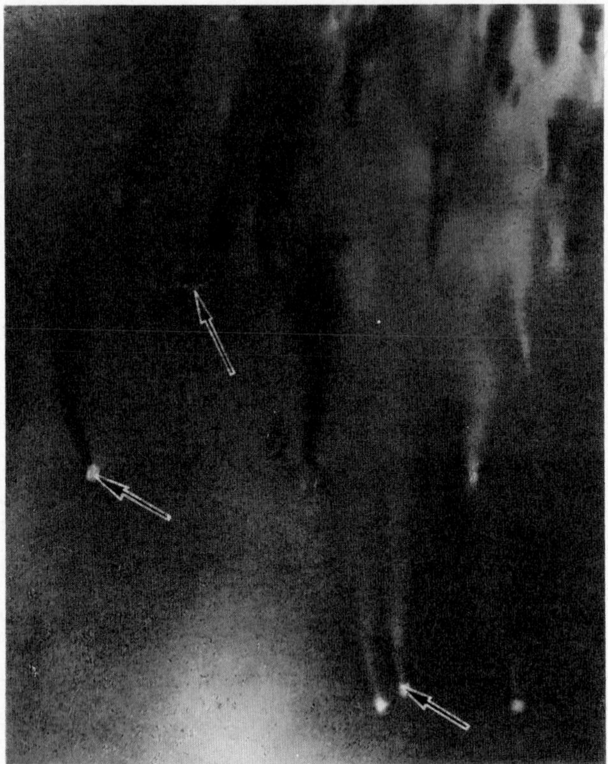

Fig.10.43. Conical grains associated with foreign particles (indicated by *arrows*) in a crystal of yttirum-aluminum garnet. Magnification × ca. 100 [10.6]

transparent crystals they are revealed by the scattering of light, for instance of a laser beam, and in optically opaque crystals, by etching and electron microscopy of thin sections. The main characteristic of the entrapment of foreign inclusions is the critical growth rate V_{cr}, above which mass trapping of particles of a given radius occurs, and below which the particles are crowded out by the crystallization front. This critical trapping rate is found from the experimental dependence of the density of foreign inclusions of a given radius on the growth rate (Fig. 10.44). The critical trapping rate can be controlled by using its dependence on the temperature gradient (Sect. 6.1).

The experimental dependence of the critical trapping rate V_{cr} on the axial temperature gradient is presented in Fig. 10.45, from which it is seen that with an increase in temperature gradient the critical trapping rate increases up to a certain value. As the axial gradient continues to increase, the experimentally observed critical trapping rate falls off. This decrease can be ascribed to the increase in the intensity of thermal convection in the melt, which in turn causes considerable temperature fluctuations (Fig. 10.46) and trapping of foreign particles.

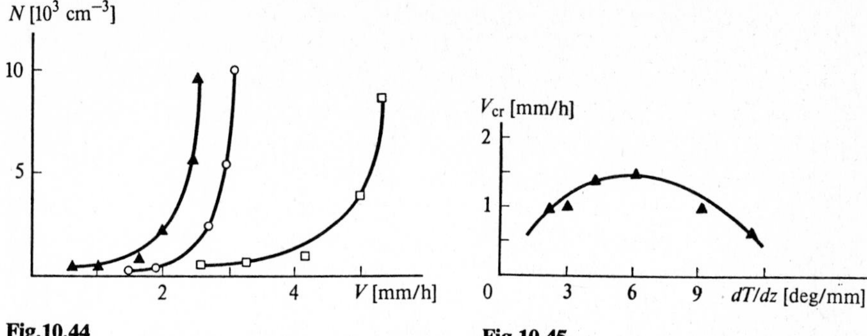

Fig.10.44. Dependence of the partide number density N of ~20-μm foreign particles on the growth rate in crystals of $Lu_3Al_5O_{12}$ with an impurity content of 3 at. % Nd^{3+} (*triangles*, critical trapping velocity $V_{cr} \simeq$ 1.5–2.5 mm/h), 2 at. % Nd^{3+} (*circles*, $V_{cr} \simeq$ 2.5–3 mm/h), and 5 at. % Yb (*squares*, $V_{cr} \simeq$ 4.5–5 mm/h) [10.22]

Fig.10.45. Dependence of the critical growth rate V_{cr} for trapping inclusions on the temperature gradient dT/dz in the vicinity of the crystallization front (see Fig. 10.16b) in unidirectional crystallization [10.22]

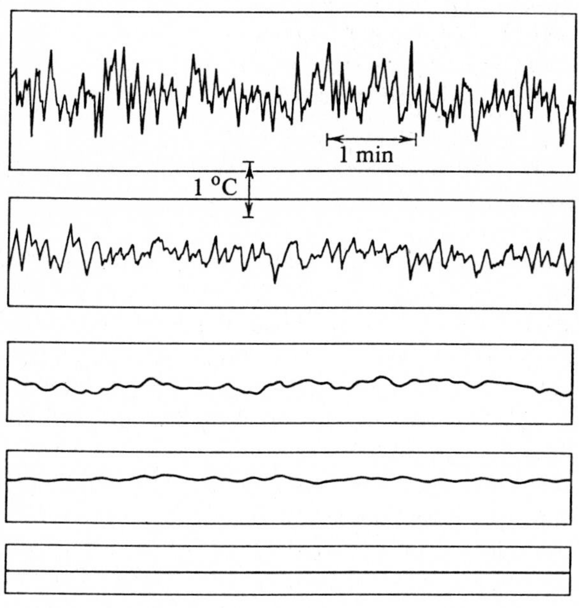

Fig.10.46. Typical temperature fluctuations in a boat containing liquid tin under stationary conditions. The fluctuations arise from natural convection at different magnitudes of horizontal temperature gradients $\partial T/\partial x$ [10.23]. The dimensions of the boat are 76 × 19 × 13 mm, the melt depth is about 10 mm. From *top* to *bottom*: $\partial T/\partial x$ = 11.5, 6.2, 2.9, 1.9, 1.1 deg/cm

10.3.2 Impurities

The problem of trapping and the distribution of impurities has two aspects. One relates to the purification of crystals from contaminants, and has been studied by many researchers (see, e.g., [9.93, 10.24]). The most suitable method of obtaining pure single crystals from the melt is by zone melting, as noted above. The other aspect of the problem has to do with intentional introduction of an activating dopant into the crystal. For instance, the production of laser crystals involves the activation of these crystals by an additive, and a similar problem is encountered in preparing impurity semiconductors and also crystals with a certain spectral transmission band. The quality and technical value of crystals doped with a specially introduced impurity are higher, the more uniformly the impurity is distributed in the crystal bulk. The role of impurities should be the guiding factor when diagnosing defects associated with them.

The distribution of impurities is estimated by different methods, of which chemical analysis is the most popular. Also widely accepted are spectral, mass spectrometric, and x-ray spectrometric analysis, as well as luminescent and adsorption analysis.

Let us consider in detail the typical cases of nonuniform impurity distribution, first along the growth direction, and then normally to it (i.e., across the crystal).

As mentioned at the beginning of Sect. 10.2, the content of impurities with distribution coefficients K that are different from unity does not remain constant along the crystal (Fig. 10.6). The regions crystallized first are enriched in impurities with $K > 1$ and contain fewer impurities with $K < 1$ than the initial charge. In the terminal regions of the crystal, the opposite relationship obtains: they have fewer impurities with $K > 1$ and more with $K < 1$ then the average value throughout the crystal.

In addition to this large-scale redistribution of impurities along the length of the crystal, a nonuniformity of a different kind is observed. It is called impurity striation, or zone structure.

Impurity striations are periodic variations in the impurity concentration along the direction of growth (Sect. 4.3.2); crystal layers, or zones, enriched in impurity, alternate with low-impurity layers. The layer width usually lies in the range of 10–100 μm.

Impurity striations can be observed with polarized light in parallel-sided specimens cut out of crystals parallel to the growth direction, and are also revealed by etching and decorating.

Among the known factors resulting in impurity striations are (1) temperature fluctuations, (2) variations in the speed of displacement of the crystal relative to the furnace, and (3) constitutional supercooling (Sect. 5.3.3).

Fluctuations in melt temperature, which promote the formation of growth zones, may be caused either by variations in the power output of the heat source or by convective flows developing in the melt. The temperature fluctuations due to the heat source are restricted with the aid of special electronic stabilization systems (Sect. 10.2.6). They ensure maintenance of the temperature within the following limits of accuracy: (2000 ± 0.5)°C, (1000 ± 0.1)°C, and (100 ± 0.001)°C. The temperature fluctuations caused by convection in the melt are eliminated by an abrupt decrease in the vertical temperature gradient or by vigorous agitation of the melt. In the former case the intensity of the convective flows diminishes because of the decrease in temperature gradients, and in the latter because they are destroyed by forced stirring.

Variations in the rate of crystal displacement can be reduced by developing reliable designs for the displacement mechanism. The best of them increase the displacement accuracy to ± 1 μm.

Variations in the growth rate due to constitutional supercooling resulting from the accumulation of an impurity or one of the components of the crystallizing substance are difficult to diagnose. Zonality caused by constitutional supercooling can usually be recognized on the basis of the morphology of the zones; they are broader, and their front boundaries are more distinct than their back ones (Fig. 10.47).

An impurity is distributed across a crystal in conformity with the shape of the growth front and hence of the crystallization isotherm. With a flat isotherm, the impurity is distributed uniformly across, and with a sloping isotherm the impurity concentration C_i increases (or decreases) from the central regions of the crystal to the periphery. The direction in which C_i changes depends on the magnitude of the trapping coefficient K compared to unity ($K > 1$ or $K < 1$) and on the sign of the curvature of the crystallization front (convex or concave). A smooth change in concentration across the crystal is observed only with a rounded growth front which coincides with the crystallization isotherm.

Faceting. The presence of one or several faces on a rounded growth front leads to a sharply nonuniform impurity distribution across the crystal. Since the coefficients of impurity trapping by the face and by the rounded front differ appreciably, the material deposited on the face contains an amount of impurity quite different from that in the material deposited on the rounded section of the front (Fig. 10.48). The crystal regions thus formed, the disposition and size of which reflect those of the faces on the growth front, can be seen in transmitted light as shadow patterns (Fig. 10.49). This method is used for crystals that are transparent in the visible and infrared regions. For opaque crystals, etching, autoradiography, and other methods are employed.

The coexistence of faceted and rounded growth forms distorts the uniformity of single crystals because of the stable fields of stresses (Fig. 10.49), which are not relieved even under conditions of prolonged high-temperature annealing.

The causes of faceting and the mechanism of its development are analyzed in detail in Sect. 5.2.5. Faceting is controlled with the aid of the axial temperature gradient, because this gradient is related to the size of the face by the expression

$$d = 2\sqrt{2R\Delta T_{\max}/\mathcal{G}}\,, \quad \mathcal{G} = \frac{\kappa_S(\partial T_S/\partial z) + \kappa_L(\partial T_L/\partial z)}{\kappa_S + \kappa_L}\,, \tag{10.6}$$

Here, d is the face diameter, $\mathcal{G}$ is the weighted temperature gradient, $\partial T_S/\partial z$

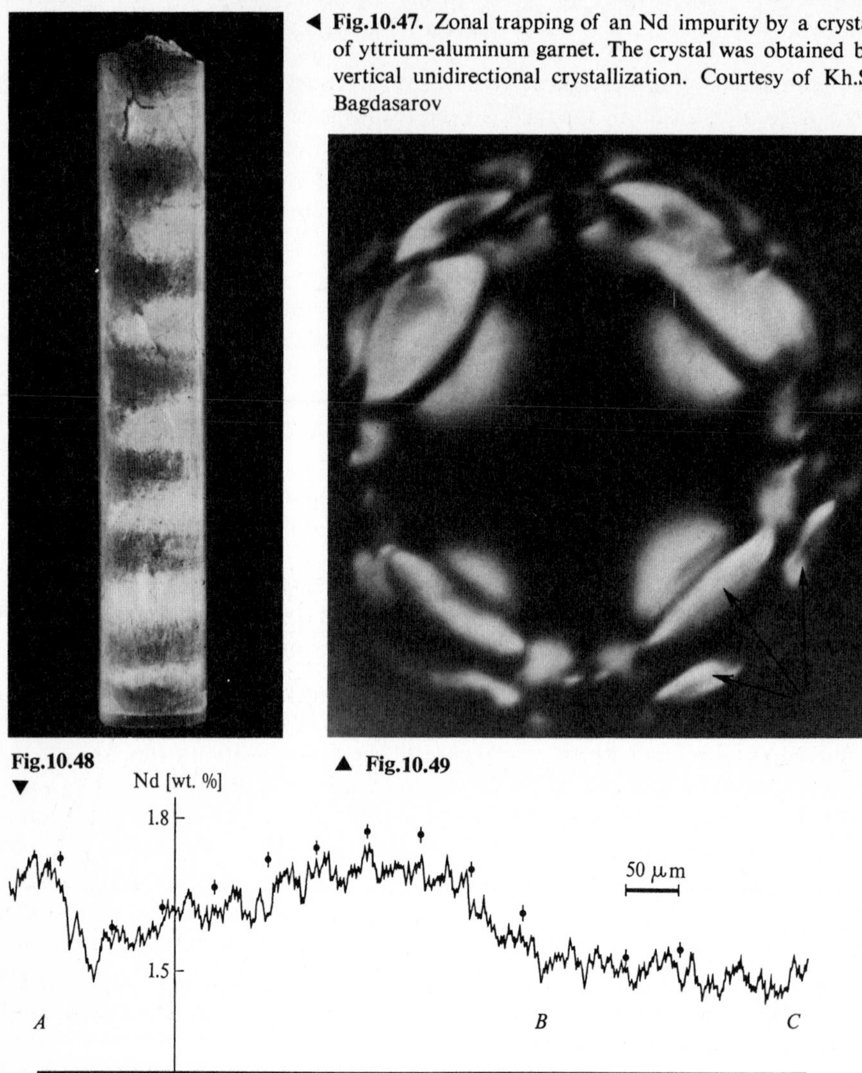

◀ **Fig.10.47.** Zonal trapping of an Nd impurity by a crystal of yttrium-aluminum garnet. The crystal was obtained by vertical unidirectional crystallization. Courtesy of Kh.S. Bagdasarov

Fig.10.48. Dopant (Nd) distribution along a platelet cut out of a crystal of yttrium-aluminum garnet perpendicular to the growth direction. Measurements were made in a scanning electron microscope microanalyzer ISM-U3. The intensity of Nd$\propto\alpha$ radiation is continuously recorded or measured at separate points of the crystal. The average impurity concentration in the material deposited in normal growth is relatively constant (BC region); in the material deposited in layer growth on the (211) face it is strongly nonuniform (AB region). The stratified impurity distribution on a microscale can also be seen. Courtesy of V.A. Meleshina

Fig.10.49. Stresses (*light areas* on the photo) in a crystal of yttrium-aluminum garnet whose growth front contains both rounded and faceted areas. Observation in polarized light; the crystal, a cylinder, is arranged with its base towards the observer. The stressed areas are located at the boundaries of the faceted and rounded growth forms. The *arrows* indicate the parts of the crystal whose material was formed by the growth of faces of an octahedron. Magnification × ca. 6. Courtesy of Kh.S. Bagdasarov

and $\partial T_L/\partial z$ are the axial temperature gradients in the crystal and melt, R is the curvature radius of the crystallization front, and ΔT_{max} is the maximum permissible supercooling on the face. If we are to obtain homogeneous crystal material, we must either minimize the size of the face or, conversely, make the face occupy the entire growing surface. According to (1.31) the normal growth mechanism is realized on the major part of the growth front (small d) at a high axial temperature gradient, and the layer growth mechanism (large d) at a low axial gradient (Fig. 10.50). But high temperature gradients are usually avoided in crystal growing, since they often unduly increase the internal stresses. Therefore to produce a homogeneous crystal it is expedient to obtain a flat crystallization front ($R \to \infty$) formed by one face. This is achieved by vigorous agitation of the melt (50–100 rpm), which results in a redistribution of the thermal fluxes near the crystallization front (Sect. 10.2.5).

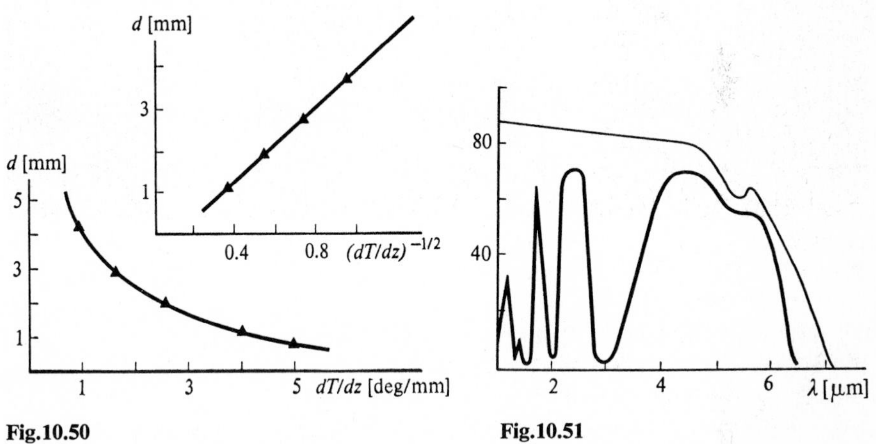

Fig.10.50 **Fig.10.51**

Fig.10.50. Dependence of the diameter d of a face forming at the growth front of a $Lu_3Al_5O_{12}$ crystal doped with Nd^{3+} on the temperature gradient dT/dz in the vicinity of the growth front during unidirectional crystallization (see Fig. 10.16b)

Fig.10.51. Absorption spectra of crystals of $Dy_3Al_5O_{12}$ (*thick line*) and $Y_3Al_5O_{12}$ (*thin line*) in the infrared region. Transmission in percent is plotted on the y axis [10.10]

The same effect can be attained by changing the thermophysical constants of the melt and the crystal by introducing impurities. It has been noted that there is no faceting of the growth front when single crystals of dysprosium-aluminum garnet ($Dy_3Al_5O_{12}$) are grown by the Czochralski method, whereas under similar conditions the growth front of yttrium-aluminum garnet ($Y_3Al_5O_{12}$) is faceted. A comparison of the absorption spectra of these crystals shows that in contrast to yttrium-aluminum garnet, single crystals of dysprosium-aluminum garnet have wide absorption bands in the range of 1.2–1.3 μm (Fig. 10.51), i.e., in the region of the maximum of the Planck function (the

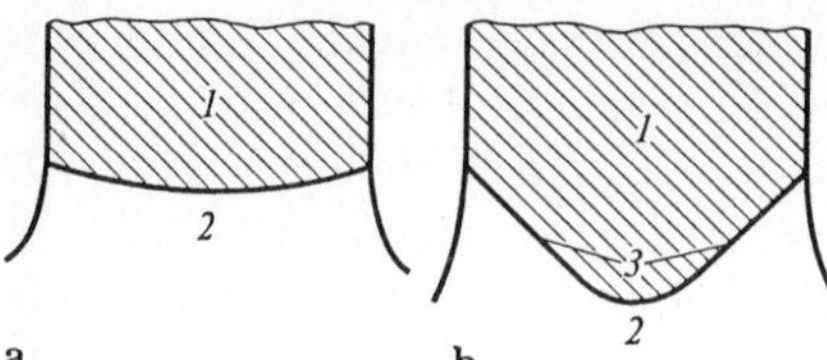

Fig.10.52a,b. Shape of the phase boundary in growing crystals of dysprosium-aluminum garent (**a**) and yttrium-aluminum garnet (**b**) by the Czochralski method. (*1*) Crystal, (*2*) melt, (*3*) faceted portions of the phase boundary [10.11]

region of maximum radiation of the heat source). Corresponding to the new thermophysical characteristics of the substance is a different distribution of thermal fluxes such that the crystallization front is flat (Fig. 10.52).

Since the dependence of the face size on the growth rate V is weaker than on the temperature gradient, $d \sim \mathcal{G}^{-1/2} V^{1/4}$ (it is assumed that $V \sim (\Delta T_{max})^2$), it is difficult to control faceting by varying the growth rate.

Thus, despite the fact that the faceting effect can be controlled with the aid of the axial temperature gradient, melt agitation, changes in the transparency of the crystal and the melt, and changes in the growth rate, only the first two factors have real value. This is because changing the transparency of the substance requires the introduction of impurities, and the face size is only slightly sensitive to changes in the growth rate.

Sectorial impurity distribution is caused by the presence of faceted growth forms (Sect. 4.3.1), and the anisotropy of the impurity distribution coefficients of the growing crystal. Sectorial distribution is controlled on the basis of the same principles as the faceting effect.

10.3.3 Residual Stresses, Dislocations, and Grain Boundaries

Residual stresses are classified into two types. The first are caused by nonlinearity of the temperature field, and the second by defects in the real structure: accumulations of point defects, dislocations, grain boundaries, impurities, and foreign inclusions. Stresses are revealed by optical and x-ray flaw detection, as well as by precision x-ray measurements of the lattice parameters [6.22]. The typical distribution of internal stresses in corundum crystals grown by the Verneuil method and by horizontal unidirectional crystallization is presented in Fig. 10.53. It can be seen that in rodlike corundum crystals grown by the Verneuil method, compression stresses operate in the central parts and tensile stresses in the peripheral parts. In crystals obtained by unidirectional crystallization, on the other hand, compression stresses act along the lateral sides of the plate, and tensile stresses in the central part, which means that the stresses in these crystals are of a quenching nature (see also Sect. 6.1).

The residual stresses depend on the temperature distribution and the growth rate. An increase in the latter increases the residual stresses, and at a certain critical value of the temperature gradient the crystal breaks. Reducing the temperature gradients by additional heating makes it possible to decrease the residual stresses to values equal to those observed after technical annealing.

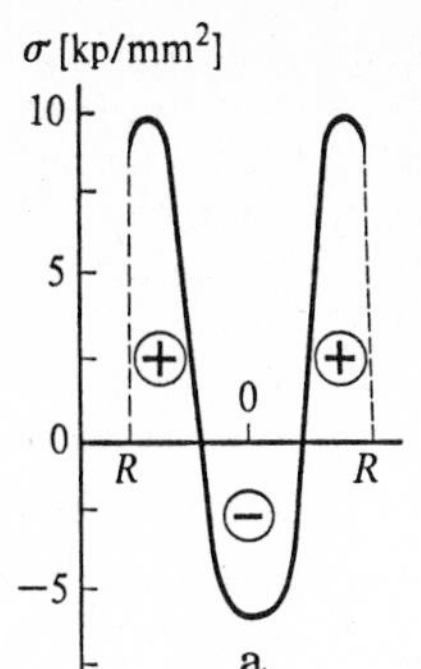

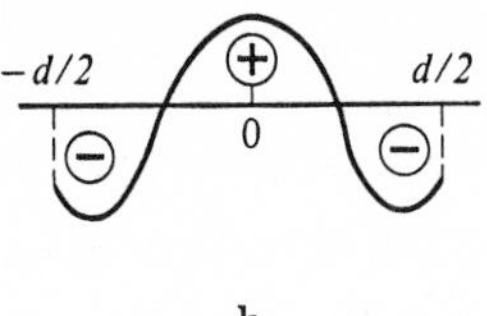

Fig.10.53. Typical stress distribution in a cylindrical crystal of radius R grown by the Verneuil method (**a**) and in a crystal platelet of thickness d grown by horizontal unidirectional crystallization (**b**). The *plus sign* denotes the compression regions and the *minus sign* the extension regions

The residual stresses in corundum, for example, can be brought down to 1.5–2 kp/mm^2 as opposed to 10–15 kp/mm^2 without additional heating.

To reduce the internal stresses it is necessary to eliminate the sources of their formation, i.e., defects in the real structure.

Dislocations may appear either directly at the crystallization front or in the bulk of the crystal when the newly grown material enters a cooler region (Sect. 6.2.4). No method of distinguishing their origin has been found as yet. Dislocations are revealed chiefly by x-ray and optical flaw detection, but also by selective etching [Ref. 4.27, Chap. 5]. Like grain boundaries, they may be inherited from the seeding crystal. Their density also depends on the impurity content in the crystal.

To obtain dislocation-free crystals from the melt, the "pinch", or "necking", method is used at reduced temperature gradients, low growth rates, and precision temperature stabilization (see Sect. 0.2.1).

Grain boundaries [Ref. 4.27, Chap. 5] may appear at the growth front and be caused by the growth processes themselves, but they can also result from subsequent temperature, mechanical, and other effects. Grains inherited from the seed are usually elongated in the growth direction. The grain boundaries are made up of dislocations, and are detected by the methods used to reveal individual dislocations.

The nature of grains has not been sufficiently studied. Their origin is intimately connected with the presence of inclusions (see Fig. 10.43) and impurities, expecially nonisomorphous ones. An increase in the concentration of these impurities reduces the grain size and consequently increases the density of grain boundaries.

The morphology of the crystal region adjoining the grain boundary shows that first a local disturbance in composition arises somewhere at the crystallization front as a result of the accumulation of impurities. This region of the growth front consequently grows more slowly, i.e., a depression is formed, in which more impurities then accumulate and a grain boundary appears. The accumulation of impurities also causes the substantial increase in dislocation density which is always observed near grain boundaries.

List of Symbols

A Component in a complex system; crystallizing substance in a hydrothermal solution
a Activity
A_α Substances involved in a chemical reaction
a Lattice parameter
a_α Activity of component α in solution
a, a_S, a_L Heat diffusivity; heat diffusivities of a crystal solid (S) and a liquid (L)
B Component in a complex system; solvent (mineralizer) in a hydrothermal solution
b Kinetic coefficient of crystallization, see Fig. 3.7c,
$\boldsymbol{b}$ Burgers vector [Ref. 4.27, Chap. 5]
C Concentration of crystallizing substance in solution
C_0 Equilibrium concentration (solubility) of crystallizing substance in solution
$\bar{C}_0$ Equilibrium concentration of crystallizing substance in solution over the infinite plane surface of a crystal
$\Delta C = C - C_0$ Deviation of solution concentration from equilibrium value (absolute supersaturation)
C_α Concentration of component α in solution
C_A Equilibrium concentration of crystallizing substance A in a hydrothermal solution
C_B Concentration of solvent (mineralizer) in a hydrothermal solution
ΔC_A Absolute supersaturation of a hydrothermal solution relative to crystallizing substance A
C_∞ Concentration of crystallizing substance in a nonstirred solution at an infinitely large distance from the crystallization front
C_{suef} Concentration of crystallizing substance in solution near the crystal surface
C_R Concentration of crystallizing substance in solution near the surface of a spherical crystal of radius R
$\langle C \rangle$ Average value of concentration in solution outside regions immediately adjacant to crystals, in mass crystallization
C_i, C_{io} Concentration and solubility of impurity in solution
C_{iS}, C_{iL} Impurity concentration in crystal and liquid
$C_{i\infty}$ Impurity concentration in medium away from the crystallization front
c Coordinate axis related to crystal lattice
$\tilde{c}$ Heat capacity
D Bulk diffusion coefficient of crystallizing substance in solution
D_s Surface diffusion coefficient
D_{iS} Diffusion coefficient of impurity in a crystal or across the phase boundary, i.e., from the surface layer of the crystal to the mother liquor or vice versa
D_{iL} Diffusion coefficient of impurity in a liquid
$D, D_0, D_{B/A}, \mathscr{D}$ Equilibrium distribution coefficients of an impurity or of component B relative to the compound A cocrystallizing from solution
d_{hkl} Spacing between neighboring atomic nets (hkl)
d Diameter
E Activation energy of surface crystallization processes

$E(N_{cr})$ Energy of formation of a plane critical nucleus in the gas phase
E_S^{conf} Total potential energy of all bonds in a crystal
e Electron charge
F Autoclave fill coefficient
ΔF Free energy of a face due to bonds parallel to the face
$f(R, t)$ Crystal size distribution function in mass crystallization from solution
$f(R, 0)$ Initial crystal size distribution in mass crystallization from solution
G Shear modulus
$\mathscr{G}$ Generalized (weighted) temperature gradient at the crystallization front
ΔH Latent heat of phase transformation (evaporation, solution, melting, crystallization) of the main component; reaction heat
ΔH_i Sublimation heat of a hypothetical polymorphous modification of a dopant crystal such that the modification has the same structure as a crystal of the main component; melting heat of crystal of impurity substance at melting point of main component T_0
h Step height; thickness of a liquid film separating a crystal from an obstacle
h, h_c Thickness of a pseudomorphous layer; its critical value
I, I_0 Flux of crystallizing substance to the surface of a crystal or substrate; equilibrium value of this flux
J Rate of nucleation (number of nuclei formed per unit time in unit volume or on unit surface)
j, j_s Fluxes of substance from the bulk of the medium and along the crystal surface which arrive at the rise of a unit-length step
j_+, j_- Number of atoms moving from the melt to the crystal or vice versa in a unit of time on one kink
K, K_0 Coefficient of impurity trapping by the crystal; equilibrium value
K_{eff} Effective coefficient of impurity trapping
$K_{st}, K_{st\,0}$ Coefficient of impurity trapping by a step; equilibrium value
$\mathscr{K}_0$ Equilibrium constant of reaction
k Boltzmann constant; absorption coefficient
L, l Linear size
L Liquid; melt
m Mass of an atom of the crystallizing substance
m Slope of liquidus line, $m = \partial T_0/\partial C_i$
m_i Mass of a dopant particle
N Number of sites in a surface net
N_1 Number of adsorbed atoms
N, N_{cr} Number of particles in an aggregate of radius R and a critical aggregate of radius R_c
N_S, N_{iS} Total number of atoms of crystallizing substance and impurity in a crystal
$\boldsymbol{n}$ vector of a unit cutward-drawn normal to the crystal surface
n Refractive index
n Gas density
n Number density of particles of solute
n Atomic number density on a step rise
n Number of grains per unit volume of a polycrystalline sample
n_0 Number density of adsorption sites on a surface
n_0 Number of "smooth" sites per unit of step length
n_0 Number of crystallization centers per unit volume of medium
n_s Number density of atoms or molecules adsorbed on a surface
$n(N_c)$ Number density of critical nuclei in N_{cr} particles on a surface
n_{sS}, n_{sV} Density of a two-dimensional gas of adsorbed particles which is in equilibrium with the crystal and the vapor
n^* Density of nucleation-active defects on a surface

n_+, n_- Numbers of positive and negative kinks per unit of step length

n_i Fraction of doped kinks relative to total kinks

P Total pressure in a system; vapor pressure in a one-component system

P_S, P_M Pressure in a crystal and in a medium

P_0 Equilibrium pressure of the vapor of a crystallizing substance

$\Delta P = P - P_0$ Deviation between the pressure of the vapor of a crystallizing substance and the equilibrium pressure (absolute supersaturation of vapor)

$P_\alpha, P_{\alpha 0}$ Partial pressure of component α of a gas mixture; equilibrium value

P_c Critical supersaturation for nucleation

p Fraction of the melt volume crystallized by a certain time

p Orientation of an interface — slope of the crystal surface to selected singular face

p^i Orientation of the ith singular face ($i = 0, 1, 2, \ldots$) relative to a selected singular face

p Average orientation of an interface relative to a selected singular face

p_0 Orientation of the slopes of vicinal growth hillocks

p_{cr} Critical value of the maximum slope of a surface at a face center. At this value, skeletal growth of a polyhedron begins.

q Heat flux density

q Probability that an arbitrary point in the volume of the initial melt is not part of the crystallized substance at given instant

q_i Density of the diffusion flux of the crystallizing substance towards the ith face of a crystal

R Gas constant

R Rate of movement of a stepped surface of a crystal along the normal to a singular face

R Characteristic size of a crystal; radius of a spherical crystal; interface curvature radius

R_c Radius of a spherical critical nucleus of a new phase

r_c Radius of a disc-shaped critical nucleus on a substrate

r_S Radius of an atom, molecule, or ion of the crystallizing substance in a crystal

r_i Radius of an atom, molecule, or ion of impurity in a crystal

S Solid

s Index showing that a quantity belongs to the surface

S, S_1 Surface area of the charge and a seed in an autoclave

$S(N), S(N_c)$ Surface area of an aggregate consisting of N particles, and of a critical aggregate consisting of N_c particles

S_{hkl} Surface area of an elementary loop in an atomic net (hkl)

S_l, S^{conf} Total configurational entropy of a step and a crystal

ΔS Entropy change in a reaction

s_S, s_V, s_M Specific entropy per particle in a crystal, in vapor, and in the initial medium (mother liquor)

$s_{iS}^{int}, s_{iM}^{int}$ Internal entropy of an impurity atom in a crystal and in a medium

$s_{iS}^{conf}, s_{iM}^{conf}$ The configurational entropy of an impurity atom associated with the number of possible sites it can occupy in the crystal and the medium

Δs_i Entropy of melting of a dopant crystal at the melting point of the main component

T Temperature

T_{cr} Critical temperature

T_0 Equilibrium temperature; melting point

T_c Critical temperature for nucleation

ΔT Deviation of the actual temperature T from the equilibrium temperature T_0, melt supercooling $\Delta T = T_0 - T$; supercooling of solution; temperature difference $\Delta T = T_1 - T_2$ between the source zone (zone of evaporation, etching, and dissolution, $T = T_1$) and the crystallization zone ($T = T_2$)

ΔT_0 Change in the equilibrium temperature when a load is placed on the crystal and when an impurity is injected

ΔT_c Critical supercooling for nucleation

T_R Critical temperature for the transition of on atomically smooth interface structure into a rough one
T_Q Ratio of the latent heat of crystallization to the heat capacity
T_S, T_L Temperatures in the solid (S) and liquid (L) phases
T_∞ Melt temperature far away from the crystallization front
T_{surf} Surface temperature of a crystal
ΔT_{max} Maximum supercooling on an interface
t Time
U, U_0 Total surface energy of a crystal at a given temperature T and at 0 K
U_1 Total step energy
$U(N_c)$ Potential energy of a critical aggregate of N particles on a substrate
U_D Potential energy barrier for surface diffusion
u Flow rate of liquid (solution) relative to crystal
$\boldsymbol{u}$ Displacement vector
u_{iK}^0 Deformation due to polymorphous transformation; residual deformation
V Vapor
V Volume
V, V_S, V_L Working volume of an autoclave; volumes of the solid (S) and liquid (L) phases in an autoclave at room temperature
V Normal rate of growth (linear growth rate of the interface along the normal to it)
V_{cr} Critical growth rate at which an extraneous particle is trapped by a crystal
v Rate of step movement
v_k Rate of kink movement along a step
w Half of binding energy between nearest neighbors
w_+ Frequency with which atoms attach to a crystal surface per atomic position
$X_{\alpha j}$ Relative molar concentration of component α in phase j
X_{AS}, X_{BS}, X_{AL}, X_{BL} Relative concentration of components A and B in a crystal (S) and a liquid (L)
X_M Molar fraction of the main component in a solution
X_{iS}, X_{iM} Molar fractions of impurity in a crystal (S) and a medium (M)
X_{iM}^0 Concentration of saturated impurity solution in molar fractions
X_{iM} Fraction of impurity molecules in a solution relative to the total amount of substances dissolved
Z Valence of an ion in a lattice
Z_1 Coordination number in a crystal or in a plane surface net
Z_i Number of neighbors of the ith order for a given atom in a crystal
Z_{1M} Coordination number in a medium
α Surface tension
α Free energy of a crystal–medium interface (surface energy)
α_{hkl} Surface energy of the (hkl) face
α_{sS}, α_{sM} Surface energies of the substrate–crystal and substrate–medium interfaces
α_s, Specific free energy of adhesion
$\Delta\alpha = \alpha + \alpha_{sS} - \alpha_{sM}$
α_l, $\alpha_{\bar{l}}$ Specific (per unit length) work of formation of opposite steps
$\hat{\alpha}$ Thermal expansion tensor
$\hat{\alpha}_C$ Lattice concentrational deformation tensor
β, β^T Kinetic coefficient of crystallization from solution and from the melt
β_{st}, β_{st}^T Kinetic coefficient for a step in crystallization from solution and from the melt
β_{tr} Coefficient characterizing the total rate of transport in a medium
β_{eff} Effective kinetic coefficient of crystallization from vapor and from the melt
γ_{iS}, γ_{iM} Activities of a dopant in a crystal (S) and a medium (M)
γ_{iM}^0 Coefficient of activity of an impurity in its saturated solution

γ_d Free energy per unit length of dislocation
δ Thickness of the diffusion boundary layer
ε_1 Binding energy between nearest neighbour in a lattice
ε_2 Binding energy between second nearest neighbours
$\varepsilon_{1/2}$ Binding energy of a particle at a kink
$\varepsilon_{SS}, \varepsilon_{MM}, \varepsilon_{SM}$ Binding energy between two neighboring atoms in a crystal, between two atoms in the melt (in the medium), and between one atom in the crystal and another in the melt
$\varepsilon_{ii}, \varepsilon_{Si}$ Binding energy of two neighboring impurity atoms in a crystal; binding energy of an atom of the main substance with an impurity atom in the crystal
$\bar{\varepsilon}$ Average work of detachment of an atom from a face of a finite crystal; chemical poetntial per particle in a finite crystal at $T = 0$
$\varepsilon_s, \varepsilon_{is}$ Adsorption energy of intrinsic atoms and impurity atoms at a crystal interface
$\varepsilon_{iS}^{int}, \varepsilon_{iM}^{int}, \varepsilon_{iS}^{conf}, \varepsilon_{iM}^{conf}$ Internal and configurational energy of an impurity atom in a crystal (S) and a medium (M)
ε^{el} Elastic energy associated with the deformation of a crystal lattice near an impurity atom
$\Delta\varepsilon_S, \Delta\varepsilon_M$ Mixing energy of the crystallizing substance and an impurity in a crystal and a medium, per bond
θ Angle of inclination of a crystal surface to a selected closepacked face
θ Wetting angle
Θ Degree of surface coverage
Θ Anisotropy of the face growth rate R and of the kinetic coefficient of crystallization b, $\Theta = (1/R)(\partial R/\partial p)_{\Delta\mu=\text{const}} = (1/b)(\partial b/\partial p)$
κ_S, κ_L Thermal conductivities of a crystal (S) and melt (L)
κ_m, κ_r Thermal conductivities: molecular (m) and radiative (r) (by reradiation)
Λ Width of the "dead"-for-nucleation zone on either side of a step
λ Step spacing
λ_k Average kink spacing in the absence of an impurity
λ_s Diffusion path length on a crystal surface
$\mu_S, \mu_V, \mu_L, \mu_M$ Chemical potential of a particle in the respective phase (in a crystalline solid, in vapor, in the melt, and in a medium in general); in an adlayer
$\mu_{AS}, \mu_{BS}, \mu_{AL}, \mu_{BL}$ Chemical potentials of components A and B in a crystalline solid (S) and in the liquid phase (L)
$\mu_{iS}, \mu_{iM}, \mu_{iS}^0, \mu_{iM}^0$ Chemical potentials of an impurity in a crystal (S) and a medium (M); their standard values
μ_i Chemical potential of a particle in an impurity crystal
$\mu^0(N)$ Chemical potential of an aggregate consisting of N atoms on a substrate; translatory, rotational, and vibrational motions of the aggregate as a whole are not taken into account.
$\Delta\mu, \Delta\mu$ Difference between the chemical potentials of phases, which serves as a measure of the deviation of a system from equilibrium (supersaturation, supercooling); value over a plane surface of an infinitely large crystal
$\Delta\mu_c$ Critical deviation from equilibrium required for a perceptible nucleation rate
$\Delta\mu_0$ Critical deviation from equilibrium, (3.31)
ν Thermal vibration frequency of an atom or molecule
ν, ν_0 Kinematic viscosity coefficient at an arbitrary temperature and at 25°C
ν Poisson ratio; total number of ions produced by dissociation
ν_α Stoichiometric coefficient for component α
ρ Curvature radius of a spiral step at a given point
$\rho, \rho_S, \rho_L, \rho_M$ Density; densities of a crystalline solid (S), melt (L), and solution medium (M)
$\Delta\rho$ Density difference between crystal and solution

σ relative supersaturation of vapor, $\sigma = P/P_0$; of solution, $\sigma = \Delta C/C_0$ and $\sigma = \Delta C_A/C_A$; and of a two-dimensional gas in the adsorption layer, $\sigma = (n_{sV} - n_{sS})/n_{Ss}$

σ_0 Threshold — or critical — supersaturation, (3.32)

$\hat{\sigma}$ Stress tensor

τ Time

τ_s Lifetime of an atom, molecule, or aggregate in a state of adsorption on the crystal surface

τ_i Lifetime of an impurity atom or molecule in the surface layer of a crystal

τ_D Lifetime of an adsorbed particle in a given potential well at the crystal surface before it shifts to a neighboring adsorption position

Φ Thermodynamic potential of a system

$\delta\Phi$ Change in the thermodynamic potential of a system

$\delta\Phi_c$, $\delta\Phi(N_c)$ Potential barrier to nucleation

$\Delta\Phi^0$ Change in thermodynamic potential during a reaction

φ Angle of deviation of an average step orientation from the orientation of the nearest PBC system

Ω, Ω_S, Ω_V, Ω_M Specific volume per particle in a crystal, in vapor, in a medium

Bibliography

Adsorption et croissance cristalline, Colloq. CNRS No.152 (Edition du Centre Nat. de la Recherche Sci., Paris 1965)

Advances in Epitaxy and Endotaxy. Physical Problems of Epitaxy, ed. by H.C. Schneider, V. Ruth (VEB Deutscher Verlag für Grundstoffindustrie, Leipzig 1971)

Advances in Epitaxy and Endotaxy. Selected Chemical Problems, ed. by H.G. Schneider, V. Ruth (with the cooperation of T. Kormány) (Akad. Kiadó, Budapest 1976)

Aleksandrov, L.N.: *Perekhodnyie oblasti epitaksialñykh poluprovodnikovykh plyonok* (Transition Regions of Epitaxial Semiconductor Films) (Nauka, Novosibirsk 1978) [in Russian]

Andreyev, V.M., Dolginov, L.M., Tretyakov, D.N.: *Zhidkostnaya epitaksiya v tekhnologii poluprovodnikovykh priborov* (Liquid Phase Epitaxy in the Technology of Semiconductor Devices) (Sov. Radio, Moscow 1975) [in Russian]

The Art and Science of Growing Crystals, ed. by J.J. Gilman (Wiley, New York 1963)

Bloem, J., Gilling, L.J.: "Mechanisms of the Chemical Vapour Deposition of Silicon," in *Current Topics in Materials Science*, Vol.1 (North-Holland, Amsterdam 1978)

Bockris, J., Razymney, G.A.: *Fundamental Aspects of Electrocrystallization* (Plenum, New York 1967)

Brice, J.C.: *The Growth of Crystals from the Melt* (North-Holland, Amsterdam 1965)

Brice, J.C.: *The Growth of Crystals from Liquids* (North-Holland, Amsterdam 1973)

Buckley, H.E.: *Crystal Growth* (Wiley, New York; Chapman & Hall, London 1951)

Chalmers, B.: *Principles of Solidification* (Wiley, New York 1964)

Cohesive Properties of Semiconductors under Laser Irradiation, ed. by L.D. Laude (NATO ASI Ser. Ser.E: Appl. Sci. No.69) (Nijhoff, The Hague 1983)

Crystal Growth, ed. by B.R. Pamplin (Pergamon, Oxford 1975)

Crystal Growth (Proc. Intern. Conf. on Crystal Growth, Boston, MA, 20-24 June, 1966), ed. by H.S. Peiser (Pergamon, Oxford 1967); see J. Phys. Chem. Solids, Suppl.1 (1967)

Crystal Growth 1968 (Proc. 2nd Intern. Conf. on Crystal Growth, Birmingham, U.K., 15-19 July 1968), ed. by F.C. Frank, J.B. Mullin, H.S. Peiser (North-Holland, Amsterdam 1968); see J. Cryst. Growth **3/4** (1968)

---

[1] Titles of the books published in Russian are given in transliteration, followed by the English translation. For all journal articles we quote their English titles. For Russian articles translated into English we usually indicate both the Russian and English editions.

Crystal Growth 1971 (Proc. 3rd Intern. Conf. on Crystal Growth, Marseille, France, 5-9 July 1971), ed. by R.A. Laudise, J.B. Mullin, B. Mutaftschiev (North-Holland, Amsterdam 1972); see J. Cryst. Growth **13/14** (1972)

Crystal Growth 1974 (Proc. 4th Intern. Conf. on Crystal Growth, Tokyo, Japan, 24-29 March 1974), ed. by K.A. Jackson, N. Kato, J.B. Mullin (North-Holland, Amsterdam 1974); see J. Cryst. Growth **24/25** (1974)

Crystal Growth 1977 (Proc. 5th Intern. Conf. on Crystal Growth, Cambridge, MA, 17-22 July 1977), ed. by R.L. Parker (North-Holland, Amsterdam 1977); see J. Cryst. Growth **42** (1977)

Crystal Growth 1980 (Proc. 6th Intern. Conf. on Crystal Growth, Moscow, USSR, 10-16 September 1980), ed. by E.I. Givargizov (North-Holland, Amsterdam 1981); see J. Cryst. Growth **52** (1981)

Crystals. Growth, Properties, and Applications, ed. by H.C. Freyhardt et al. (Springer, Berlin, Heidelberg, New York)
Vol.1: Crystals for Magnetic Applications (1978)
Vol.2: Growth and Properties (1980)
Vol.3: III-Y Semiconductors (1980)
Vol.4: Organic Crystals, Germanates, Semiconductors (1980)
Vol.5: Silicon, ed. by J. Grabmaier (1981)
Vol.6: R. Wagner: Field-Ion Microscopy (1982)
Vol.7: Analytical Methods. High-Melting Metals (1982)
Vol.8: Silicon Chemical Etching, ed. by J. Grabmaier (1982)
Vol.9: Modern Theory of Crystal Growth I, ed. by A.A. Chernov, H. Müller-Krumbhaar (1983)

Crystal Growth: An Introduction, ed. by P. Hartman (North-Holland, Amsterdam 1973)

Crystal Growth. Theory and Techniques, Vols.1,2, ed. by C.H.L. Goodman (Plenum, New York 1974, 1978)

Crystal Growth: A Tutorial Approach, ed. by W. Bardsley, D.T.J. Hurle, J.B. Mullin (North-Holland, Amsterdam 1979)

Crystal Growth and Characterization, ed. by R. Ueda, J.B. Mullin (North-Holland, Amsterdam 1975)

Crystal Growth and Epitaxy from the Vapour Phase (Proc. 1st Intern. Conf. on Crystal Growth and Epitaxy from the Vapour Phase, Zürich, Switzerland, 23-26 September 1970), ed. by E. Kaldis, M. Schieber (North-Holland, Amsterdam 1971); see J. Cryst. Growth **9** (1971)

Current Topics in Materials Science, ed. by E. Kaldis et al. (North-Holland, Amsterdam); Vol.1 (1978), Vol.2 (1977), Vols.3,4 (1979), Vols.5,6 (1980) Vol.7 (1981), Vols.8-10 (1982)

Dorfman, V.F.: *Mikrometallurgiya v mikroelektronike* (Micrometallurgy in Microelectronics) (Metallurgiya, Moscow 1978) [in Russian]

Elwell, D., Scheel, H.J.: *Crystal Growth from High-Temperature Solutions* (Academic, London 1975)

Epitaxial Growth, ed. by J.W. Matthews (Academic, New York 1975)

Factor, M.M., Garret, I.: *Growth of Crystals from the Vapour* (Chapman & Hall, London 1974)

Flemings, M.C.: *Solidification Processing* (McGraw-Hill, New York 1974)

Gidrotermal'nyi sintez kristallov, ed. by A.N. Lobachev (Nauka, Moscow 1968) [English transl.: *Hydrothermal Synthesis of Crystals* (Consultants Bureau, New York 1971)]

Givargizov, E.I.: *Rost nitevidnykh i plastinchatykh kristallov iz para* (Growth of Whiskers and Platelets from the Vapour) (Nauka, Moscow 1977) [in Russian]

Gorelik, S.S., Dashevsky, M.Ya.: *Materialovedeniye poluprovodnikov i metallovedeniye* (Semiconductor Materials Science and Physical Metallurgy) (Metallurgizdat, Moscow 1973) [in Russian]

Growth and Perfection of Crystals (Proc. Intern. Conf. on Crystal Growth, Cooperstown, NY, 27-29 August 1958), ed. by R.H. Doremus, B.W. Roberts, D. Turnbull (Wiley, New York; Chapman & Hall, London 1958)

Henish, H.K.: *Crystal Growth in Gels* (The Pennsylvania State University Press, University Park 1970)

Hirth, J.P., Pound, G.M.: *Condensation and Evaporation. Nucleation and Growth Kinetics*, Progr. Mat. Sci. **11** (Pergamon, Oxford 1963)

Holland, L.: *Vacuum Deposition of Thin Films* (Chapman & Hall, London 1956)

Honigman, B.: *Gleichgewichts- und Wachstumsformen von Kristallen* (Steinkopff, Darmstadt 1958)

Hurle, D.T.J.: *Mechanisms of Growth of Metal Single Crystals from the Melt*, Progr. Mat. Sci. **10**, 79 (Pergamon, Oxford 1962)

Ikornikova, N.Yu.: *Gidrotermal'nyi sintez kristallov v khloridnykh sistemakh* (Hydrothermal Synthesis of Crystals in Chloride Systems) (Nauka, Moscow 1975) [in Russian]

Issledovaniye protsessov kristallizatsii v gidrotermal'nykh usloviyakh, ed. by A.N. Lobachev (Nauka, Moscow 1970) [English transl.: *Crystallization Processes under Hydrothermal Conditions* (Consultants Bureau, New York 1973)]

Industrial Crystallization (Proc. 6th Sympos. on Industrial Crystallization, Usti nad Laben, Czechoslovakia, 1-3 September 1975), ed. by J.W. Mullin (Plenum, New York 1976)

Jackson, K.A., Uhlman, D.R., Hunt, J.D.: On the nature of crystal growth from the melt. J. Cryst. Growth **1**, 1 (1967)

R. Kern, G. Le Lay, J.J. Métois: "Basic Mechanisms in the Early Stages of the Epitaxy," in *Current Topics in Materials Science*, Vol.3 (North-Holland, Amsterdam 1979) p.132

Kinetika i mekhanizm kristallizatsii, ed. by N.N. Sirota (Crystallization Kinetics and Mechanism) (Nauka i Tekhnika, Minsk 1973) [in Russian]

Kozlova, O.G.: *Rost i morfologiya kristallov* (Crystal Growth and Morphology) (Izd-vo MGU, Moscow 1980) [in Russian]

Kristallizatsiya i fazovyie perekhody, ed. by N.N. Sirota (Crystalization and Phase Transitions) (Izd-vo Akad. Nauk Belorus. SSR, Minsk 1962) [in Russian]

Kristallizatsiya i fazovyie prevrashchniya, ed. by N.N. Sirota (Crystallization and Phase Transformations) (Nauka i Tekhnika, Minsk 1971) [in Russian]

Kuznetsov, V.D.: *Kristally i kristallizatsiya* (Crystals and Crystallization) (Gostekhizdat, Moscow 1953) [in Russian]

Laboratory Manual on Crystal Growth, ed. by I. Tarján, M. Mâtroi (Akad. Kiadó, Budapest 1972)

Laudise, R.A.: *The Growth of Single Crystals* (Prentice Hall, Englewood Cliffs, NJ 1970)

Lawson, W.D., Nielsen, S.: *Preparation of Single Crystals* (Butterworths, London 1958)

Liquid Phase Epitaxy, ed. by G.M. Blom (North-Holland, Amsterdam 1974); see J. Cryst. Growth **27** (1974)

Lyubov, B.Ya.: *Kineticheskaya teoriya fazovykh prevrashchenii* (Kinetic Theory of Phase Transformations) (Metallurgiya, Moscow 1969) [in Russian]

Lyubov, B.Ya.: *Teoriya kristallitatsii v bol'shikh obyomakh* (Theory of Crystallization in Large Volumes) (Nauka, Moscow 1975) [in Russian]

Material Sciences in Space (Proc. 2nd Europ. Sympos. on Material Sciences in Space, Frascati, Italy, 6-8 April 1976) (ESA, Noordwijk, The Netherlands 1976)

Material Sciences in Space (Proc. 3rd Europ. Sympos. on Material Sciences in Space, Grenoble, France, 24-27 April 1979) (ESA, Paris 1979)

Mekhanizm i kinetika kristallizatsii, ed. by N.N. Sirota (Crystallization Mechanism and Kinetics) (Nauka i Tekhnika, Minsk 1964) [in Russian]

Mullin, J.W.: *Crystallization*, 2nd ed. (Butterworths, London 1972)
Müller-Krumbhaar, H.: "Kinetics of Crystal Growth. Microscopic and Phenomenological Theories," in *Current Topics in Materials Science*, Vol.1 (North-Holland, Amsterdam 1978)
Naumovets, A.G.: Investigation of surface structure by LEED: Progress and perspectives. Ukr. Fiz. Zh. **23**, 1585 (1978) [in Russian]
Nucleation, ed. by A.C. Zettlemoyer (Dekker, New York 1969)
Nucleation Phenomena, ed. by A.C. Zettlemoyer (Elsevier, Amsterdam 1977)
Nývlt, J.: *Kryštalizácia z roztokov* (Slovenské vyd-vo technïckej literatúry, Bratislava 1967) [English transl.: *Industrial Crystallization from Solutions* (Butterworths, London 1971)]
Obrazovaniye kristallov (Crystal Growth). Bibliographical Index 1945-1968 (Nauka, Moscow 1970)
Palatnik, L.S., Sorokin, V.K.: *Materialovedeniye v mikroelektronike* (Materials Science in Microelectronics) (Energiya, Moscow 1978) [in Russian]
Parker, R.L.: "Crystal Growth Mechanisms: Energetics, Kinetics and Transport," in *Solid State Physics. Advances in Research and Applications*, Vol.25 (Academic, New York 1970) p.151
Pfann, W.G.: *Zone Melting* (Wiley, New York 1963)
Protsessy real'nogo kristalloobrazovaniya, ed. by N.V. Belov (Processes of Real Crystal Formation) (Nauka, Moscow 1977) [in Russian]
Rosenberger, F.: *Fundamentals of Crystal Growth. I. Macroscopic Equilibrium and Transport Concepts*, Springer Ser. Solid-State Sci., Vol.5 (Springer, Berlin, Heidelberg, New York 1979)
Rost kristallov, Vols.1-14 (Izd-vo Akad. Nauk SSSR, Moscow 1957-1983) [English transl.: *Growth of Crystals* (Consultants Bureau, New York) Vols.1,2 (1959), Vol.3 (1962), Vol.4 (1966), Vols.5A,5B (1968), Vols.6-8 (1969), Vol.9 (1975), Vol.10 (1976), Vol.11 (1979), Vol.12 (1981), Vol.13 (in print)]
Rost i defekty metallicheskikh kristallov, ed. by D.E. Ovsienko (Growth and Defects of Metallic Crystals) (Naukova Dumka, Kiev 1972) [in Russian]
Rost i nesovershenstva metallicheskikh kristallov, ed. by D.E. Ovsienko (Naukova Dumka, Kiev 1966) [English transl.: *Growth and Imperfections of Metallic Crystals* (Consultants Bureau, New York 1968)]
Rost kristallov iz vysokotemperaturnykh vodnykh rastvorov, ed. by A.N. Lobacher (Growth of Crystals from High-Temperature Aqueous Solutions) (Nauka, Moscow 1977) [in Russian]
II-YI Compounds 1982 (Proc. Intern. Conf. on II-VI Compounds, Durham, UK, 21-23 April 1982), ed. by S.J.C. Irvine (North-Holland, Amsterdam 1982); see J. Cryst. Growth **59**, N1/2 (1982)
Shafranovsky, I.I.: *Kristally mineralov* (Mineral Crystals), Part 1: Ploskogrannyie formy (Faceted Forms) (Izd-vo LGU, Leningrad 1957); Part 2: Krivogrannyie, skeletnyie i zernistyie formy (Round-Shaped, Skeletal and Granular Forms) (Gosgeoltekhizdat, Moscow 1961) [in Russian]
Shaped Crystal Growth, ed. by G.W. Surek (in collaboration with P.I. Antonov) (North-Holland, Amsterdam 1980); see J. Cryst. Growth **50** (1980)
Shubnikov, A.V.: *Obrazovaniye kristallov* (Crystallization) (Izd-vo Akad. Nauk SSSR, Moscow 1947) [in Russian]
Shubnikov, A.V.: *Kak rastut kristally* (How Crystals Grow) (Izd-vo Akad. Nauk SSSR, Moscow 1935) [in Russian]
Shubnikov, A.V., Parvov, V.F.: *Zarozhdeniye i rost kristallov* (Crystal Nucleation and Growth) (Nauka, Moscow 1969) [in Russian]
Single-Crystal Films (Proc. Intern. Conf. Philco Sci. Lab., Blue Bell, PA, May 1963), ed. by M.H. Francombe, H. Sato (Pergamon, New York 1964)
Strickland-Constable, R.: *Kinetics and Mechanism of Crystallization* (Academic, London 1968)

Third Intern. Conf. on Thin Films, Budapest, Hungary, 25-29 August 1975 (Elsevier, Lausanne 1976); see Thin Solid Films **32** (1976)
Timofeyeva, V.A.: *Rost kristallov iz rastvorov-rasplavov* (Crystal Growth from Flux) (Nauka, Moscow 1978) [in Russian]
Treivus, E.B.: *Kinetika rosta i rastvoreniya kristallov* (The Kinetics of Crystal Growth and Dissolution) (Izd-vo LGU, Leningrad 1979) [in Russian]
Trusov, L.I., Kholmyansky, V.A.: *Ostrovkovyie metallicheskiye plyonki* (Island Metallic Films) (Metallurgiya, Moscow 1973) [in Russian]
Van Hook, A.: *Crystallization. Theory and Practice* (Reinhold, New York; Chapman & Hall, London 1961)
Van Hove, M.A., Tong, S.Y.: *Surface Crystallography by LEED*, Springer Ser. Chem. Phys., Vol.2 (Springer, Berlin, Heidelberg, New York 1979)
Vapour Growth and Epitaxy (Proc. 2nd Intern. Conf. on Vapour Growth and Epitaxy, Jerusalem, Israel, 21-25 May 1972), ed. by G.W. Cullen, E. Kaldis, R.L. Parker, M. Schieber (North-Holland, Amsterdam 1972); see J. Cryst. Growth **17** (1972)
Vapour Growth and Epitaxy (Proc. 3rd Intern. Conf. on Vapour Growth and Epitaxy, Amsterdam, The Netherlands, 18-21 August 1975), ed. by G.W. Cullen, E. Kaldis, R.L. Parker, C.J.M. Rooymans (North-Holland, Amsterdam 1975); see J. Cryst. Growth **31** (1975)
Vapour Growth and Epitaxy (Proc. 4th Intern. Conf. on Vapour Growth and Epitaxy, Nagoya, Japan, 9-13 July 1978), ed. by K. Takahashi (North-Holland, Amsterdam 1978); see J. Cryst. Growth **45** (1978)
Verma, A.R.: *Crystal Growth and Dislocations* (Butterworths, London 1953)
Volmer, M.: *Kinetik der Phasenbildung* (Steinkopf, Dresden 1939)
Wilke, K.-Th. (in collaboration with I. Böhm): *Kristallzüchtung* (VEB Deutscher Verlag der Wissenschaften, Berlin 1973)

References

Chapter 1 [1]

1.1 L.D. Landau, E.M. Lifshits: *Statisticheskaya fizika* (Nauka, Moscow 1964) [English transl.: *Statistical Physics*, 2nd ed. (Pergamon, London 1969)]

1.2 a) F. Rosenberger: *Fundamentals of Crystal Growth I. Macroscopic Equilibrium and Transport Concepts;* Springer Ser. Solid-State Sci., Vol.5 (Springer, Berlin, Heidelberg, New York 1981)
b) O. Söhnel, J. Garside: J. Cryst. Growth *46*, 238 (1979)
c) J.W. Mullin: The Chemical Engineer, June, 1973, p.316

1.3 a) A.V. Shubnikov, N.N. Sheftal' (eds.): *Rost kristallov, T.2* (Izd-vo Akad. Nauk SSSR, Moscow 1959) [English transl.: *Growth of Crystals. Vol.2* (Consultants Bureau, New York 1959)]
b) V.Ya. Khaimov-Mal'kov: In [Ref.1.3a, p.5] [English transl.: "The Thermodynamics of Crystallization Pressure", in Ref.1.3a, p.3]
c) V.Ya. Khaimov-Mal'kov: In [Ref.1.3a, p.17] [English transl.: "Experimental Measurement of Crystallization Pressure", in Ref.1.3a, p.14]
d) V.Ya. Khaimov-Mal'kov: In [Ref.1.3a, p.26] [English transl.: "The Growth Conditions of Crystals in Contact with Large Obstacles", in Ref.1.3a, p.20]

1.4 Yu.S. Barâsh, V.L. Ginzburg: Usp. Fiz. Nauk *116*, 5 (1975) [English transl.: Electromagnetic fluctuations in matter and molecular (Van der Waals) forces between them. Sov. Phys. Usp.*18*, 305 (1975)]

1.5 B.V. Derjaguin, I.I. Abrikosova, E.M. Lifshits: Molekulyarnoe prityazhenie kondensirovannykh tel (Molecular attraction of condensed bodies). Usp. Fiz. Nauk *64*, 493 (1958)

1.6 I.E. Dzyaloshinsky, E.M. Lifshits, L.P. Pitayevsky: Usp. Fiz. Nauk *73*, 381 (1961) [English transl.: General theory of Van der Waals forces. Sov. Phys. Usp.*4*, 153 (1961)]

1.7 a) E. Kaldis, H.J. Scheel (eds.): 1976 *Crystal Growth and Materials*: Review Papers of the First European Conference on Crystal Growth ECCG-1, Zurich, Sept. 1976 (North-Holland, Amsterdam 1977)
b) A.A. Chernov, D.E. Temkin: "Capture of Inclusions in Crystal Growth", in [Ref.1.7a, p.3]
c) R.G. Horn, J.N. Israelashvili: J. Chem. Phys.*75*, 1400 (1981)

1.8 C.W. Correns: Discuss. Faraday Soc. No. *5*, 267 (1949)

1.9 P. Hartman: In *Rost kristallov, T.7*, ed. by N.N. Sheftal' (Nauka, Moscow 1967) p.8 [English transl.: "The Dependence of Crystal Morphology on Crystal Structure", in *Growth of Crystals, Vol.7* (Consultants Bureau, New York 1969) p.3]

1.10 B. Honigmann: *Gleichgewichts- und Wachstumsformen von Kristallen* (Steinkopf, Darmstadt 1958)

1.11 a) P. Hartman: "Le coté cristallographique de l'adsorption vu par le changement de faciés", in *Adsorption et croissance cristalline*, Colloq. Int. CNRS, No.152 (Edition du Centre National de la Recherche Scientifique, Paris 1965) p.477, 506

[1] For papers of review nature and for some others the titles are included in references to Chaps.1-7.

b) P. Hartman: "Structure and Morphology", in *Crystal Growth: An Introduction*, ed. by P. Hartman (North-Holland, Amsterdam 1973) p.367

1.12 a) Ya.E. Geguzin, N.N. Ovcharenko: Usp. Fiz. Nauk *76*, 283 (1962) [English transl.: Surface energy and processes on solid surfaces. Sov. Phys. Usp.*5*, 129 (1962)]
b) N. Eustatopulos, J.-C. Joud: "Interfacial Tension and Adsorption of Metallic Systems", *Current Topics Mat. Sci.*, Vol.4 (North-Holland, Amsterdam 1979) p.281

1.13 V.I. Marchenko, A.Ya. Parshin: Zh. Eksp. Teor. Fiz.*79*, 257 (1980) [English transl.: Sov. Phys. JETP *52*, 129 (1980)]

1.14 A.F. Andreyev, Yu.A. Kosevich: Zh. Eksp. Teor. Fiz.*81*, 1435 (1981)

1.15 I.N. Stransky, R. Kaischew: K teorii rosta kristallov i obrazovaniya kristallicheskikh zarodyshei (On the theory of crystal growth and nucleation), Usp. Fiz. Nauk *21*, 408 (1939)

1.16 Ch. Kittel: *Introduction to Solid State Physics*, 4th ed. (Wiley, New York 1971)

1.17 W.K. Burton, N. Cabrera, F.C. Frank: The growth of crystals and the equilibrium structure of their surfaces. Philos. Trans. R. Soc. London A*243*, 299 (1951)

1.18 A.A. Chernov: Usp. Fiz. Nauk *73*, 277 (1961) [English transl.: The spiral growth of crystals. Sov. Phys. Usp.*4*, 116 (1961)]

1.19 a) S.F. Bedair: Surf. Sci.*42*, 595 (1974)
b) S. Hok, M. Drechsler: Surf. Sci.*107*, 262 (1981)
c) B.M. Bulakh, A.A. Chernov: J. Cryst. Growth *52*, 39 (1981)

1.20 T.N. Kompaniyets: Zh. Tekh. Fiz.*46*, 1361 (1976)

1.21 R.C. Weast (ed.): *Handbook of Chemistry and Physics*, 56th ed. (CRC, Cleveland, Ohio 1975)

1.22 a) A.A. Chernov, N.S. Papkov: Kristallografiya *25*, 1002 (1980) [English transl.: Sov. Phys. Crystallogr.*25*, 572 (1980)]
b) A.A. Chernov, M.P. Ruzaikin: J. Cryst. Growth *45*, 73 (1978); *52*, 185 (1981)
c) A.A. Chernov: Equilibrium adsorption and some interfacial growth processes in gaseous solutions and CVD systems. J. Jp. Assoc. Cryst. Growt *5*, 227 (1978)
d) R. Cadoret: "Application of the Theory of the Rate Processes in the CVD of GaAs", *Current Topics Mat. Sci.*, Vol.5 (North-Holland, Amsterdam 1980) p.218
e) A.A. Chernov, M.P. Ruzaikin, N.S. Papkov: Poverkhn. Fiz. Khim. Mekh.*I*, 94 (1982)

1.23 a) F. Hottier, R. Cadoret: J. Cryst. Growth *52*, 199 (1981)
b) F. Hottier, J.B. Theeten: J. Cryst. Growth *48*, 644 (1980)

1.24 a) J. Bloem, L.J. Gilling: "Chemical Vapour Deposition of Silicon", *Curren Topics Mat. Sci.*, Vol.1 (North-Holland, Amsterdam 1978) p.147
b) J. Bloem, W.A.P. Claassen: J. Cryst. Growth *49*, 435 (1980)
c) W.A.P. Claassen, J. Bloem: J. Cryst. Growth *50*, 803 (1980); *51*, 443 (1981)
d) J. Bloem, W.A.P. Claassen, W.G.J.N. Valkenburg: J. Cryst. Growth *57*, 177 (1982)
e) J. Bloem: J. Cryst. Growth *50*, 581 (1980)

1.25 a) V.V. Voronkov: Kristallografiya *11*, 284 (1966) [English transl.: Sov. Phys. Crystallogr.*11*, 259 (1966)]
b) G.H. Gilmer, J.D. Weeks: J. Chem. Phys.*68*, 950 (1978)

1.26 a) H.J. Leamy, G.H. Gilmer, K.A. Jackson: "Statistical Thermodynamics of Clean Surfaces", in *Surface of Materials*, Vol.1, ed. by J.M. Blakely (Academic, New York 1975) p.121
b) H.J. Leamy, K.A. Jackson: J. Appl. Phys.*42*, 2121 (1971)
c) G.H. Gilmer: J. Cryst. Growth *42*, 3 (1977)

1.27 a) J.Q. Broughton, G.H. Gilmer, J.D. Weeks: J. Chem. Phys.*75*, 5128 (1981); Phys. Rev. B*25*, 4651 (1982)

b) J.M. Phillips, L.W. Bruch, R.D. Murphy: J. Chem. Phys.*75*, 5097 (1981)
c) M.R. Mruzik, S.H. Garofalini, G.M. Pound: Surf. Sci.*103*, 353 (1981)
d) Y. Hiwatari, E. Stoll, T. Schneider: J. Chem. Phys.*68*, 3401 (1978)
e) J.Q. Brougton, A. Bonissent, F.F. Abraham: J. Chem. Phys.*74*, 4029 (1981)
f) A. Bonissent, F. Abraham: J. Chem. Phys.*74*, 1306 (1981)
g) A. Bonissent: Structure of Solid-Liquid Interface. *Crystals*, Vol.9 (Springer, Berlin, Heidelberg, New York 1983) p.1

1.28 D.E. Temkin: "O molekulyarnoi sherokhovatosti granitsy kristall—rasplav" (On Molecular Roughness of the Crystal—Melt Interface), in *Mekhanizm i kinetika kristallizatsii* (Mechanism and Kinetics of Crystallization), ed. by N.N. Sirota (Nauka i Tekhnika, Minsk 1964) p.86

1.29 a) R.H. Doremus, B.W. Roberts, D. Turnbull (eds.): *Growth and Perfection of Crystals* (Wiley, New York 1958)
b) K.A. Jackson: "Interface Structure", in [Ref.1.29a, pp.319, 324]

1.30 a) A.A. Chernov (ed.): *Rost kristallov, T.11* (Izd-vo EGU, Erevan 1975) [English transl.: *Growth of Crystals, Vol.11* (Consultants Bureau, New York 1979)]
b) K.A. Jackson: In [Ref.1.30a, p.116] [English transl.: "Computer Modelling of Crystal Growth Processes", in Ref.1.30a, p.115]
c) D. Turnbull: J. Appl. Phys.*21*, 1022 (1950)

1.31 a) V.V. Voronkov, A.A. Chernov: Structure of crystal ideal solution interface. J. Phys. Chem. Solids Suppl.*1*, 593 (1967); Kristallografiya *11*, 662 (1966) [English transl.: Sov. Phys. Cryst. *11*, 571 (1967)]
b) H.W. Kerr, W.C. Winegard: Eutectic solidification. J. Phys. Chem. Solids Suppl.*1*, 179 (1967)
c) V.V. Podolinsky, V.G. Drykin: 6th Int. Conf. Cryst. Growth, Extended Abstracts, Vol.2, USSR Academy of Sciences, Moscow 1980, p.77

1.32 a) A.F. Andreev, A.Ya. Parshin: Zh. Eksp. Teor. Fiz.*75*, 1511 (1978) [English transl.: Sov. Phys. JETP *48*, 763 (1978)]
b) K.O. Keshishev, A.Ya. Parshin, A.V. Babkin: Zh. Eksp. Teor. Fiz.*80*, 716 (1981) [English transl.: Sov. Phys. JETP *53*, 362 (1981)]; Zh. Eksp. Teor. Fiz. Pis'ma Red.*30*, 63 (1979) [English transl.: Sov. Phys. JETP Lett.*30*, 56 (1979)]
c) S. Balibar, D.O. Edward, C. Laroche: Phys. Rev. Lett.*42*, 782 (1979)
d) J. Landau, S.G. Lipson, M.M. Määttänen, L.S. Balfour, D.O. Edwards: Phys. Rev. Lett.*45*, 31 (1980)

1.33 a) C. Herring: "The Use of Classical Macroscopic Concepts in Surface-Energy Problems", in *Structure and Properties of Solid Surfaces*, ed. by E.R. Gomer, C.J. Smith (University of Chicago Press, Chicago 1953) pp. 5, 72
b) C. Herring: "Surface Tension as a Motivation for Sintering", in *The Physics on Powder Metallurgy* (McGraw-Hill, New York 1951) p.143

1.34 a) G.G. Lemmlein: Dokl. Akad. Nauk SSSR *98*, 973 (1954)
b) G.G. Lemmlein: *Morfologiya i genezis kristallov* (Crystal Morphology and Genesis) (Nauka, Moscow 1973)

1.35 M.O. Kliya: Kristallografiya *1*, 577 (1956)

1.36 a) J.C. Heyraud, J.J. Metois: J. Cryst. Growth *50*, 571 (1980)
b) J.J. Metois, J.C. Heyraud: J. Cryst. Growth *57*, 487 (1982)
c) J.C. Heyraud, J.J. Metois: Acta Metall.*28*, 1789 (1980)
d) M. Drechsler, J.F. Nicholas: J. Phys. Chem. Solids *28*, 2609 (1967)

Chapter 2

2.1 L. Bosio, A. Defrain, I. Epelboin: J. Phys. (Paris) *27*, 61 (1966)
2.2 D.H. Rasmussen: Thermodynamics and nucleation phenomena – a set of experimental observations. J. Cryst. Growth *56*, 45 (1982)
2.3 S.S. Stoyanov: "Nucleation Theory for High and Low Supersaturations", *Current Topics Mat. Sci.*, Vol.3 (North-Holland, Amsterdam 1979) p.421
2.4 a) Ya.B. Zel'dovich: Zh. Eksp. Teor. Fiz.*12*, 525 (1942)
b) S. Toshev: "Homogeneous Nucleation", in *Crystal Growth: An Introduction* ed. by P. Hartman (North-Holland, Amsterdam 1973) p.1
2.5 J. Hirth, G.M. Pound: *Condensation and Evaporation. Nucleation and Growth Kinetics* (Pergamon, Oxford 1963)
2.6 a) M. Volmer: *Kinetik der Phasenbildung* (Steinkopf, Dresden 1939)
b) M. Volmer, H. Flood: Z. Phys. Chem.*170*, 273 (1934)
2.7 J. Lothe, G.M. Pound: J. Chem. Phys.*36*, 2080 (1962)
2.8 J. Miyazaki, G.M. Pound, F.F. Abraham, J.A. Barker: J. Chem. Phys.*67*, 3851 (1977)
2.9 a) K. Nishioka, K.C. Russel: Surf. Sci.*104*, 213 (1981)
b) M. Blauder, J.L. Katz: Surf. Sci.*104*, 217 (1981)
c) A.C. Zettlemoyer (ed.): *Nucleation Phenomena* (Elsevier, Amsterdam 1977)
2.10 a) D. Turnbull: J. Appl. Phys.*21*, 1022 (1950)
b) D.E. Ovsienko, G.A. Alfintsev: Crystal Growth from the Melt. Experimental Investigation of Kinetics and Morphology, *Crystals*, Vol.2 (Springer Berlin, Heidelberg, New York 1980) p.119
2.11 G. Matz: *Die Kristallization in der Verfahrenstechnik* (Springer, Berlin, Göttingen, Heidelberg 1954)
2.12 a) A.G. Khachaturyan: *Theory of Phase Transformations and the Structure of Solid Solutions* (Amerind, New Delhi, to be published)
b) B.Ya. Lyubov: *Kineticheskaya teoriya fazovykh prevrashchenii* (Kinetic Theory of Phase Transformations) (Metallurgiya, Moscow 1969)
c) V.V. Voronkov: Kristallografiya *15*, 1120 (1970) [English transl.: Sov. Phys. Crystallogr. *15*, 979 (1971)]
d) E.A. Brener, V.I. Marchenko, C.V. Meshkov: Zh. Eksp. Teor. Fiz. (1983) (in print)
2.13 a) L.E. Murr, O.T. Inal, H.P. Singh: Thin Solid Films *9*, 241 (1972)
b) T.T. Tsong, P.L. Cowan: "Behavior and Properties of Single Atoms on Metal Surfaces", in *Chemistry and Physics of Metal Surfaces, Vol.2*, ed. by R. Vanselow (CRC, Boca Raton, Florida 1979) p.209
c) V.N. Shrednik, G.A. Odisharia, O.L. Golubev: J. Cryst. Growth *2*, 249 (1971)
d) K.L. Moazed, G.M. Pound: Trans. Met. Soc. AIME *230*, 2341 (1964)
e) S.C. Hardy: In *Proceedings of the International Conference on Crystal Growth*, Boston 1966, ed. by H.S. Peiser (Pergamon, Oxford 1967) p.287; Also in: J. Phys. Chem. Solids Suppl. I
2.14 a) V.I. Danilov: *Stroyeniye i kristallizatsiya zhidkostei* (Structure and Crystallization of Liquids) (Izd-vo Akad. Nauk Ukr. SSR, Kiev 1956) p.311
b) J. Nyvlt, V. Pekarek: Z. Phys. Chem. Neue Folge *22*, 199 (1980)
2.15 D. Turnbull: J. Chem. Phys.*20*, 411 (1952)
2.16 D. Turnbull, R.E. Cech: J. Appl. Phys.*21*, 804 (1950)
2.17 J.A. Koutsky, A.G. Walton, E. Baer: J. Appl. Phys.*38*, 1832 (1967)
2.18 V.P. Skripov, V.P. Koverda, G.T. Butorin: Kristallografiya *15*, 1219 (1970) [English transl.: Sov. Phys. Crystallogr.*15*, 1065 (1971)]
2.19 a) S. Toshev, I. Gutzov: Krist. Tech.*7*, 43 (1972)
b) I. Gutzov: Contemp. Phys.*21*, 121, 243 (1980)
2.20 V.P. Skripov: "Homogeneous Nucleation in Melts and Amorphous Films", in [Ref.1.7a, p.327]

2.21 V.P. Skripov, V.P. Koverda, G.T. Butorin: In [Ref.1.30a. p.25] [English transl.: "Crystal Nucleation Kinetics in Small Volumes", in Ref.1.30a, p.22]
2.22 L. Bosio: Mêt. Corros. Ind.*40*, 421, 451 (1965)
2.23 J. Fehling, E. Scheil: Z. Metallkd.*53*, 593 (1962)
2.24 G.L.F. Powell, L.M. Hogan: Trans. Metall. Soc. AIME *242*, 2133 (1968)
2.25 D.E. Ovsienko, V.V. Maslov, V.P. Kostyuchenko: Kristallografiya *16*, 405 (1971) [English transl.: Sov. Phys. Crystallogr.*16*, 331 (1971)]
2.26 J.L. Walker: "The Influence of Large Amounts of Undercooling on the Grain Size of Nickel", in *Physical Chemistry of Process Metallurgy, Metallurgical Society Conferences*, Vol.8, ed. by G.R.S. Pierre (Interscience, New York 1961) Pt.2, p.845
2.27 a) D.W. Gomersall, S.Y. Shiraishi, R.G. Ward: J. Aust. Inst. Met.*10*, 220 (1965)
b) N.J. Pugh, D.A. Jefferson: 12th Int. Congr. Crystallography, Ottawa, Canada (1981), Collected Abstracts, p. C-302
2.28 V.M. Fokin, A.M. Kalinina, V.N. Filipovich: J. Cryst. Growth *52*, 115, (1981)
2.29 S. Stoyanov. D. Kashchiev, "Thin Film Nucleation and Growth Theories: A Confrontation with Experiment", *Current Topics Mat. Sci.*, Vol.7 (North-Holland, Amsterdam 1981) p.69
2.30 a) M.R. Hoare, P. Pal: J. Cryst. Growth *17*, 77 (1972)
b) A. Bonissent, B. Mutaftschiev: J. Chem. Phys.*58*, 372 (1973)
c) J.K. Lee, J.A. Barker, F.F. Abraham: J. Chem. Phys.*58*, 3166 (1973)
d) C.L. Briant, J.J. Burton: J. Chem. Phys.*63*, 2045, 332 (1975)
e) J.J. Burton, C.L. Briant: "Atomistic Models for Microclusters; Implications for Nucleation Theory", in *Nucleation Phenomena*, ed. by A.C. Zettlemoyer (Elsevier, Amsterdam 1977) p.131
2.31 a) K. Sattler: Diagnostics of Clusters in Molecular Beams, Preprint
b) K. Sattler: "Clusters in Molecular Beams", Preprint, should be published in *Current Topics Mat. Sci.* (North-Holland, Amsterdam 1983)
c) K. Sattler, J. Mühlbach, E. Recknagel: Phys. Rev. Lett.*45*, 821 (1980)
d) O. Echt, K. Sattler, E. Recknagel: Phys. Rev. Lett.*47*, 1121 (1981)
e) J. Mülbach, K. Sattler, P. Pfau, E. Recknagel: in print
f) K. Sattler, J. Mülbach, P. Pfau, E. Recknagel: Phys. Lett. A*87*, 418 (1982)
2.32 a) J.-P. Borel, J. Buttet (eds.): *Small Particles and Inorganic Clusters* (North-Holland, Amsterdam 1981); Surf. Sci.*106*, (1981)
b) D. Dreyfuss, H.Y. Wachman: J. Chem. Phys.*76*, 2031 (1981)
2.33 a) M.R. Hoare: "Structure and Dynamics of Simple Microclusters", in *Advances in Chemical Physics*, ed. by I. Prigogine, S.A. Rice (Wiley, New York 1979) p.49
b) M.R. Hoare, P. Pal, P.P. Wegener: J. Colloid Interface Sci.*75*, 126 (1980)
2.34 a) B.G. de Boer, G.D. Stein: Surf. Sci.*106*, 84 (1981)
b) A. Renou, M. Gillet: Surf. Sci.*106*, 27 (1981)
2.35 D. Walton: J. Chem. Phys.*37*, 2182 (1962)
2.36 D. Walton: "Condensation of Metals on Substrates", in *Nucleation*, ed. by A.C. Zettlemoyer (Marcel Dekker, New York 1969) p.379
2.37 T.N. Rhodin, D. Walton: "Nucleation of Oriented Films". in *Single-Crystal Films*, ed. by M.H. Francombe, H. Sato (Pergamon, New York 1964) p.31
2.38 P.W. Palmberg, T.N. Rhodin, C.J. Todd: Appl. Phys. Lett.*10*, 122 (1967)
2.39 H. Bethge: "Surface Structures and Molecular Processes", in *Molecular Processes on Solid Surfaces*, ed. by E. Drauglis, R.D. Gretz, R.I Jaffee (McGraw-Hill, New York 1969) pp.569, 585
2.40 G.R. Henning: Appl. Phys. Lett.*4*, 52 (1964)

2.41 D.J. Stirland: Appl. Phys. Lett.*8*, 326 (1966)
2.42 P.W. Palmberg, C.J. Todd, T.N. Rhodin: J. Appl. Phys.*39*, 4650 (1968)
2.43 G.I. Distler: J. Cryst. Growth *3/4*, 175 (1968)
2.44 L.S. Palatnik, M.Ya. Fuks, V.M. Kosevich: *Mekhanizm obrazovaniya i substructura kondensirovannykh plyonok* (Mechanism of Formation and Substructure of Condensed Films) (Nauka, Moscow 1972)
2.45 R.A. Sigsbee: J. Appl. Phys.*42*, 3904 (1971)
2.46 R.A. Sigsbee: J. Cryst. Growth *13/14*, 135 (1972)
2.47 M. Krohn: In [Ref.1.30a, p.192] [English transl.: "Estimation of the Mean Displacement of Adsorbed Molecules from the Growth of Lochkeims", in Ref.1.30a, p.191]
2.48 G. Ehrlich: Direct observation of individual atoms on metals. Surf. Sci. *63*, 422 (1977)
2.49 G.L. Kellog, T.T. Tsong, P. Cowan: Direct observation of surface diffusion and atomic interaction on metal surfaces. Surf. Sci.*70*, 485 (1978)
2.50 M. Klaua: In [Ref.1.30a, p.65] [English transl.: "Electron Microscopic Investigations of Surface Diffusion and Nucleation of Au on Ag (111)", in Ref.1.30a, p.60]
2.51 V.N.E. Robinson, J.L. Robins: Thin Solid Films *20*, 155 (1974)
2.52 M.J. Stowell: J. Cryst. Growth *24/25*, 45 (1974)
2.53 S.S. Stoyanov: J. Cryst. Growth *24/25*, 293 (1974)
2.54 I. Markov: Thin Solid Films *35*, 11 (1976)
2.55 I. Markov, D. Kashchiev: J. Cryst. Growth *13/14*, 131 (1972)
2.56 M. Gebhardt, A. Neuhaus: In *Landolt-Börnstein*. Numerical Data and Functional Relationship in Science and Technology, New Series, Group 3: Crystal and Solid State Physics, Vol.8, Structure Data of Organic Compounds (Springer, Berlin, Heidelberg, New York 1972)
2.57 L.S. Palatnik, I.I. Papirov: *Epitaksial'nyie plyonki* (Epitaxial Films) (Nauka, Moscow 1971)
2.58 J.W. Matthews (ed.): *Epitaxial Growth* (Academic, New York 1975)
2.59 a) R. Kern, G. Le Lay, J.J. Métois: "Basic Mechanisms in the Early Stages of Epitaxy", *Current Topicis Mat. Sci.*, Vol.3 (North-Holland, Amsterdam 1979) p.132
b) G. Honjo, K. Yagi: "Studies of Epitaxial Growth of Thin Films by *in situ* Electron Microscopy", *Current Topics Mat. Sci.*, Vol.4 (North-Holland, Amsterdam 1980) p.195
2.60 J.J. Métois, M. Gauch, A. Masson, R. Kern: Thin Solid Films *11*, 205 (1972)
2.61 G.I. Distler, V.P. Vlasov: Kristallografiya *14*, 872 (1969) [English transl.: Sov. Phys. Crystallogr.*14*, 747 (1970)]
2.62 a) R. Kern, A. Masson, J.J. Métois: Surf. Sci.*27*, 483 (1971)
b) A. Masson, J.J. Métois, R. Kern: Surf. Sci.*27*, 463 (1971)
c) J.C. Zanghi, J.J. Métois, R. Kern: Philos. Mag.*31*, 743 (1975)
2.63 J.J. Métois, K. Heinemann, H. Poppa: Appl. Phys. Lett.*29*, 134 (1976)
2.64 a) C. Chapon, C.R. Henry: Surf. Sci.*106*, 152 (1981)
b) K. Kinosita: Thin Solid Films *85*, 223 (1981)
2.65 B.A. Malyukov, V.E. Korolev, V.S. Papkov, M.A. Vorkunova: Kristallografiya *25*, 444 (1980) [English transl.: Sov. Phys. Crystallogr.*25*, 257 (1980)]
2.66 G. Le Lay, R. Kern: J. Cryst. Growth *44*, 197 (1978)
2.67 P.B. Barraclough, P.G. Hall: Surf. Sci.*46*, 393 (1974)
2.68 J. Estel, H. Hoinkes, H. Kaarmann, H. Nahr, H. Wilsch: Surf. Sci.*54*, 393 (1976)
2.69 a) I.A. Kotzé, J.C. Lombaard: Thin Solid Films *23*, 221 (1974)
b) E. Gillet, B. Gruzza: Surf. Sci.*97*, 553 (1980)
c) E. Bauer, H. Poppa, C. Todd, P. Davis: J. Appl. Phys.*48*, 3773 (1977)

d) E. Bauer, H. Poppa, C. Todd, F. Benczek: J. Appl. Phys.*45*, 5164 (1974)
e) J.W. Jesser, J.W. Matthews: Philos. Mag.*17*, 595 (1968)
f) J.W. Jesser, J.W. Matthews: Philos. Mag.*17*, 461 (1968)
g) C.T. Herng, R.W. Vook: J. Vac. Sci. Technol.*2*, 140 (1974)
h) U. Gradman, W. Kümmerle, P. Tillmans: Thin Solid Films *34*, 249 (1976)
i) S. Stoyanov, I. Markov: Surf. Sci.*116*, 313 (1982)

2.70 a) A.A. Chernov: Growth Kinetics and Capture of Impurities During Gas Phase Crystallization. J. Cryst. Growth *42*, 55 (1977)
b) A.A. Chernov, S.S. Stoyanov: Kristallografiya *22*, 248 (1977) [English transl.: Sov. Phys. Crystallogr.*22*, 141 (1977)]

2.71 J. Hölzl, F.K. Schulte: "Work Function of Metals", *Springer Tracts Mod. Phys.*, Vol.85 (Springer, Berlin, Heidelberg, New York 1979) p.1

2.72 H. Wagner: "Physical and Chemical Properties of Stepped Surfaces", *Springer Tracts Mod. Phys.*, Vol.85 (Springer, Berlin, Heidelberg, New York 1979) p.151

2.73 E.V. Klimenko, A.G. Naumovets: Surf. Sci.*14*, 141 (1969)

2.74 J. Woltersdorf: Thin Solid Films *32*, 277 (1976)

2.75 J.P. Hirth, J. Lothe: *Theory of Dislocations* (McGraw-Hill, New York 1968)

2.76 B.K. Vainshtein, V.M. Fridkin, V.L. Indenbom: *Sovremennaya Kristallografiya. T.2* Struktura kristallov, ed. by B.K. Vainshtein (Nauka, Moscow 1979) [English transl.: *Modern Crystallography II*, Springer Ser. Solid-State Sci., Vol.21 (Springer, Berlin, Heidelberg, New York 1982)]

2.77 a) M.S. Abrahams: "Epitaxy, Heteroepitaxy, and Misfit Dislocations", in *Crystal Growth and Characterization*, ed. by R. Ueda, J.B. Mullin (North-Holland, Amsterdam 1975) p.187
b) F. Hila, M. Gillet: Thin Solid Films *87*, L7 (1982)
c) J. Woltersdorf: Interface problems in relation to epitaxy. Thin Solid Films *85*, 241 (1981)

2.78 R. Wakkernagel: Kästners Archiv f.d. gesamte Naturlehre *5*, 295 (1825)

2.79 M.L. Frankenheim: Ann. Phys. Chem.*37*, 516 (1836)

2.80 R.S. Bradley: Z. Kristallogr.*96*, 499 (1937)

2.81 G.I. Distler: In [Ref.1.30a, p.47] [English transl.: "Crystallization as a Matrix Replication Process", in Ref.1,30a, p.44]

2.82 G.I. Distler, V.P. Vlasov, Yu.M. Gerasimov, S.A. Kobzareva, E.I. Kortukova, V.N. Lebedeva, V.V. Moskvin, L.A. Shenyavskaya: *Dekorirovaniye poverchnosti tvyordykh tel* (Decoration of Solid Surfaces) (Nauka, Moscow 1976)

2.83 V.L. Indenbom: In [Ref.1.30a, p.62] [English transl.: "Elastic Interaction in Epitaxial Effects", in Ref.1.30a, p.57]

Chapter 3

3.1 K.A. Jackson: The present state of the theory of crystal growth from the Melt. J. Cryst. Growth *24/25*, 130 (1974)

3.2 V. Bostanov, R. Rusinova, E. Budevski: In [Ref.1.30a, p.131] [English transl.: "Rate of Propagation of Monatomic Layers and the Mechanism of Electrodeposition of Silver", in Ref.1.30a, p.130]

3.3 A.A. Ballman, R.A. Laudise: "Hydrothermal Growth", in *The Art and Science of Growing Crystals*, ed. by J.J. Gilman (Wiley, New York 1963) p.231

3.4 G.G. Lemmlein, M.O. Kliya, A.A. Chernov: Kristallografiya *9*, 231 (1964) [English transl.: Sov. Phys. Crystallogr.*9*, 181 (1964)]

3.5 G.G. Lemmlein, E.D. Dukova: Kristallografiya *1*, 112 (1956)
3.6 S.K. Tung: J. Electrochem. Soc.*112*, 436 (1968)
3.7 a) A.A. Chernov (ed.): *Rost kristallov. T.12* (Izd-vo EGU, Erevan 1977) [English transl.: *Growth of Crystals. Vol.12* (Consultants Bureau, New York 1981]
b) L.I. Tsinober, V.E. Khadzhi, L.A. Gordienko, L.I. Litvin: In [Ref. 3.7a, p.75] [English transl.: "Growth Conditions and Real Structure of Quartz Crystals", in Ref.3.7a]
3.8 A.V. Belyustin, I.M. Levina: Private communication
3.9 M. Volmer, I. Estermann: Z. Phys.*7*, 13 (1921)
3.10 M. Volmer, W. Schultze: Z. Phys. Chem. A*156*, 1 (1931)
3.11 M.I. Kozlovsky, G.G. Lemmlein: Kristallografiya *3*, 351 (1958) [English transl.: Sov. Phys. Crystallogr.*3*, 352 (1958)]
3.12 V.G. Levich: *Physicochemical Hydrodynamics* (Prentice Hall, Englewood Cliffs, New Jersey 1962)
3.13 P. Bennema: J. Cryst. Growth *24/25*, 76 (1974)
3.14 a) N.N. Sheftal' (ed.): *Rost kristallov, T.5* (Nauka, Moscow 1965) [English transl.: *Growth of Crystals, Vol.5A* (Consultants Bureau, New York 1968)]
b) A.A. Chernov, B.Ya. Lyubov: In [Ref.3.14a, p.11] [English transl.: "Aspects of Crystal Growth Theory", in Ref.3.14a, p.7]
c) A.E. Nielsen: *Kinetics of Precipitation* (Pergamon, New York 1964)
d) W.B. Hillig: Acta Metall.*14*, 1868 (1966)
e) H.J. Meyer, H. Dabringhaus: "Molecular Processes of Condensation and Evaporation of Alkali Halides", *Current Topics Mat. Sci.*, Vol.1 (North-Holland, Amsterdam 1978) p.47
f) G.G. Lemmlein: *Morfoligiya i genesis kristallov* (Morphology and Genesis of Crystals, Collected Papers of G.G. Lemmlein) (Nauka, Moscow 1973) p.278
g) G.G. Lemmlein: Fine structure of crystal relief. Report to the General Meeting of the Phys. Div. of the USSR Academy of Sciences, Vestn. Akad. Nauk SSSR *4*, 119 (1945)
3.15 V. Bostanov: J. Cryst. Growth *42*, 194 (1977)
3.16 G. Staikov, W. Obretenov, V. Bostanov, E. Budevski, H. Bort: Electrochim. Acta *25*, 1619 (1980)
3.17 V. Bostanov, W. Obretenov, G. Staikov, D.K. Roe, E. Budevski: J. Cryst. Growth *52*, 761 (1981)
3.18 G.H. Gilmer: J. Cryst. Growth *49*, 465 (1980)
3.19 A.A. Chernov, I.L. Smol'sky, V.F. Parvov, Yu.G. Kuznetsov, V.N. Rozhansky: Kristallografiya *25*, 821 (1980) [English transl.: Sov. Phys. Crystallogr. *25*, 469 (1980)]
3.20 A.A. Chernov: "Struktura poverchnosti i rost kristallov" (Surface Structure and Crystal Growth), in *Fiziko-khimicheskiye problemy kristallizatsii* (Physical-Chemical Problems of Crystallization) (Izd-vo KGU, Alma-Ata 1969) p.8
3.21 A.A. Chernov: "Crystallization", in *Annual Review of Materials Science*, Vol.3, ed. by R.A. Huggins, R.H. Bube, R.W. Roberts (Annual Reviews, Palo Alto, CA 1973) p.397
3.22 C.E. Miller: J. Cryst. Growth *42*, 357 (1977)
3.23 a) N. Cabrera, M. Levine: Philos. Mag.*1*, 450 (1956)
b) H. Müller-Krumbhaar, T.W. Burkhardt, D.M. Kroll: J. Cryst. Growth *38*, 13 (1977)
c) H. Müller-Krumbhaar: "Kinetics of Crystal Growth. Microscopic and Phenomenological Theories", *Current Topics Mat. Sci.*, Vol.1 (North-Holland, Amsterdam 1978) p.1
d) R.Kaishev: J. Cryst. Growth *3*, 15 (1962)

e) B. van der Hoek, J.P. van der Eerden, P. Bennema: J. Cryst. Growth *56*, 108 (1982)
f) K. Tsukamoto, B. van der Hoek: J. Cryst. Growth *57*, 131 (1982)
g) F.C. Frank: J. Cryst. Growth *51*, 367 (1982)
h) B. van der Hoek, J.P. van der Eerden, P. Bennema, I. Sunagawa: J. Cryst. Growth *58*, 365 (1982)
i) J.P. van der Eerden: "Surface and Volume Diffusion Controlling Step Movement", in *Crystals*, Vol.9 (Springer, Berlin, Heidelberg, New York 1983) p.112

3.24 N. Cabrera, R.V. Coleman: "Theory of Crystal Growth from the Vapour", in *The Art and Science of Growing Crystals*, ed. by J.J. Gilman (Wiley, New York 1963) p.3

3.25 A.A. Chernov: "Surface Morphology and Growth Kinetics", in *Crystal Growth and Characterization*, ed. by R. Ueda, J.-B. Mullin (North-Holland, Amsterdam 1975) p.33

3.26 M. Hayashi, T. Shichiri: J. Cryst. Growth *21*, 254 (1974)

3.27 A.A. Chernov: Kristallografiya *16*, 842 (1971) [English transl.: Sov. Phys. Crystallogr.*16*, 741 (1971)]

3.28 M. Quivi: "Contribution à l'étude du role de la diffusion dans la croissance des cristaux à partir de solution"; These, Université Strasbourg (1965)

3.29 A.A. Chernov, E.D. Dukova: Kristallografiya *14*, 169 (1969) [English transl.: Sov. Phys. Crystallogr.*14*, 150 (1969)]

3.30 A.A. Chernov, N.S. Papkov, A.F. Volkov: Kristallografiya *25*, 997 (1980) [English transl.: Sov. Phys. Crystallogr.*25*, 572 (1980)]

3.31 A.A. Chernov, N.S. Papkov: Kristallografiya *25*, 1002 (1980) [English transl.: Sov. Phys. Crystallogr.*25*, 575 (1980)]

3.32 a) R. Kaishev, E. Budevski: Surface properties in electrocrystallization. Contemp. Phys. *8*, 489 (1967)
b) E. Budevski, E. Bostanov, G. Staikov: Electrocrystallization. Annu. Rev. Mater. Sci.*10*, 85 (1980)

3.33 a) G.A. Alfintsev, D.E. Ovsienko: Investigation of the mechanism of growth of some metallic crystals from the melt. J. Phys. Chem. Solids Suppl.*1*, 757 (1967)
b) D.E. Ovsienko, G.A. Alfintsev: "Crystal Growth from the Melt. Experimental Investigation of Kinetics and Morphology", *Crystals*, Vol.2 (Springer, Berlin, Heidelberg, New York 1980) p.119

3.34 Z. Gyulai: Z. Phys.*138*, 317 (1954)

3.35 V.A. Kuznetsov: In *Gidrotermal'nyi sintez kristallov*, ed. by A.N. Lobachev (Nauka, Moscow 1968) p.77 [English transl.: "Kinetics of the Crystallization of Corundum, Quartz and Zincite", in *Hydrothermal Synthesis of Crystals* (Consultants Bureau, New York 1971) p.52]

3.36 O.K. Mel'nikov, B.N. Litvin, N.S. Triodina: In *Issledovaniye protsessov kristallizatsii v gidrotermal'nykh usloviyakh*, ed. by A.N. Lobachev (Nauka, Moscow 1970) p.134 [English transl.: "Crystallization of Sodalite on a Seed", in *Crystallization Processes under Hydrothermal Conditions* (Consultants Bureau, New York 1973) p.151]

3.37 L.A. Gordienko, V.F. Miuskov, V.E. Khadzhi, L.I. Tsinober: Kristallografiya *14*, 539 (1969) [English transl.: Sov. Phys. Crystallogr.*14*, 454 (1969)]

3.38 R.L. Barns, P.E. Freeland, E.D. Kolb, R.A. Laudise, J.R. Patel: J. Cryst. Growth *43*, 676 (1978)

3.39 Yu.M. Fishman: Kristallografiya *17*, 607 (1972) [English transl.: Sov. Phys. Crystallogr.*17*, 524 (1972)]

3.40 a) Yu.G. Kuznetsov, I.L. Smol'sky, A.A. Chernov, V.N. Rozhansky: Dokl. Akad. Nauk SSSR, *260*, 864 (1981)

b) A.A. Chernov, V.F. Parvov, M.O. Kliya, D.V. Kostomarov, Yu.G. Kuznetsov: Kristallografiya *26*, 1125 (1981) [English transl.: Sov. Phys. Crystallogr.*26*, 640 (1981)
3.41 a) M. Shimbo, J. Nishisawa, T. Terasaki: J. Cryst. Growth *23*, 267 (1974)
b) K.W. Keller: "Surface Microstructure and Processes of Crystal Growth Observed by Electron Microscope", in *Crystal Growth and Characterization*, ed. by R. Ueda, J.B. Mullin (North-Holland, Amsterdam 1975) p.361; 6th Int. Conf. Crystal Growth, Moscow, 1980, p.62
c) A.A. Chernov: J. Cryst. Growth *42*, 55 (1977)
d) A.A. Chernov, G.F. Kopylova: Kristallografiya *22*, 1247 (1977) [English transl.: Sov. Phys. Crystallogr.*22*, 709 (1977)]
e) J. Nishizawa, H. Tadano, Y. Oyama, M. Shimbo: J. Cryst. Growth *55*, 402 (1981)
f) E. Bauser, H. Strunk: J. Cryst. Growth *51*, 362 (1981)
g) E. Bauser: In *III-V Optoelectronics Epitaxy and Device Related Processes*, ed. by V.C. Keramidas, S. Mahajan (Electrochem. Society, Pennington, NJ 1983)
3.42 T.A. Carlson: *Photoelectron and Auger Spectroscopy* (Plenum Press, New York 1975)
3.43 G. Margaritondo, J.E. Rowe: J. Vac. Sci. Technol.*17*, 561 (1980)
3.44 M.A. van Hove, S.Y. Tong: *Surface Crystallography by LEED*, Springer Ser. Chem. Phys., Vol.2 (Springer, Berlin, Heidelberg, New York 1979)
3.45 a) A. Benninghoven, C.A. Evans, Jr., R.A. Powell, R. Shimizu, H.A. Storms (eds.): *Secondary Ion Mass Spectrometry SIMS II*, Springer Ser. Chem. Phys., Vol.9 (Springer, Berlin, Heidelberg, New York 1980)
b) A. Benninghoven, J. Giber, J. Lászlő, M. Riedel, H.W. Werner (eds.): *Secondary Ion Mass Spectrometry SIMS III*, Springer Ser. Chem. Phys., Vol.19 (Springer, Berlin, Heidelberg, New York 1982)
c) R. Wagner: *Field Ion Microscopy in Materials Science*, Crystals: Growth, Properties and Applications, Vol.6 (Springer, Berlin, Heidelberg, New York 1982)
3.46 E.W. Müller, T.T. Tsong: *Field Ion Microscopy. Field Ionization and Field Evaporation* (Pergamon, Oxford 1973)
3.47 G. Thomas: *Transmission Electron Microscopy of Metals* (Wiley, New York 1962)
3.48 P.B. Hirsch, A. Howie, R.B. Nicholson, D.W. Pashley, M.J. Whelan: *Electron Microscopy of Thin Crystals* (Butterworths, London 1965)
3.49 G. Schimmel: *Elektronenmikroskopische Methodik* (Springer, Berlin, Heidelberg, New York 1969)
3.50 L.M. Utevsky: *Difraktsionnaya elektronnaya mikroskopiya v metallovedenii* (Electron Microscopy in Physical Metallurgy) (Metallurgiya, Moscow 1973)
3.51 L. Reimer, *Transmissions Electron Microscopy*, Springer Ser. Opt. Sci., Vol.36 (Springer, Berlin, Heidelberg, New York 1983)
3.52 a) M. Born, E. Wolf: *Principles of Optics* (Pergamon, Oxford 1964)
b) G.S. Landsberg: *Optika* (Optics) (Nauka, Moscow 1976)
3.53 V.B. Tatarsky: *Kristallooptika i immersionnyi metod* (Crystal Optics and Immersion Method) (Nedra, Moscow 1965)
3.54 F. Rinne, M. Berek: *Anleitung zu optischen Untersuchungen mit dem Polarizationsmikroskop* (Dr. Max Jänecke Verlagsbuchhandlung, Leipzig 1934)
3.55 G.R. Bartini, E.D. Dukova, I.P. Korshunov, A.A. Chernov: Kristallografiya *8*, 758 (1963) [English transl.: Sov. Phys. Crystallogr.*8*, 605 (1964)]
3.56 H.E.L. Madsen: J. Cryst. Growth *32*, 84 (1976)
3.57 a) M.E. Glicksman, R.J. Schaefer, J.A. Blodgett: J. Cryst. Growth *13/14*, 68 (1972)
b) F. Bedarida, L. Zefiro: Acta Crystallogr. A*31*, 212 (1975)
3.58 a) W. Schumann, M. Dubas: *Holographic Interferometry*, Springer Ser. Opt. Sci., Vol.16 (Springer, Berlin, Heidelberg, New York 1979)

b) Yu.I. Ostrovsky, M.M. Butusov, G.V. Ostrovskaya: *Interferometry by Holography*, Springer Ser. Opt. Sci., Vol.20 (Springer, Berlin, Heidelberg, New York 1980)

3.59 a) I.N. Guseva, V.M. Ginzburg, V.A. Kramarenko: In [Ref.1.30a, p.216] [English transl.: "Concentration Inhomogeneity in a Solution During Crystal Growth and Dissolution", in Ref.1.30a, p.215]
b) L.N. Rashkovich, A.N. Israilenko, V.T. Leschenko, Z.S. Pashina: In *6th Int. Conf. Cryst. Growth, Extended Abstracts*, Vol.4 (Moscow 1980) p.49
c) I.A. Berstein, V.P. Ershov, V.I. Katsman, V.A. Rogachev: ibid., p.11
d) L.N. Rashkovich, V.T. Leschenko, A.T. Amandosov, V.A. Koptsik: Kristallografiya *28*, 768 (1983)

3.60 M. Françon: *Le microscope à contraste de phase et le microscope interférentiel* (Edition du Centre National de la Recherche Scientifique, Paris 1954)

3.61 S. Tolansky: *High Resolution Spectroscopy* (Methuen, New York 1947)

3.62 a) M.M. Gorshkov: *Ellipsometriya* (Ellipsometry) (Sovetskoye Radio, Moscow 1974)
b) R.M.A. Azzam, N.M. Bashara: *Ellipsometry and Polarized Light* (North-Holland, Amsterdam 1977)
c) P.S. Theocaris, E.E. Gdoutos, *Matrix Theory of Photoelasticity*, Springer Ser. Opt. Sci., Vol.11 (Springer, Berlin, Heidelberg, New York 1979)

3.63 D.E. Anderson: Interface Ellipsometry: Surf. Sci.*101*, 84 (1980)

3.64 a) J.B. Theeten, L. Hollan, R. Cadoret: "Growth Mechanisms in CVD of GaAs" in [Ref.1.7a, p.196]
b) M. Gauch, G. Quentel: Surf. Sci.*108*, 617 (1981)

3.65 G.A. Basset: Philos. Mag.*3*, 1042 (1958)

3.66 R.A. Rabadanov, S.A. Semiletov, Z.A. Magomedov: Fiz. Tverd. Tela *12*, 1430 (1970)

3.67 A.A. Chernov: J. Cryst. Growth *42*, 55 (1977)

3.68 G.G. Lemmlein, E.D. Dukova: Kristallografiya *1*, 352 (1956)

3.69 M.I. Kozlovsky: Kristallografiya *3*, 209 (1958) [English transl.: Sov. Phys. Crystallogr.*3*, 206 (1958)]

3.70 G.G. Lemmlein, E.D. Dukova, A.A. Chernov: Kristallografiya *5*, 662 (1960) [English transl.: Sov. Phys. Crystallogr.*5*, 634 (1961)]

3.71 E.D. Dukova: Kristallografiya *16*, 200 (1971) [English transl.: Sov. Phys. Crystallogr.*16*, 160 (1971)]

3.72 T.G. Petrov, E.B. Treivus, A.P. Kasatkin: *Vyrashchivaniye kristallov iz rastvorov*, 2nd ed. (Growing of Crystals from Solutions) (Nedra, Leningrad 1983)

3.73 F.C. Frank: "On the Kinematic Theory of Crystal Growth from Solution", in [Ref.1.29a, pp.411, 418]

3.74 N. Cabrera, D.A. Vermilyea: "The Growth of Crystals from Solution", in [Ref.1.29a, pp.393, 408]

3.75 D. Nenov, A. Pavlovska, E. Dukova, V. Stoyanova: Surface Melting and Appearence of Non-Singular Crystalline Faces. Commun. Dep. Chem. Bulg. Acad. Sci.*13*, 526 (1980)

3.76 E. Dukova, D. Nenov, Kristallografiya *23*, 816 (1978) [English transl.: Sov. Phys. Crystallogr.*23*, 457 (1978)]

3.77 A. Pavlovska, D. Nenov: J. Cryst. Growth *12*, 9 (1972)

3.78 A. Pavlovska: J. Cryst. Growth *46*, 551 (1979)

3.79 M. McLean, H. Mykura: Surf. Sci.*5*, 466 (1966)

3.80 V.I. Kvilividze, V.F. Kiselev, A.B. Kurzaev, L.A. Ushakova: Surf. Sci. *44*, 60 (1974)

3.81 S. Grande, St. Limmer, A. Lösche: Phys. Lett. A*54*, 69 (1975)

3.82 D. Beaglehole, D. Nason: Surf. Sci.*96*, 357 (1980)

3.83 T. Kuroda, R. Lackmann: J. Cryst. Growth *56*, 189 (1982)

3.84 R. Goodman, S. Somorjai: J. Chem. Phys.*52*, 6325 (1970)

Chapter 4

4.1 E.B. Treivus: *Kinetika rosta i rastvoreniya kristallov* (Kinetics of Crystal Growth and Dissolution) (Izd-vo LGU, Leningrad 1979)

4.2 A.A. Chernov, V.V. Sipyagin: "Peculiarities in Crystal Growth from Aqueous Solutions Connected with Their Structures", *Current Topics Mat. Sci.*, Vol.5 (North-Holland, Amsterdam 1980) p.279

4.3 a) S. Troost: J. Cryst. Growth *3/4*, 340 (1968)
b) S. Troost: J. Cryst. Growth *13/14*, 449 (1972)

4.4 a) L.M. Belyaev, M.G. Vassilieva, L.V. Soboleva: Kristallografiya *25*, 871 (1980) [English transl.: Sov. Phys. Crystallogr.*25*, 499 (1980)]
b) M.G. Vassilieva, L.V. Soboleva: Zh. Neorg. Khim.*23*, 2795 (1979)
c) L.M. Belyaev, M.G. Vassilieva, L.V. Soboleva: Kristallografiya *26*, 373 (1981) [English transl.: Sov. Phys. Crystallogr.*26*, 212 (1981)]
d) A.F. Banishev, Yu.K. Voron'ko, A.B. Kudryavtsev, V.V. Osiko, A.A. Sobol': Kristallografiya *27*, 618 (1982) [English transl.: Sov. Phys. Crystallogr. *27*, 374 (1982)]
e) W. van Erk: J. Cryst. Growth *46*, 539 (1979)

4.5 T. Yamamoto: Bull. Inst. Phys. Chem. Res.*11*, 1083 (1933)

4.6 J.W. Mullin: *Crystallization* (Butterworths, London 1972)

4.7 *Adsorption et croissance cristalline*, Colloq. Int. CNRS, Vol.152 (Edition du Centre National de la Recherche Scientifique, Paris 1965)

4.8 R. Boistelle, M. Mathieu, B. Simon: Surf. Sci.*42*, 373 (1974)

4.9 A.A. Chernov, V.A. Kuznetsov: Kristallografiya *14*, 879 (1969) [English transl.: Sov. Phys. Crystallogr.*14*, 753 (1970)]

4.10 J. Dugua, B. Simon: J. Cryst. Growth *44*, 280 (1978)

4.11 B. Simon, R. Boistelle: J. Cryst. Growth *52*, 779 (1981)

4.12 J. Gilman, W. Johnston, G. Sears: J. Appl. Phys.*29*, 747 (1958)

4.13 P. Bennema, G. Brouwer, G.M. van Rosmalen: "Growth Kinetics of Gypsum", in Int. Conf. on Crystal Growth, MIT, 17-22 July 1977, p.263

4.14 G.N. Nancollas: Private communication

4.15 G. Bliznakov: Fortschr. Mineral.*36*, 149 (1958)

4.16 V.N. Rozhansky, E.V. Parvova, V.M. Stepanova, A.A. Predvoditelev: Kristallografiya *6*, 704 (1961) [English transl.: Sov. Phys. Crystallogr. *6*, 564 (1962)]

4.17 L.M. Kolganova, A.M. Ovrutsky, E.V. Finagina: In [Ref.1.29a, p.289] [English transl.: "Factors Governing Crystal Growth and Dissolution Shapes in Molten Metals", in Ref.1.29a, p.295]

4.18 V.V. Podolinsky, V.G. Drykin: "Changes in Surface Structure of Crystals in Solutions", in 6th Int. Conf. on Crystal Growth, Moscow, USSR, 10-16 September 1980. Extended Abstrcts, Vol.2, p.77

4.19 a) G. Bliznakov: "Sur le méchanisme de l'action des additifs adsorbants dans la croissance cristalline", in *Adsorption et croissance cristalline*, Colloq. Int. CNRS, No.152 (Edition du Centre National de la Recherche Scientifique, Paris 1965) pp.291, 300
b) R. Boistelle, M. Mathieu, B. Simon: Surf. Sci.*42*, 373 (1974)
c) R.J. Davey, J.W. Mullin: J. Cryst. Growth *23*, 89 (1974)
d) A.A. Chernov, V.F. Parvov, M.O. Kliya, D.V. Kostomarov, Yu.G. Kuznetsov: Kristallografiya *26*, 1125 (1981) [English transl.: Sov. Phys. Crystallogr. *26*, 640 (1981)]
e) Yu.G. Kuznetsov, I.L. Smol'sky, A.A. Chernov, V.N. Rozhansky: Dokl. Akad. Nauk SSSR *260*, 864 (1981)

4.20 A. Glasner, M. Zidon: J. Cryst. Growth *21*, 294 (1974)

4.21 G.G. Lemmlein: Dokl. Akad. Nauk SSSR *84*, 1167 (1952)

4.22 R. Kern: In *Rost kristallov, T.8*, ed. by N.N. Sheftal' (Nauka, Moscow 1968) p.5 [English transl.: "Crystal Growth and Adsorption", in *Growth of Crystals, Vol.8* (Consultants Bureau, New York, London 1969) p.3]

4.23 C.W. Bunn, H. Emmett: Discuss. Faraday Soc.*5*, 119 (1949)
4.24 W. Kleber, S. Schiemann: Krist. Tech.*1*, 553 (1966)
4.25 H.E. Buckley: *Crystal Growth* (Wiley, New York 1951)
4.26 L.I. Tsinober, M.I. Samoilovich: "Raspredeleniye strukturnykh defektov i anomal'naya opticheskaya simmetriya v kristallakh kvartsa" (Distribution of Structural Defects and Anomalous Optical Symmetry in Quartz Crystals), in *Problemy sovremennoi kristallografii* (Problems of Modern Crystallography) (Nauka, Moscow 1975) p.207
4.27 B.K. Vainshtein, V.M. Fridkin, V.L. Indenbom: *Sovremennaya Kristallografiya. T.2.* Struktura kristallov, ed. by B.K. Vainshtein (Nauka, Moscow 1979) [English transl.: *Modern Crystallography II*, Springer Ser. Solid-State Sci., Vol.21 (Springer, Berlin, Heidelberg, New York 1982)]
4.28 T.L. Allen: J. Chem. Phys.*27*, 810 (1957)
4.29 K. Weiser: J. Phys. Chem. Solids *7*, 118 (1958)
4.30 K.M. Rozin, O.L. Kreinin, M.P. Shaskol'skaya: Izv. Akad. Nauk SSSR, Neorg, Mater.*7*, 1105 (1971)
4.31 C.D. Thurmond: "Control of Composition in Semiconductors by Freezing Methods", in *Semiconductors*, ed. by N.B. Hannay (Reinhold, New York 1959) p.145
4.32 C.D. Thurmond: J. Phys. Chem.*57*, 827 (1953)
4.33 C.D. Thurmond, J.D. Struthers: J. Phys. Chem.*57*, 831 (1953)
4.34 F.A. Trubmore: Bell Syst. Tech. J.*39*, 205 (1960)
4.35 A.P. Ratner: "K teorii raspredeleniya elektrolitov mezhdu tverdoi kristallicheskoi i zhidkoi fazami" (On the Theory of Electrolyte Distribution Between Solid and Liquid Phases), in *Trudy Gosudarstvennogo radievogo instituta*, Vol.2 (Collected Works of the State Radium Institute, Vol.2) (Goskhimtekhizdat, Leningrad 1933)
4.36 a) G.I. Gorshtein: Zh. Neorg. Khim.*3*, 51 (1958)
b) V.S. Urusov: Energeticheskaya formulirovka zadachi ravnovesnoi sokristallizatsii iz vodnogo rastvora (Energetic formulation of equilibrium cocrystallization from aqueous solutions). Geokhimiya *5*, 627 (1980)
c) C. Balarev: "Inclusion of Isomorphous Admixtures in Crystal Hydrate Salts", in *Industrial Crystallization 1981*, ed. by S.J. Jancič, E.J. de Jong (North-Holland, Amsterdam 1982) p. 117
d) V.S. Urusov: *Teoriya isomorfnoi smessimosti* (The Theory of Isomorphous Mixing) (Nauka, Moscow 1977)
4.37 J.O. McKaldin: J. Appl. Phys.*34*, 1748 (1963)
4.38 O.B. Pelevin, V.V. Voronkov, B.G. Girich, M.G. Mil'vidsky: Izv. Akad. Nauk SSSR, Neorg. Mater.*8*, 57 (1972)
4.39 J.B. Mullin: J. Cryst. Growth *42*, 77 (1977)
4.40 R.A. Laudise, E.D. Kolb, N.C. Lias, E.E. Grudensky: In [Ref.1.29a, p.352] [English transl.: "The Distribution Constant in Hydrothermal Quartz Growth", in Ref.1.29a, p.355]
4.41 A.A. Chernov: Rost tsepei sopolimerov i smeshannykh kristallov – statistika prob i oshibok. Usp. Fiz. Nauk *100*, 277 (1970) [English transl.: Growth of copolymer chains and mixed crystals – trial and error statistics. Sov. Phys. Usp.*13*, 101 (1970)]
4.42 a) V.V. Voronkov: In [Ref.1.30a, p.357] [English transl.: "Dope Uptake Factor in Relation to Growth Rate and Surface Inclination", in Ref. 1.30a, p.364]
b) W. Haubenreisser, H. Pfeiffer: Microscopic Theory of the Growth of Two-Component Crystals. *Crystals*, Vol.9 (Springer, Berlin, Heidelberg, New York 1983) p.43
c) T.A. Cherepanova: J. Cryst. Growth *52*, 319 (1981)
4.43 I.V. Melikhov: In [Ref.1.30a, p.302] [English transl.: "Trapping of Impurities During Growth from Solution", in Ref.1.30a, p.309]
4.44 I.V. Melikhov, M.S. Merkulova: *Sokristallizatsiya* (Cocrystallization) (Khimiya, Moscow 1975)

4.45 J.B. Mullin: "Segregation in InSb", in *Compound Semiconductors*, Vol.1 Preparation of III-V Compounds, ed. by R.K. Willard, H.L. Goering (Reinhold, New York 1962) p.365
4.46 J.C. Brice: *The Growth of Crystals from Liquids* (North-Holland, Amsterdam 1973)
4.47 I.S. Miroshnichenko: In: *Rost i nesovershenstva metallicheskikh kristallov* ed. by D.E. Ovsienko (Naukova Dumka, Kiev 1966) p.320 [English transl.: "Effect of the Cooling Rate During Crystallization on the Composition of Axial Regions of Dendritic Branches", in *Growth and Imperfections of Metallic Crystals* (Consultants Bureau, New York 1968)]; "Vlianiye skorosti okhlazhdeniya na protsessy kristallizatsii metallicheskikh splavov" (Effect of the Cooling Rate on Crystallization Processes of Metallic Alloys), in *Rost i defekty metallicheskikh kristallov*, (Growth and Defects of Metallic Crystals) ed. by D.E. Ovsienko (Naukova Dumka, Kiev 1972)
4.48 D.E. Temkin: Kristallografiya *15*, 884 (1970) [English transl.: Sov. Phys. Crystallogr.*15*, 773 (1971)]
4.49 D.E. Polk, B.C. Giessen: "Overview of Principles and Applications", in *Metallic Glasses. ASM Materials Science Seminar Series* (American Society for Metals, Metals Park, Ohio 1977) Chap.1
4.50 J.J. Gilman: "Metallic Glasses — A New Technology", in [Ref.1.6a, p.727]
4.51 H.-J. Güntherodt, H. Beck (eds.): *Glassy Metals I*, Topics Appl. Phys. Vol.46 (Springer, Berlin, Heidelberg, New York 1981)
4.52 R. Hasegawa (ed.): *The Magnetic, Chemical and Structural Properties of Glassy Metallic Alloys* (CRC, Boca Raton, Florida 1981)
4.53 I.B. Khaibullin, E.I. Styrkov, M.M. Zaripov, M.F. Galyautdinov, G.G. Zakirov: Sov. Phys. Semicond.*11*, 190 (1977)
4.54 A.Kh. Antonenko, N.N. Gerasimenko, A.V. Dvurechensky, L.S. Smirnov, G.M. Tseitlin: Sov. Phys. Semicond.*10*, 81 (1976)
4.55 G.A. Kachurin, E.V. Nidaev, A.V. Khodyachikh, L.A. Kovaleva: Sov. Phys. Semicond.*10*, 1128 (1976)
4.56 A.V. Dvurechensky, G.A. Kachurin, T.N. Mustafin, L.S. Smirnov: "Laser Annealing of Ion-Implanted Semiconductors", in *Laser-Solid Interactions and Laser Processing — 1978*, ed. by S.D. Ferris, H.J. Leamy, J.M. Poate (American Institute of Physics, New York 1979) p.245
4.57 S.D. Ferris, H.J. Leamy, J.M. Poate (eds.): *Laser-Solid Interactions and Laser Processing — 1978* (American Institute of Physics, New York 1979); A.I.P. Conf. Proc. No.50
4.58 C.W. White, P.S. Peercy (eds.): *Laser and Electron Beam Processing of Materials* (Academic, New York 1980)
4.59 J.F. Gibbons, L.D. Hess, T.W. Sigmon (eds.): *Laser and Electron Beam Solid Interactions and Materials Processing* (North-Holland, New York 1981)
4.60 D.S. Gnanamathu, C.B. Shaw, Jr., W.E. Lawrence, M.R. Mitchell: "Laser Transformation Hardening", in [Ref.4.57, p.173]
4.61 J.A. van Vechten: "Evidence for and Nature of a Nonthermal Mechanism of Pulsed Laser Annealing of Si", in [Ref.4.59, p.53]
4.62 W.L. Brown: "Fundamental Mechanisms in Laser and Electron Beam Processing of Semiconductors", in [Ref.4.58, p.1]
4.63 W.L. Brown: "Transient Laser-Induced Processes in Semiconductors", in [Ref.4.59, p.20]
4.64 B.P. Fairand, A.H. Clauer: "Laser-Generated Stress Waves: Their Characteristics and Their Effects on Materials", in [Ref.4.57, p.27]
4.65 N. Bloembergen: "Fundamentals of Laser-Solid Interactions", in [Ref. 4.57, p.1]
4.66 P. Baeri, S.U. Campisano, G. Foti, E. Rimini: J. Appl. Phys.*50*, 788 (1979)

4.67 P.L. Liu, R. Yen, N. Bloembergen, R.T. Hodgson: "Picosecond Laser Pulse-Induced Melting and Resolidification Morphology on Silicon", in [Ref.4.59, p.156]
4.68 D.H. Auston, J.A. Golovchenko, A.L. Simons, R.E. Slusher, P.R. Smith, C.M. Surko: "Dynamics of Laser Annealing", in [Ref.4.57, p.11]
4.69 G.L. Olson, S.A. Kokorowski, J.A. Roth, L.D. Hess: "Direct Measurements of CW Laser-Induced Crystal Growth Dynamics by Time-Resolved Optical Reflectivity", in [Ref.4.59, p.125]
4.70 C.W. White, B.R. Appleton, B. Stritzker, D.M. Zehner, S.R. Wilson: "Kinetic Effects and Mechanisms Limiting Substitutional Solubility in the Formation of Supersaturated Alloys by Pulse Laser Annealing", in [Ref.4.59, p.59]
4.71 C.W. White, S.R. Wilson, B.R. Appleton, F.W. Young, Jr.: J. Appl. Phys. *51*, 738 (1980)
4.72 J.C. Baker, J.W. Cahn: Acta Metall.*17*, 575 (1969)
4.73 J.W. Cahn, S.R. Coriell, W. Boettinger: "Rapid Solidification", in [Ref.4.58, p.89]
4.74 K.A. Jackson, G.H. Gilmer, H.J. Leamy: "Solute Trapping", in [Ref.4.58, p.104]
4.75 S.E. Bradshaw, J. Goeerissen: J. Cryst. Growth *48*, 514 (1980)
4.76 B. Strocka, P. Willich: J. Cryst. Growth *56*, 606 (1982)
4.77 A.M.J.G. van Run: J. Cryst. Growth *54*, 195 (1981)
4.78 D.T.J. Hurle: "Melt Growth", in *Crystal Growth, An Introduction*, ed. by P. Hartman (North-Holland, Amsterdam 1973) p.210
4.79 R.L. Parker: "Results of Crystal Growth in Skylab (and ASTP)", in [Ref. 1.7a, p.851]
4.80 L.H. Brixner: J. Electrochem. Soc. *114*, 108 (1967)
4.81 A. Murgai, H.C. Gatosand, A.F. Witt: J. Electrochem. Soc. *123*, 224 (1976)
4.82 E.P. Martin, A.F. Witt, J.R. Carruthers: J. Electrochem. Soc. *126*, 284 (1979)
4.83 D.E. Holmes, H.C. Gatos: J. Electrochem. Soc. *128*, 429 (1981)

Chapter 5

5.1 a) B.Ya. Lyubov: *Teoriya kristallizatsii v bol'shikh obyomakh* (Theory of Crystallization in Large Volumes) (Nauka, Moscow 1975)
b) B.Ya. Lyubov: *Kineticheskaya teoriya fazovykh prevrashchenii* (Kinetic Theory of Phase Transformations) (Metallurgiya, Moscow 1969)
5.2 S.M. Pimputkar, S. Ostrach: Connective effects in crystals grown from the melt. J. Cryst. Growth *55*, 614 (1981)
5.3 R.L. Parker (ed.): *Crystal Growth 1977* (North-Holland, Amsterdam 1977) Sect. VIII, pp.377-410
Also in: J. Cryst. Growth *42*, 377-410 (1977)
5.4 E.I. Givargizov (ed.): *Crystal Growth 1980* (North-Holland, Amsterdam 1980) Sect. VIII, pp.423-492
Also in: J. Cryst. Growth *52*, 423-492 (1981)
5.5 D.T.J. Hurle: "Hydrodynamics in Crystal Growth", in [Ref.1.7a, p.549]
5.6 R.L. Parker: "Crystal Growth Mechanisms: Energetics, Kinetics and Transport", in Solid State Physics, Vol.25 (Academic, New York 1970) p.151
5.7 J.R. Carruthers: J. Cryst. Growth *32*, 13 (1976)
5.8 D.E. Temkin, V.B. Polyakov: Kristallografiya *21*, 661 (1976) [English transl.: Sov. Phys. Crystallogr.*21*, 374 (1976)]
5.9 V.I. Kuznetsov, G.G. Kharin: Dokl. Akad. Nauk SSSR *180*, 1354 (1968)
5.10 A. Seeger: Philos. Mag.*44*, 1 (1953)
5.11 A.A. Chernov: Kristallografiya *16*, 842 (1971) [English transl.: Sov. Phys. Crystallogr.*16*, 741 (1971)]

5.12 A.A. Chernov: J. Cryst. Growth *24/25*, 11 (1974)
5.13 S.P.F. Humphreys-Owen: Proc. R. Soc. London A*197*, 218 (1949)
5.14 a) I. Sunagawa (ed.): *Morphology of Crystals* (in preparation)
b) A.A. Chernov: Kristallografiya *7*, 895 (1962) [English transl.: Sov. Phys. Crystallogr.*7*, 728 (1963)]
5.15 D.A. Petrov, A.A. Bukhanova: "Izucheniye form pervichnoi kristallizatsii metallov" (Studying the Forms of Primary Crystallization of Metals), in *Trudy MATI* (Collected Works of the Moscow Institute of Aviation Technology), No.7 (Oborongiz, Moscow 1949) p.3
5.16 a) A.V. Shubnikov, N.N. Sheftal' (eds.): *Rost kristallov. T.1* (Izd-vo Akad. Nauk SSSR, Moscow 1957) [English transl.: *Growth of Crystals. Vol.1* (Consultants Bureau, New York 1958)]
b) G.P. Ivantsov: In [Ref.5.16a, p.98] [English transl.: "Thermal and Diffusion Processes in Crystal Growth", in Ref.5.16a, p.76]
c) V.V. Voronkov: Statistics of Surfaces, Steps and Two-Dimensional Nuclei: A Macroscopic Approach, *Crystals*, Vol.9 (Springer, Berlin, Heidelberg, New York 1983) p.74
d) V.V. Voronkov: Theory of Crystal Surface Formation in the Pulling Process. J. Cryst. Growth *52*, 311 (1980)
e) T. Surek, S.R. Coriell, B. Chalmers: The Growth of Shaped Crystals from the Melt. In [Ref.5.16g, p.21]
f) V.A. Tatarchenko, E.A. Brener, G.I. Babkin: Crystallization Stability During Capillary Shaping I, II. In [Ref. 5.16g, p.33]
g) G.W. Cullen, T. Surek, P.I. Antonov (eds.): Shaped Crystal Growth. J. Cryst. Growth *50* (1980)
h) P.I. Antonov: J. Cryst. Growth *23*, 318 (1974)
i) W. Bardsley, F.C. Frank, G.W. Green, D.T.J. Hurle: J. Cryst. Growth *23*, 341 (1974)
j) W. Bardsley, D.T.J. Hurle, G.C. Joyce: J. Cryst. Growth *40*, 21 (1977)
5.17 D.E. Temkin: Dokl. Akad. Nauk SSSR *132*, 1307 (1960)
5.18 a) J.S. Langer, H. Müller-Krumbhaar: J. Cryst. Growth *42*, 11 (1977)
b) J.S. Langer: Rev. Mod. Phys. *52*, 1 (1980)
c) H. Müller-Krumbhaar: "Theory of Crystal Growth", in *Cohesive Properties of Semiconductors Under Laser Irradiation*, Ser. E: Appl. Sciences, N 69, ed. by L.D. Laude (Nijhoff, The Hague 1983) p.197
5.19 M.E. Glicksman, R.J. Schaefer, J.D. Ayers: Metall. Trans. A*7*, 1747 (1976)
5.20 R.F. Sekerka: "Morphological Stability", in *Crystal Growth: An Introduction*, ed. by P. Hartman (North-Holland, Amsterdam 1973) p.403
5.21 R. Glardon, W. Kurz: "Optimizing the Properties of Cobalt – Rare-Earth Permanent Magnets. Brittle Matrix/Ductile Dendrite Composites, in *New Developments and Applications in Composites*, ed. by D. Kuhlmann-Wilsdorf, W.C. Harrigan, Jr. (Metallurgical Society of AIME, Warrendale, Pennsylvania 1979) p.85
5.22 W.W. Mullins, R.F. Sekerka: J. Appl. Phys.*34*, 323 (1963)
5.23 W.W. Mullins. R.F. Sekerka: J. Appl. Phys.*35*, 444 (1964)
5.24 D.E. Ovsienko, G.A. Alfintsev, V.V. Maslov: J. Cryst. Growth *26*, 233 (1974)
5.25 S. Goldsztaub, R. Itti, F. Mussard: J. Cryst. Growth *6*, 130 (1970)
5.26 A. Glasner, M. Zidon: J. Cryst. Growth *21*, 294 (1974)
5.27 A.A. Chernov, E.D Dukova: Kristallografiya *14*, 169 (1969) [English transl.: Sov. Phys. Crystallogr.*14*, 150 (1969)]
5.28 C. Nanev, D. Iwanov: J. Cryst. Growth *3/4*, 530 (1968)
5.29 D. Nenov, V. Stoyanova: J. Cryst. Growth *41*, 73 (1977)
5.30 A. Papapetrou: Z. Kristallogr.*92*, 89 (1935)
5.31 G.A. Alfintsev, D.E. Ovsienko: In *Rost i nesovershenstva metallicheskikh kristallov*, ed. by D.E. Ovsienko (Naukova Dumka, Kiev 1966) p.40 [English transl.: "Investigation of the Growth Mechanism of Some Metallic Crystals

Growing from the Melt", in *Growth and Imperfections of Metallic Crystals* (Consultants Bureau, New York 1968)]
5.32 P.G. Shewmon: Trans. Metall. Soc. AIME *233*, 736 (1965)
5.33 A.A. Chernov, D.E. Temkin: "Capture of Inclusions in Crystal Growth", in [Ref.1.7a, p.3]
5.34 R.F. Sekerka: J. Cryst. Growth *3/4*, 71 (1968)
5.35 T. Takahashi, A. Kamio, N.A. Trung: J. Cryst. Growth *24/25*, 477 (1974)
5.36 A.A. Chernov: In [Ref.1.30a, p.221] [English transl.: "Stability of a Planar Growth Front for Anisotropic Surface Kinetics", in Ref.1.30a, p.223]

Chapter 6

6.1 N.N. Sheftal': In [Ref.5.16a, p.5] [English transl.: "Real Crystal Formation", in Ref.5.16a, p.5]
6.2 A.A. Chernov, D.E. Temkin: "Capture of Inclusions in Crystal Growth", in [Ref.1.7a, p.3]
6.3 V.M. Kuznetsov, B.A. Lugovskoi, E.I. Sher: Prikl. Mekh. Tekhn. Fiz. No.*1*, 124 (1966)
6.4 A.A. Chernov, D.E. Temkin, A.M. Mel'nikova: Kristallografiya *21*, 652 (1976) [English transl.: Sov. Phys. Crystallogr.*21*, 369 (1976)]
6.5 D.R. Uhlmann, B. Chalmers, K.A. Jackson: J. Appl. Phys.*35*, 2986 (1964)
6.6 Yu.M. Polukarov, L.I. Lyamina, V.V. Grinina, N.I. Tarasova, V.P. Chernov: Elektrokhimiya *14*, 1635 (1978)
6.7 Yu.M. Polukarov, L.I. Lyamina, N.I. Tarasova: Elektrokhimiya *14*, 1468 (1978)
6.8 M.O. Kliya, I.G. Sokolova: Kristallografiya *3*, 219 (1958) [English transl.: Sov. Phys. Crystallogr.*3*, 217 (1958)]
6.9 V.Ya. Khaimov-Mal'kov: In [Ref.1.3a, p.26] [English transl.: "The Growth Conditions of Crystals in Contact with Large Obstacles", in Ref.1.3a, p.20]
6.10 A.A. Chernov, V.E. Khadzhi: J. Cryst. Growth *3/4*, 641 (1968)
6.11 A.J.R. de Kock, S.D. Ferris, L.C. Kimmerling, H.J. Leamy: J. Appl. Phys. *48*, 301 (1977)
6.12 a) H. Föll, U. Gösele, B.O. Kolbesen: J. Cryst. Growth *40*, 90 (1977)
b) W. Dietze, W. Keller, A. Mühlbauer: "Float-Zone Grown Silicon", in *Silicon*, ed. by J. Grabmaier, *Crystals*, Vol.5 (Springer, Berlin, Heidelberg, New York 1981)
6.13 A.A. Shternberg: Kristallografiya *7*, 114 (1962) [English transl.: Sov. Phys. Crystallogr.*7*, 92 (1962)]
6.14 Yu.M. Fishman: Kristallografiya *17*, 607 (1972) [English transl.: Sov. Phys. Crystallogr.*17*, 524 (1972)]
6.15 V.L. Indenbom: "Growth Dislocations in Nonplastic Crystals", in 5th All-Union Conf. on Crystal Growth, Tbilisi 1977. Abstracts, Vol.2, Crystal Growth and Structure, p.260
6.16 H. Klapper, Yu.M. Fishman, V. Lutzau: Phys. Status Solidi A*21*, 115 (1974)
6.17 E. Billig: Proc. R. Soc. London A*235*, 37 (1956)
6.18 V.L. Indenbom: Izv. Akad. Nauk SSSR, Ser. Fiz.*37*, 2258 (1973)
6.19 V.L. Indenbom: I.S. Zhitomirsky, T.S. Chebanova: Kristallografiya *18*, 39 (1973) [English transl.: Sov. Phys. Crystallogr.*18*, 24 (1973)]
6.20 M.G. Mil'vidsky, V.B. Osvensky: "Polucheniye sovershennykh monokristallov" (Preparation of Perfect Single Crystals) in *Problemy sovremennoi kristallografii* (Problems of Modern Crystallography), ed. by B.K. Vainshtein, A.A. Chernov (Nauka, Moscow 1975) p.79
6.21 M.G. Mil'vidsky, V.B. Osvensky, S.S. Shifrin: J. Cryst. Growth *52*, 396 (1981)

6.22 B.K. Vainshtein: *Sovremennaya kristallografiya. T.1.* Simmetriya kristallov. Metody strukturnoi kristallografii (Nauka, Moscow 1979) [English transl.: *Modern Crystallography I*, Springer Ser. Solid-State Sci., Vol.15 (Springer, Berlin, Heidelberg, New York 1981)]
6.23 L.A. Shuvalov, A.A. Urusovskaya, I.S. Zheludev, A.V. Zalesskii, B.N. Grechushnikov, I.G. Chistyakov, S.A. Semiletov: *Sovremennaya kristallografiya. T.4.* Fizicheskiye svoistva kristallov, ed. by B.K. Vainshtein (Nauka, Moscow 1981) [English transl.: *Modern Crystallography IV*, Springer Ser. Solid-State Sci., Vol.37 (Springer, Berlin, Heidelberg, New York 1984)]
6.24 a) W.C. Dash: "Growth of Silicon Crystals Free from Dislocations", in Ref.1.29a, p.361, 382
Also in: J. Appl. Phys.*30*, 459 (1959)
b) M. Duseaux, G. Jacob: Appl. Phys. Lett.*40*, 790 (1982)
c) V.L. Indenbom: Ein Beitrag zur Entstehung von Spannungen und Versetzungen beim Kristallwachstum. Krist. Tech.*14*, 493 (1979)
d) W. Geil, K. Schmugge: Krist. Tech. *14*, 343 (1979)
e) A.A. Chernov, S.N. Maximovsky, L.A. Vlasenko, E.N. Kholina, V.P. Martovitsky, V.L. Levtov: Dokl. Akad. Nauk SSSR *271*, 106 (1983)
f) V.V. Voronkov: J. Cryst. Growth *59*, 625 (1982)
6.25 J. Barthel: In [Ref.1.30a, p.315] [English transl.: "Trapping During Growth from a Melt", in Ref.1.30a, p.322]
6.26 A.F. Witt, H.C. Gatos: J. Electrochem. Soc.*113*, 808 (1966)
6.27 A. Murgai, H.C. Gatos, W.A. Westdorp: J. Electrochem. Soc.*126*, 2240 (1979)
6.28 M.O. Kliya: Kristallografiya *13*, 667 (1968) [English transl.: Sov. Phys. Crystallogr.*13*, 565 (1969)]

Chapter 7

7.1 J.W. Mullin: *Crystallization*, 2nd ed. (Butterworths, London 1972)
7.2 a) L.N. Matusevich: *Kristallizatsiya iz rastvorov v khimicheskoi promyshlennosti* (Crystallization from Solutions in the Chemical Industry) (Khimiya, Moscow 1968)
b) E.V. Khamsky: *Kristallizatsiya iz rastvorov* (Crystallization from Solutions) (Nauka, Leningrad 1967)
c) E.V. Khamsky: *Peresyshchennyie rastvory* (Supersaturated Solutions) (Nauka, Leningrad 1975)
7.3 a) J. Nývlt: *Industrial Crystallization from Solutions* (Butterworths, London 1971)
b) A.D. Randolf, M.A. Larson: *Theory of Particulate Processes. Analysis and Techniques of Continuous Crystallization* (Academic, New York 1971)
c) A.W. Bamforth: *Industrial Crystallization* (Leonard Hill, London 1965)
7.4 S.J. Jancić: Industrial Crystallization. Part I: Fundamentals of Crystallization from Solution; Part II: Crystallizer Design (submitted for publication)
7.5 a) E.J. de Jong, S.J. Jancić (eds.): *Industrial Crystallization '78* (North-Holland, Amsterdam 1979)
b) S.J. Jancić, E.J. de Jong (eds.): *Industrial Crystallization '81* (North-Holland, Amsterdam 1982)
7.6 a) A.N. Kolmogorov: Izv. Akad. Nauk SSSR, Ser. Matem. No.*3*, 355 (1937)
b) W. Johnson, R. Mehl: Trans. Am. Inst. Min. Metall. Eng.*135*, 416 (1939)
c) M. Avrami: J. Chem. Phys.*7*, 1103 (1939); *8*, 212 (1940); *9*, 117 (1941)

7.7 a) V.Z. Belen'ky: *Geometriko-veroyatnostnyie modeli kristallizatsii* (Geometrical Probability Models of Crystallization) (Nauka, Moscow 1980)
b) L.I. Trusov, V.A. Kholmyansky: *Ostrovkovye metallicheskiye plyonki* (Island Metallic Films) (Metallurgiya, Moscow 1973)
7.8 J.P. Hirth, J. Lothe: *Theory of Dislocations* (McGraw-Hill, New York 1968)
7.9 G.G. Lemmlein: Dokl. Akad. Nauk SSSR *48*, 177 (1945)
7.10 A.N. Kolmogorov: Dokl. Akad. Nauk SSSR *65*, 681 (1949)
7.11 M.C. Flemings: *Solidification Processing* (McGraw-Hill, New York 1974)
7.12 G.A. Hughmark: Chem. Eng. Sci.*24*, 291 (1969)
7.13 A.W. Nienow, P.D.B. Bujac, J.W. Mullin: J. Cryst. Growth *13/14*, 488 (1972)
7.14 A.W. Nienow: Chem. Eng. Sci.*23*, 1459 (1968)
7.15 O.T. Todes: "Kinetika koagulyatsii i ukrupneniya chastits v zolakh" (Kinetics of Coagulation and Coarsening of Particles in Sols), in *Problemy kinetiki i kataliza, T.7, Statisticheskiye yavleniya v geterogennykh sistemakh* (Problems of Kinetics and Catalysis, Vol.7, Statistical Phenomena in Heterogeneous Systems) (Izd-vo Akad. Nauk SSSR, Moscow 1949) p.137
7.16 I.M. Lifshits, V.V. Slyozov: Zh. Eksp. Teor. Fiz.*35*, 479 (1958)
7.17 a) R.I. Garber, V.S. Kogan, L.M. Polyakov: Zh. Eksp. Teor. Fiz.*35*, 1364 (1958)
b) M.A. Belyshev, A.A. Chernov: In *Industrial Crystallization 1981*, ed. by S.J. Jancic, E.J. De Jong (North-Holland, Amsterdam 1982) p.315
7.18 N.V. Gordeyeva, A.V. Shubnikov: Kristallografiya *12*, 186 (1967) [English transl.: Sov. Phys. Crystallogr.*12*, 154 (1967)]
7.19 I.G. Bazhal: Kristallografiya *14*, 1106 (1969) [English transl.: Sov. Phys. Crystallogr.*14*, 127 (1970)]
7.20 I.G. Bazhal: "Issledovaniye mekhanizma rekristallizatsii v dispersnykh sistemakh" (An Investigation of the Recrystallization Mechanism in Disperse Systems), Ph.D. Thesis, Ukrainian SSR Institute of Colloid Chemistry and the Chemistry of Water (Kiev 1972)
7.21 E.G. Denk, G.D. Botsaris: J. Cryst. Growth *13/14*, 493 (1972)
7.22 H. Garabedian, R.F. Strickland-Constable: J. Cryst. Growth *13/14*, 506 (1972)
7.23 J. Estrin, M.L. Wang, G.R. Youngquist: AIChE J.*21*, 392 (1975)

Chapter 8

8.1 G.R. Booker, B.A. Joyce: Philos. Mag.*14*, 301 (1966)
8.2 Yu. Khariton, A.I. Shal'nikov: *Mekhanizm kondensatsii i obrazovaniye kolloidov* (Condensation Mechanism and the Formation of Colloids) (Gostekhteorizdat, Leningrad 1934)
8.3 M.H. Francombe, J.E. Johnson: "The Preparation and Properties of Semiconductor Films", in *Physics of Thin Films*, Vol.5, ed. by G. Hass, R.E. Thun (Academic, New York 1969) p.143
8.4 a) H.M. Manasevit: J. Cryst. Growth *22*, 125 (1974)
b) G.W. Cullen: In *Heteroepitaxial Semiconductors for Electronic Devices*, ed. by G.W. Cullen, C.C. Wang (Springer, Berlin, Heidelberg, New York 1978)
c) J. Cryst. Growth (special issue) *58* (1982)
8.5 a) E.I. Givargizov, N.N. Sheftal', V.I. Klykov: "Oriented Crystallization on Amorphous Substrates", *Current Topics Mat. Sci.*, Vol.10 (North-Holland, Amsterdam 1982), pp.1-53
b) N.N. Sheftal': "Trends in Real Crystal Formation and Some Principles for Single-Crystal Growth", in *Growth of Crystals*, Vol.10, ed. by N.N. Sheftal' (Consultants Bureau, New York 1976) pp.185-210
c) V.I. Klykov, N.N. Sheftal', E. Hartmann: Acta Phys. Acad. Sci. Hung. *47*, 167 (1979)

d) M.W. Geis, D.C. Flanders, H.I. Smith: Appl. Phys. Lett.*35*, 71 (1979)
e) M.W. Geis, D.A. Antoniadis, D.J. Silversmith, R.W. Mountain, H.I. Smith: Appl. Phys. Lett.*37*, 454 (1980)
f) C. Weissmantel: J. Vac. Sci. Technol.*18*, 179 (1981)
g) R. Anton, H. Poppa, D.C. Flanders: J. Cryst. Growth *56*, 433 (1982)
h) L.S. Darken, D.H. Lowndes: 161st Electrochem. Soc. Meeting, Montreal, Canada, May 1982, Extended Abstracts
i) J.C.C. Fan, M.W. Geis, B.-Y. Tsaur: Appl. Phys. Lett.*38*, 365 (1981)
j) B.-Y. Tsaur, J.C.C. Fan, M.W. Geis, D.J. Silversmith, R.W. Mountain: Appl. Phys. Lett.*39*, 561 (1981)
k) R.W. McClelland, C.O. Bozler, J.C.C. Fan: Appl. Phys. Lett.*37*, 560 (1980)
l) P. Vohl, C.O. Bozler, R.W. McClelland, A. Chu, A.J. Strauss: J. Cryst. Growth *56*, 410 (1982)

8.6 a) L. Holland: *Vacuum Deposition of Thin Films* (Chapman & Hall, London 1956)
b) P. Archibald, E. Parent: Solid State Technol.*19*, N7, 32 (1976)

8.7 B.J. Miller, J.H. McFee: J. Electrochem. Soc.*125*, 1310 (1978)

8.8 K.-G. Günther: "Vaporization and Reaction of the Elements", in *Compound Semiconductors*, Vol.1, Preparation of III-V Compounds, ed. by R.K. Willardson, H.L. Goering (Reinhold, New York 1962) p.313

8.9 H. Freller, K.G. Günther: Thin Solid Films *88*, 291 (1982)

8.10 K. Morimoto, H. Watanabe, S. Itoh: J. Cryst. Growth *45*, 334 (1978)

8.11 R.D. Gretz, C.M. Jackson, J.P. Hirth: Surf. Sci.*6*, 171 (1967)

8.12 a) A.Y. Cho, J.R. Arthur: "Molecular Beam Epitaxy", in *Progress in Solid State Chemistry*, Vol.10, ed. by G. Somorjai, J. McCaldin (Pergamon, New York, 1975)
b) A.C. Gossard, P.M. Petroff, W. Weigman, R. Dingle, A. Savage: Appl. Phys. Lett.*29*, 323 (1976)
c) *Progr. Cryst. Growth and Characterization* (special issue), *2*, p.1 (1979)
d) R.Z. Bachrach: "MBE – Molecular Beam Epitaxial Evaporative Growth", in *Crystal Growth*, ed. by B. Pamplin (Pergamon, London 1980)
e) R.F.C. Farrow: J. Vac. Sci. Technol.*19*, 150 (1981)

8.13 E. Kasper: Appl. Phys. A*28*, 1 (1982)

8.14 P.E. Luscher: Thin Solid Films *83*, 125 (1981)

8.15 a) T. Yao, S. Maekawa: J. Cryst. Growth *53*, 423 (1981)
b) F. Kitagawa, T. Mishima, K. Takahashi: J. Electrochem. Soc.*127*, 937 (1980)
c) J.P. Faurie, A. Millon: J. Cryst. Growth *54*, 577, 582 (1981)

8.16 A.A. Tikhonova: Kristallografiya *20*, 615 (1975) [English transl.: Sov. Phys. Crystallogr.*20*, 375 (1975)]

8.17 L.I. Maissel: "The Deposition of Thin Films by Cathode Sputtering", in *Physics of Thin Films*, Vol.3, ed. by G. Hass, R.E. Thun (Academic, New York 1966) p.61

8.18 J.E. Greene, S.A. Barnett, K.C. Cadien, M.A. Ray: J. Cryst. Growth *56*, 389 (1982)

8.19 a) V. Hoffman: Solid State Technol.*19*, N12, 57 (1976)
b) L. Holland: Thin Solid Films *86*, 227 (1981)
c) A.R. Nyaiesh: Thin Solid Films *86*, 267 (1981)
d) K. Urbanek: Solid State Technol.*20*, N7, 87 (1977)
e) R. Adachi, K. Takashita: J. Vac. Sci. Technol.*20*, 98 (1982)

8.20 a) J.E. Varga, W.A. Bailey: Solid State Technol.*20*, N12, 67 (1973)
b) A.E.T. Kuiper, G.E. Thomas, W.J. Schouten: J. Cryst. Growth *45*, 332 (1978)
c) T. Spalvins: J. Vac. Sci. Technol.*17*, 315 (1980)
d) A. Matthews, D.G. Teer: Thin Solid Films *80*, 41 (1981)

8.21 a) J.N. Avaritsiotis, R.P. Howson: Thin Solid Films *77*, 351 (1981)

b) H.S. Randhava, M.D. Matthews, R.F. Bunshah: Thin Solid Films *83*, 267 (1981)

8.22 a) E. Kaldis: "Principles of the Vapour Growth of Single Crystals", in *Crystal Growth. Theory and Techniques*, Vol.1, ed. by C.H.L. Goodman (Plenum, New York 1974) p.49
b) M.M. Factor, I. Garrett: *Growth of Crystals from the Vapour* (Chapham & Hall, London 1974)
c) E. Schönherr: "The Growth of Large Crystals from the Vapour Phase", *Crystals*, Vol.2 (Springer, Berlin, Heidelberg, New York 1980)
d) V.P. Zlomanov, E.V. Masyakin, A.V. Novoselova: J. Cryst. Growth *26*, 261 (1974)
e) S.J.C. Irvine, J.B. Mullin: J. Cryst. Growth *53*, 458 (1981)
f) K. Mochizuki: J. Cryst. Growth *53*, 355 (1981)
g) K. Kinoshita, S. Miyazawa: J. Cryst. Growth *57*, 141 (1982)
h) J. Morimoto, T. Ito, T. Yoshioka, T. Miyakawa: J. Cryst. Growth *57*, 362 (1982)
i) E. Schönherr: J. Cryst. Growth *57*, 493 (1982)
j) C. Barta, E. Kostal, A. Triska: Cryst. Res. Technol.*17*, 411 (1982)
k) Yu.M. Tairov, V.F. Tsvetkov: J. Cryst. Growth *52*, 146 (1981)

8.23 a) R. Triboulet: Rev. Phys. Appl.*12*, 123 (1977)
b) T. Taguchi, S. Fujita, Y. Inuishi: J. Cryst. Growth *45*, 204 (1978)
c) R. Triboulet, Y. Marfaing: J. Cryst. Growth *51*, 89 (1981)
d) H. Kezuka, K. Iwamura, T. Masaki: Thin Solid Films *83*, 47 (1981)
e) Yu.A. Vodakov, E.N. Mokhov, M.G. Ramm, A.D. Roenkov: Krist. Tech.*14*, 729 (1979);
E.N. Mokhov, I.L. Shulpina, A.S. Tregulova, Yu.A. Vodakov: Cryst. Res. Technol.*16*, 879 (1981)

8.24 A. Lopez-Otero: Thin Solid Films *49*, 3 (1978)

8.25 W.M. Yim, E.J. Stofko: J. Electrochem. Soc.*119*, 381 (1972)

8.26 A. Lely: Ber. Dtsch. Keram. Ges.*32*, 229 (1955)

8.27 E.V. Markov, A.A. Davydov: Izv. Akad. Nauk SSSR, Neorg. Mater.*7*, 575 (1971)

8.28 H. Schäfer: *Chemische Transportreaktionen* (Verlag Chemie, Weinheim 1961)

8.29 a) R. Nitsche: Fortschr. Mineral *44*, 231 (1966)
b) J. Mercier: J. Cryst. Growth *56*, 235 (1982)

8.30 H. Scholz, R. Kluckow: J. Phys. Chem. Solids, Suppl.*1*, 475 (1967);
H. Scholz: Acta Electron.*17*, 69 (1974)

8.31 M. Schieber, M.F. Schnepple, L. Van den Berg: J. Cryst. Growth *33*, 125 (1976)

8.32 L.A. Zadorozhnaya, V.A. Lyakhovitskaya, E.I. Givargizov, L.M. Belyaev: J. Cryst. Growth *41*, 61 (1977)

8.33 a) L.S. Gagara, P.A. Gashin, G.G. Dvornik, V.V. Leondar, P.S. Paskal, A.V. Simashkevich: Cryst. Res. Technol.*17*, 345 (1981)
b) M. Aoki, K. Tada, T. Murai, T. Inoue: Thin Solid Films *83*, 283 (1981)
c) M. Shimizu, T. Shiozaki, A. Kawabata: J. Cryst. Growth *57*, 94 (1982)

8.34 F.H. Nicoll: J. Electrochem. Soc.*110*, 1165 (1963)

8.35 J. Nishizawa: "Aspects of Silicon Epitaxy", in *Crystal Growth. Theory and Techniques*, Vol.2, ed. by C.H.L. Goodman (Plenum, New York 1978) p.57

8.36 N.N. Sheftal', N.P. Kokorish, A.V. Krasilov: Izv. Akad. Nauk SSSR, Ser. Fiz.*21*, 146 (1957)

8.37 H.C. Theuerer: J. Electrochem. Soc.*108*, 649 (1961)

8.38 J. Bloem, L.J. Gilling: "Mechanisms of the Chemical Vapour Deposition of Silicon", *Current Topics Mat. Sci.*, Vol.1 (North-Holland, Amsterdam 1978) p.147

8.39 a) J. Bloem: J. Cryst. Growth *50*, 581 (1980)
b) J. Bloem, W.A.P. Claassen, W.C.J.N. Valkenburg: J. Cryst. Growth *57*, 177 (1982)
c) K.E. Bean: Thin Solid Films *83*, 173 (1981)

8.40 W. Steinmaier: Philips Res. Rep.*18*, 75 (1963)

8.41 E. Sirtl, L.P. Hunt, D.H. Sawyer: J. Electrochem. Soc.*121*, 919 (1974)
8.42 V.S. Ban: J. Electrochem. Soc.*125*, 317 (1978)
8.43 F. Eversteyn, P. Severin, C.H. Brekel: J. Electrochem. Soc.*117*, 925 (1970)
8.44 E.I. Givargizov: Fiz. Tverd. Tela *6*, 1804 (1964) [English transl.: Sov. Phys. Solid State *6*, 1415 (1964)]
8.45 D.C. Gupta: Solid State Technol.*14*, 33 (1978)
8.46 M.L. Hammond: Solid State Technol.*21*, 68 (1978)
8.47 a) Y.S. Chiang, G.W. Looney: J. Electrochem. Soc.*120*, 550 (1973)
b) W.G. Townsend, M.E. Uddin: Solid State Technol.*16*, N3, 39 (1973)
8.48 M.J.-P. Duchemin, M.M. Bonnet, M.F. Koelsch: J. Electrochem. Soc.*125*, 637 (1978)
8.49 J.L. Gentner, C. Bernard, R. Cadoret: J. Cryst. Growth *56*, 332 (1982)
8.50 a) R.S. Rosler: Solid State Technol.*20*, N4, 63 (1977)
b) C.H.J. van den Brekel, L.J.M. Bollen: J. Cryst. Growth *54*, 310 (1981)
c) W.A.P. Claassen, J. Bloem, W.G.J.N. Valkenburg, C.H.J. van den Brekel: J. Cryst. Growth *57*, 259 (1982)
8.51 C.F. Powell, J.H. Oxley, J.M. Blocher (eds.): *Vapor Deposition* (Wiley, New York 1966)
8.52 a) B.V. Derjaguin, B.V. Spitsyn, A.E. Gorodetsky, A.P. Zakharov, L.L. Bouilov, A.E. Alekseenko: J. Cryst. Growth *31*, 44 (1975)
b) B.V. Spitsyn, L.L. Bouilov, B.V. Derjaguin: J. Cryst. Growth *52*, 219 (1981)
8.53 W. von Muench, E. Pettenpaul: J. Electrochem. Soc.*125*, 294 (1978)
8.54 W.M. Feist, S.R. Steel, D W Readey: "The Preparation of Films by Chemical Vapour Deposition", in *Physics of Thin Films*, Vol.5 (Academic, New York 1969) p.237
8.55 a) J.R. Knight, D. Effer, P.R. Evans: Solid State Electron.*8*, 178 (1965)
b) J.V. DiLorenzo: J. Cryst. Growth *17*, 189 (1972)
8.56 a) T. Mizutani, M. Yoshida, A. Usui, H. Watanabe, T. Yuasa, I. Hayashi: Jpn. J. Appl. Phys.*19*, L113 (1980)
b) P. Vohl: J. Cryst. Growth *54*, 101 (1981)
8.57 a) G.H. Olsen, C.J. Nuese, M. Ettenberg: Appl. Phys. Lett.*34*, 262 (1979)
b) H. Seki, A. Koukitu, M. Matsumara: J. Cryst. Growth *54*, 615 (1981)
8.58 a) J. Cryst. Growth (special issue: "Metalorganic Vapor Phase Epitaxy") *55* (1981)
b) M. Morita, N. Uesugi, S. Isogai, K. Tsubouchi, N. Mikoshiba: Jpn. J. Appl. Phys.*20*, 17 (1981)
c) T. Fukui, Y. Horikoshi: Jpn. J. Appl. Phys.*20*, 587 (1981)
d) R.M. Biefeld: J. Cryst. Growth *56*, 382 (1982)
e) S.J.C. Irvine, J.B. Mullin, A. Royle: J. Cryst. Growth *57*, 15 (1982)
8.59 a) K. Shohno, H. Ohtake, J. Bloem: J. Cryst. Growth *45*, 187 (1978)
b) M. Sano, M. Aoki: Thin Solid Films *83*, 247 (1981)
8.60 a) T. Muranoi, M. Furukoshi: J. Electrochem. Soc.*127*, 2295 (1980)
b) T. Matsumoto, T. Morita, T. Ishida: J. Cryst. Growth *53*, 225 (1981); M.D. Scott, J.O. Williams, R.C. Goodfellow: J. Cryst. Growth *51*, 267 (1981)
8.61 A.E.T. Kuiper, G E Thomas, W J Schouten: J. Cryst. Growth *51*, 17 (1981)
8.62 S.N. Maximovsky, I.P. Revocatova, M.A. Selezneva: J. Cryst. Growth *52*, 141 (1981)
8.63 I.W. Boyd, J.I.B. Wilson, J.L. West: Thin Solid Films *83*, L173 (1981)
8.64 a) M. Hanabusa, A. Namiki, K. Yoshihara: Appl. Phys. Lett.*35*, 626 (1979)
b) V. Baranauskas, C.I.Z. Mammana, R.E. Klinger, J.E. Greene: Appl. Phys. Lett.*36*, 930 (1980)
c) G. Leyendecker, D. Bauerle, P. Geitner, H. Lydtin: Appl. Phys. Lett. *39*, 921 (1981)
d) T.F. Deutsch, D.J. Ehrlich, R.M. Osgood: Appl. Phys. Lett.*35*, 175 (1979); *38*, 1018 (1981); *39*, 957 (1981); J. Electrochem. Soc.*128*, 2039 (1981)
8.65 R.S. Wagner, W.C. Ellis: Appl. Phys. Lett.*4*, 89 (1964)
8.66 R.S. Wagner: "Growth of Crystals by the Vapour–Liquid–Solid Mechanism", in *Whisker Technology*, ed. by A.P. Levitt (Wiley, New York 1970) p.47

8.67 G.A. Bootsma, H.J. Gassen: J. Cryst. Growth *10*, 223 (1971)
8.68 E.I. Givargizov: "Growth of Whiskers by the Vapour–Liquid–Solid Mechanism", *Current Topics Mat. Sci.*, Vol.1 (North-Holland, Amsterdam 1978) p.79
8.69 E.I. Givargizov, A.A. Chernov: Kristallografiya *18*, 147 (1973) [English transl.: Sov. Phys. Crystallogr.*18*, 89 (1973)]
8.70 E.I. Givargizov: J. Cryst. Growth *31*, 20 (1975)
8.71 O. Nittono, H, Hasegawa, S. Nagakura: J. Cryst. Growth *42*, 175 (1977)
8.72 a) R.R. Hasiguti, T. Ishibashi, H. Yumoto: J. Cryst. Growth *45*, 13 (1978)
b) R.R. Hasiguti, H. Yumoto, Y. Kuriyama: J. Cryst. Growth *52*, 135 (1981)
8.73 R.S. Wagner, W.C. Ellis: Trans. Metall. Soc. AIME *233*, 1053 (1965)
8.74 E.N. Sickafus, D.B. Barker: J. Cryst. Growth *1*, 93 (1967)
8.75 D. Nenov, E D Dukova: Krist. Tech.*7*, 779 (1972)
8.76 R.S. Wagner: J. Cryst. Growth *3/4*, 159 (1968)
8.77 C.Y. Lou, G.A. Somorjai: J. Chem. Phys.*55*, 4554 (1971)
8.78 E.I. Givargizov, R.A. Babasyan: J. Cryst. Growth *37*, 140 (1977)
8.79 E. Kaldis: "Liquid Layers on Vapour Grown Crystals", in *Crystal Growth and Characterization*, ed. by R. Ueda, J.B. Mullin (North-Holland, Amsterdam 1975) p.225

Chapter 9

9.1 M.I. Ravich: Zh. Neorg. Khim.*15*, 2019 (1970)
9.2 M.I. Ravich, F.E. Borovaya: Zh. Neorg. Khim.*9*, 952 (1964)
9.3 S. Sourirajan, G. Kennedy: Am. J. Sci.*260*, 115 (1962)
9.4 V.K. Yanovsky, V.I. Voronkova, V.A. Koptsik: Kristallografiya *15*, 362 (1970) [English transl.: Sov. Phys. Crystallogr.*15*, 302 (1970)]
9.5 T.G. Petrov, E.B. Treivus, A.P. Kasatkin: *Vyrashchivaniye kristallov iz vodnykh rastvorov* (Crystal Growth from Aqueous Solutions) (Nedra, Leningrad 1967)
9.6 a) A.V. Shubnikov, N.N. Sheftal' (eds.): *Rost kristallov. T.3* (Izd-vo Akad. Nauk SSSR, Moscow 1961) [English transl.: *Growth of Crystals. Vol.3* (Consultants Bureau, New York 1962)]
b) V. Šip, V. Vaniček: In [Ref.9.6a, p.265] [English transl.: "New Items of Equipment for the Production of Monocrystals", in Ref.9.6a, p.191]
9.7 R.W. Moore: J. Am. Chem. Soc.*41*, 1060 (1919)
9.8 M.F. Koldobskaya, I.V. Gavrilova: In [Ref.9.6a, p.278] [English transl.: "Growth of Large Regular Crystals of Triglycine Sulphate under Laboratory Conditions", in Ref.9.6a, p.199]
9.9 H.E. Buckley: *Crystal Growth* (Wiley, New York 1951)
9.10 A.C. Walker, G.T. Kohman: Trans. Am. Inst. Electr. Eng.*67*, 565 (1948)
9.11 V.F. Parvov: Kristallografiya *9*, 584 (1964) [English transl.: Sov. Phys. Crystallogr.*9*, 499 (1965)]
9.12 M.I. Kozlovsky, A.V. Kotorobai, I.I. Melent'yev, M.F. Burchakov: In *Rost kristallov, T.6*, ed. by N.N. Sheftal' (Nauka, Moscow 1965) p.9 [English transl.: "Apparatus and Procedure for Study of the Effects of the Crystallization Conditions on the Growth and Properties of Crystals", in *Growth of Crystals, Vol. 6A* (Consultants Bureau, New York 1968) p.7]
9.13 a) A.I. Munchayev: Kristallografiya *18*, 894 (1973) [English transl.: Sov. Phys. Crystallogr.*18*, 560 (1974)]
b) R.A. Laudise: *The Growth of Single Crystals* (Prentice-Hall, Englewood Cliffs 1970)
9.14 a) H.K. Henish: Helv. Phys. Acta *41*, 888 (1968)
b) B. Brezina, J. Horvath: J. Cryst. Growth *52*, 858 (1981)
c) F. le Faucheux, M.C. Robert, E. Manghi: J. Cryst. Growth *56*, 141 (1982)
d) M. Abdulkhadar, M.A. Ittyachen: J. Cryst. Growth *48*, 149 (1980)
9.15 P. Kratochvil, B. Sprusil: J. Cryst. Growth *3/4*, 360 (1968)
9.16 J.C. Murphy, H.A. Kues, J. Bohandy: Nature *218*, 165 (1968)

9.17 Z. Blank, W. Brenner, Y. Okamoto: Mater. Res. Bull.*3*, 555 (1968)
9.18 A.F. Armington, J.J. O'Connor: J. Cryst. Growth *3/4*, 367 (1968)
9.19 R.G. Robins: J. Nucl. Mater.*2*, 189 (1960)
9.20 R.G. Robins: J. Nucl. Mater.*3*, 294 (1961)
9.21 R.C. De Mattei, R.A. Huggins, R.S. Feigelson: J. Cryst. Growth *34*, 1 (1976)
9.22 a) J. Bockris, G.A. Razymney: *Fundamental Aspects of Electrocrystallization* (Plenum, New York 1967)
b) D. Elwell: J. Cryst. Growth *52*, 741 (1981)
c) V. Bostanov, W. Obretenov, G. Staikov, D.K. Roc, E. Budevski: J. Cryst. Growth *52*, 761 (1981)
9.23 H. Jaffe, B.R.F. Kjellgren: Discuss. Faraday Soc. No.*5*, 319 (1949)
9.24 a) A.V. Shubnikov, N.N. Sheftal' (eds.): *Rost kristallov, T.4* (Nauka, Moscow 1964) [English transl.: *Growth of Crystals, Vol.4* (Consultants Bureau, New York 1966)]
b) I.M. Byteva: In [Ref.9.24a, p.22] [English transl.: "Effect of pH on the Shape of Ammonium Dihydrogen Phosphate Crystals", in Ref.9.24a, p.16]
9.25 I.V. Gavrilova, L.I. Kuznetsov: In [Ref.9.24a, p.85] [English transl.: "Aspects of the Growth of Monocrystals of Potassium Dihydrogen Phosphate", in Ref.9.24a, p.69]
9.26 I.M. Byteva: In [Ref.3.14a, p.219] [English transl.: "Effect of pH on the Growth of ADP Crystals in the Presence of Fe^{3+} and Cr^{3+}", in Ref.3.14a, Vol. B, p.26]
9.27 A.V. Belyustin, N.S. Stepanova: Kristallografiya *10*, 743 (1965) [English transl.: Sov. Phys. Crystallogr.*10*, 624 (1966)]
9.28 A.V. Belyustin, A.V. Kolina: Kristallografiya *20*, 206 (1975) [English transl.: Sov. Phys. Crystallogr.*20*, 126 (1975)]
9.29 Yu.M. Fishman: "Rentgenovskoye topograficheskoye issledovaniye defektov i difraktsionnogo kontrasta v kristallakh gruppy KH_2PO_4 (KDP)" (X-Ray Topographic Investigation of Defects and Diffraction Contrast in Crystals of Group KH_2PO_4 (KDP); Candidate's Thesis, Nauchno-issledovatel'skii institut mashinovedeniya, Moscow 1972)
9.30 a) H.J. Kolb, J.J. Gomer: J. Am. Chem. Soc.*67*, 894 (1945)
b) I.A. Batyreva, V.I. Bespalov, V.I. Bredikhin, G.L. Galushkina, V.P. Ershov, V.I. Katsman, S.P. Kuznetsov, L.A. Lavrov, M.A. Novikov, N.R. Shvetsova: J. Cryst. Growth *52*, 832 (1981)
9.31 R.J. Davey, J.W. Mullin: J. Cryst. Growth *23*, 89 (1974); *26*, 45 (1974)
9.32 V.B. Treivus, Yu.O. Punin, T.V. Ushakovskaya, O.I. Artamonova: Kristallografiya *20*, 199 (1975) [English transl.: Sov. Phys. Crystallogr.*20*, 121 (1975)]
9.33 W. Kibalczyc, W. Kolasinski: Zesz. Nauk Plodz. No.*271*, 51 (1977); 1st Europ. Conf. on Crystal Growth, Zürich, Switzerland, 12-18 September 1976, p.125
9.34 B. Wojciechowski, J. Karniewicz: "Influence of Properties of the Solvent on the Growth of Crystal from Aqueous Solutions", in 1st. Europ. Conf. on Crystal Growth, Zürich, Switzerland, 12-18 September 1976, p.110
9.35 A.A. Chernov, V.V. Sipyagin: "Peculiarities in Crystal Growth from Aqueous Solutions Connected with Their Structures", *Current Topics Mat. Sci.*, Vol.5 (North-Holland, Amsterdam 1980) p.281
9.36 C. Belouet, M. Monnier, J.C. Verplanke: J. Cryst. Growth *29*, 109 (1975)
9.37 C. Belouet, E. Dunia, T.F. Pëtroff: J. Cryst. Growth *23*, 243 (1974)
9.38 R.J. Davey, J.W. Mullin: J. Cryst. Growth *26*, 45 (1974)
9.39 P.P. Chirvinsky: *Iskusstvennoye poluchenyie mineralov v XIX stoletii* (Man-Made Minerals in the Nineteenth Century) (Kiev 1903-1906)
9.40 V.A. Kuznetsov, A.N. Lobachev: Kristallografiya *17*, 878 (1972) [English transl.: Sov. Phys. Crystallogr.*17*, 775 (1973)]
9.41 a) A.N. Lobachev (ed.): *Issledovaniye protsessov kristallizatsii v gidrotermal'nykh usloviyakh* (Nauka, Moscow 1970) [English transl.: *Crystallization Processes under Hydrothermal Conditions* (Consultants Bureau, New York 1973)]

b) A.A. Shternberg: In [Ref.9.41a, p.199] [English transl.: "Controlling the Growth of Crystals in Autoclaves", in Ref.9.41a, p.225]
9.42 R.A. Laudise, A.A. Ballman: J. Phys. Chem.*64*, 688 (1960)
9.43 L.N. Demyanets, A.N. Lobachev: In [Ref.9.41a, p.7] [English transl.: "Some Problems of Hydrothermal Crystallization", in Ref.9.41a, p.1]
9.44 a) N.N. Sheftal', E.I. Givargizov (eds.): *Rost kristallov. T.9* (Nauka, Moscow 1972) [English transl.: *Growth of Crystals. Vol.9* (Consultants Bureau, New York 1975)]
b) V.P. Butuzov, A.N. Lobachev: In [Ref.9.44a, p.13] [English transl.: "Some Results of Research on Hydrothermal Crystal Systems and Growth", in Ref.9.44a, p.11]
9.45 N.Yu. Ikornikova: *Gidrotermal'nyi sintez kristallov v khloridnykh sistemakh* (Hydrothermal Synthesis of Crystals in Cloride Systems) (Nauka, Moscow 1975)
9.46 B.N. Litvin: *Gidrotermal'nyi sintez neorganicheskikh soyedinenii. Annotirovannyi ukazatel' faktograficheskikh dannykh* (Hydrothermal Synthesis of Inorganic Compounds. Annotated Index of Literature Data) (VNIIKI, Moscow 1971)
9.47 R.A. Laudise, J.W. Nielsen: "Hydrothermal Crystal Growth", in *Solid State Physics. Advances in Research and Applications*. Vol.12, ed. by F. Seitz, D. Turnbull (Academic, New York 1961) p.149
9.48 R.A. Laudise, E.D. Kolb: Endeavour *28*, 114 (1969)
9.49 C.J.M. Rooymans, W.F.Th. Langenhofft: J. Cryst. Growth *3/4*, 411 (1968)
9.50 R.A. Laudise, A.A. Ballman: J. Am. Chem. Soc.*80*, 2655 (1958)
9.51 A.A. Sternberg, V.A. Kuznetsov: Kristallografiya *13*, 745 (1968) [English transl.: Sov. Phys. Crystallogr.*13*, 647 (1969)]
9.52 R.C. Puttbach, R.R. Manchamp, J.W. Nielsen: J. Phys. Chem. Solids, Suppl.*1*, 569 (1967)
9.53 V.I. Popolitov, A.N. Lobachev, M.N. Tseitlin: "Kristallizatsiya ortoantimonata sur'my v gidrotermal'nykh usloviyakh" (Crystallization of Antimony Orthoantimonate under Hydrothermal Conditions), in *Rost kristallov iz vysokotemperaturnykh vodnykh rastvorov* (Crystal Growth in High Temperature Aqueous Solutions), ed. by A.N. Lobachev (Nauka, Moscow 1977) p.198
9.54 E.D. Kolb, R.A. Laudise: J. Appl. Phys.*42*, 1552 (1971)
9.55 C.T. Ashby, J.W. Berry: Am. Mineral.*55*, 1800 (1970)
9.56 A.N. Lobachev (ed.) *Gidrotermal'nyi sintez kristallov* (Nauka, Moscow 1968) [English transl.: *Hydrothermal Synthesis of Crystals* (Consultants Bureau, New York 1971)]
9.57 N.Yu. Ikornikova, A.N. Lobachev, A.R. Vasenin, V.M. Egorov, V.M. Antoshin: In [Ref.9.41a, p.212] [English transl.: "Apparatus for Precision Research in Hydrothermal Experiments", in Ref.9.41a, p.241]
9.58 A.A. Shternberg: In [Ref.9.56, p.203] [English transl.: "Double Autoclave for Operation at 700°C and 3000 kgf/cm^2", in Ref.9.56, p.147]
9.59 A.N. Lobachev, L.N. Demyanets, I.P. Kuz'mina, E.N. Emelyanova: J. Cryst. Growth *13/14*, 540 (1972)
9.60 V.A. Kuznetsov: Kristallografiya *9*, 123 (1964) [English transl.: Sov. Phys. Crystallogr.*9*, 103 (1965)]
9.61 G.C. Kennedy: Am. J. Sci.*248*, 540 (1950)
9.62 N.Yu.Ikornikova, V.M. Egorov: In [Ref.9.56, p.58] [English transl.: "Experimentally Determined PTFC Diagrams of Aqueous Solutions of Li, Na, K, and Cs Chlorides", in Ref.9.56, p.34]
9.63 L.A. Samoilovich: *Zavisimost' mezhdu davleniyem, temperaturoi i plotnost'yu vodno-solevykh rastvorov* (Pressure–Temperature–Density Relation for Aqueous Salt Solutions) (Vsesoyuznyi nauchno-issledovatel'skii institut sinteza mineral'nogo syr'ya, Moscow 1969)
9.64 H. Rau: High Temp. High Pressures *6*, 671 (1974)
9.65 T.B. Kosova, L.N. Demyanets: "Izucheniye rastvorimosti kankrinita v rastvorakh NaOH pri temperaturakh 200°–400°C" (Solubility of Cancrinite

in NaOH Solutions at 200°–400°C), in *Rost kristallov iz vysokotemperaturnykh vodnykh rastvorov* (Crystal Growth in High Temperature Aqueous Solutions), ed. by A.N. Lobachev (Nauka, Moscow 1977) p.43
9.66 M.I. Ravich: *Vodno-solevyie sistemy pri povyshennykh temperaturakh i davleniyakh* (Water–Salt Systems at Elevated Temperatures and Pressures) (Nauka, Moscow 1974)
9.67 G.B. Boky, I.N. Anikin: Zh. Neorg. Khim.*1*, 1926 (1956)
9.68 H.L. Barnes, S.B. Romberger, M. Stemprok: Econ. Geol.*62*, 957 (1967)
9.69 B.M. Ryzhenko: Geokhimiya *2*, 229 (1976)
9.70 H.C. Helgeson: *Complexing and Hydrothermal Ore Deposition* (Pergamon, Oxford 1964)
9.71 N.G. Duderov, L.N. Demianets, A.N. Lobachev: Krist. Tech.*10*, 37 (1975)
9.72 V.V. Badikov, A.A. Godovikov: "Temperaturnyi rezhim avtoklavov i massoperenos galenita v gidrotermal'nykh usloviyakh" (Temperature Regime of Autoclaves and Mass Transfer of Galenite under Hydrothermal Conditions), in *Eksperimental'nyie issledovaniya po mineralogii* (Experimental Investigations in Mineralogy) (Inst. geologii i geofiziki Sib. Otd. Akad. Nauk SSSR, Novosibirsk 1969) p.154
9.73 R.A. Laudise: J. Am. Chem. Soc.*81*, 362 (1959)
9.74 G.G. Lemmlein, L.I. Tsinober: "Nekotoryie osobennosti morfologii kristallov iskusstvennogo kvartsa" (Some Peculiarities in the Morphology of Artificial Quartz Crystals), in *Trudy Vsesoyuznogo nauchno-issledovatel'skogo instituta piezoopticheskogo mineralogicheskogo cyr'ya, T.6, Materialy po izucheniyu iskusstvennogo kvartsa* (Collected Works of the All-Union Scientific Research Institute of Piezooptical Raw Materials, Vol.6, Materials on Synthetic Quartz) (Gosgeoltekhizdat, Moscow 1962) p.13
9.75 O.K. Mel'nikov, B.N. Litvin, N.S. Triodina: In [Ref.9.41, p.134] [English transl.: "Crystallization of Sodalite on a Seed", in Ref.9.41, p.151]
9.76 A.A. Chernov, V.A. Kuznetsov: Kristallografiya *14*, 879 (1969) [English transl.: Sov. Phys. Crystallogr.*14*, 753 (1970)]
9.77 A.A. Shternberg: Kristallografiya *7*, 114 (1962) [English transl.: Sov. Phys. Crystallogr.*7*, 92 (1962)]
9.78 E.D. Kolb, D.L. Wood, E.G. Spenser, R.A. Laudise: J. Appl. Phys.*38*, 1027 (1967)
9.79 A.A. Ballman, D.M. Dodd, N.A. Kuebler, R.A. Laudise, D.L. Wood, D.W. Rudd: Appl. Opt.*7*, 1387 (1968)
9.80 I.P. Kuz'mina, A.N. Lobachev, N.S. Triodina: In [Ref.9.41a, p.187] [English transl.: "Crystallization Kinetics of Sodium Zincogermanate", in Ref.9.41a, p.211]
9.81 V.E. Khadzhi, M.V. Lelyakova: In *Rost kristallov. T.8*, ed. by N.N. Sheftal' (Nauka, Moscow 1968) p.51 [English transl.: "Effects of Temperature and Supersaturation on the Entry of Aluminium into Synthetic Quartz", in *Growth of Crystals. Vol.8* (Consultants Bureau, New York 1969) p.43]
9.82 L.I. Tsinober, I.E. Kamentsev: Kristallografiya *9*, 448 (1969) [English transl.: Sov. Phys. Crystallogr.*9*, 374 (1969)]
9.83 K.V. Shalimova, I.K. Morozova, M.M. Malov, V.A. Kuznetsov, A.A. Shternberg, A.N. Lobachev: Kristallografiya *19*, 147 (1974) [English transl.: Sov. Phys. Crystallogr.*19*, 86 (1975)]
9.84 A.N. Lobachev, L.M. Belyayev, I.M. Sil'vestrova, O.K. Mel'nikov, Yu.V. Pisarevsky, N.S. Triodina: Kristallografiya *19*, 126 (1974) [English transl.: Sov. Phys. Crystallogr.*19*, 72 (1974)]
9.85 G.I. Distler, A.N. Lobachev, V.P. Vlasov, O.K. Mel'nikov, N.S. Triodina: Dokl. Akad. Nauk SSSR *215*, 91 (1974)
9.86 B. Ferrand, J. Daval, J.O. Joubert: J. Cryst. Growth *17*, 312 (1972)
9.87 a) L.N. Demianets: "Hydrothermal Crystallization of Magnetic Oxides", *Crystals* (Springer, Berlin, Heidelberg, New York 1978) p.98
b) V.A. Timofeyeva: *Rost kristallov iz rastvorov-rasplavov* (Crystal Growth from Flux) (Nauka, Moscow 1978)

c) R.A. Laudise: *The Growth of Single Crystals* (Prentice-Hall, Englewood Cliffs 1970)
9.88 V.A. Timofeyeva, N.I. Lukyanova: In [Ref.9.44a, p.104] [English transl.: "Effects of the Phase Boundaries of the Compounds Formed in Molten Solutions Used to Grow Garnets", in Ref.9.44a, p.116]
9.89 P.K. Konakov, G.E. Verevochkin, L.A. Goryainov, L.A. Zaruvinskaya, Yu.P. Konakov, V.V. Kudryavtsev, G.A. Tretyakov: *Teplo- i massoobmen pri poluchenii monokristallov* (Heat and Mass Exchange in the Preparation of Single Crystals) (Metallurgiya, Moscow 1971)
9.90 D. D. Elwell, H.J. Scheel: *Crystal Growth from High-Temperature Solutions* (Academic, London 1975)
9.91 J.W. Nielsen: In [Ref.3.7, p.139] [English transl.: "Recent Progress in Flux Growth", in Ref.3.7]
9.92 G.A. Wolff, A.I. Mlavsky: "Travelling Solvent Techniques", in *Crystal Growth, Theory and Techniques*, ed. by C.H.L. Goodman (Plenum, New York 1974) p.193
9.93 W.G. Pfann: *Zone Melting* (Wiley, New York 1966)
9.94 U.M. Kulish: *Rost i elektrofizicheskiye svoistva plyonok poluprovodnikov (zhidkostnaya epitaksiya,* (Growth and Electrophysical Properties of Semiconductor Films (Liquid Phase Epitaxy), (Kalmyk Izd-vo, Elista 1976)
9.95 P.A. Arseniev, Kh. S. Bagdasarov, V.V. Fenin: "Vyrashchivaniye monokristallicheskikh plyonok dlya kvantovoi elektroniki" (Growing of Single Crystal Films for Quantum Electronics), Review No. 1495-76, All-Union Institute for Scientific and Technical Information, Moscow 1977

Chapter 10

10.1 I.S. Kulikov: *Termodinamicheskaya dissotsiatsiya soyedinenii* (Thermodynamic Dissociation of Compounds) (Metallurgiya, Moscow 1966)
10.2 Kh.S. Bagdasarov: In [Ref.3.7, p.179] [English transl.: "Problems of Synthesis of Refractory Optical Single Crystals", in Ref.3.7]
10.3 S.S. Gorelik, M.Ya. Dashevsky: *Materialovedeniye poluprovodnikov i metallovedeniye* (Materials Science of Semiconductors and Physical Metallurgy) (Metallurgiya, Moscow 1973)
10.4 V.P. Butuzov: "Metody polucheniya iskusstvennykh almazov" (Methods of Preparation of Synthetic Diamonds), in *Issledovaniye prirodnogo i tekhnicheskogo mineraloobrazovaniya* (Study of Natural and Artificial Formation of Diamonds) (Nauka, Moscow 1966) p.10
10.5 T.P. Markholiya, B.F. Yudin, N.I. Voronin: "O vzaimodeistvii tugoplavkikh metallov s glinozemom i kremnezemom" (On the Interaction of Refractory Metals with Alumina and Silica), in *Khimia vysokotemperaturnykh materialov* (Chemistry of Refractory Materials) (Nauka, Leningrad 1967) p.203
10.6 L.M. Belyayev (ed.): *Rubin i sapfir* (Ruby and Sapphire) (Nauka, Moscow 1974)
10.7 N.A. Toropov, I.A. Bondar', F.Ya. Galakhov, Kh.S. Nikogosyan, N.V. Vinogradova: Izv. Akad. Nauk SSSR, Ser. Khim. 1158 (1964)
10.8 F. Schmidt, D. Viechnicki: J. Am. Ceram. Soc.*53*, 528 (1970)
10.9 W.C. Dash: "The Growth of Silicon Crystals Free from Dislocations", in [Ref.1.29a, pp.361, 382]
10.10 B. Cockayne: J. Cryst. Growth *3/4*, 60 (1968)
10.11 B. Cockayne, M. Chesswas, D.B. Gasson: J. Mater. Sci.*4*, 450 (1969)
10.12 H.F. Sterling, R.W. Warren: Br. J. Met.*67*, 404 (1963)
10.13 V.I. Aleksandrov, V.V. Osiko, V.M. Tatarintsev: Prib. Tekh. Eksp.*5*, 222 (1970)
10.14 A.V. Stepanov: Izv. Akad. Nauk SSSR, Ser. Fiz.*33*, 1946 (1969)
10.15 H.E. LaBelle, Jr., A.J. Mlavsky: Nature *216*, 574 (1968)
10.16 T.R. Kyle, G. Zudzik: Mater. Res. Bull.*8*, 443 (1973)

10.17 K.-Th. Wilke: Kristallzüchtung (VEB Deutscher Verlag der Wissenschaften, Berlin 1973)
10.18 R.E. De la Rue, F.A. Halden: Rev. Sci. Instrum. *31*, 35 (1960)
10.19 E.M. Sparrow, R.D. Cess: *Radiation Heat Transfer*, 3rd ed. (Brooks/Cole, Belmont, California 1970)
10.20 E.I. Givargizov (ed.): *Crystal Growth 1980*, Proc. 6th Int. Conf. on Crystal Growth, Moscow, USSR, 10-16 September 1980 (North-Holland, Amsterdam 1981)
10.21 D. Schwabe, A. Scharmann, F. Preisser, R. Oeder: J. Cryst. Growth *43*, 305 (1978)
10.22 A.G. Petrosyan: Issledovaniye uslovii vyrashchivaniya lazernykh kristallov $Lu_3Al_5O_{12}$, aktivirovannykh ionami redkozemel'nykh elementov (Study of Growth Conditions of Laser Crystals $Lu_3Al_5O_{12}$ Activated by Rare Earth Ions), Candidate's Thesis, Institute for Physical Research, Academy of Sciences of the Armenian SSR, Erevan (1974)
10.23 H.P. Utech, M.C. Flemings: J. Phys. Chem. Solids Suppl. *1*, 651 (1967)
10.24 J.C. Brice: *The Growth of Crystals from Liquids* (North-Holland, Amsterdam 1973)

Materials Index

Subject Index

S